THE ECONOMIES OF AFRICA

The
ECONOMIES OF AFRICA

Geography Population History Stability Structure Performance Forecasts

Michael Hodd
University of London

G.K.HALL&CO.

© Michael Hodd 1991

All rights reserved. No part of this publication may be reproduced, stored in a retrieval system, or transmitted in any form or by any means, electronic, mechanical, photocopying, recording, or otherwise without the prior permission of the publisher.

Published by

G.K. Hall & Co.
70 Lincoln Street
Boston
Massachusetts 02111
USA

Library of Congress Cataloging-in-Publication data forthcoming.

HC
800
.H633
1991

ISBN   0-8161-7357-5

CONTENTS

Acknowledgements *viii*

PART 1
THE AFRICAN ECONOMIES

1 Introduction *3*

2 Africa and the World *7*

3 The African Economies *26*

   References *39*

PART 2
COUNTRY ENTRIES

1 Angola *43*

2 Benin *50*

3 Botswana *57*

4 Burkina Faso *64*

5 Burundi *71*

6 Cameroon *78*

7 Cape Verde *85*

Contents

8 Central African Republic *91*

9 Chad *97*

10 Comoros *103*

11 Congo *110*

12 Côte d'Ivoire *116*

13 Djibouti *123*

14 Equatorial Guinea *130*

15 Ethiopia *137*

16 Gabon *144*

17 Gambia *150*

18 Ghana *156*

19 Guinea *163*

20 Guinea-Bissau *170*

21 Kenya *176*

22 Lesotho *183*

23 Liberia *189*

24 Madagascar *196*

25 Malawi *202*

26 Mali *209*

27 Mauritania *216*

28 Mauritius *223*

29 Mozambique *229*

30 Namibia *236*

31 Niger *243*

32 Nigeria *250*

33 Réunion *257*

34 Rwanda *263*

Contents

35 São Tomé and Príncipe *269*

36 Senegal *275*

37 Seychelles *281*

38 Sierra Leone *287*

39 Somalia *293*

40 South Africa *299*

41 Sudan *307*

42 Swaziland *315*

43 Tanzania *322*

44 Togo *329*

45 Uganda *335*

46 Zaïre *342*

47 Zambia *350*

48 Zimbabwe *357*

Acknowledgements

Dartmouth Publishing has been very patient in waiting for this book as other work has interfered with its completion. Research assistance was provided by Ann Collins, Hania el Farhan, Muleganga Tembo, Mushiba Nyamazana, Mugisha Maliyamkono and Simon Wilson. I would like to think they are responsible for all the errors that remain.

PART 1

THE
AFRICAN ECONOMIES

1
INTRODUCTION

This book is divided into two parts. The first looks at the African economies generally. Chapter 2 examines the African economies in a world context, and Chapter 3 goes on to consider the African economies as a group. The second part of the book deals with individual countries.

Forty-eight economies in sub-Saharan Africa are covered. Forty-seven are sovereign states, and Réunion, administered by France, is also included. The North African countries of Morroco (and the former Spanish Sahara), Algeria, Tunisia, Libya and Egypt are not included. South Africa is covered, but except when otherwise indicated, the term 'Africa' is taken to comprise the forty-seven economies south of the Sahara, excluding South Africa.

Considerable use is made of comparisons. These are made between the African economies and other world groupings, between various sub-groupings within Africa, and between individual African economies and African averages. This is done on the grounds that it is easier to appreciate orders of magnitude, and to discern unusual economic features in a comparative context.

Africa provides the unique opportunity of a large sample of countries, and comparisons can be used to draw conclusions as to cause and effect in explaining economic performance in Africa.

Africa and the World

Chapter 2 compares Africa with other developing countries, the high-income countries and, where data is available, the 'non-reporting' group (mostly Eastern bloc or former Eastern bloc) countries. Considerable use is made of diagrams and charts to point up the main differences between these groups of countries.

The chapter begins with sections on economic structure, education and health, and economic performance. The final section is economic outlook and forecasts. The forecasts presented for Africa are aggregated from the forecasts made for forty-eight individual sets of forecasts for the African countries, and presented in the Country Entries part of the book. A description of the forecasting basis is given in the Country Entries section in this chapter. The forecasts for Africa are compared with the forecasts for developing countries and for the high-income countries generated by the World Bank (1989b).

The African Economies

Chapter 3 attempts to give something of an overall picture of the African economies by grouping them on the basis of geographical region, population and economic size, income level, economic zone, stability record, economic performance, recent economic policy initiatives, and future prospects. The basis for the stability assessments is described in the Country Entries section of this chapter

Considerable use is made of tables listing the African economies by their particular characteristics. The classifications are made on the basis of the analysis for individual economies contained in the Country Entries part of the book.

Introduction

Chapter 3 also contains a discussion of the evolution of economic policy in Africa, in particular the enormous changes brought about by the liberalising economic reforms of the 1980s

Country Entries

The coverage of each individual economy in Africa begins with a short description of physical geography and climate, a section on the composition and characteristics of the population, and a brief history. These sections provide background for the subsequent sections on the economy.

There then follows a section on stability, which is seen as as comprising both political and economic factors.

The political set of factors relates to events such as the incidence of coups, attempted coups, riots, civil and cross-border wars, strikes, and civil disorder. The economic set of factors relates to continuity in the policy environment in which economy has operated. Abrupt changes in economic strategy which involve the transfer of large sections of the economy from private ownership to state ownership or control, or *vice-versa*, are judged to have an adverse affect on stability.

Countries are ranked into five categories from bad though poor, moderate, and fair to good. Absence of any seriously de-stabilising examples of the listed factors will generally result in a rating of good, while virtually continual incidence of factors will result in a bad rating. Isolated political disruption or one abrupt change in economic strategy will generally lead to a fair or moderate ranking, depending on the severity and extent of the incident or change. Several incidents interspersed with some tranquil periods will lead to a poor ranking. These classifications are not rigid, however, and a certain degree of subjective judgement is employed.

These assessments are reasonably well correlated with other more rigidly compiled stability indices (see El Farhan and Hodd 1989). These are variously based on numbers of years with incidents (Wheeler 1984), an explicit weighting of incidents (Jackman 1978, Johnson *et al* 1984, Johnson *et al*. 1985) and on success in raising sovereign loans (Euromoney 1989). While the weighting of incidents in the assessments used in this study is subjective, it has the advantage of being able to include the economic policy factors. It also allows an element of discretion when there may be some doubt as to the genuineness of a reported coup attempt, say, or in judging the severity of an event such as rioting The indices based on loan-raising are susceptible to the criticism that a stable country might have a low ranking because of a policy to avoid external indebtedness.

Each country entry then has a section dealing with the country's economic structure, and the emphasis is on comparison with the rest of Africa and the high-income OECD countries. Production, expenditure and international trade are covered, and the section concludes with a review of social factors such as education and health.

The section on economic performance covers overall growth rates, growth of the major producing sectors, growth of merchandise trade volumes and inflation.

A section on recent economic developments reviews the evolution of the country's economic policies, and goes on to cover such factors as exchange rate changes, the external debt situation, and aid commitments.

Each country entry concludes with a section on economic outlook and forecasts. There is a brief discussion of the country's stability prospects and of the economic policies being pursued. Prospects for the main commodity exports are considered, together with any possible changes in the flows of aid and foreign investment. Expected prices for commodity exports are drawn, in the main, from the World Bank (1989a).

The forecasts are generated as follows. The forecasts of export volumes are obtained from supply response equations for major export commodities, normalised for each country. The dollar value of export earnings is generated from the volume projections and the commodity price forecasts. Import volumes are generated from export receipts and estimated net capital inflow. GDP is generated from a two-gap (foreign exchange and savings) model (see for example World Bank 1975) normalised for each economy. Population growth forecasts from the World Bank (1989b) are used to convert the GDP forecasts to GDP per head forecasts. The money supply is projected from

Introduction

the budget deficit, and, with the real GDP forecast, is used in a quantity equation, again normalised for each economy, to forecast inflation.

The parameters for the behavioural relations used in the forecasting process are adjusted for two factors. Firstly, to allow for any changes in economic policy which improve the productivity of resources, and secondly to allow for any changes in stability conditions.

Base forecasts are obtained by taking the most likely value for a country's net capital inflow. High forecasts are generated by making more optimistic assumptions regarding net capital inflows. Thus the high forecasts are generated on a different set of assumptions to those in World Bank (1989b). The latter are generated by assuming more rapid adjustment in the world economy to eliminate macroeconomic imbalances. These differences should be borne in mind when comparing the more optimistic forecasts.

There is considerable variation in year-to-year economic growth in African countries. The main factors are abrupt changes in stability conditions, the effect of climatic conditions on the agricultural sector, and variations in world commodity prices. Fairly successful forecasts can be made for individual countries for very short periods, up to a year ahead, as is done by the Economist Intelligence Unit (quarterly) *Country Reports*s. These are based on harvest projections, recent rainfall, and current world commodity prices.

The medium-term forecasts presented here are taken to be averages for the five-year forecast period and implicitly assume that climatic variations will cancel out. However, some African countries have experienced drought cycles in recent years which are clearly not random, and will not cancel out over a five-year span. Thus forecasts for such countries assume that normal (that is, long-term average) climatic conditions are experienced.

At the end of each country entry is a compilation of statistical data and charts. The first of these is a comparison of economic data for each country with the regional, African and high-income (OECD) country averages. The dates for the data in the tables refer to the country in question. There are certain limitations in the comparisons, with some countries not having such recent economic data as others. In the event, the main single-date comparisons were centred on 1986, and the period comparisons on 1980-86. The basic population data, for which there are more recent data for all countries, is based on 1989.

There is then a set of leading economic indicators since 1970 for each country, covering annual changes in real GDP, GDP per head, and inflation, together with the exchange rate, and merchandise exports and imports valued in $US. These series are aligned in the six charts to show up any correspondences in fluctuations.

Finally, there is a table of the data used in the leading indicator charts, together with further charts which give the commodity composition of the country's exports and imports, the destination of exports, and the origins of imports.

Statistical Sources

Data is for the most part taken from the International Monetary Fund *International Financial Statistics*, the World Bank *World Development Report*, the World Bank *World Tables*, the United Nations *Monthly Bulletin of Statistics* and the United Nations Economic Commission for Africa *African Statistical Abstract*. These sources are supplemented, particularly for the smaller countries, by national statistical sources. Recent data is collected from preliminary reports released by ministries and reported in the British Broadcasting Corporation *Summary of World Broadcasts: Economic Report*.

There are considerable measurement problems with the compilation of economic data arising from difficulties of definition, sampling error, construction of indices and inaccurate reporting. Some of these problems are more severe for the African economies, when small, unrepresentative samples are taken, particularly in the agriculture sector and for domestic prices Data is often haphazardly collected and processed. A major problem is set by the variability of international commodity prices, with different base years sometimes giving markedly different weightings and growth rates. The problems are most severe in GDP growth rate data and inflation figures. Data

Introduction

on trade flows and exchange rates tend to be of much better quality.

Other Sources on the African Economies

The present book attempts to provide concise summaries of the stability structure, performance and prospects for African economies within an analytical framework. As much of the data as possible is presented in comparative form, or in charts. As such, it is complimentary to other sources of information on African economies, the most useful being listed below.

Africa South of the Sahara (annual), published by Europa gives longer historical accounts for each country, with considerable detail on recent political devleopments. The sections on the economy contain helpful descriptive material on the main producing sectors of the economy. One strong feature of the volume is the section which deals with commodities produced in Africa and their importance in world markets.

Africa Confidential (fortnightly) presents information and analysis of political events that are generally available from other sources.

Africa Analysis (fortnightly) contains accounts of economic events and developments, many of which are assembled from inside sources. There is a particularly useful listing of black market exchange rates in each issue.

Africa Economic Digest (weekly) contains up-to-date information on developments in African economies, with the emphasis on immediacy rather than reflective analysis. A particular strength is the summaries of deliberations of international bodies and the implications of these for Africa.

Africa Business (monthly) has less emphasis on up-to the-minute information, but has good reports and analysis from some of its country specialists.

Country Reports (quarterly) from the Economist Intelligence Unit have the great virtue of providing coverage for all the African economies. There are sections on political developments as well as economic affairs. All of the reports contain an outlook section, and some of the reports have good short-term forecasts for up to a year ahead.

Country Profiles (annual) from the Economist Intelligence Unit are succinct summaries of the political and economic circumstances in African countries, and provide sound background material on each economy.

2

AFRICA AND THE WORLD

The African economies located south of the Sahara are considered in comparison with world groupings based on the classifications of the World Bank 1989. These are the other developing countries (DCs) group, comprising the World Bank low-income and middle-income countries less the African countries; the high-income countries; and the non-reporting (mostly Eastern bloc or former Eastern bloc) countries.

Countries with populations of 1m or less are not included in any of the groupings - their inclusion would have a negligible influence on the totals and averages that are computed.

In economic terms, South Africa has a considerable influence on some averages and totals, and where it is considered helpful, calculations are made both with and without South Africa.

Five countries in North Africa - Morocco, Algeria, Tunisia, Libya and Egypt - as well as Spanish Morocco are not included in the Africa group, but are included in the other DCs group. Two African countries, Angola and Namibia, are included in the non-reporting group.

There are 48 territories included in sub-Saharan Africa. Eleven have populations below 1m, Angola and Namibia are excluded as non-reporting, and South Africa is omitted as well. Thus the Africa group comprises 34 countries.

The survey begins by comparing the structures of the economies of the various groups, then looks at comparative economic performance, and concludes with forecasts for the period up to 1995.

Averages are computed using the same weighting systems as employed by the World Bank (1989b). For GDP, its components, and economic items such as inflation, GDP weights are used. For trade items, export and import values are employed as weights, and for items expressed on a per head basis, population weights are used

Economic Structure

Population and Land

Figure 2.1 gives population sizes and percentages of world totals.

Africa contains roughly 9.0% of the world's people, other DCs make up another 69%, the high-income 15% and non-reporting countries comprise the remaining 7%. The World Bank sub-divides the developing countries into low, lower middle-income and upper middle-income groups. Africa's population makes up 10% of the total population of the low income group, this group being dominated by India and China which together comprise 75%. Of the lower middle-income group, African countries comprise 21%. Of the upper middle-income group, Africa, including South Africa, makes up 6.4%. South Africa is the dominant African country in this group with a population of 33.8 million in 1987, the only others being Gabon (1 m), Réunion (600 thousand) and Seychelles (65 thousand).

It should be noted that some of these

Africa and the World

Figure 2.1 Population 1989

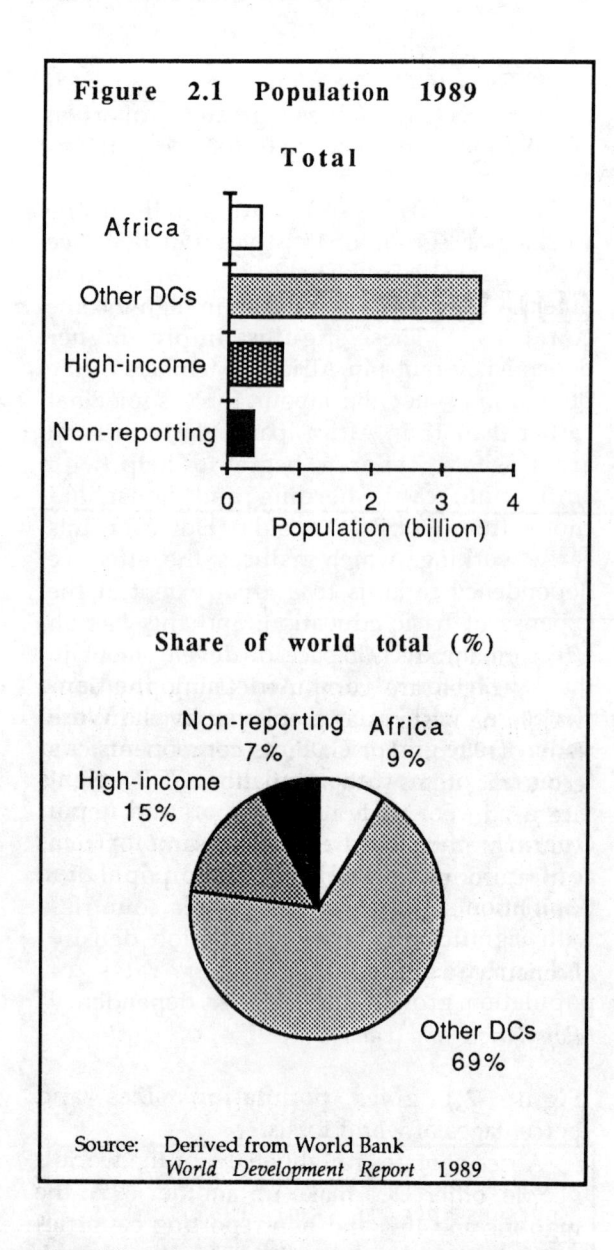

Total

[Bar chart showing population in billions for Africa, Other DCs, High-income, Non-reporting, with x-axis from 0 to 4 billion]

Share of world total (%)

[Pie chart: Non-reporting 7%, Africa 9%, High-income 15%, Other DCs 69%]

Source: Derived from World Bank *World Development Report* 1989

classifications are somewhat unstable, depending as they do on GNP in local currency being converted to $US at official exchange rates. As exchange rates vary, so do the classifications. South Africa is a case in point, being variously classified as upper middle-income (World Bank 1988b), lower-middle-income (World Bank 1989b) and upper middle-income (World Bank 1990). This problem is returned to later in this chapter, when alternative approaches to making inter-country comparisons are discussed.

Population growth rates in Africa, shown in Figure 2.2, are substantially higher than elsewhere.

What is striking is to compare Africa's rates of population growth at 3.2% a year with other DCs at 1.9% a year. Africa has a crude birth rate of 47 per thousand whereas for other DCs the rate is 28 per thousand. Crude death rates in Africa are 16 per thousand, and they are 9 per thousand for other DCs. These figures imply large family sizes, with Africa showing a fertility rate of 6.6, while for other DCs, the rate is 3.6. High birth rates in Africa which are only partly offset by higher death rates imply a population that will grow by 41% in the 11 years up to the year 2000. By contrast, the population of other DCs will rise by 23% in this period.

Figure 2.3 gives land areas, and percentages of total world land area. These are reflected in population densities, which are shown in Figure 2.4.

Despite fast rates of population growth, Africa has a low overall population density, 22 persons per square kilometre, with only the non-reporting countries significantly lower at 14 per square kilometre. Africa's density is under a third of that recorded for other DCs at 68 persons a square kilometre. However, the nature of the economic structure in Africa, which is highly dependent on agriculture, and the percentage of the land area which is fertile, need to be taken into account.

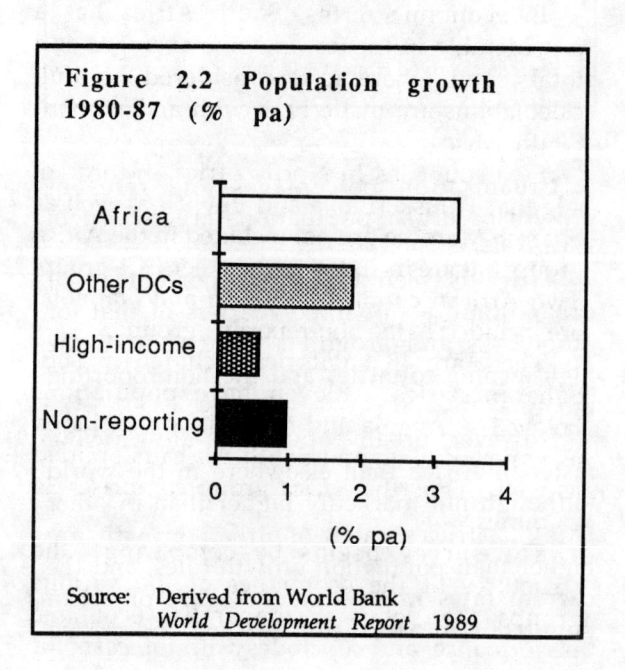

Figure 2.2 Population growth 1980-87 (% pa)

[Bar chart showing % pa for Africa, Other DCs, High-income, Non-reporting, with x-axis from 0 to 4]

(% pa)

Source: Derived from World Bank *World Development Report* 1989

Africa and the World

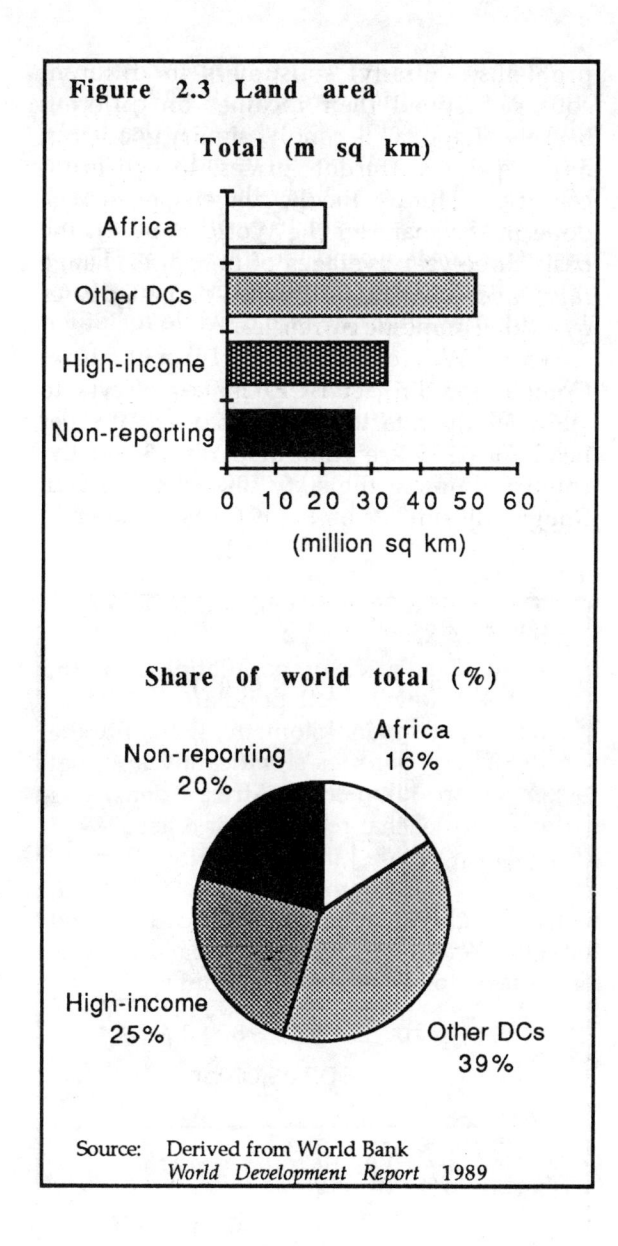

Figure 2.3 Land area

Total (m sq km)

[Bar chart showing:
- Africa: ~22
- Other DCs: ~52
- High-income: ~33
- Non-reporting: ~25
(million sq km)]

Share of world total (%)

[Pie chart:
- Africa 16%
- Non-reporting 20%
- High-income 25%
- Other DCs 39%]

Source: Derived from World Bank *World Development Report* 1989

Urbanisation percentages and rates of growth of urban populations are shown in Figure 2.5.

Africa is seen to have a percentage of urbanisation about three-quarters of that for other DCs and around a third that of the high-income countries and the non-reporting group.

However, urban populations are growing faster in Africa than elsewhere in the world (although not markedly higher than in other DCs). Africa's rates of urban growth are higher than population growth rates, and the current rates imply that the share of urban population in Africa rises by just under 1% a year. The rate of expansion of the urban areas can be expected to fall as a greater proportion of the population comes to be located in towns and cities.

50% of Africa's population fell in the working age-group of 15-64 years in 1985 (see World Bank 1988b). This is 8% lower than in other DCs, and 17% lower than in high-income countries. These figures imply higher dependency ratios in Africa. Members of the population enter the labour force somewhat earlier than 15 in Africa, particularly in rural areas where children begin to help with cultivation, cattle herding and household chores from 6 or 7 years old. However, this early working, which reduces the effective dependency ratio, is to a great extent at the expense of basic education, and this has an effect on the overall pace of development in the longer term. The African population-structure, with greater proportions in the under-15 age group, is a direct result of Africa's faster rate of population growth.

Overall, the picture emerges of Africa containing a small proportion of the population of the world's poorer countries with significantly lower population density and urbanisation, but higher rates of population growth and higher dependency ratios.

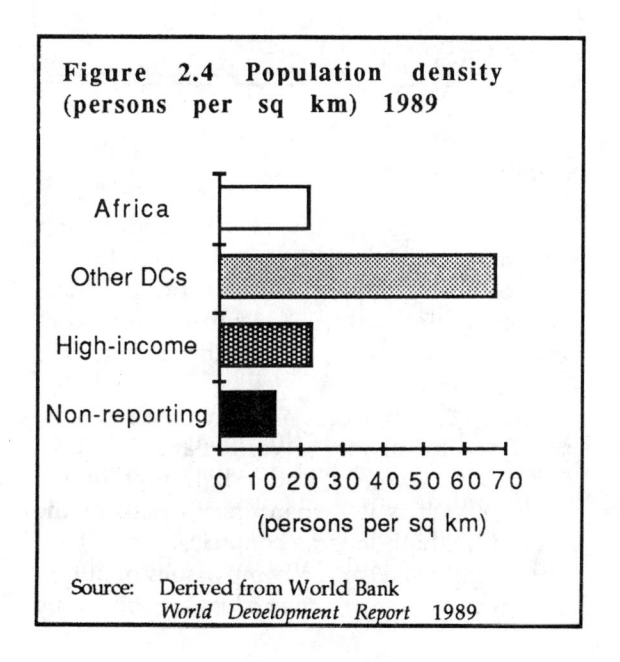

Figure 2.4 Population density (persons per sq km) 1989

[Bar chart showing:
- Africa: ~20
- Other DCs: ~68
- High-income: ~22
- Non-reporting: ~10
(persons per sq km)]

Source: Derived from World Bank *World Development Report* 1989

Africa and the World

Figure 2.5 Urbanisation 1987

Percent of total population

(bar chart showing Africa, Other DCs, High-income, Non-reporting on scale 0-80%)

Urban population growth 1980-87 (% pa)

(bar chart showing Africa, Other DCs, High-income, Non-reporting on scale 0-7 % pa)

Source: Derived from World Bank *World Development Report* 1989

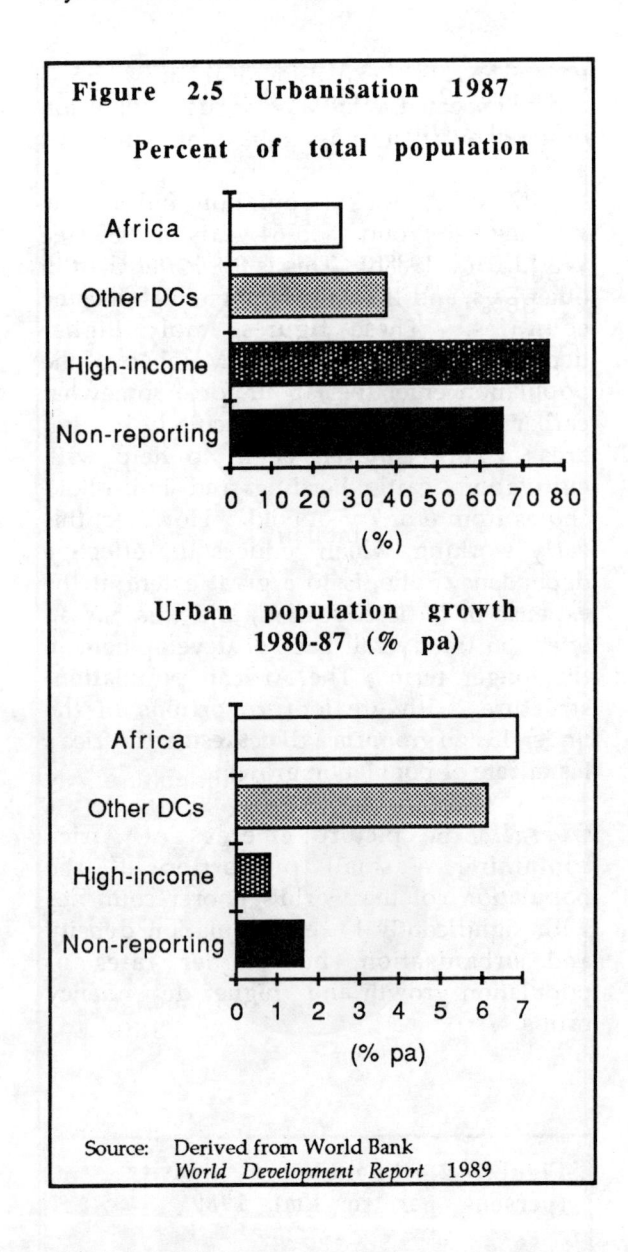

Output and Income

Figure 2.6 shows levels of output and percentage shares of the world total for the main country groupings. Thus Africa produces 1% of world output, (excluding the non-reporting countries), with the other DCs recording 17%.

Figure 2.7 shows levels of income per head calculated by standard World Bank methods. The Africa level, is under half that of the other DCs. It is less than one fortieth of the level in the high-income countries.

As mentioned earlier, comparisons of output and income across countries are beset by problems. Firstly, consumers in different countries spend their incomes on different baskets of goods. Secondly, similar goods can have quite different prices in different countries. Thirdly, the the conversions to $US, done in the main by the World Bank on the basis three-year averages of official exchange rates, over-estimate GDP when exchange rates overvalue domestic currencies.

The World Bank's International Comparison Project (ICP) makes efforts to allow for these factors. Adjusted incomes per head for 1985 are computed for 13 African countries among the sixty countries covered. Rough adjustments on the ICP basis, shown in

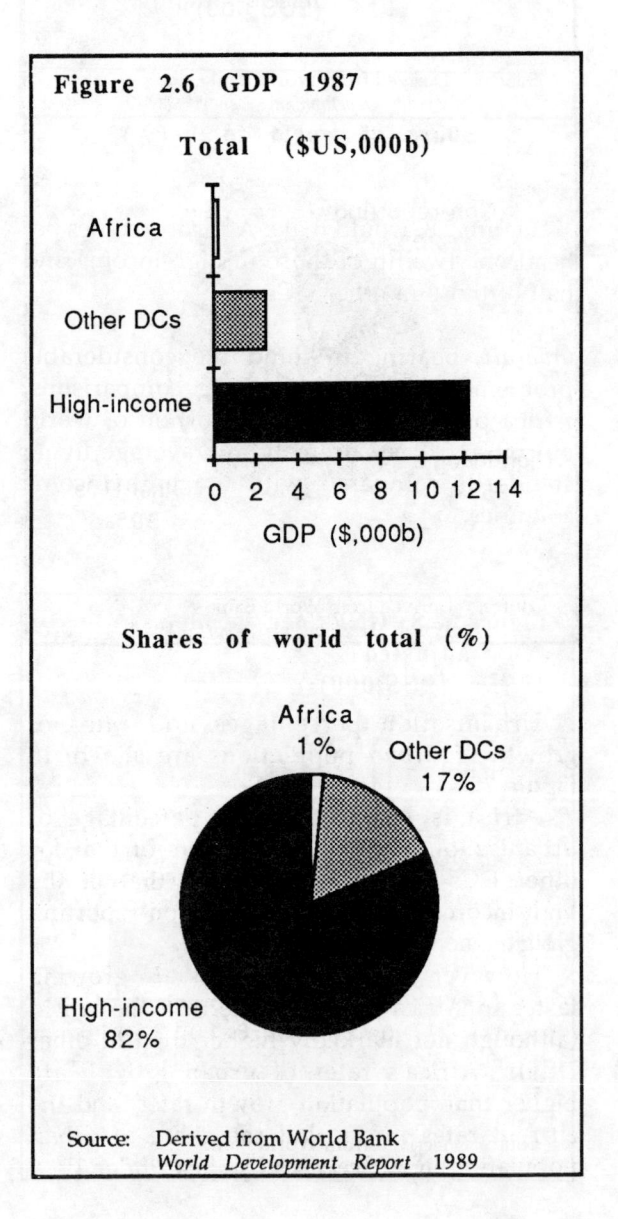

Figure 2.6 GDP 1987

Total ($US,000b)

(bar chart showing Africa, Other DCs, High-income on scale 0-14 GDP ($,000b))

Shares of world total (%)

(pie chart: Africa 1%, Other DCs 17%, High-income 82%)

Source: Derived from World Bank *World Development Report* 1989

Africa and the World

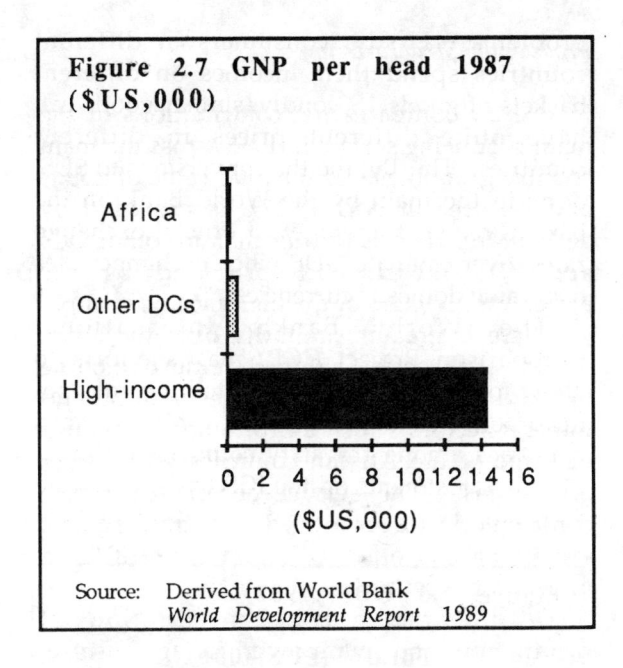

Figure 2.7 GNP per head 1987 ($US,000)

- Africa
- Other DCs
- High-income

($US,000)

Source: Derived from World Bank *World Development Report* 1989

Figure 2.8, would make African incomes per head one-twelfth of those in high-income and half of those in other DCs.

Overall, bearing in mind the considerable problems involved in making comparisons, Africa produces a small proportion of world output and has extremely low average living standards compared with the high income countries.

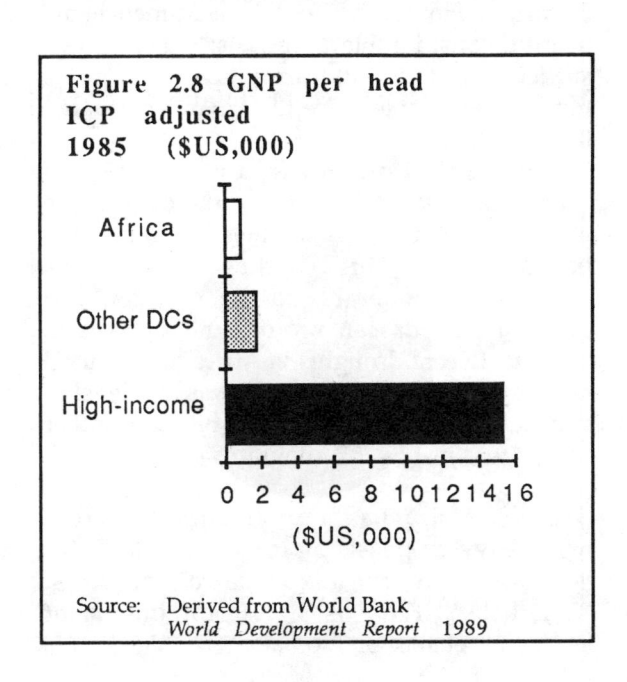

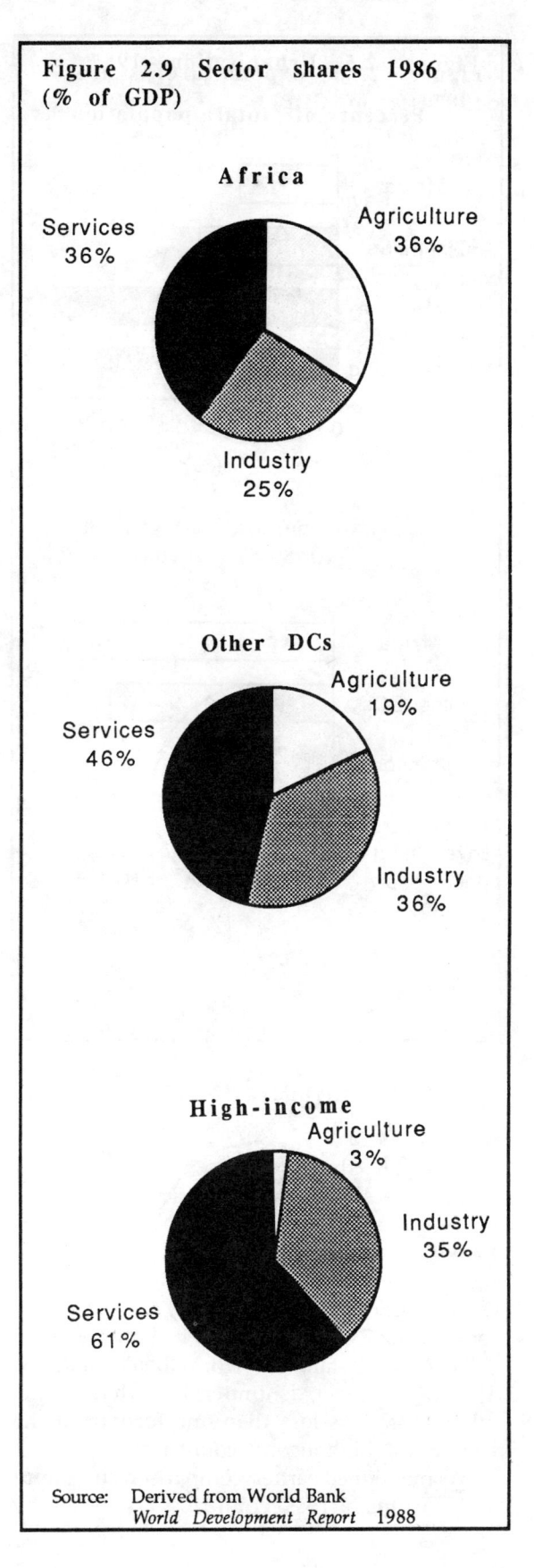

Figure 2.8 GNP per head ICP adjusted 1985 ($US,000)

- Africa
- Other DCs
- High-income

($US,000)

Source: Derived from World Bank *World Development Report* 1989

Figure 2.9 Sector shares 1986 (% of GDP)

Africa
- Services 36%
- Agriculture 36%
- Industry 25%

Other DCs
- Services 46%
- Agriculture 19%
- Industry 36%

High-income
- Services 61%
- Agriculture 3%
- Industry 35%

Source: Derived from World Bank *World Development Report* 1988

Africa and the World

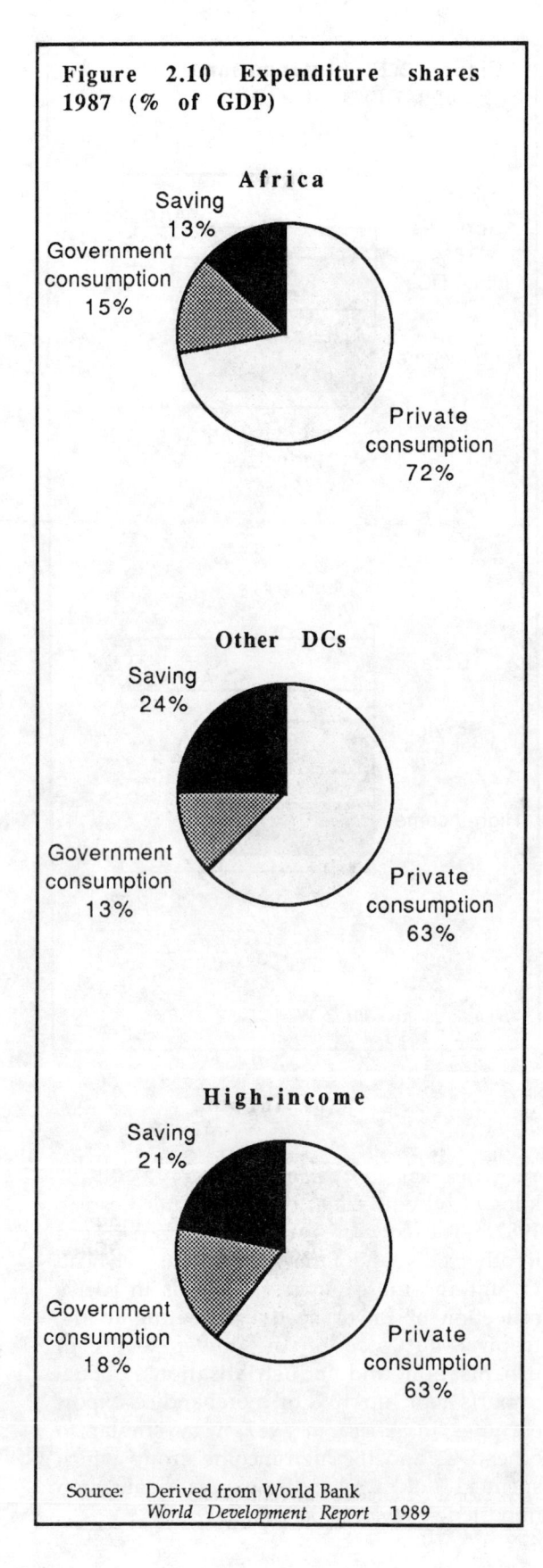

Figure 2.10 Expenditure shares 1987 (% of GDP)

Africa
- Saving 13%
- Government consumption 15%
- Private consumption 72%

Other DCs
- Saving 24%
- Government consumption 13%
- Private consumption 63%

High-income
- Saving 21%
- Government consumption 18%
- Private consumption 63%

Source: Derived from World Bank *World Development Report* 1989

Production Structure

Figure 2.9 compares the contributions of the main producing sectors in 1986 across the main country groupings. Of the total value of goods and services in Africa, 36% is provided by agriculture. This is greater than for other DCs at 19%, and the 3% provided by the agricultural sectors of the high-income group.

There is greater similarity over the share of industrial sectors, although it must be borne in mind that this sector includes both oil and mineral sectors which are prominent features of some of the African countries with larger economies. The service sectors in Africa contribute just over a third of output, against almost half in other DCs and three-fifths in high-income countries.

Despite the heavy concentration of production and manpower in agriculture, Africa imported more than 17 kilograms of cereals per head in 1987. However, this was less than in other DCs, where 26 kilograms of food per head were imported. There is no reason why food imports should indicate an inadequate ability on the part of the economies in providing for domestic needs. Indeed, it is argued that for many African countries the best use of agricultural resources is in producing valuable tropical crops for export and using the revenue to purchase cheap cereals produced in temperate, high-income countries. However, African governments tend to give considerable emphasis to being self-sufficient in food, and are prepared to pay a not inconsiderable price for the independence this represents.

Most agriculture in Africa is rain-fed, and compared with other parts of the developing world, particularly Asia, there is very little irrigation. This makes agriculture particularly vulnerable to the weather, and drought periods can produce heavy falls in output. Recent droughts in Africa have caused famines, and in 1987, food aid was at a level of almost 7 kilograms per head, whereas in other DCs it was just over 2 kilograms per head.

Overall, Africa has a production structure with heavy emphasis on agriculture, and this is mainly at the expense of services. Nonetheless, there is significant import of food, and reliance on food aid.

Africa and the World

Demand Structure

Figure 2.10 shows expenditure as a percentage of GDP in 1987 for the main country groups.

Africa commits a higher proportion of GDP to private consumption, 72%, whereas in other DCs and in the high-income group, it is close to 60%. The higher commitment to consumption is largely at the expense of saving. African saving was 13% of GDP in 1987, it was almost twice as large in other DCs at 24%, and in the high-income countries it was 21% of GDP. Government consumption as a percentage of GDP was 15% in Africa, higher than in other DCs at 13%, but lower than the high-income countries, where government consumption comprises 18% of GDP.

Poor savings performance is reflected in Africa in the low investment ratio in 1987, as shown in Figure 2.11. Investment was 16% of GDP in Africa, it was half as much again in other DCs, while it stood at 21% in high-income countries.

Figure 2.12 shows trade dependence. Africa exports a quarter of all its output, and this is a higher level of dependence on overseas markets than other DCs which export 20%, and the high-income group which exports 19%. Africa obtains 24% of domestic absorption of resources from overseas, whereas this is lower at 17% in other DCs and 16% in high-income countries.

Figure 2.11 Investment 1987 (% of GDP)

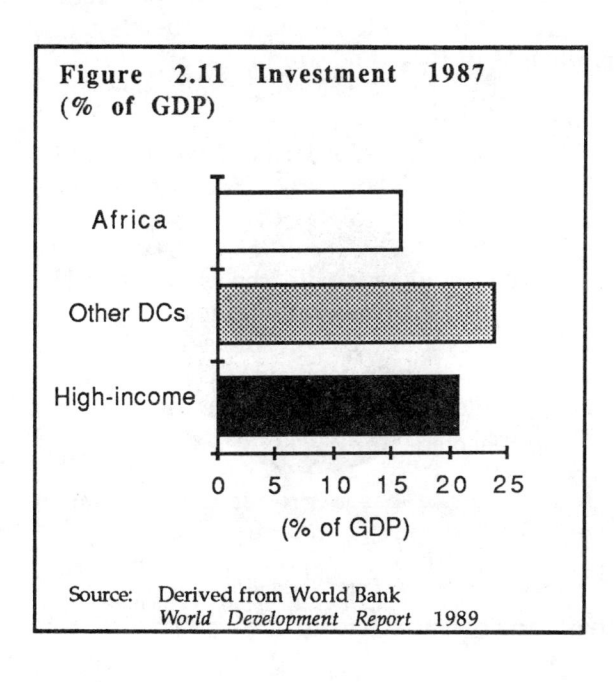

Source: Derived from World Bank *World Development Report* 1989

Figure 2.12 International trade 1987 (% of GDP)

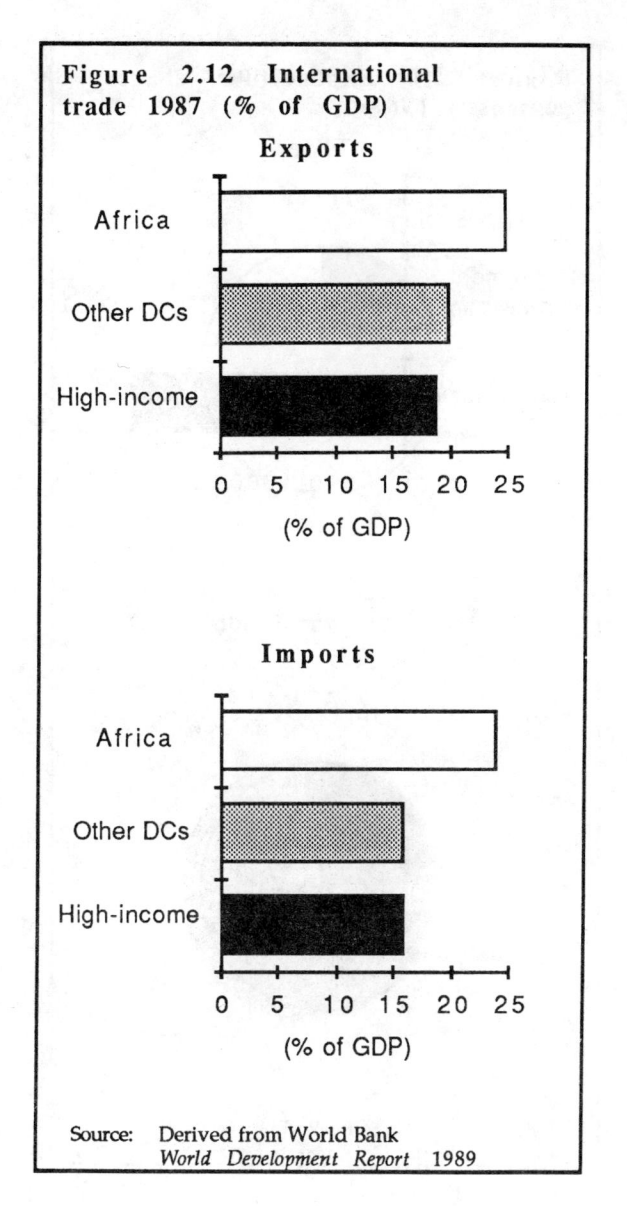

Exports

Imports

Source: Derived from World Bank *World Development Report* 1989

Africa has low usage of energy, with 82 kilos of oil equivalent consumed per head in 1987. This compares with 560 kilos per head in other DCs, and almost 5,000 kilos per head in high-income countries. This is in part a reflection of lower need for heating in the tropics, but also of the lower levels of urbanisation and industrialisation. Fuel imports took up 10% of merchandise export earnings in Africa in 1987, very similar to other DCs and the high-income group which spent 11% of merchandise export earnings on imported fuels.

Africa and the World

Figure 2.13 Merchandise exports 1987

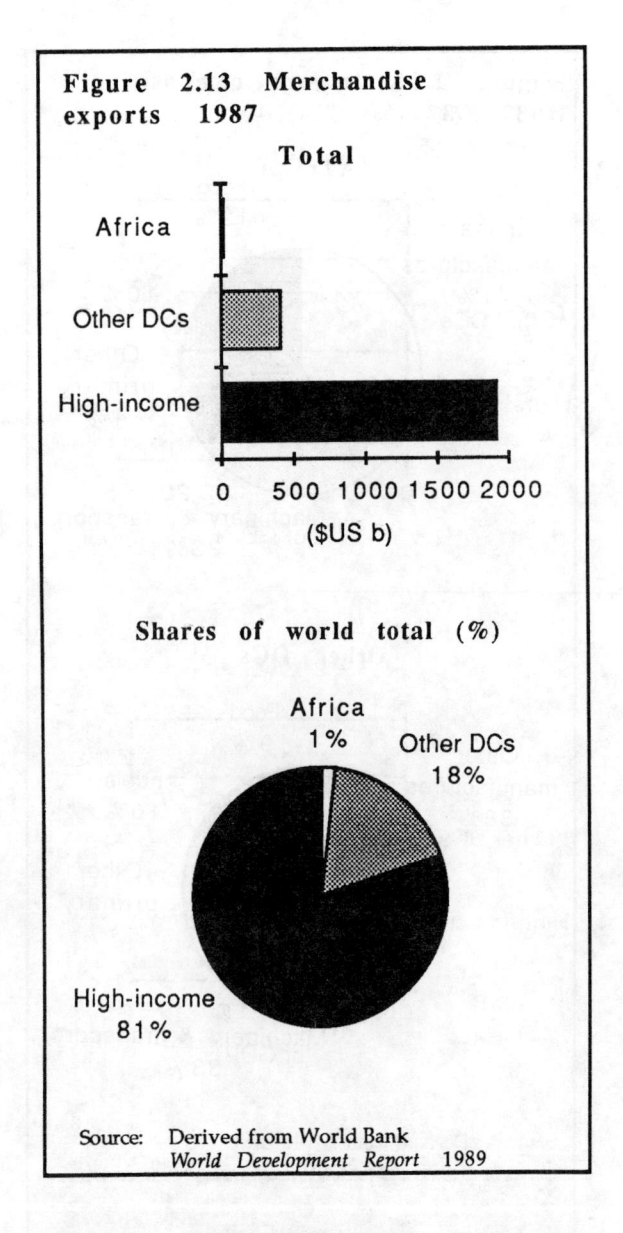

Total ($US b)

Shares of world total (%)
- Africa 1%
- Other DCs 18%
- High-income 81%

Source: Derived from World Bank *World Development Report* 1989

Overall, Africa's expenditure emphasises private consumption, at the expense of saving. Africa is noticeably more dependent on overseas trade than other country groupings, and has relatively low levels of energy usage.

International Trade

Africa was responsible for $US 29b (1%) of world merchandise exports in 1987 (see Figure 2.13).

Other DCs were responsible for $US 437b (18%) of world exports, and the high-income group $US 1,925b (81%). These figures are closely matched by expenditure on imports as shown in Figure 2.14.

48% of Africa's exports were fuels, minerals and metals, and 40% were other primary commodities, comprising agricultural, forestry and fishery products. Manufactures are 12% of Africa's exports. This primary product dependence is in marked contrast to the other DCs where 58% of exports are manufactures, and to the high-income group, where the proportion is 78% (see Figure 2.15).

In 1987, food, fuels and other primary products combined made up 26% of Africa's imports, the other 74% being machinery and manufactures (see Figure 2.16).

Figure 2.14 Merchandise imports 1987

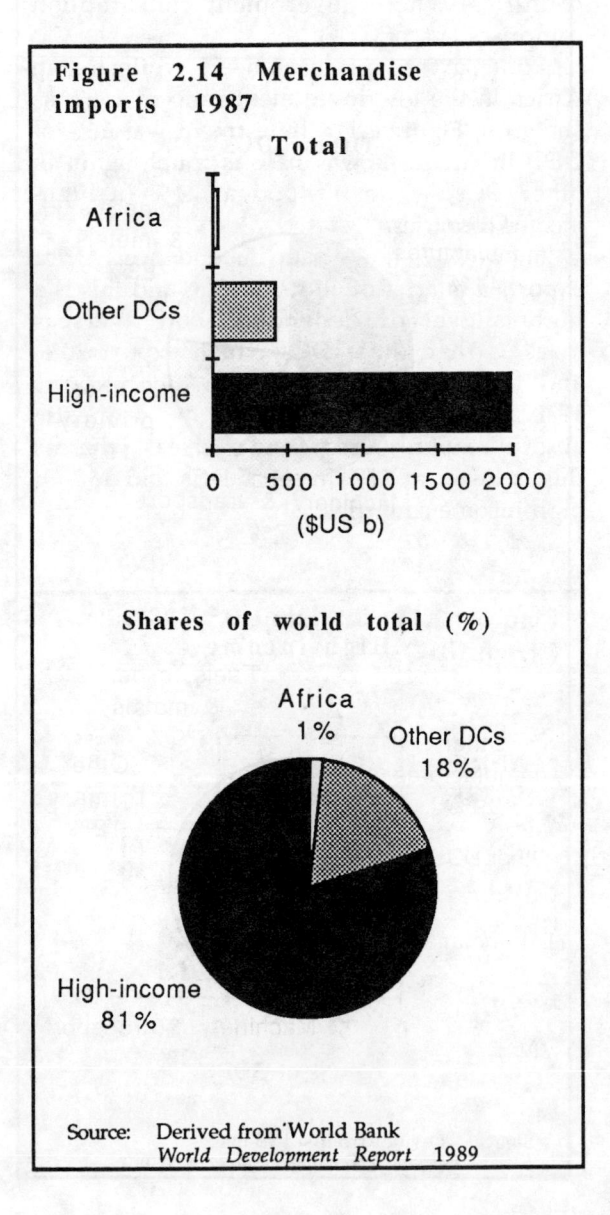

Total ($US b)

Shares of world total (%)
- Africa 1%
- Other DCs 18%
- High-income 81%

Source: Derived from World Bank *World Development Report* 1989

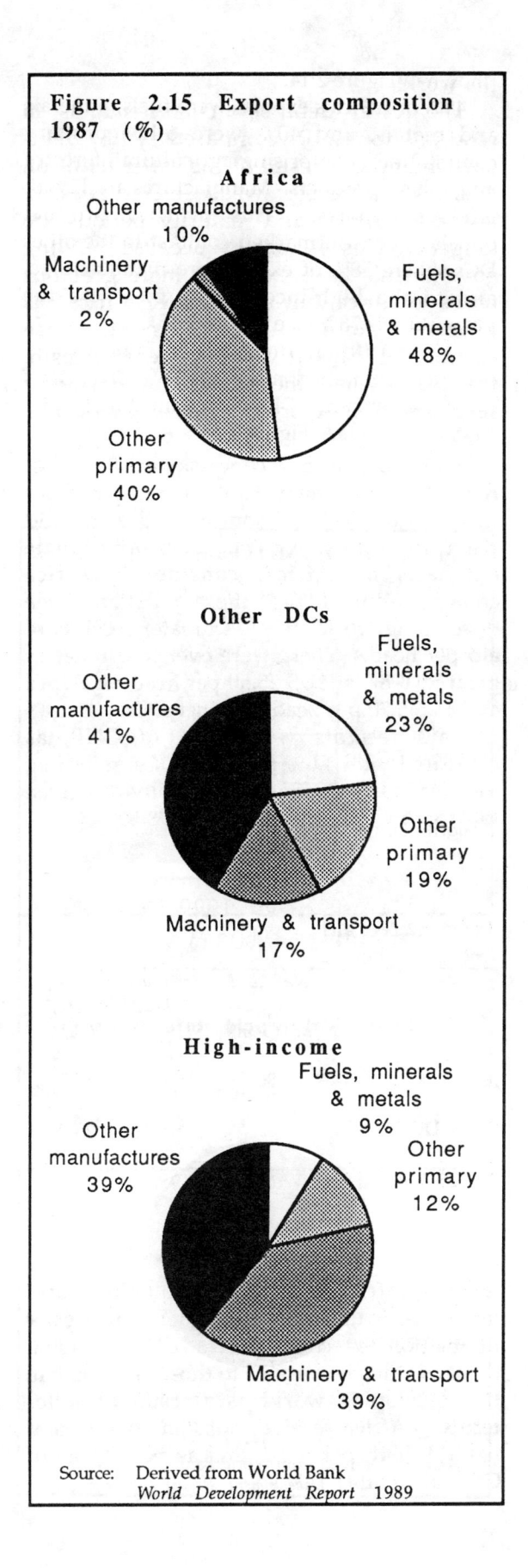

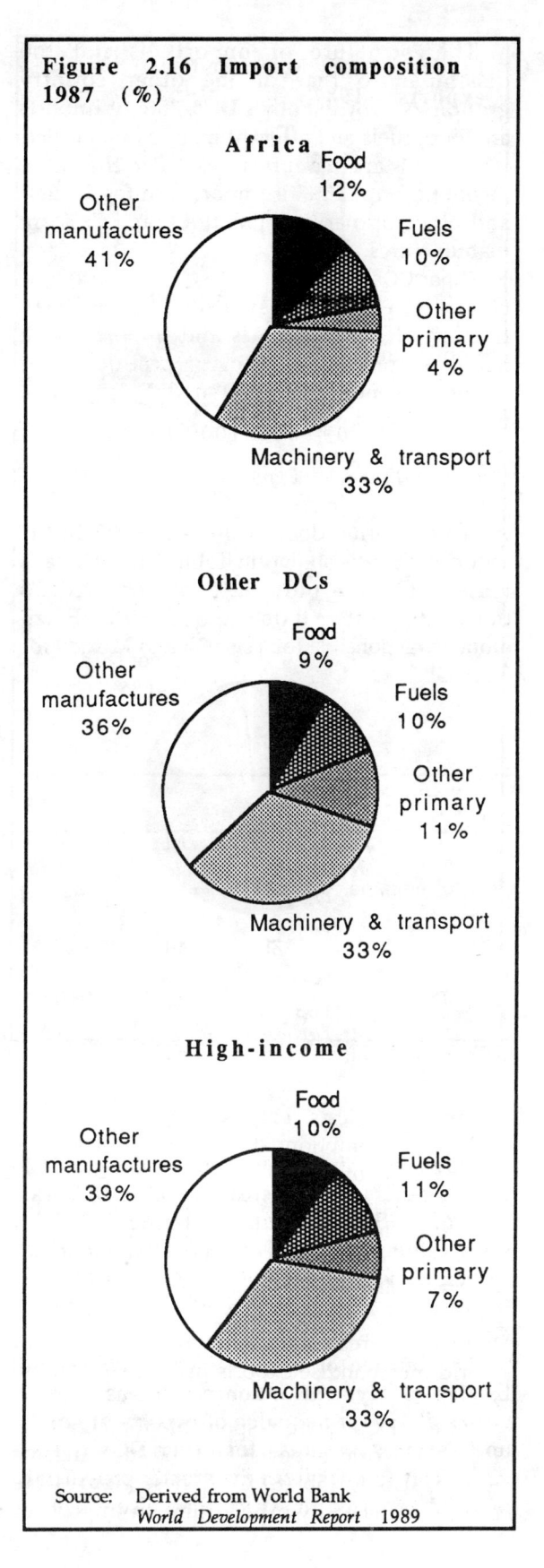

Africa and the World

Figure 2.15 Export composition 1987 (%)

Africa
- Other manufactures 10%
- Machinery & transport 2%
- Fuels, minerals & metals 48%
- Other primary 40%

Other DCs
- Other manufactures 41%
- Fuels, minerals & metals 23%
- Other primary 19%
- Machinery & transport 17%

High-income
- Other manufactures 39%
- Fuels, minerals & metals 9%
- Other primary 12%
- Machinery & transport 39%

Source: Derived from World Bank *World Development Report* 1989

Figure 2.16 Import composition 1987 (%)

Africa
- Other manufactures 41%
- Food 12%
- Fuels 10%
- Other primary 4%
- Machinery & transport 33%

Other DCs
- Other manufactures 36%
- Food 9%
- Fuels 10%
- Other primary 11%
- Machinery & transport 33%

High-income
- Other manufactures 39%
- Food 10%
- Fuels 11%
- Other primary 7%
- Machinery & transport 33%

Source: Derived from World Bank *World Development Report* 1989

Africa and the World

This structure of imports is not too dissimilar to that of the other country groupings. For the other DCs, 30% of imports are food, fuels and other primary commodities, while 69% are manufactures. For the high-income group, 28% of imports are food, fuels and other primary commodities, and 72% are manufactures.

Overall, Africa is responsible for a small fraction of world exports and imports, with heavy emphasis on primary products for export revenue. Imports mainly comprise machinery and manufactures.

Debt and Financial Flows

Africa's external debt position in 1987 in the world context is shown in Table 2.1. Africa is responsible for only 12% of the world's outstanding external debt at $US 104b. Brazil alone is responsible for $US 92b and Mexico for $US 93b.

Table 2.1 External debt
1987

	Total debt ($US b)	Debt to GDP (%)	Debt service to GDP (%)	Debt service to exports (%)
Africa	104	81	4.1	14.7
Other DCs	782	31	4.5	23.0

Source: Derived from World Bank 1989b

However, Africas outstanding debt is 81% of one year's GNP, whereas for other DCs it is 31% of GNP. Overall the burden of debt service (interest payments and repayments of principle) was about the same in Africa and the rest of the developing world at about 4% of GNP.

The burden of debt service as far as the balance of payments is concerned was less for Africa at 15% of the value of exports of goods and services, whereas for other DCs it was 23%. This follows from the greater proportion of GDP exported in Africa than elsewhere in the world.

The lower debt servicing demands in relation to GNP compared with debt-outstanding is a result of the softer terms on which Africa has borrowed. Interest rates are lower for Africa and maturity periods are longer. This is partly because lending governments have extended more generous terms to Africa, but also because Africa has generally found it less easy to raise commercial loans at variable interest rates in the 1970s when the balance of payments surpluses of oil exporters were being re-lent by banks.

Table 2.2 gives new concessionary flows less repayments of principle and interest payments (net aid) in 1987. About a third of net aid went to Africa. This disproportionate emphasis on Africa, considering Africa comprises only 12% of the population of the developing world, implies greater receipts of aid per head. These were over four times as great in 1987, at $US 25.50 per head for Africa and $US 5.70 per head for other DCs. Africa's net aid receipts were 8.3% of GNP, as compared with 1.0% for other DCs, reflecting the emphasis the donor community has given to Africa's circumstances in the 1980s.

Table 2.2 Net aid
1987

	Net aid ($US m)	Net aid per head ($US m)	Net aid to GNP (%)
Africa	81	4.1	14.7
Other DCs	31	4.5	23.0

Source: Derived from World Bank 1989b

Overall, Africa is a minor debtor in world terms. Although outstanding debts are greater in relation to economy size, debt servicing demands are comparable to those elsewhere in the developing world as a result of softer terms. Africa receives substantially greater net aid, both per head, and as percentage of GNP, than other LDCs.

Africa and the World

Figure 2.17 Central government budgets 1985-87 (% of GNP)

Government expenditure

- Africa (1985)
- Other DCs (1987)
- High-income (1987)

(% of GNP)

Government revenue

- Africa (1985)
- Other DCs (1987)
- High-income (1987)

(% of GNP)

Budget deficits

- Africa (1985)
- Other DCs (1987)
- High-income (1987)

(% of GNP)

Source: Derived from World Bank *World Development Report* 1989

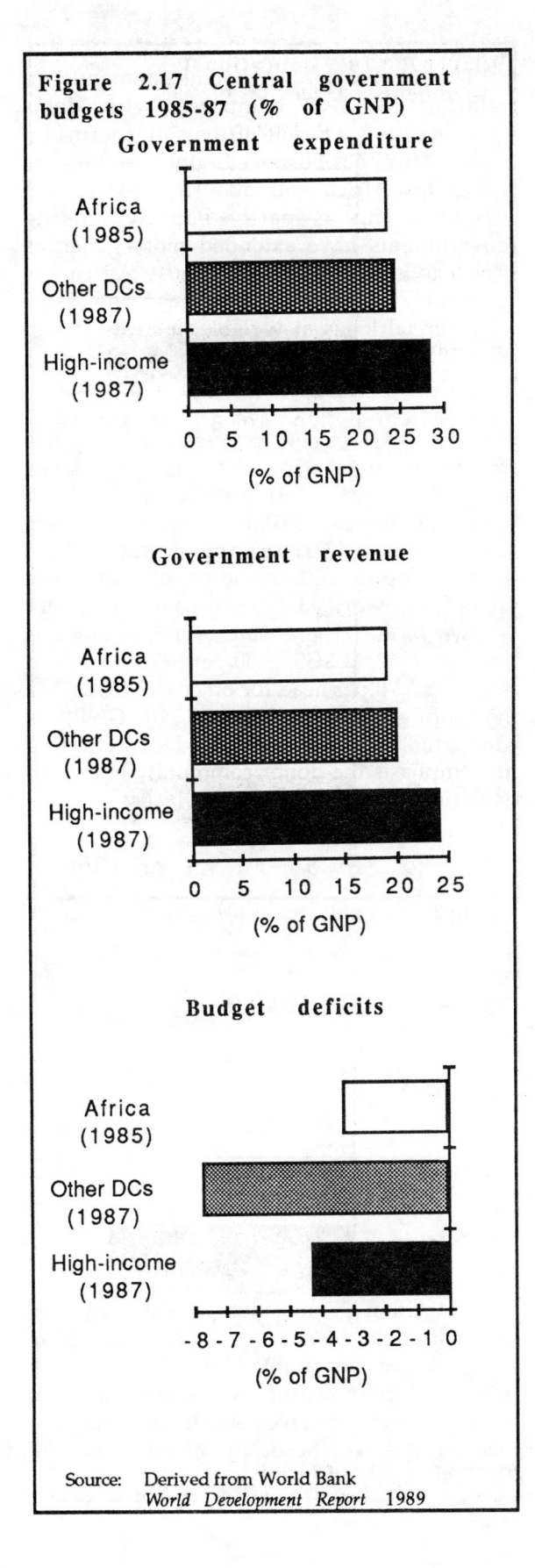

Government Sector

Figure 2.17 shows government expenditure as a percentage of GNP. The governments of Africa and other LDCs are responsible for similar expenditures as proportions of GNP, at around 24%. Governments do a greater share of spending in high-income countries at close to 29%.

Central government spending is a poor reflection of total public sector spending in comparisons made across country groupings as local government is responsible for a greater degree of educational and housing expenditure in high-income countries than in the developing world. Certain comparisons are perhaps instructive, however, with the African countries devoting almost 40% of central government spending to general administration in 1985, whereas this is 25% in other DCs (1985) and high-income countries (1987). A quarter of Africa's central government spending was on economic services in 1985. This was 22% in other DCs (1987), and under 10% in the high-income group (1987).

Government current revenue is shown in Figure 2.17. Africa (1985) and other DCs (1987) raised close to 20% of GNP in current revenue. The high-income group raised 24% (1987).

On the revenue side, Africa raised almost a quarter of total revenue from taxes or international trade in 1985. This was only 1% in high-income countries. Africa raised only 1.5% of total revenue from social security contributions, other LDCs raised 10% and high-income countries 30%. To a great extent, these differences reflect the constraints placed on revenue raising in Africa where most of the population are engaged in small farm agriculture with a substantial subsistence component.

Budget deficits as a percentage of GDP are shown in Figure 2.17. They are lowest in Africa (1985) at -3.3%, highest in other DCs (1987) at -7.7%, and stood at -4.3% in high-income countries (1987). It should be noted that the difference between government current revenue and expenditure does not correspond to the budget deficit. The deficit includes current and capital revenue and grants received, less total expenditure plus lending less repayments. Africa receives a net inflow

Africa and the World

on items other than expenditure and current revenue, and this lowers the budget deficit. The converse is true for other DCs, while for high-income countries these other items balance.

Overall, Africa has relatively modest levels of government spending and budget deficits, with heavy emphasis on general administrative spending by central government and on revenue raised from taxes on international trade.

Education and Health

Adult literacy rates for 1980 are given in Figure 2.18.

The low emphasis given to education in colonial Africa is reflected in the low overall literacy rate, where over three-fifths of adults were unable to read or write in 1980.

Literacy rates were almost 50% better in other DCs at 58%, and there was almost complete literacy in the high-income group at 97%. It must be noted that these literacy rates were compiled a decade ago, and the enrolment in primary education in this period in Africa is expected to have raised the current level of African literacy closer to 50% by 1990.

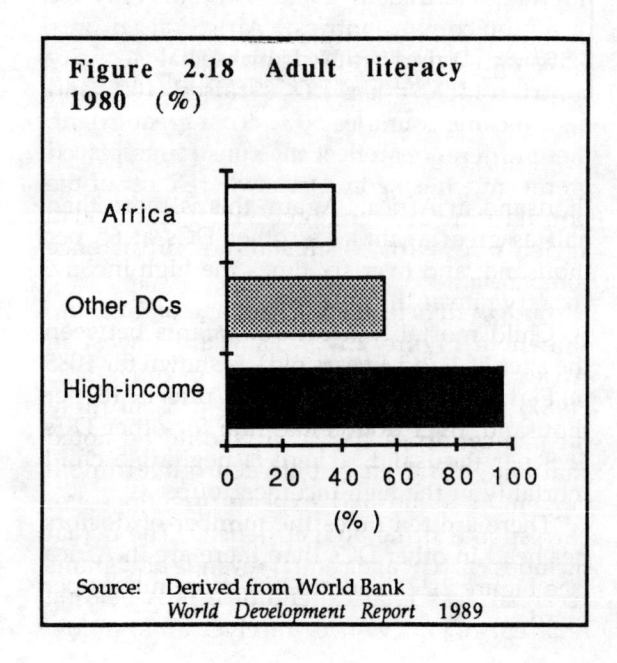

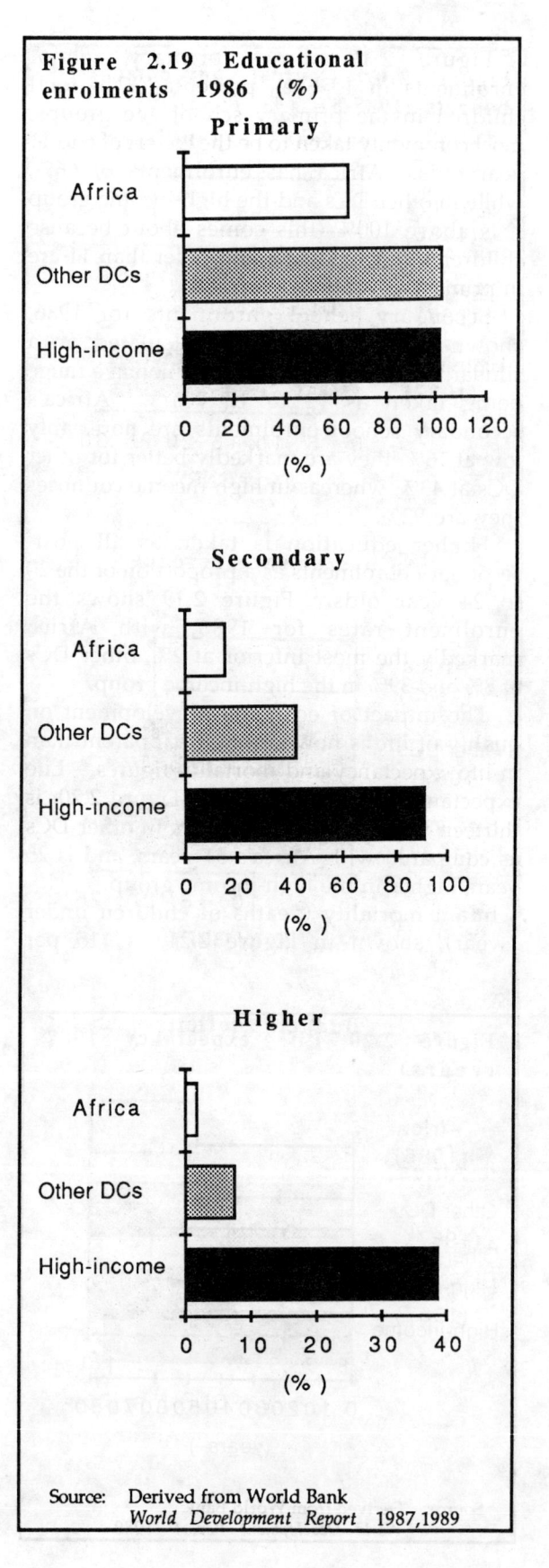

Figure 2.18 Adult literacy 1980 (%)

Source: Derived from World Bank *World Development Report* 1989

Figure 2.19 Educational enrolments 1986 (%)

Primary

Secondary

Higher

Source: Derived from World Bank *World Development Report* 1987, 1989

Africa and the World

Figure 2.19 shows primary school enrolments in 1986 as percentages of total children in the primary school age groups, most commonly taken to be the cohort of 6 to 11 year olds. Africa has enrolments of 66%, while in other DCs and the high-income group it is above 100% (this comes about because children younger than 6 and older than 11 are in primary education).

Secondary school enrolments for 1986, shown in Figure 2.19, are calculated on a similar basis, with the most common age range being taken as 12 to 17 years. Africa's secondary school enrolments are noticeably low at 16%, they are markedly better for other DCs at 43%, whereas in high-income countries they are 92%.

Higher education is taken as all post-secondary enrolments as a proportion of the 20 to 24 year olds. Figure 2.19 shows the enrolment rates for 1986, with Africa markedly the most inferior at 2%, other DCs at 8% and 39% in the high-income group.

The impact of economic development on quality of life is nowhere more apparent than in life expectancy and mortality figures. Life expectancy in 1987, shown in Figure 2.20, is thirteen years higher at 64 years in other DCs as compared with Africa's 51 years, and is 25 years higher in the high-income group.

Infant mortality (deaths of children under 1 year), shown in Figure 2.21, is 115 per thousand in Africa. Again, this is more than half as great again as in other DCs at 65 per thousand, and over six times the high-income country rate at 10 per thousand.

Child mortality (death of infants between the ages of 1 and 4 years old), is shown for 1985 in Figure 2.21. In Africa the rate is 18 per thousand, over double the rate for other DCs at 9 per thousand. There is negligible child mortality in the high-income groups.

There are ten times the number of doctors per head in other DCs than there are in Africa (see Figure 2.22), and are 17% more nurses per head.

Figure 2.21 Mortality rates (per thousand)

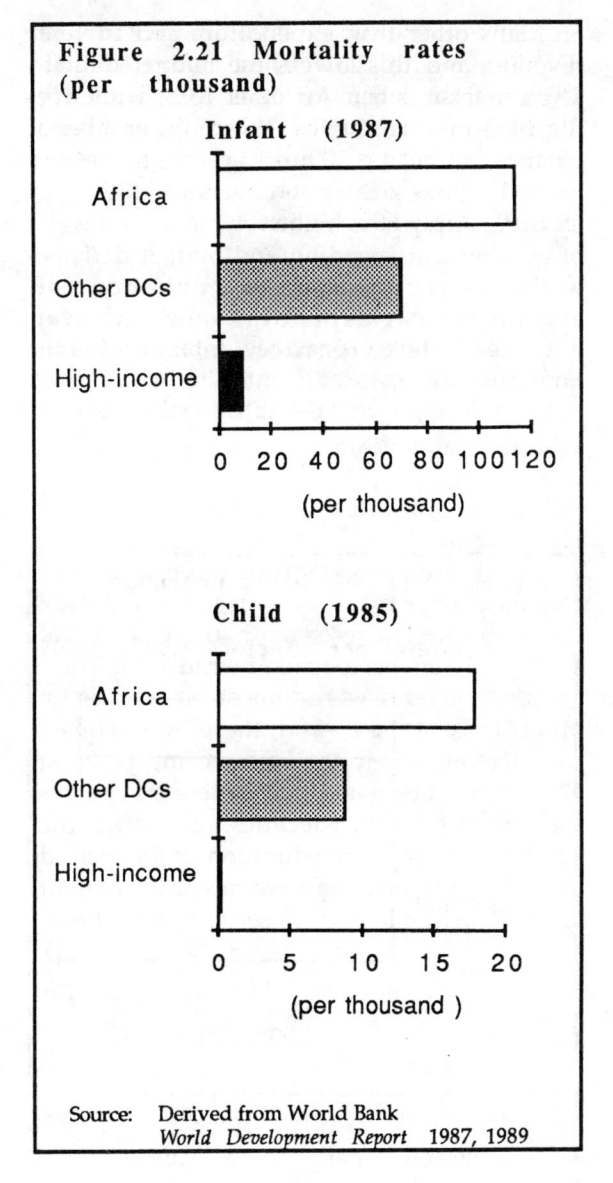

Infant (1987)

Child (1985)

Source: Derived from World Bank *World Development Report* 1987, 1989

Figure 2.20 Life expectancy 1987 (years)

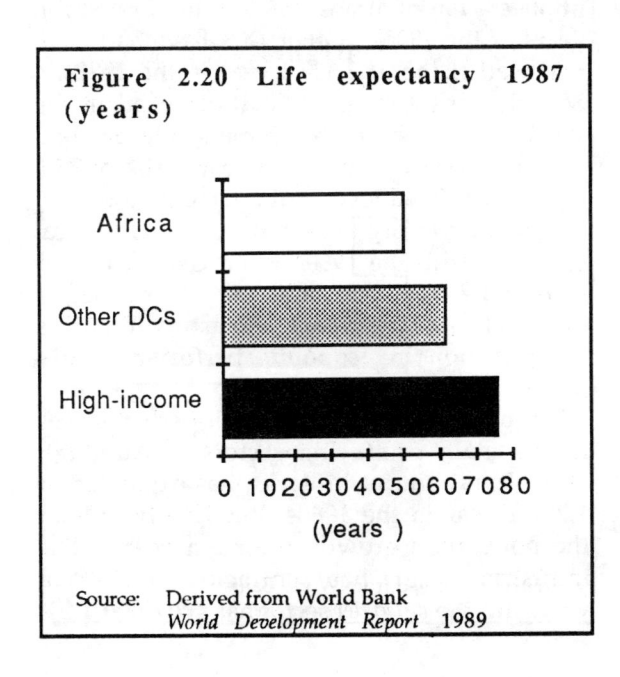

Source: Derived from World Bank *World Development Report* 1989

Africa and the World

Daily calorie supply is 26% higher in other DCs (see Figure 2.23). These comparisons are even starker when Africa is compared with the high-income countries where the number of doctors per head is 50 times greater, nurses per head 16 times greater and average daily calorie supply 60% higher.

Overall, Africa has lower literacy and educational enrolment rates than elsewhere in the world. Life expectancy, infant and child mortality are markedly inferior in Africa as are physician, nurse and nutritional provision.

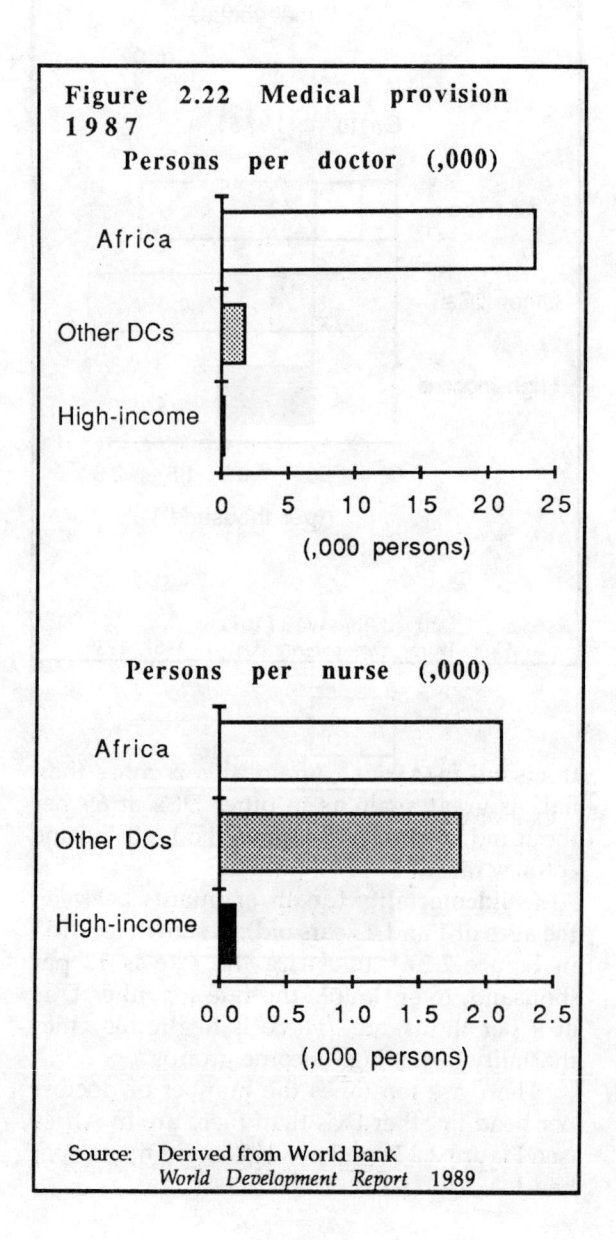

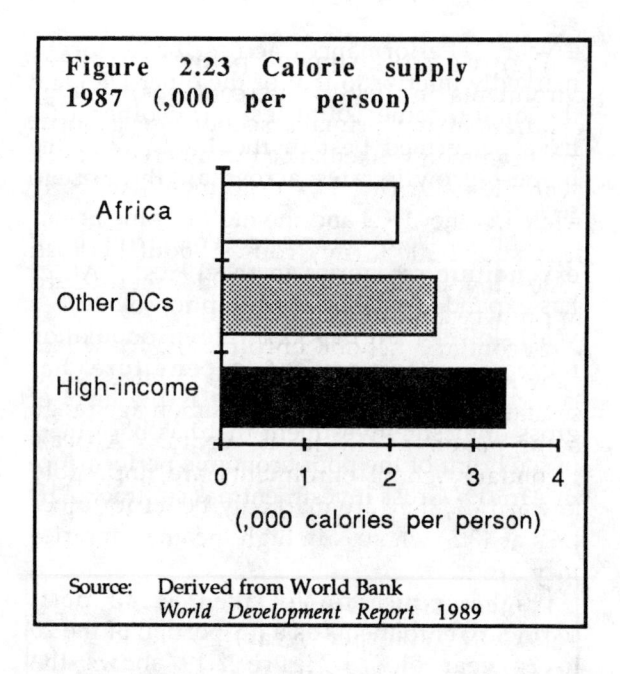

Figure 2.22 Medical provision 1987
Persons per doctor (,000)

Persons per nurse (,000)

Source: Derived from World Bank
World Development Report 1989

Figure 2.23 Calorie supply 1987 (,000 per person)

Source: Derived from World Bank
World Development Report 1989

Economic Performance

Figure 2.24 shows GDP growth in the 1980s. Africa has had falling GDP over the period 1980-87, at a rate of -1.3% a year. Other DCs have expanded GDP at almost 4.5% a year in the same period, and the high-income countries have had economies growing at 3.4%.

When population growth rates are taken into account, to give GDP per head growth rates, Africa has had average output per head falling at -4.5% a year (see Figure 2.24). This implies a fall of almost -28% in the first seven years of the 1980s. Other DCs have had GDP per head rising at 3.1% a year in the 1980s, a 24% rise in the level of GDP per head in the seven years. The high-income group has had GDP per head rise at 2.7% a year, a rise of 20% in GDP per head level in the same period.

Growth performance in the 1980s for Africa, contrasts with the 1960s when GDP per head rose at 1.4 % a year, and the 1970s when it grew at 0.2% a year. Reasons for this deterioration in economic performance are taken up in Chapter 3

Figure 2.25 gives the growth performances in the main producing sectors. Although agricultural output in Africa has expanded at 1.2% a year in the 1980s, this is slower than the population growth of 3.2% a year. The industrial sector has contracted at -1.2% a year, and the services sector has grown at 1.2%

Africa and the World

a year. Performances across the sectors is markedly inferior to that of the other DCs and the high-income countries. The other DCs have performed best in the 1980s, with the highest growth rates across all three main sectors.

Figures 2.26 shows recent growth of expenditure categories in the 1980s. Africa has expanded private consumption by 1.1% a year, slower than the 3.2% rate of population increase. Government expenditure has contracted at -1.0% a year. It is the level of gross domestic investment that has born most of the brunt of the poor economic performance in Africa. Gross investment has contracted by

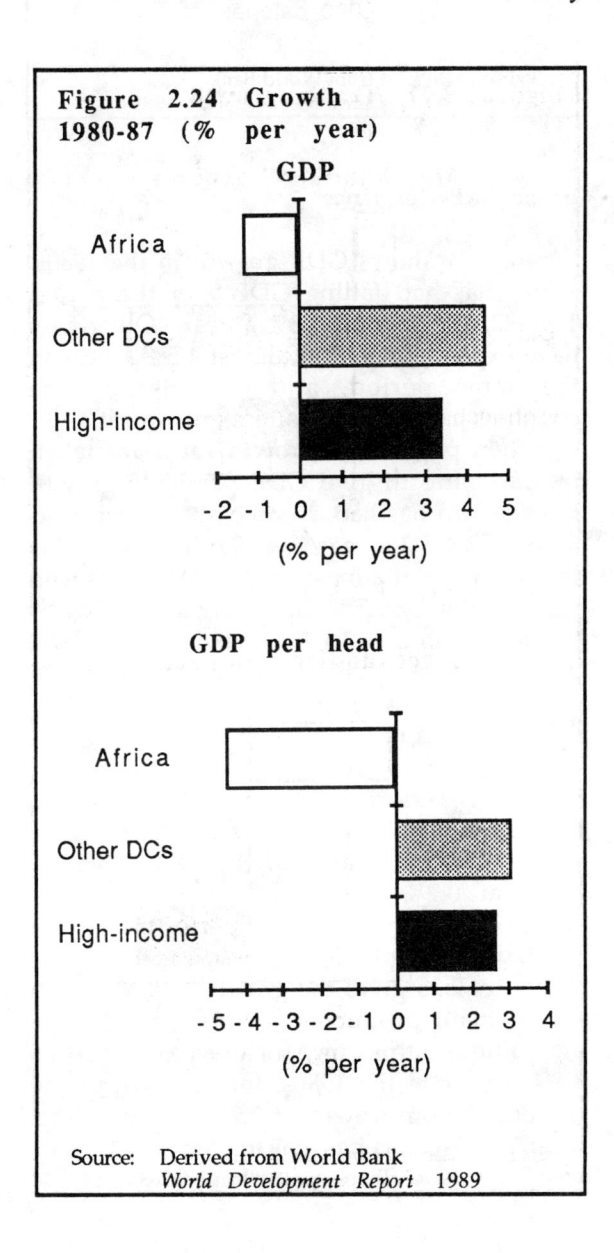

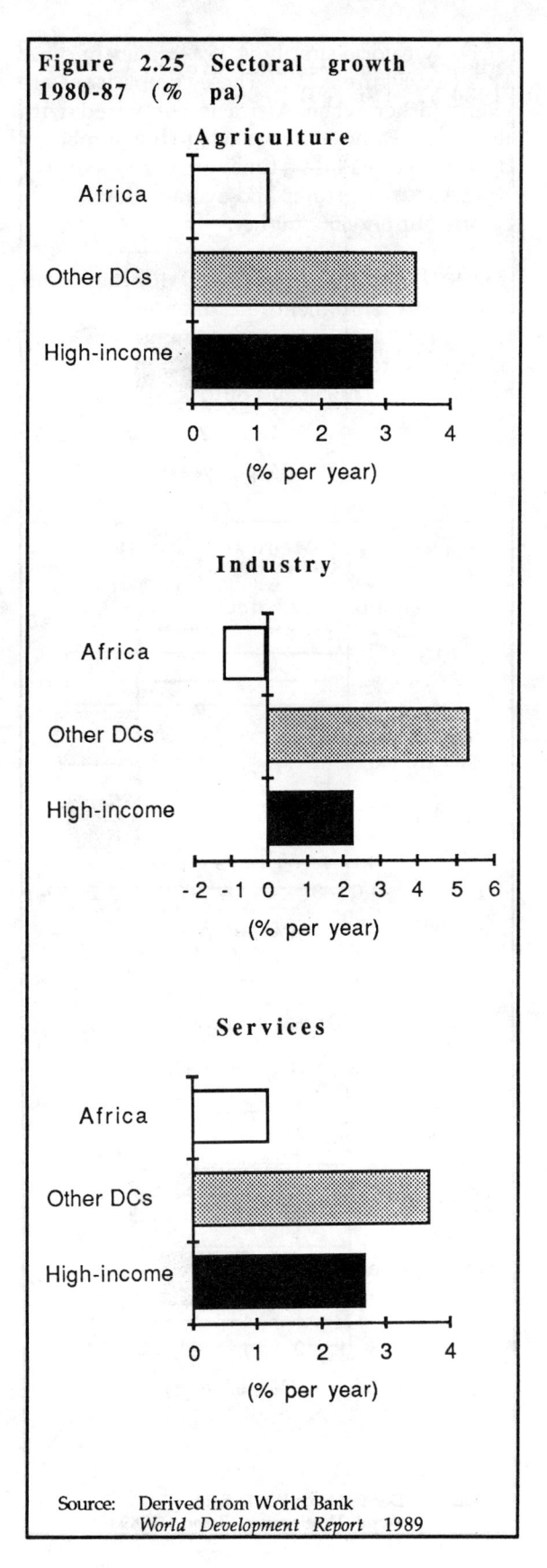

Figure 2.24 Growth 1980-87 (% per year)

GDP

GDP per head

Source: Derived from World Bank *World Development Report* 1989

Figure 2.25 Sectoral growth 1980-87 (% pa)

Agriculture

Industry

Services

Source: Derived from World Bank *World Development Report* 1989

Africa and the World

Figure 2.26 Demand growth 1980-87 (% pa)

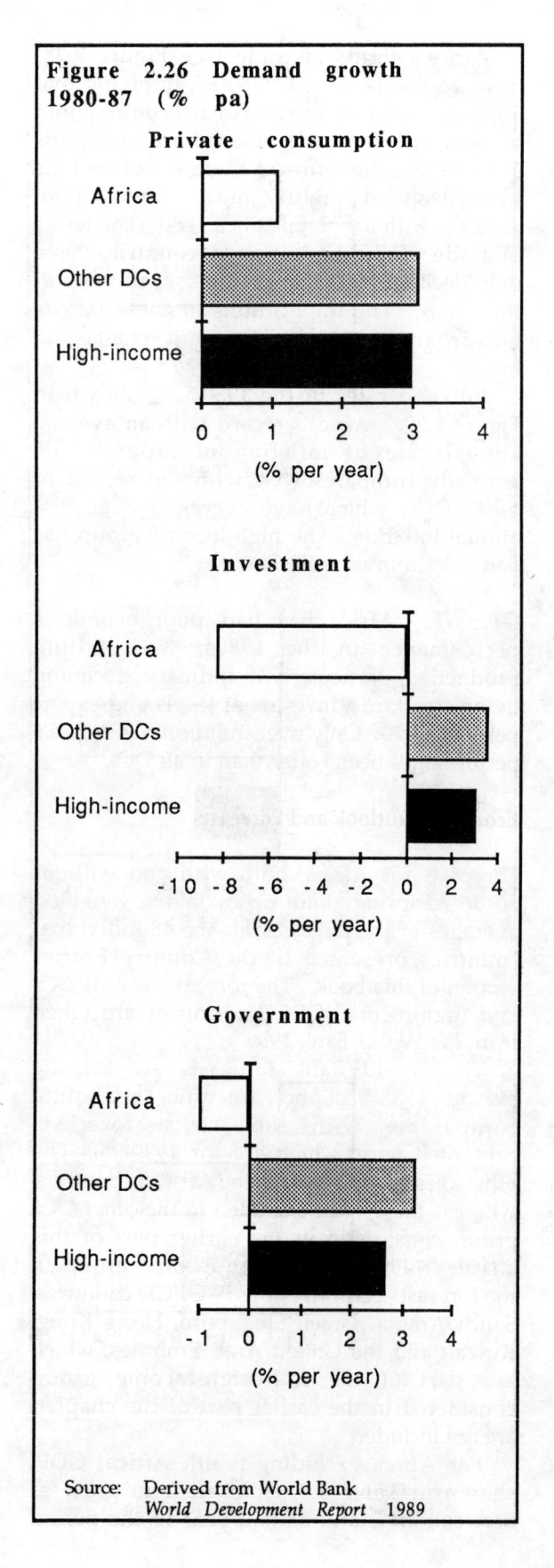

Private consumption
- Africa: ~1
- Other DCs: ~3
- High-income: ~3
(% per year)

Investment
- Africa: ~-8
- Other DCs: ~3
- High-income: ~3
(% per year)

Government
- Africa: ~-1
- Other DCs: ~3
- High-income: ~2.5
(% per year)

Source: Derived from World Bank *World Development Report* 1989

-8.3% a year. Investment levels in 1987 were 55% of their levels in 1980.

Ability to increase expenditure has been noticeably inferior in Africa compared with other DCs and the high-income countries, where better economic performance has enabled expenditure to rise in all categories by around 3.0% a year.

Trade performance is shown in Figure 2.27 for 1980-87. Africa's export volumes have fallen at -1.0% a year, and this, with declining terms of trade (see Figure 2.28) and rising levels of debt service as a proportion of export earnings has compressed import

Figure 2.27 Trade growth 1980-87 (% pa)

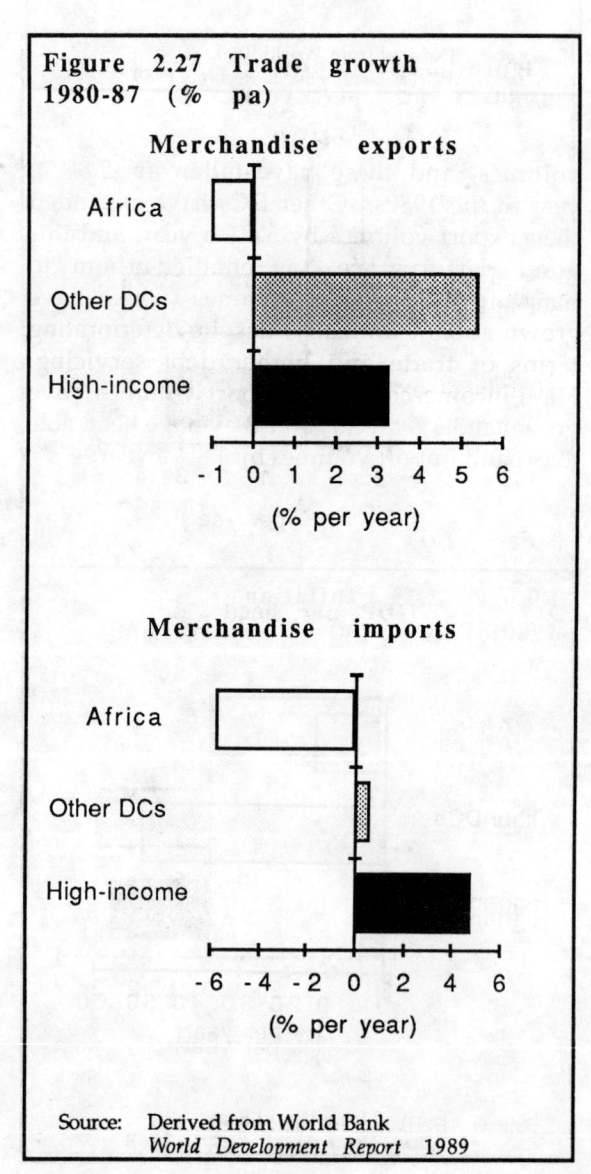

Merchandise exports
- Africa: ~-1
- Other DCs: ~5
- High-income: ~3
(% per year)

Merchandise imports
- Africa: ~-5
- Other DCs: ~0
- High-income: ~4
(% per year)

Source: Derived from World Bank *World Development Report* 1989

Africa and the World

Figure 2.28 Terms of trade 1987 (1980 = 100)

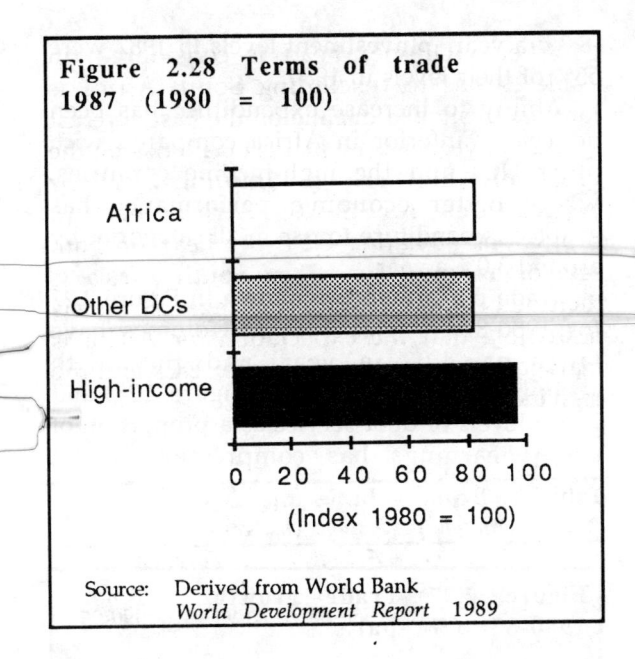

(Index 1980 = 100)

Source: Derived from World Bank *World Development Report* 1989

volumes, and these have fallen at -5.8% a year in the 1980s. Other DCs have expanded their export volumes by 5.4% a year, and this good performance has enabled them to maintain their import volumes (which have grown at 0.6% annually) despite deteriorating terms of trade and higher debt servicing. High-income countries' export volumes have grown at 3.3% a year, and they have been able to expand import volumes by 4.8% a year.

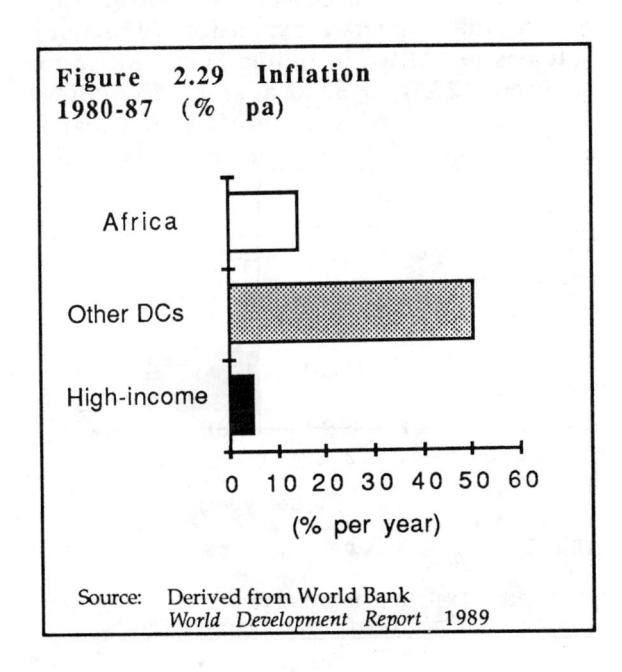

Figure 2.29 Inflation 1980-87 (% pa)

(% per year)

Source: Derived from World Bank *World Development Report* 1989

Africa's terms of trade (see Figure 2.28) have declined by -16% 1980-87, implying that a greater volume of exports has been required in 1987 to purchase the same volume of imports as in 1980. The terms of trade for other DCs have declined slightly more rapidly than Africa's, with a 17% fall since 1980. The terms of trade of the high-income countries have only declined by 3% in the first seven years of the 1980s. The implications of these factors for African economic development are taken up later.

Inflation rates in the 1980s are shown in Figure 2.29. Africa's record with an average annual rate of inflation of around 15% annually compares well with the record of other DCs, which have averaged over 50% annual inflation. The high-income group has had 5.2% annual inflation.

Overall, Africa has had poor economic performance in the 1980s, with falling production, particularly in industry, declining living standards, investment levels and export performance. Only over inflation has Africa's performance been better than in all DCs.

Economic Outlook and Forecasts

Forecasts for Africa, both with and without South Africa, are built up by taking weighted averages of the forecasts for the 48 individual countries presented in the Country Entries section of this book. The forecasts for all DCs and high-income OECD countries are taken from the World Bank 1989.

The World Bank forecasts are for the period 1988-95, and are thus not quite comparable with the Africa forecasts presented here, which are for 1990-95. The other DCs group includes the African countries, whereas these were excluded in the other DCs group considered in the earlier part of this chapter. Finally, the high-income group in the forecasts comprises the 19 OECD countries. Saudi Arabia, Israel, Singapore, Hong Kong, Kuwait and the United Arab Emirates, which are part of the full high-income group considered in the earlier part of the chapter, are not included.

For Africa excluding South Africa, GDP, shown in Table 2.3, is forecast to grow at between 2.9% and 3.5% a year in the period

Africa and the World

1990-95. If realised, this will be a considerable improvement on the record for the 1980-87 period when GDP expanded at 0.4% a year. The World Bank low forecasts for 1988-95 for this group are slightly more optimistic, projecting 3.1% GDP growth, but the high forecasts are more pessimistic at 3.2%.

Table 2.3 GDP forecasts
(average annual percentage change)

	Actual 1980-87	Forecast Base 1990-95	Forecast High 1990-95
Africa (excl. S. A.)	0.4	2.9	3.5
Africa (incl. S. A.)	0.6	2.4	2.9
All DCs	4.0	3.7[a]	4.6[a]
High-income OECD	2.7	2.4[a]	2.6[a]

[a] 1988-95 World Bank 1989b

When South Africa is included, the base forecast is lower at 2.5% a year, and the high forecast is 2.9%. Despite the improvements projected compared with the 1980s, the African countries are still expected to grow less rapidly than the other DCs, although their growth is forecast to be faster than the OECD group.

Table 2.4 GDP per head forecasts
(average annual percentage change)

	Actual 1980-87	Forecast Base 1990-95	Forecast High 1990-95
Africa (excl. S. A.)	-2.7	-0.2	0.4
Africa (incl. S. A.)	-2.4	-0.6	-0.1
All DCs	2.0	1.8[a]	2.7[a]
High-income OECD	2.0	1.9[a]	2.1[a]

[a] 1988-95 World Bank 1989b

When GDP per head is considered, in Table 2.4, Africa excluding South Africa is forecast to show a decline on the base forecast of -0.2% a year, with a modest 0.4% rise on the basis of the high assumptions. Poor expectations for South Africa lead to projections of falling GDP per head on both sets of assumptions when South Africa is included in the Africa aggregates. It is noticeable that the expectation for Africa is markedly inferior than for other DCs and the high income group.

Table 2.5 Export volume forecasts
(average annual percentage change)

	Actual 1980-87	Forecast Base 1990-95	Forecast High 1990-95
Africa (excl. S. A.)	-1.0	3.6	4.8
Africa (incl. S. A.)	-0.6	2.2	3.0
All DCs	5.0	4.1[a]	5.1[a]

[a] 1988-95 World Bank 1989b

Export volume forecasts, shown in Table 2.5, have Africa excluding South Africa expanding export volumes at between 3.6% and 4.8% a year. Again poor prospects for South Africa reduce the expected expansion of export volumes for Africa including South Africa to between 2.2% and 3.0% a year. Excluding South Africa, Africa is forecast to expand export volumes at rates that are reasonably

Table 2.6 Import volume forecasts
(average annual percentage change)

	Actual 1980-87	Forecast Base 1990-95	Forecast High 1990-95
Africa (excl. S. A.)	-5.8	3.0	3.8
Africa (incl. S. A.)	-6.9	1.7	2.3
All DCs	0.1	4.6[a]	5.7[a]

[a] 1988-95 World Bank 1989b

Africa and the World

comparable with those projected for the all DCs group by the World Bank for 1988-95.

Import volumes for Africa excluding South Africa, shown in Table 2.6 are forecast to expand at between 3.0% and 3.8% a year, and these rates are lower, at between 1.7% and 2.3% when South Africa is included. All DCs are forecast by the World Bank to expand their imports rather faster at between 4.6% and 5.7%.

Inflation rates for Africa excluding South Africa are forecast to accelerate from around 15% a year in the 1980s to between 19% and 23% a year, and these rates are slightly lower when South Africa is included. It is noticeable that African inflation rates are expected to be more than four times greater than in the high-income OECD countries. In general this can be expected to put downward pressure on African exchange rates where these are adjusted to reflect market valuations, or to widen the divergence between parallel and official rates when governments are reluctant to devalue.

Table 2.7 Inflation forecasts
(average annual percentage change)

	Actual 1980-87	Forecast Base 1990-95	Forecast High 1990-95
Africa (excl. S. A.)	15.2	19.6	23.2
Africa (incl. S. A.)	14.7	18.0	20.7
High-income OECD	5.0	4.1[a]	4.2[a]

[a] 1988-95 World Bank 1989b

3
THE AFRICAN ECONOMIES

The forty-eight economies of Africa are considered in various groupings based on geographical region, population and economic size, income level, economic zone, and stability. Each section is summarised at its conclusion in an endeavour to discern patterns and causality in the structure and performance of the various groups.

There is then a section discussing economic policies and performance in the past three decades, followed by an examination of the liberalising economic reform programmes that have been introduced in the 1980s. The chapter concludes with a summary of economic prospects.

Geographical Regions

Groupings of the African countries by geographical region are given in Table 3.1. These groupings are somewhat arbitrary, but they have become a common point of reference. Some institutions, particularly organisations dealing with intra-Africa trade, regional markets and transport have based policies on geographical groupings. Certain regional features, such as average population size, are of little significance, being largely the reflection boundaries established in the colonial period. However, there are some significant differences in in the economic structures and economic performance of the regional groups.

Central Africa is the most urbanised region with 39% of the population living in towns in 1985, with Southern Africa the least at 24%. East and West Africa both have urbanisation rates close to 30%. Population density is highest in West Africa, 31 persons per square kilometre, followed by East Africa (24), while in Central and Southern Africa densities are significantly lower at around 11 persons per kilometre. Population growth rates are close to 3% in all regions.

In terms of income, West Africa had the highest average in 1986 at $US 510 per person, followed by Central and Southern Africa at close to $US 400, and these levels were significantly higher than those in East Africa, which averaged $US 250 per person

On the supply side, the East and West African economies tend to have large agriculture sectors (40% or more of GDP) and smaller industrial sectors when compared with Central and Southern Africa.

On the demand side, the most noticeable difference was that in 1986 Central Africa had a markedly higher investment ratio at 25% of GDP, compared with the other regions, where it was around 15%.

The Central and Southern regions were more dependent on trade in 1986, with around 35% of GDP exported and imported. In East and West Africa, the ratio of traded goods to GDP was close to 20%.

Central and Southern Africa appear to have higher literacy rates, and to make better provision for primary education than in East and West Africa. However, secondary and higher education provision is better in West and Central Africa, and rather poorer in East and Southern Africa.

Life expectancies are more or less the same in all regions, with expectancy in Central

The African Economies

Africa slightly higher than in the others. Calorie provision per person per day is fairly even, too, but with Southern Africa slightly lower than the other three regions. Supply of doctors per head of population is best in Central Africa and West Africa, while it is noticeably less good in Southern Africa, and poorest in East Africa.

In terms of overall economic performance in the period 1980-86, positive annual expansion of GDP was achieved by Central (4.2% a year) and East Africa (1.6% a year), while declines were experienced in West (-2.0% per year) and Southern Africa (-3.7% a year). When population growth is taken into account, only Central Africa has managed to increase GDP per head, with the other regions recording declines, which have been fastest in Southern Africa and slowest in East Africa.

All sectors (agriculture, industry and services) have grown in the period 1980-86 in East Africa and Central Africa. In West Africa, agriculture has grown, but the others have declined, and in Southern Africa all sectors have declined.

Volumes of traded goods have grown in East and Central Africa in the 1980-86 period, whereas they have declined in Central and Southern Africa.

Inflation has been highest, 1980-86, in East Africa (24% a year) and lowest in West Africa (13% a year), while Central and Southern Africa experienced around 17% a year.

In terms of economic outlook, all regions are forecast to have positive GDP growth in the period 1990-95. It is expected to be highest in Central Africa at 3.9% a year, followed by Southern Africa at 3.9%. Lower rates are forecast for West Africa at 2.7% a year and East Africa at 2.2% a year.

When population growth rates are taken into account, the projections show increases in GDP per head in Central Africa (9% a year) and Southern Africa (0.4%), but falls in GDP are forecast for West Africa (-0.6%) and East Africa (-0.9%)

Overall, West Africa has the highest income levels and East Africa the lowest. Central Africa has had the best economic performance in the 1980s, with increasing GDP per head, while East Africa has had the worst, with falling GDP per head.

Table 3.1 Geographical regions

#	Country	East	West	Central	Southern
1	Angola				Southern
2	Benin		West		
3	Botswana				Southern
4	Burkina F		West		
5	Burundi	East			
6	Cameroon			Central	
7	Cape Verde		West		
8	Central A R			Central	
9	Chad			Central	
10	Comoros	East			
11	Congo			Central	
12	Côte d'Ivoire		West		
13	Djibouti	East			
14	Equatorial G			Central	
15	Ethiopia	East			
16	Gabon			Central	
17	Gambia		West		
18	Ghana		West		
19	Guinea		West		
20	G Bissau		West		
21	Kenya	East			
22	Lesotho				Southern
23	Liberia		West		
24	Madagascar	East			
25	Malawi				Southern
26	Mali		West		
27	Mauritania		West		
28	Mauritius	East			
29	Mozambique				Southern
30	Namibia				Southern
31	Niger		West		
32	Nigeria		West		
33	Réunion	East			
34	Rwanda	East			
35	São Tomé & P			Central	
36	Senegal		West		
37	Seychelles	East			
38	Sierra Leone		West		
39	Somalia	East			
40	South Africa				Southern
41	Sudan	East			
42	Swaziland				Southern
43	Tanzania	East			
44	Togo		West		
45	Uganda	East			
46	Zaïre			Central	
47	Zambia				Southern
48	Zimbabwe				Southern
	Totals	14	16	8	10

The African Economies

Population Size

A summary of African countries by population size is given in Table 3.2. There are eleven countries with populations in 1989 of under one million, twenty-seven with populations of between one and ten million, and ten with populations above ten million.

The small countries have quite the highest average level of GDP per head at $US 1,190 in 1987, while the medium-sized countries had GDP per head of $US 464. The large countries had the lowest GDP per head at $US 213.

The small countries have the lowest rates of population growth at 2.8% a year 1980-87, the medium-sized countries 3.1%, and the large countries 3.2%.

The medium-sized countries have had the fastest rate of growth of GDP at 2.9% a year 1980-87, well ahead of the small countries at 0.6%, while GDP fell by -0.2% a year in the large countries. Although small size does not preclude a high income level, contrary to the view that argues that a large domestic market is necessary to generate scale economies, it would appear to leave small countries vulnerable to economic fluctuations.

Outlook is judged best for the small countries, which are forecast to experience GDP growth of 4.4% a year for 1990-95. GDP growth for the medium-sized countries is projected at 3.0% a year, and at 2.7% a year for the large countries. This leads to forecasts of GDP per head rising at 1.6% a year in the small countries, but falling at -0.1% a year in the medium-sized group and at -0.4% in the large countries.

Overall, the small population countries have the highest average income levels, and the large countries the lowest. However, the medium-sized countries have had the best economic performance in the 1980s, with the large countries the worst.

Economic Size

Countries are grouped according to their total GDP in Table 3.3. GDP is measured in $US, and, as argued in Chapter 2, there are considerable problems in converting local values to dollars. This is done by using three

Table 3.2 Population size 1989

	Country	Small	Medium	Large
1	Angola		Medium	
2	Benin		Medium	
3	Botswana		Medium	
4	Burkina F		Medium	
5	Burundi		Medium	
6	Cameroon		Medium	
7	Cape Verde	Small		
8	Central A R		Medium	
9	Chad		Medium	
10	Comoros	Small		
11	Congo		Medium	
12	Côte d'Ivoire		Medium	
13	Djibouti	Small		
14	Equatorial G	Small		
15	Ethiopia			Large
16	Gabon	Small		
17	Gambia	Small		
18	Ghana			Large
19	Guinea		Medium	
20	G Bissau	Small		
21	Kenya			Large
22	Lesotho		Medium	
23	Liberia		Medium	
24	Madagascar		Medium	
25	Malawi		Medium	
26	Mali		Medium	
27	Mauritania		Medium	
28	Mauritius		Medium	
29	Mozambique			Large
30	Namibia		Medium	
31	Niger		Medium	
32	Nigeria			Large
33	Réunion	Small		
34	Rwanda		Medium	
35	São Tomé & P	Small		
36	Senegal		Medium	
37	Seychelles	Small		
38	Sierra Leone		Medium	
39	Somalia		Medium	
40	South Africa			Large
41	Sudan			Large
42	Swaziland	Small		
43	Tanzania			Large
44	Togo		Medium	
45	Uganda			Large
46	Zaïre			Large
47	Zambia		Medium	
48	Zimbabwe		Medium	
	Totals	11	27	10

Key: Small = under 1 million
Medium = 1 million to 10 million
Large = over 10 million

The African Economies

year averages of offical exchange rates, but in many cases these exchange rates are poor reflections of the real puchasing powers of the respective currencies. Nevertheless, the figures do give some idea of orders of magnitude.

There are fourteen economies with total GDP of less than $US 1b, twenty-five economies with GDP between $US 1b and $US 5b, and nine economies with GDP above $US 5b.

No significant differences in levels of GDP per head are observed, with all economy sizes having average GDP per head at close to $US 300 in 1987.

The medium-sized economies had the fastest rate of GDP expansion at 1.7% a year 1980-87, with the small and the large economies expanding at 0.8% a year. These rates led to falls in GDP per head in all three groups.

The medium-sized economies are projected to grow fastest in the 1990-95 period at 3.4% a year, followed by the large economy group at 2.7% and the small economies at 1.6% a year.

When population growth is taken into account, the medium-sized economies are forecast to show increasing GDP per head at 0.5% a year, but there are projected falls at -1.0% a year, and at -0.7% a year for the small and large economies respectively.

Overall, there are no significant differences in income level for the various economy-size groups. The medium-sized economies have had the best economic performance in the 1980s.

Income Levels

The World Bank groups countries according to average level of income, expressed as GNP per head in $US. As with the inter-country comparisons of GDP, there are considerable problems in converting local values to dollars.

Table 3.4 shows that Africa has thirty-four countries classified as low-income in 1987, ten in the lower-middle income group, and three upper-middle income countries (excluding South Africa).

The low-income countries had the equal lowest rate of population growth 1980-87 at

Table 3.3 Economic size 1988

	Country	Small	Medium	Large
1	Angola		Medium	
2	Benin		Medium	
3	Botswana		Medium	
4	Burkina F		Medium	
5	Burundi		Medium	
6	Cameroon			Large
7	Cape Verde	Small		
8	Central A R		Medium	
9	Chad	Small		
10	Comoros	Small		
11	Congo		Medium	
12	Côte d'Ivoire			Large
13	Djibouti	Small		
14	Equatorial G	Small		
15	Ethiopia		Medium	
16	Gabon		Medium	
17	Gambia	Small		
18	Ghana			Large
19	Guinea		Medium	
20	G Bissau	Small		
21	Kenya			Large
22	Lesotho	Small		
23	Liberia	Small		
24	Madagascar		Medium	
25	Malawi		Medium	
26	Mali		Medium	
27	Mauritania	Small		
28	Mauritius		Medium	
29	Mozambique		Medium	
30	Namibia		Medium	
31	Niger		Medium	
32	Nigeria			Large
33	Réunion		Medium	
34	Rwanda		Medium	
35	São Tomé & P	Small		
36	Senegal		Medium	
37	Seychelles	Small		
38	Sierra Leone	Small		
39	Somalia		Medium	
40	South Africa			Large
41	Sudan			Large
42	Swaziland	Small		
43	Tanzania		Medium	
44	Togo		Medium	
45	Uganda		Medium	
46	Zaïre			Large
47	Zambia		Medium	
48	Zimbabwe			Large
	Totals	14	25	9

Key: Small = GDP under $US 1b
      Medium = GDP $US 1 b to $US 5b
      Large = GDP over $US 5b

The African Economies

3.1% a year, the lower-middle-income group had the highest at 3.4% a year, and the upper-middle-income group experienced 3.1% annual population expansion. This is consistent with the explanation of changes in population growth rates based on a demographic transition. This explanation argues that population growth rates first rise with increased income per head as death rates fall as a result of better nutrition and health provision. Later, population growth rates fall as increased income is accompanied by greater urbanisation and better educational and employment prospects for women. These factors serve to reduce family sizes as they tend to increase the costs and sacrifices involved in having children.

The economies of the lower-middle-income group have grown fastest with a 4.2% a year expansion of GDP 1980-87, with the low-income group only expanding GDP at 0.3%. The upper-middle-income group, comprising as it does Gabon, Réunion, and Seychelles, is really too small and atypical of Africa to make any significant comparisons on the basis of economic factors.

Population growth led to falls in GDP per head at -2.8% a year 1980-87 for the low-income group. The lower-middle-income countries expanded GDP per head at 0.8% a year

The lower-middle-income countries had good price stability with 8.5% a year inflation 1980-87. In the low-income group, prices rose rather faster at 21.8% a year. In the low-income countries, there is continual pressure to increase government spending, which, in the context of a narrow tax base invariably leads to an inflationary increase in the money supply.

The low-income countries are forecast to expand GDP at 2.7% a year 1990-95, with the lower-middle-income group expanding GDP at 3.2%. However, population growth leads to projections of GDP per head falling at -0.4% a year in the low-income group, and at -0.2% in the lower-middle-income group. The low-income group are forecast to expand exports at 2.6% a year, and the lower-middle-income group at 4.0% a year. Inflation is projected at 25% a year in the low-income countries, and at under 8% a year for the upper-middle-income group.

Table 3.4 Income level 1987

1	Angola	Low		
2	Benin	Low		
3	Botswana		L-middle	
4	Burkina F	Low		
5	Burundi	Low		
6	Cameroon		L-middle	
7	Cape Verde	Low		
8	Central A R	Low		
9	Chad	Low		
10	Comoros	Low		
11	Congo		L-middle	
12	Côte d'Ivoire		L-middle	
13	Djibouti		L-middle	
14	Equatorial G	Low		
15	Ethiopia	Low		
16	Gabon			U-middle
17	Gambia	Low		
18	Ghana	Low		
19	Guinea	Low		
20	G Bissau	Low		
21	Kenya	Low		
22	Lesotho	Low		
23	Liberia	Low		
24	Madagascar	Low		
25	Malawi	Low		
26	Mali	Low		
27	Mauritania	Low		
28	Mauritius		L-middle	
29	Mozambique	Low		
30	Namibia		L-middle	
31	Niger	Low		
32	Nigeria	Low		
33	Réunion			U-middle
34	Rwanda	Low		
35	São Tomé & P	Low		
36	Senegal		L-middle	
37	Seychelles			U-middle
38	Sierra Leone	Low		
39	Somalia	Low		
40	South Africa			U-middle
41	Sudan	Low		
42	Swaziland		L-middle	
43	Tanzania	Low		
44	Togo	Low		
45	Uganda	Low		
46	Zaïre	Low		
47	Zambia	Low		
48	Zimbabwe		L-middle	
	Totals	34	10	4

Key:
Low = GDP per head under $US 480
L-middle = GDP per head $US 480 to $US 2,000
U-middle = GDP per head $US 2,000 to $US 6,000

Overall, the lower-middle-income countries have had markedly better economic performance in the 1980s than the low-income countries, implying a widening of inequality within Africa.

Economic Zones

Table 3.5 groups countries according to economic zone. The main grouping centres around the members of the Franc zone which is distinguished by convertability of the currency, the CFA franc, and the budgetary and monetary discipline imposed as a condition of the convertability, by metropolitan France. Currently there are fifteen countries in the zone - this includes Réunion in which the French franc circulates, but to all intents and purposes the monetary arrangements are the same as for the other fourteen countries..

The second group, the Anglophone zone, comprises those countries administered or protected by the British during the colonial period. There is less coherence to this group, apart from the generally greater flow of British aid and technical assistance. The third group is made up of the remaining countries. For the most part, comparisons are made between the Franc zone countries and the rest.

It needs to be observed that in terms of population size, the Franc zone is quite the smallest. The Franc zone contains 16% of Africa's population, the Anglophone zone 53% (excluding South Africa), and the Other zone 31%.

There is less difference in terms of economic size. The Franc zone contributes 32% of Africa's GDP, the Anglophone zone 47%, and the Other zone 21%.

The Franc zone is significantly more prosperous than the two other zones, with a level of GDP per head of $US 598 in 1987, whereas in the Anglophone countries it was $US 275 and in the Other group $US 214.

It is in economic performance that the difference between the Franc zone countries and the rest of Africa is most apparent. GDP expanded at 3.1% a year 1980-87 in the Franc zone, at 1.4% a year in the other group, and at 0.1% in the Anglophone countries.

Table 3.5 Economic zone 1989

	Country	Franc	Anglophone	Other
1	Angola			Other
2	Benin	Franc		
3	Botswana		Anglophone	
4	Burkina F	Franc		
5	Burundi			Other
6	Cameroon	Franc		
7	Cape Verde			Other
8	Central A R	Franc		
9	Chad	Franc		
10	Comoros	Franc		
11	Congo	Franc		
12	Côte d'Ivoire	Franc		
13	Djibouti			Other
14	Equatorial G	Franc		
15	Ethiopia			Other
16	Gabon	Franc		
17	Gambia		Anglophone	
18	Ghana		Anglophone	
19	Guinea			Other
20	G Bissau			Other
21	Kenya		Anglophone	
22	Lesotho		Anglophone	
23	Liberia			Other
24	Madagascar			Other
25	Malawi		Anglophone	
26	Mali	Franc		
27	Mauritania			Other
28	Mauritius		Anglophone	
29	Mozambique			Other
30	Namibia		Anglophone	
31	Niger	Franc		
32	Nigeria		Anglophone	
33	Réunion	Franc		
34	Rwanda			Other
35	São Tomé & P			Other
36	Senegal	Franc		
37	Seychelles		Anglophone	
38	Sierra Leone		Anglophone	
39	Somalia			Other
40	South Africa		Anglophone	
41	Sudan		Anglophone	
42	Swaziland		Anglophone	
43	Tanzania		Anglophone	
44	Togo	Franc		
45	Uganda		Anglophone	
46	Zaïre			Other
47	Zambia		Anglophone	
48	Zimbabwe		Anglophone	
	Totals	15	18	15

The African Economies

The Franc zone maintained unchanged levels of GDP per head in the 1980s. GDP per head fell at -1.4% a year in the Other group, and at -3.3% in the Anglophone countries.

Price stability was markedly better in the Franc zone at 6.5% a year 1980-87, while it was over 20% a year in the Anglophone and Other groups.

Prospects are judged to be slightly better for the Franc zone group with GDP forecast to expand at 3.2% a year 1990-95, allowing an increase in GDP per head of 0.2% a year. The Anglophone countries are forecast to expand GDP at 3.1% a year, giving falling GDP per head at -0.3% a year. The Other group, GDP is projected to grow at 2.7% a year, with GDP per head declining at -0.2% a year.

The forecasts indicate a narrowing of the difference in performance between the Franc zone countries and the rest. This is explained in part by the adoption of policies by the non-Franc zone countries that bring them closer to the Franc zone's exchange rate, budgetary, and monetary arrangements.

Overall, the Franc zone is the smallest population grouping, but is the most prosperous with levels of GDP per head twice the Africa average. The Franc zone has had markedly better economic performance in the 1980s in terms of GDP growth, GDP per head growth, and price stability.

Stability

Stability is judged on the record over the past two decades. It is a subjective assessment of factors that adversely affect economic performance. Thus non-constitutional changes of government, threats of such changes, internal and cross-border armed conflicts, riots and disruptions, and sudden changes in economic policy are all seen as having adverse effects. Details of stability assessment for each economy are given in the Country Entries section of this book A summary is given in Table 3.6 Sixteen countries are judged to have had bad stability records, eight are judged poor, four moderate, nine fair and eleven good.

The division into five stability groups is rather too detailed for comparative purposes, and in what follows, the twenty countries in

Table 3.6 Stability record 1970-89					
1	Angola	Bad			
2	Benin		Moderate		
3	Botswana				Good
4	Burkina F	Bad			
5	Burundi	Bad			
6	Cameroon				Good
7	Cape Verde			Fair	
8	Central A R	Bad			
9	Chad	Bad			
10	Comoros	Bad			
11	Congo		Moderate		
12	Côte d'Ivoire				Good
13	Djibouti				Good
14	Equatorial G	Bad			
15	Ethiopia	Bad			
16	Gabon				Good
17	Gambia			Fair	
18	Ghana	Bad			
19	Guinea		Poor		
20	G Bissau	Bad			
21	Kenya				Good
22	Lesotho		Moderate		
23	Liberia		Poor		
24	Madagascar		Poor		
25	Malawi				Good
26	Mali		Poor		
27	Mauritania	Bad			
28	Mauritius				Good
29	Mozambique	Bad			
30	Namibia	Bad			
31	Niger		Moderate		
32	Nigeria	Bad			
33	Réunion				Good
34	Rwanda			Fair	
35	São Tomé & P			Fair	
36	Senegal				Good
37	Seychelles		Poor		
38	Sierra Leone		Poor		
39	Somalia		Poor		
40	South Africa			Fair	
41	Sudan	Bad			
42	Swaziland			Fair	
43	Tanzania				Good
44	Togo			Fair	
45	Uganda	Bad			
46	Zaïre			Fair	
47	Zambia			Fair	
48	Zimbabwe		Poor		
	Totals	16	8	4	9	11

the poor, moderate and fair groups (excluding South Africa) are considered together as a medium stability group.

The bulk of Africa, 56% of the total population is made up of countries with bad stability records. The middle group comprises 25% of Africa's population, and the countries with good stability, 19%.

The countries with good stability had GDP per head of $US 513 in 1987, double the level for the rest of Africa.

The good stability countries have experienced population growth at 3.6% a year, higher than the rest of Africa at 3.0% a year.

It is in terms of economic performance that the correspondence with stability is most striking. The bad stability group has grown at -0.2% a year 1980-87, the middle group at 1.4%, and the good stability group at 3.7% a year. Thus the bad stability group has seen GDP per head decline at -3.2% a year, the middle group at -1.7%, while the good stability countries have managed to maintain GDP per head virtually unchanged with 0.1% a year growth.

The bad stability group has seen export volumes decline at 4.9% a year 1980-87, the middle group -1.1%, while the good stability group has expanded export volumes at 2.8% a year. Finally, the bad and middle stability countries have experienced 21.5% a year inflation 1980-87, while the good stability countries have had 9.2% annual inflation.

The strong association between stability on the one hand and income and economic performance on the other raises the question as to which is the cause, and which is the effect. Tentative evidence in El Farhan and Hodd 1989 indicates that stability conditions affect economic performance, but economic performance does not strongly affect stability.

It is invariably difficult to improve material well-being if there is not internal security, stable government and continuity in economic policy, and the attainment of stability is generally a pre-requisite for economic progress. Only Burkina Faso and Chad have experienced good economic performance in the 1980s despite poor stability records.

On the other hand, it is quite possible to have stability, and through adverse external circumstances or poor domestic economic policies, experience poor economic performance. Côte d'Ivoire, Djibouti, Gabon, Malawi, and Tanzania would come into this category.

Overall, the majority of African people live in countries with poor stability record. Those countries with poor stability records have lower income levels and inferior economic performance.

Economic Performance

Countries that have managed, more or less, to maintain average living standards in the period 1980-87, by maintaining GDP per head growth within half of one percent of zero growth in GDP per head, are judged to have had moderate economic performance. The record is summarised in Table 3.7 Thirty-one African countries have had poor economic performance, eight countries have managed to hold their own in expanding output at roughly the same pace as population growth. Nine countries have had good performance in the early 1980s.

Overall, only a third of African nations have maintained average living standards or better and this contrasts with the two decades 1960-1979, when this was achieved by over 80% of African countries.

Analysis of Economic Performance

Explanations of the deterioration in African economic performance in recent years have centred around assessments of the relative impact of external events as against domestic policies pursued.

The first set of explanations emphasizes the poor educational preparation of the labour force during the Colonial period, the impact of the two major oil price rises in the 1970s, droughts and declining terms of trade. Some instability has external origins particularly in Southern African states as the result of South Africa's activities. There has also been the recession in the high-income countries in the early 1980s which has affected volumes of Africa's exports as well as their prices. High

The African Economies

Figure 3.7 Economic performance 1980-87		
1 Angola	Moderate	
2 Benin	Moderate	
3 Botswana		Good
4 Burkina F		Good
5 Burundi	Moderate	
6 Cameroon		Good
7 Cape Verde		Good
8 Central A R	Poor	
9 Chad		Good
10 Comoros	Poor	
11 Congo		Good
12 Côte d'Ivoire	Poor	
13 Djibouti	Poor	
14 Equatorial G	Poor	
15 Ethiopia	Poor	
16 Gabon	Poor	
17 Gambia	Moderate	
18 Ghana	Poor	
19 Guinea	Poor	
20 G Bissau	Moderate	
21 Kenya	Moderate	
22 Lesotho	Moderate	
23 Liberia	Poor	
24 Madagascar	Poor	
25 Malawi	Poor	
26 Mali		Good
27 Mauritania	Poor	
28 Mauritius		Good
29 Mozambique	Poor	
30 Namibia	Poor	
31 Niger	Poor	
32 Nigeria	Poor	
33 Réunion		Good
34 Rwanda	Poor	
35 São Tomé & P	Poor	
36 Senegal	Moderate	
37 Seychelles	Poor	
38 Sierra Leone	Poor	
39 Somalia	Poor	
40 South Africa	Poor	
41 Sudan	Poor	
42 Swaziland	Poor	
43 Tanzania	Poor	
44 Togo	Poor	
45 Uganda	Poor	
46 Zaïre	Poor	
47 Zambia	Poor	
48 Zimbabwe	Poor	
Totals	31	8 9

Key: Poor = Falling GDP per head
Moderate = Unchanged GDP per head
Good = Rising GDP per head

interest rates have increased the burden of servicing external debts, especially for those middle-income oil and mineral exporters who have borrowed from commercial sources.

The other view, while acknowledging the adverse impact of some of the external factors, emphasizes poor domestic economic policies in Africa.

In the 1950s and 1960s, the period in which most African countries were approaching and attaining political independence, the consensus among economists was that the market mechanism was not adequate in meeting the development needs of poor countries. There were a number of theoretical reasons to support this, in particular the arguments that restricted competition and monopolies in markets, imperfect knowledge and poorly established private property rights led to impaired economic efficiency. Markets were thought to result in distributions of income that did not meet with the social values of society, and it was feared that, in the early stages of a development strategy reliant on markets, there would be a movement toward a more unequal distribution of income.

These views were reinforced by an impression that planning had performed well in the Soviet Union in the inter-war period when the market economies of the West had experienced a debilitating depression. During the Second World War, Western economies were subject to greater government control, overall planning, and rationing regulations which distributed goods according to social priorities rather than via the price mechanism. This experience encouraged the view that government intervention was necessary if national objectives were to be achieved.

The Colonial period had established regulations, controls and institutionalised monopolies to protect settler and Colonial interests in Africa. The creation of export crop marketing boards, and restricting the growing of the more valuable cash crops to settlers, were notable examples. The performance of China after 1949 in successfully collectivising agriculture and introducing socialist planning provided a model which several African countries admired. In the Colonial period the ownership of most capitalist enterprises was in the hands of the Colonial rulers. The new

The African Economies

African leaders at that time, with little vested interest in supporting large scale capitalist enterprises, pursued a set of policies which emphasized the role of the state rather than corporate business.

The 1960s saw independent African countries increasing their role in their economies. Marketing boards with monopolies over cash crop purchasing were retained, and were seen as mechanisms whereby governments could effectively tax agriculture by passing on less than the eventual sale price of the crops to domestic producers. Industrialization rather than primary product production was seen as essential for fast development. Consequently, African countries began setting up manufacturing plants to produce goods previously imported, invariably with monopolies in domestic markets, tariff protection and state ownership. Foreign investment was discouraged and subjected to controls, and in many instances existing foreign-owned enterprises were nationalised. Prices of goods in domestic markets were often regulated. Official exchange rates were kept fixed and when balance of payments problems arose, governments reduced imports by imposing licences and introducing restrictions on foreign exchange movements. Concern to be self-sufficient in food led to pricing policies which favoured foodstuffs rather than cash-crop production in agriculture. Governments undertook substantial public sector programmes in health, education, and the provision of the infrastructure necessary for industrialisation.

A change in view among economists began to emerge in the 1970s. However it was almost a decade before international institutions began to press for changes in policy, and African governments began to receive similar advice from their own advisors.

It gradually became clear that developing countries that had relied on international markets had experienced good economic performance in the post-war period. The best use of their resources was judged to be in producing goods in which they either had a natural advantage, such as tropical agricultural products or minerals, or in which they had built up experience and expertise, and in trading these for goods more efficiently manufactured by other countries.

At independence there were few university graduates in Africa, and those that there were had mostly trained in medicine arts or natural science subjects, and they lacked the ability to implement comprehensive economic plans. Planning failure began to be seen as more serious than market failure. A new generation of economic advisors began to man government ministries, many of whom had received postgraduate training in economics in the West where there is heavy emphasis on the efficiency of competitive markets.

Short-comings of the interventionist policies in Africa also became apparent. The export crop-marketing boards often passed on very low percentages of the realised world price to domestic producers. And instead of the margins between the world price and the producer price being used to augment government revenue for productive investment, it was often defrayed in paying salaries in over-staffed institutions. Many export marketing boards, far from adding to government revenue succeeded only in incurring losses which were funded from other government receipts or increased budget deficits. The marketing boards responsible for domestic foods paid high prices to encourage self-sufficiency, but sold goods at subsidised prices to urban consumers. Marketing boards made losses as a result of the negative trading margins to which were added their operating costs. These policies conspired to reduce export crop output as farmers moved out of export crops into domestic food production.

The exchange rate policies pursued exacerbated the adverse effect of marketing boards on exports. Heavy levels of domestic public sector spending eventually led to acceleration in domestic inflation. With fixed exchange rates, this led to over-valued domestic currencies. The official rates could only be maintained by controls on foreign exchange dealings and restrictions on imports by licencing. Black markets in foreign exchange emerged in most countries. In the 1970s Africa's black market exchange rates were on average 60% higher than official rates. Over-valued currencies further reduced receipts by producers of export-crops in domestic currencies, and encouraged smuggling. Allied to this was the effect of cheaper imports in discouraging domestic producers of

The African Economies

manufactures. These production effects added to the pressure on the balance of payments, and in the absence of corrective means led to even greater domestic currency over-valuation.

Those import-substituting industrial projects with tariff protection or domestic monopolies, used their market position to produce high cost goods from inefficient, over-manned enterprises. Lack of experience and expertise in manufacturing, particularly shortage of managerial skills added to poor production performance in these sectors. Control over prices of domestic goods led to shortages and the emergence of black markets in basic commodities.

With the rises in oil prices in 1974 and 1979, and increased burden of debt-servicing resulting from higher interest rates, African countries found the shortage of foreign exchange caused by poor export performance was intensified. Import licences restricted the availability of fuel, industrial and agricultural imputs and machinery spare parts. Manufacturing began to operate at low levels of capacity, agriculture grew slowly, and the transport system began to fall into disrepair.

Although the effects of government intervention were apparent, it was argued (see Bates 1981) that the predominantly one-party states in Africa, acting as self-perpetuating oligarchies, had no incentive to improve efficiency. Political power required the support of the urban population, a minority which comprised 20% of the total.

Ruling groups discriminated through the exchange rate policy and marketing board operations to favour urban communities. Thus the low prices to rural export producers, subsidised food prices for urban consumers, and cheap imports of manufactures, all served to distribute income toward town-dwellers. Inflated staffs of marketing boards and state-owned enterprises mainly served to provide high-income employment for urban residents. In addition, those lucky enough to receive import licences were in a position to resell the goods at black market prices (see Kreuger 1974). The plethora of regulations, licences and (often discretionary) controls involved in standard business transactions opened the way to substantial opportunities for corruption.

The Reform Programmes of the 1980s

In the face of political factors which have favoured intervention in African economies it requires some explanation as to why the 1980s have witnessed a series of reforms in African economies which have seen the beginnings of returns to market-orientated policies. Table 3.8 summarises the position at the end of 1989, when three economies had introduced some reforms, and thirty-two had instituted fairly comprehensive changes to their economic policies. Thirteen countries had not made any substantial changes. Of these, it could be argued that Botswana, Namibia, Réunion and South Africa have fairly market-orientated economies, with substantial foreign investment, making the need for liberalising reforms less pressing

The reforms involve devaluation to close to the market valuation of currencies, and in some case, exchange rates set by weekly auctions. Marketing boards have had their monopolies ended and in some cases have been wound-up or had their operations severely curtailed. Private trading has re-emerged in these markets. Subsidies have been reduced or abolished on foodstuffs, and controls have been lifted on domestic prices. Foreign investment has been encouraged and state-owned enterprises are in the process of being closed down or sold-off.

The changes have been more comprehensive and introduced more rapidly than observers of the entrenched political positions in the 1960s and 1970s considered possible. Several influences appear to have been important.

In the first place, the interventions in the economy produced redistributions in favour of ruling élites and urban populations at the expense of long-run development of the economy. After two decades of intervention the new ruling elites of Africa have now accumulated substantial assets, and now have more to gain by making these assets more productive by more efficient economic policies, than by continuing redistribution in their favour.

Secondly, there has been a change in the pressure from the IMF and World Bank. Up to 1979, the IMF lent on the condition that borrowing countries undertook policies to

The African Economies

Table 3.7 Liberalising reforms				
1	Angola	No		
2	Benin			Yes
3	Botswana	No		
4	Burkina F	No		
5	Burundi			Yes
6	Cameroon			Yes
7	Cape Verde	No		
8	Central A R			Yes
9	Chad			Yes
10	Comoros	No		
11	Congo			Yes
12	Côte d'Ivoire			Yes
13	Djibouti	No		
14	Equatorial G			Yes
15	Ethiopia	No		
16	Gabon			Yes
17	Gambia			Yes
18	Ghana			Yes
19	Guinea			Yes
20	G Bissau			Yes
21	Kenya			Yes
22	Lesotho			Yes
23	Liberia		Some	
24	Madagascar			Yes
25	Malawi			Yes
26	Mali			Yes
27	Mauritania			Yes
28	Mauritius			Yes
29	Mozambique			Yes
30	Namibia	No		
31	Niger			Yes
32	Nigeria			Yes
33	Réunion	No		
34	Rwanda	No		
35	São Tomé & P			Yes
36	Senegal			Yes
37	Seychelles		Some	
38	Sierra Leone			yes
39	Somalia			Yes
40	South Africa	No		
41	Sudan		Some	
42	Swaziland	No		
43	Tanzania			Yes
44	Togo			Yes
45	Uganda			Yes
46	Zaïre			Yes
47	Zambia			Yes
48	Zimbabwe	No		
	Totals	13	3	32

correct the balance of payments, usually by reducing government spending. From 1979 on, they have demanded more wide-ranging reforms to improve efficiency of resources use. In this they have worked in co-operation with the World Bank.

More important than the financial sanctions that the IMF and World Bank have been able to impose via their own lending has been their influence on bilateral vendors and policy-advisors in African countries (see Hodd 1987). They have succeeded in persuading these groups to support their market-orientated policies by arguing that markets work more efficiently than planned economies, and that, as a result, all sections of the community can be better-off.

Experience of the reform programmes have been mixed. Hurried implementation in Zambia and the Sudan caused urban unrest at the ending of food subsidies and the programmes were aborted. Elsewhere results have been encouraging, with return to positive growth rates and improved living standards, most notably in Ghana, Malawi and Tanzania. Critics argue that it is not the policies changes that have been responsible for improved performance but the restoration of aid flows withheld until reforms were adopted.

Most of these reform programmes are in their early stages. Reforming South American countries have experienced deteriorating performance after initial improvements. There is thus reason to be cautious about the overall impact of reforms in Africa.

Future prospects

Table 3.9 summarises the forecasts developed for individual countries up to 1995.

Poor prospects are interpreted in the limited sense of falling GDP per head. Moderate prospects are taken to be GDP increasing roughly in line with population, and good prospects imply a forecast of rising GDP per head.

Thirty-one countries have poor prospects, and they are invariably countries with poor stability or showing unwillingness to introduce reforms, or both.

Eight countries have moderate prospects and for eight the prospects are good. The

The African Economies

Table 3.9 Ecomomic Prospects 1990-95				
1	Angola		Moderate	
2	Benin		Moderate	
3	Botswana			Good
4	Burkina F			Good
5	Burundi		Moderate	
6	Cameroon			Good
7	Cape Verde			Good
8	Central A R	Poor		
9	Chad			Good
10	Comoros	Poor		
11	Congo			Good
12	Côte d'Ivoire	Poor		
13	Djibouti	Poor		
14	Equatorial G	Poor		
15	Ethiopia	Poor		
16	Gabon	Poor		
17	Gambia		Moderate	
18	Ghana	Poor		
19	Guinea	Poor		
20	G Bissau		Moderate	
21	Kenya		Moderate	
22	Lesotho		Moderate	
23	Liberia	Poor		
24	Madagascar	Poor		
25	Malawi	Poor		
26	Mali			Good
27	Mauritania	Poor		
28	Mauritius			Good
29	Mozambique	Poor		
30	Namibia	Poor		
31	Niger	Poor		
32	Nigeria	Poor		
33	Réunion			Good
34	Rwanda	Poor		
35	São Tomé & P	Poor		
36	Senegal		Moderate	
37	Seychelles	Poor		
38	Sierra Leone	Poor		
39	Somalia	Poor		
40	South Africa	Poor		
41	Sudan	Poor		
42	Swaziland	Poor		
43	Tanzania	Poor		
44	Togo	Poor		
45	Uganda	Poor		
46	Zaïre	Poor		
47	Zambia	Poor		
48	Zimbabwe	Poor		
	Totals	31	8	9

Key: Poor = Falling GDP per head
Moderate = Unchanged GDP per head
Good = Rising GDP per head

countries with good prospects are invariably stable with economic reform programmes in place.

REFERENCES

Bates, R. L. 1981. *Markets and States in Tropical Africa*. San Francisco: University of California Press.

El Farhan, H. and Hodd, M. 1989. 'Stability and Growth in Africa, 1960-89'. *(mimeo)*

Euromoney. 1989. 'Country Risk Rating'.

Hodd, M. 1986. *African Economic Handbook* London: Euromonitor.

Hodd, M. 1987. 'Africa, the IMF and the World Bank'. *African Affairs*.

Jackman, R. W. 1978. 'The Predictability of Coups d'Etat: a Model with African Data' *American Political Science Review*.

Johnson, T. H., Slater, R. O., and McGowan, P. 1984. 'Explaining African Military Coups d'Etat 1960-82'. *American Political Science Review*.

Johnson, T.H., McGowen, P. 1985. 'Forecasting African Military Coups' Paper prepared for delivery at the 1985 annual meeting of the American Political Science Association. *(mimeo)*.

Kreuger, A. O. 1974. 'The Political Economy of the Rent-Seeking Society' *American Economic Review*.

Wheeler, D. 1984. 'Sources of Stagnation in Sub-Saharan Africa'. *World Devleopment*.

World Bank. 1975. *Kenya: Into the Second Decade*. Washington D. C.: Johns Hopkins University Press.

World Bank. 1981. *Accelerated Development in Sub-Saharan Africa*. Washington D.C: World Bank.

World Bank. 1988a. *World Debt Tables..* Washington D.C: World Bank.

World Bank. 1988b. *World Development Report*. New York: Oxford University Press.

World Bank 1989a. *Price Prospects for Major Primary Commodities 1988-2000*. 2 Volumes. Washington D.C.:World Bank.

World Bank. 1989b. *World Development Report*. New York: Oxford University Press.

World Bank. 1990. *World Development Report*. New York: Oxford University Press.

PART 2

COUNTRY ENTRIES

1 ANGOLA

Physical Geography and Climate

Angola, in the Southern Africa region, is the third largest sub-Saharan country, located on the south western Atlantic coast and bounded by Zaïre, Zambia and Namibia, while the Cabinda enclave to the north of the Zaïre River shares a border with Congo-Brazzaville. It has a tropical climate, modified by both altitude and the effect of the cold Benguela current. Two-thirds of the country consists of a high plateau at 1000 metres divided by the Rivers Cuanze, Cuango, Kasai running south to north and the Rivers Zambezi Cubango and Kunene running north to south. The coastal region is arid, whilst the interior uplands have a Mediterranean climate. The border areas experience high temperatures and heavy rainfall, with jungle covered mountains in the north west and equatorial jungle around Cabinda. The River Kwanza is navigable, but the other rivers do not give access to the interior from the coast, however they are used for irrigation and hydro-electricity generation. There are dams at Cambambe on the River Cuanza and at Lobito-Benguela, Lomaum and Biópio, Matala and Ruacaná Falls. The main ports are Lobito, Luanda, Namibe and Cabinda. Mineral deposits include oil, diamonds, copper, manganese, phosphates, salt, and iron ore.

Population

Estimates for 1989 suggest a population of 9.7 million, which for such a large country is relatively under populated. The civil war has made a census impossible since independence. The official language is Portuguese with 38% speaking Umbundu, 27% Kimbundu, 12% Kikongo and 9% Luanda-Quieco. The main ethnic groups being 38% Ovimbandu, 23% Mbundu, 14% Kongo, 9% Lunada-Quieco and 7% Ngangula with other smaller groups (1960 census). The majority of the population are animists. With an area of 1,247,000 sq. km, the density of population is 7.6 per square kilometre, which is low compared with the regional average of 11.3 per square kilometre. Following the exodus of the Portuguese prior to independence in 1975 there has been a decline in urbanisation, it being 25% compared to the regional average of 23.8%. The rate of population growth was 2.6% for the period 1980-86, while the regional average was 3.1%. The growth rate has been limited by the armed conflicts which has led to displacement and starvation in some areas.

History

Angola was a former Portuguese colony first settled in the fifteenth century. By the mid-eighteenth century Portuguese colonial governors were trying to halt the slave trade and secure the interior for Portuguese immigrants. This policy was ineffectively supported and by the mid-nineteenth century the main source of income was the the supply of slaves to Brazil. In 1822 Brazil became independent and there was a mutiny by troops in Luanda. Stability returned in 1834, the slave trade was made illegal, but its volume did not diminish. Following the Brazilian revolt of 1848 there was a substantial increase in Portuguese immigration and by the 1880s the intention was to occupy the interior between the twelth and eighteenth parallels. There were tribal uprisings in 1902 by the Ovimbundu and by the Bakongo in 1913, but up until 1960 the Portuguese administrative structures remained intact.

Angola

Since 1960 Angola has experienced armed struggle between either the Portuguese colonial government, or between various liberation forces. A number of resistance movements grew up in the 1960s, notably Movimento Popular de Libertação de Angola (MPLA), Frente Nacional de Libertação de Angola (FNLA) and União Nacional para a Independência Total de Angola (UNITA). They fought the Portuguese army around the northern and eastern borders of the country, though there was little unified action between the three groups.

The 1974 coup in Portugal paved the way to independence for all the Portuguese colonies as moral in that country continued to decline with the wars draining the Portuguese economy. Independence was finally achieved in 1975.

When the MPLA government came to power it was faced with the threat of invasion from South Africa who supported the FNLA and UNITA forces. While the MPLA received large scale military assistance from Cuba, USSR and Eastern Europe, UNITA and the FNLA were able to count on open military support from South Africa, as well as covert supplies from the United States, but the western European countries, worried about the movements' links with South Africa, preferred not to get involved in the conflict.

Relations with Zaïre have been poor since independence, with Zaïre accusing Angola of aiding rebels in the Shaba crises of 1977 and 1978, while Angola continued to complain of Zaïre's support for UNITA. Cross-border incursions by the South African armed forces have also continued since indepencence. However, 1989 saw South Africa finally withdraw from Angola in the run up to Namibian independence. With South Africa removed from the war zone, UNITA's position has been greatly weakened, forcing it to negotiate peace with the MPLA government on ending the civil war. However, Angola continues to have security problems in the north with the FNLA and the movement in Cabinda province. The Angolan civil war has had devastating effects on a potentially prosperous economy with large scale infrastructural damage and the diversion of resources towards the armed forces. This is in addition to the problems of the exodus of skilled labour at independence and the major economic restructuring of the economy on Marxist-Leninist lines. In 1986 FAO declared the famine to be seriously affecting Angola. President dos Santos had been in power for ten years by 1989.

Stability

There has been armed struggle in Angola involving first the liberation struggle against the Portuguese colonial government and then against the South African-backed UNITA forces. UNITA has been able to disrupt communications, prevent development in agriculture and even the heavily protected oil sector has been the subject of UNITA attacks.

Since Independence in 1975, the ruling MPLA have attempted to restructure the economy on Marxist-Leninist lines. The overwhelming majority of the Portuguese community, who supplied almost half the skilled and professional manpower in the economy, have departed. There are tentative signs that the government might be considering a second reversal in development strategy, having applied to join the IMF and initiating more contacts with the West.

Overall, Angola has a very poor stability record, characterised by an absence of internal security and a lack of any continuity in development strategy. As a result of recent initiatives involving talks with UNITA and South African withdrawal, there is some prospect of a modest improvement in stability.

Economic Structure

Recent economic figures are not available. In 1982, the economy generated $US 3.7 m of GDP, about as much as the average for the 48 countries in the region, but almost double the average for Africa, where only Zimbabwe has a larger economy. In terms of GNP per head, however, although it is classified as a middle-income country, only Malawi and Mozambique in the region have lower per capita incomes.

Despite an important oil sector, agriculture generates 47% of Angola's GDP and only Mozambique in the region, a similarly unstable country where the population has also reverted to subsistence agriculture, is the relative size of the agricultural sector comparable at 43%. The industrial sector contributes 24% of GDP and services 29%, both these percentages being smaller than for Africa and the region. 48% of GDP is exported, higher than for any large country in the region and double the export dependence of Africa. Imports comprise 36% of total spending, roughly the regional level, but indicating more openness to imports than in Africa generally where 23% of spending is on overseas goods.

Petroleum and its products made up 93% of exports in 1987, coffee and diamonds the rest. Imports are mostly industrial goods, machinery, manufactures and chemicals comprising 60%, with food only 14% and fuels 4%.

Angola's refusal to join the IMF until recently and her socialist development strategy have limited the build-up of external debt. In 1984 debt service was estimated to be under 1% of export earnings.

There are no recent figures for adult literacy or the provision of primary education, though these have improved in recent years, they are thought to be amongst the poorest in the region. Secondary education provision compares better, with 12% of the relevant age-groups enrolled, although this is worse than the regional average. 2% of the 20-24 age-group are in higher education, which compares favourably with the regional and African averages of around 1.4%.

Life expectancy of 43 years is lower than in any other country except for Namibia, although provision of doctors per head of population is twice as good as in the region generally.

Overall, the economic structure is one of a low level of mainly subsistence development for the bulk of the population, but with the presence of oil providing some higher incomes in the modern sector. There is heavy dependence on oil for export earnings and for the higher modern sector incomes, but little of this benefits the 75% of the population in the rural areas.

Economic Performance

The performance of the economy has been one of rapid decline over the period 1970 - 81, and with the GDP contracting at -7.3% a year, with GDP per head falling at -10.4% a year. Agriculture and services sector output have both fallen rapidly at over -8.0% a year and only the industry sector, which includes the oil sector, has shown any positive growth at 0.6% a year.

Export volumes have fallen at -6.7% a year since 1973, but the favourable price movements in Angola's main export, oil, has allowed import volumes to rise at 1.2% over the period.

Gross domestic investment was 9% of GDP in 1982 and such a low level is not enough to maintain the size of the capital stock in the face of depreciation. Lack of stability and heavy commitment of government expenditure toward the military costs of fighting the UNITA guerillas are mostly responsible for the low investment performance. Only Mozambique in the region, similarly affected by destabilising guerila activity, has a comparable low rate of investment.

In 1981, the government was running a budget deficit of 9.9% of GDP. Heavy military expenditure and an inability to raise revenue anywhere other than in the oil sector continued to lead to high deficit spending. High deficit spending is reflected in an inflation rate of 21% in the 1980s and this underestimates actual inflation as parallel market transactions at markedly higher prices were not included in the price indices. Angola's inflation has been higher than any country in the region and higher than the African average of 16% for the past decade.

Overall, economic performance is dominated by the war with UNITA and the destabilizing activities of South Africa. The oil sector, with considerable involvement of foreign companies, continues to provide the main source of foreign exchange and government revenue. Improved economic performance and more widespread development will not occur until there is an end to UNITA's activities inside Angola's borders.

Recent Economic Developments

Angola has shown some signs of wishing to relax the socialist development strategy which involves pervasive state ownership and control in the economy. In October of 1987, Angola formally applied for IMF membership, but the application has been hindered by US opposition. Early in 1988 there were pledges made to industrialisation of the economy and the encouragement of foreign investment, but few actual reforms appear to have been implemented. Price controls have been

Angola

abolished on fruit and vegetables in September 1988 and in future only a small number of basic commodities were expected to be subject to regulation. Despite speculation that the exchange rate was to be devalued, no changes had occurred by the end of 1989. More comprehensive reform measures are only likely to occur when Angolian application to join the IMF is accepted, but there is little indication as to when this might be.

There are no figures for growth of GDP after 1982, and the figures prior to that date are very unreliable. There is a marked duality between the oil sector and other enclaves such as diamond mining which can be protected against the prevalent insecurity in the countryside. In oil and diamonds Angola has been able to expand production, while in agriculture there is evidence of continuing falls in output. There are conflicting reports on current levels of oil production but there are plans to expand production from around 350 thousand barrels-a-day to closer to 450 thousand in 1989, a 30% rise, to follow 17% achieved in 1987. Diamond earnings have risen to $US 180 m in 1987 from $US 15 m in 1986. By way of contrast, coffee production at 38,000 tons in the 1986-7 season was at its lowest level since the mid 1960s, a massive fall compared with 215,000 tons produced in 1974.

Inflation figures represent very little as official, controlled prices affect only a small proportion of the sales of basic commodities, with much trading being at black market rates. Black market prices appear to have doubled in 1988, indicating a current rate of inflation of close to 100% a year.

The trade sector depends crucially on the world oil price, with almost 90% of foreign exchange earnings coming from the petroleum sector. Angola has attempted to maintain earnings by increasing production. It is hoped by this means that export receipts can be restored to close to the $US 2,000 m realised in 1985, before the oil price fall. If this is achieved, it will allow merchandise imports to increase closer to $US 1,500 m. Angola's external debts are estimated at $US 4,000 m, and there have been problems in servicing this debt since the fall in oil prices in 1986. In October 1988 there was an informal Paris Club re-scheduling of $US 320 m of debts.

There has been no adjustment to the exchange rate and this has remained at around AK30 = $US 1 since 1970. The absence of adjustment has been reflected in steady depreciation of the black market value of the Kwanza, and this stood at AK 2,800 = $US 1 in 1989 as compared with AK 1,500 in 1987. The 1988 budget anticipated total expenditure of AK 95,000m ($US 3,167 at the official exchange rate), with receipts estimated at AK 80,000m ($US 2,667m). This implies a budget deficit of AK 15,000m ($US 500m), which is probably over 10% of GDP and with GDP unlikely to show any growth, implies considerable inflationary pressure unless export revenues rise as a result of higher world oil prices.

Economic Outlook and Forecasts

Angola's economic prospects depend on the world price of oil and on the prospect of peace and stability in the region. Oil prices are expected to fall slightly in 1990. There will be no economic development outside the oil sector until the security situation improves, and despite optimism in 1989, this is expected to be a slow process.

The forecasts assume moderate stability improvements and a slow rate of implementation of policy reforms. Both scenarios indicate positive GDP and GDP per head growth. Inflation is expected to remain above 20% annually. Growth rates for export and import volumes are expected to fall as the figures for 1980-87 reflect recovery from the falls in the 1970s.

Angola: Economic Forecasts
(average annual percentage change)

	Actual 1980-87	Forecast Base 1990-95	Forecast High 1990-95
GDP	3.0	3.3	4.0
GDP per Head	0.4	0.3	1.0
Inflation Rate	21.0	25.0	20.0
Export Volumes	20.0 [a]	5.0	7.0
Import Volumes	16.5 [a]	4.0	5.0

Note : [a] 1970-82

ANGOLA: Comparative Data

	ANGOLA	SOUTHERN AFRICA	AFRICA	INDUSTRIAL COUNTRIES
POPULATION & LAND				
Population, mid year, millions, 1989	9.7	6.2	10.2	40.0
Urban Population, %, 1985	25	23.8	30	75
Population Growth Rate, % per year, 1980-86	2.6	3.1	3.1	0.8
Land Area, thou. sq. kilom.	1,247	531	486	1,628
Population Density, persons per sq kilom., 1988	7.6	11.3	20.4	24.3
ECONOMY: PRODUCTION & INCOME				
GDP, $US millions, 1982	3,700	2,121	3,561	550,099
GNP per head, $US, 1982	460	383	389	12,960
ECONOMY: SUPPLY STRUCTURE				
Agriculture, % of GDP, 1982	47	25	35	3
Industry, % of GDP, 1982	24	33	27	35
Services, % of GDP, 1982	29	42	38	61
ECONOMY: DEMAND STRUCTURE				
Private Consumption, % of GDP, 1982	54	67	73	62
Gross Domestic Investment, % of GDP, 1982	9	14	16	21
Government Consumption, % of GDP, 1982	25	21	14	17
Exports, % of GDP, 1982	48	34	23	17
Imports, % of GDP, 1982	36	36	26	17
ECONOMY: PERFORMANCE				
GDP growth, % per year, 1970-81	-7.3	-3.7	-0.6	2.5
GDP per head growth, % per year, 1970-81	-10.4	-6.6	-3.7	1.7
Agriculture growth, % per year, 1970-81	-8.3	-6.8	0.0	2.5
Industry growth, % per year, 1970-81	0.6	-3.8	-1.0	2.5
Services growth, % per year, 1970-81	-8.9	-0.9	-0.5	2.6
Exports growth, % per year, 1973-84	-6.7	-5.1	-1.9	3.3
Imports growth, % per year, 1973-84	1.2	-2.5	-6.9	4.3
ECONOMY: OTHER				
Inflation Rate, % per year, 1970-81	21.0	18.1	16.7	5.3
Aid, net inflow, % of GDP, 1985	3.6	8.6	6.3	-
Debt Service, % of Exports, 1984	0.7	7.8	20.6	-
Budget Surplus (+), Deficit (-), % of GDP, 1981	-9.9	-4.1	-2.8	-5.1
EDUCATION				
Primary, % of 6-11 group, 1985	93	95	76	102
Secondary, % of 12-17 group, 1985	13	18	22	93
Higher, % of 20-24 group, 1985	1	1.3	1.9	39
Adult Literacy Rate, %, 1960	5	47	39	99
HEALTH & NUTRITION				
Life Expectancy, years, 1986	44	50	50	76
Calorie Supply, daily per head, 1983	1,926	1,998	2,096	3,357
Population per doctor, 1965	13,150	26,118	24,185	550

Notes: 'Southern Africa' and 'Africa' exclude South Africa. Dates are for the country in question, and do not always correspond with the Regional, African and Industrial averages.

ANGOLA: Leading Indicators

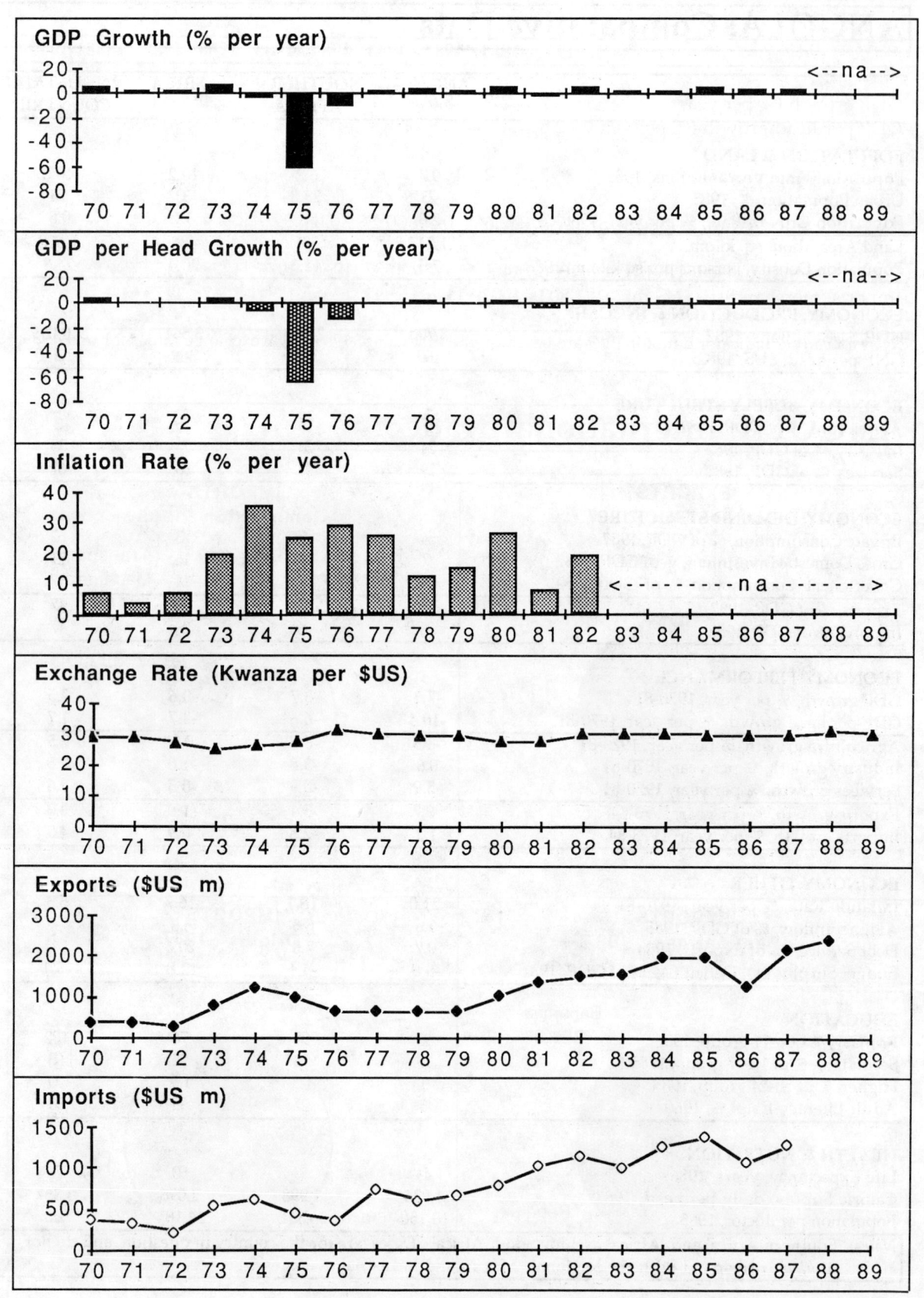

GDP Growth (% per year)

GDP per Head Growth (% per year)

Inflation Rate (% per year)

Exchange Rate (Kwanza per $US)

Exports ($US m)

Imports ($US m)

Note: Leading indicator data for 1989 are based on the first half of 1989. 1989 exchange rate is for mid-1989.

ANGOLA: Leading Indicator Data

	1980	1981	1982	1983	1984	1985	1986	1987	1988	1989
GDP growth (% per year)	5.7	-2.2	5.6	1.4	2.5	5.4	3.4	3.6		
GDP per head growth (% per year)	2.8	-4.9	2.9	-1.3	-0.2	2.7	0.7	0.9		
Inflation (% per year)	26.2	8.2	19.1							
Exchange rate (Kwanzas per $US)	27.6	27.6	30.2	30.2	30.0	29.9	29.9	29.9	30.6	29.7
Exports, merchandise ($US m)	1036	1400	1517	1587	1960	1976	1278	2150	2400	
Imports, merchandise ($US m)	803	1002	1126	993	1265	1384	1062	1275		

Note: Leading indicator data for 1989 are based on the first half of 1989. 1989 exchange rate is for mid-1989.

ANGOLA: International Trade

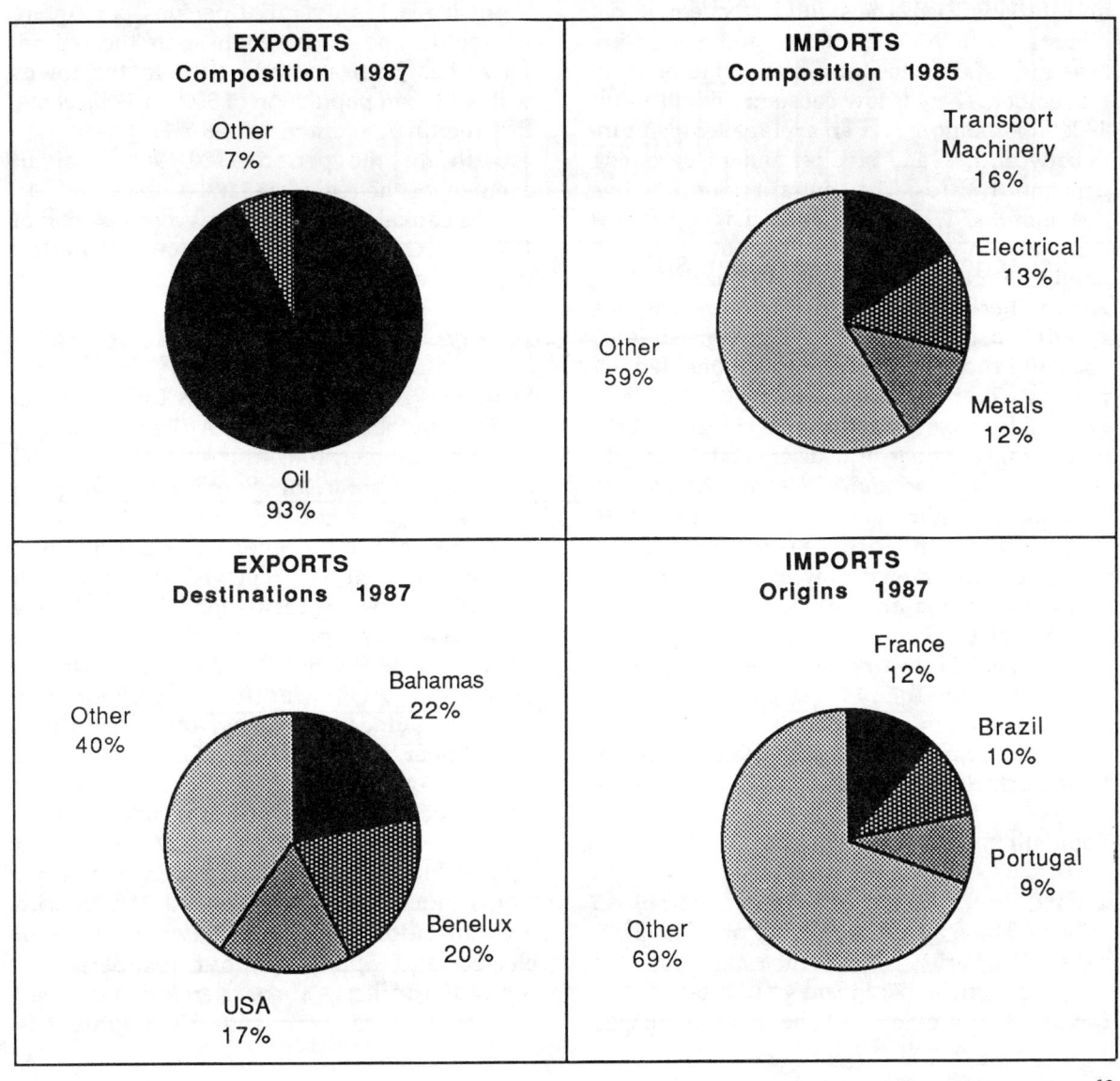

EXPORTS Composition 1987
- Oil 93%
- Other 7%

IMPORTS Composition 1985
- Transport Machinery 16%
- Electrical 13%
- Metals 12%
- Other 59%

EXPORTS Destinations 1987
- Bahamas 22%
- Benelux 20%
- USA 17%
- Other 40%

IMPORTS Origins 1987
- France 12%
- Brazil 10%
- Portugal 9%
- Other 69%

2 BENIN

Physical Geography and Climate

Benin, in the West-African region, is a relatively small, poor, densely populated coastal state bounded by Niger, Nigeria, Togo and Burkina Faso. The south has an equatorial climate, with high humidity and two rainy seasons, March to July and September to November. There is low constant rainfall about 130cm per annum and an average temperature of between 25°C and 34°C per annum, becoming drier northwards. The central region has five dry months, whilst the north has a tropical climate with eight dry months, but retains an annual rainfall of 130cm. As is typical to West Africa there are three distinct zones to the country: a coastal strip, which consists of a coastal sand bar enclosing lagoons fed by marshy river estuaries, swamp, forest and farmland; the south and centre, which become increasingly higher and drier, and the north which is dry and more barren. Due to the volume of rain many villages are only accessible in the dry season. There is a navigable lagoon waterway link to the port of Lagos in Nigeria and a moderate deepwater harbour at Cotonou. A hydro-electric scheme has been established on the River Mono in co-operation with Togo. There are off-shore oil fields and mineral deposits of iron ore, chromium and phosphates, but these have yet to be worked commercially.

Population

Estimates for 1989 suggest a population of 4.6 million. The main ethnic groups are: Fon 47%; Adja 12%; Bariba 10%; Yoruba and Mali 9%; Aizo 5%; Somba 5%; Fulani 4%; Coto-Coli 3%; Dendi 2% and others 4%. The main languages are French and Hausa with 26% speaking Fon, 14% Nago, 13% Bariba, 12% Goun and 11% Adjo-Walchi. The density of population is high at 40 persons per square kilometre for 1988, which is above the regional average of 31 persons per square kilometre. In parts of the south it reaches over 120 persons per square kilometre, one of the highest in the region. There has been a steady drift to the towns with an urban population of 35% in 1985, above the regional average of 28.7%. Population growth for the period 1980-1986 is about average for the region at 3.2%.

The capital, Porto Novo has a population of 132,000 though Cotonou has a higher population of 350,000.

History

Benin was previously known as Dahomey and was one of the ancient West African kingdoms. By the eighteenth century it was a well organized militaristic Fon state. European traders played a complex role in local power struggles but the Dahomey economy was devastated when the French banned slavery in 1794. Following the restrictions elsewhere, the Dahomey slave ports prospered until the 1850s. In 1904 the French West African Federation (AOF) was founded comprising Senegal, Guinea, Dahomey, the Ivory Coast and Upper Senegal-Niger. Politically there was a north/south split which reflected relative economic development. Independence was acheived in 1960 and Prime Minister Maga became President. There have been three army coups since independence and the interim civilian administrations have been both elected and appointed, but hampered by regional conflicts. Major Kerekou has been President since 1972 and his regime has introduced some political continuity. From 1974 there has been a move towards more public ownership and by 1975 relations had deteriorated so much with France that technical assistance and aid was ended. In 1978 this policy was reversed and French loans and expertise have been redeployed accompanied by a tolerance towards the private sector. There seems to be a response to short term economic difficulties rather than a change of philosophy. In 1985 there was student unrest leading to the closure of schools and colleges in May and June.

Stability

Since Independence in 1960, Benin has had three coups by the military. Interspersed between these military take-overs have been various elected and appointed civilian governments, but invariably civilian rule has been hampered by regional conflicts.

Brigadier-General Mathieu Kerekou has ruled as President since assuming office in 1972, being confirmed by elections in 1980 and 1984. There has therefore been some fifteen years of political continuity after twelve years of frequent changes of government.

On the economic front, Kerekou's government initiated more public ownership in 1974. Relations with France deteriorated in 1975, and French aid and technical assistance was terminated. Collaboration was restored in 1978, and loans and skilled personnel have subsequently been received from France. From 1981 there was more sympathy toward the private sector, although more from the infeasibility of further state involvement rather than commitment to market forces. In 1986 the government prepared to adopt a programme of economic liberalisation, including privatisation and reform of public sector operations and anticipated a deal with the IMF which was finally concluded in 1989. Continued membership of the Franc zone has provided a fully convertible currency.

Overall, Benin has a moderate stability record. Enthusiastic adoption of the IMF programme and several more years of encouragement of the private sector will be necessary before increased investment is forthcoming from other than French firms.

Economic Structure

Benin's population, estimated in 1989 to be 4.6 m, makes it smaller than the average in the region, and in Africa, where the average population is around 10 m, 35% of the population live in urban areas, and this is above the average level of urbanisation for the region and Africa. Benin is small, in terms of land area, with only 113 thousand square kilometres, about one third of the average regional size. Population density, at 39.8 per square kilometre is almost twice the African average, and 30% higher than in the region. Population, growing at 3.2% a year, is comparable with the regional and African average, while agricultural growth 1980-86 was slower at 3.0%.

In 1987, the economy generated $US 1,570m of GDP, and this makes the economy small in West African terms where the average is three times larger. In GNP per head terms at $US 310 per capita, Benin is a low-income country, and average living standards are below regional and African averages.

Agriculture generates 46% of GDP and this is above average for the region and Africa. The size of the industry sector at 14% of GDP, though only 4% is manufacturing, is half the size typical in both the region and Africa, while services at 39% is similar to average African service sector sizes.

15% of GDP is exported, and this is more or less typical of West Africa and Africa. Imports comprise 25% of total spending, and this is a fifth higher than for West Africa generally. Benin's exports are predominantly agricultural, and are affected by fluctuating world commodity prices, with cocoa and cotton each comprising 20% of exports, with vegetable oil also being significant. Hydrocarbons are increasingly making an important contribution to exports. There is considerable unofficial trade with neighbouring countries, especially Nigeria, though efforts are being made to curb it. Imports are mostly industrial goods, with machinery, manufactures and chemicals comprising 63%, with food imports high at 17% and fuels making up 4%.

External debt servicing payments took up 15.9% of export earnings in 1987, well below the average regional figure of 21.4%.

Benin

Adult literacy was 28% in 1980, about the regional average, which is lower than that for Africa which stands at 36%. Literacy rates can be expected to have improved since the last literacy survey, as 67% of the relevant age group are in primary education, which is low compared with averages in the region, although the percentages in secondary education (22%) and higher education are equivalent or better.

Life expectancy at 49 years is typical of the region and Africa, average nutrition levels are inferior, provision of doctors is slightly worse, and the availability of nurses better.

Overall, the economic structure is one of a low level of development, with heavy dependence on agriculture for generation of GDP and export earnings. The educational system gives preference to secondary and higher education, and is in that sense élitist.

Economic Performance

GDP has shown positive growth at 2.8% per year over the period 1980-87, and this is substantially better than the averages of the region and Africa where there were negative growths of -2.0% and -0.6% respectively. Only Burkina Faso, Mali and Senegal in the region have grown faster. This has enabled rises in average living standards of only 0.2% per year, but elsewhere in Africa, on average, there have been falls of around -3.6% per year. Over the period 1980-87, the agriculture sector expanded output at 2.5% per year, roughly the same rate as population growth, and the services sector growth at 1.3% but it is the industry sector with 8.3% which has contributed most to the good performance.

Export volumes have fallen at -0.1% a year, though not as fast as they have declined elsewhere in Africa, while import volumes have grown minimally at 0.4% a year, though this is compared with a decline of -12.9% in West Africa generally over the same period.

Investment as a percentage of GDP was fairly low in 1987 at 14%, and it has been considerably higher in previous years, being 24% in 1980. Aid receipts have been high, being 10% of GDP in 1985, and this is double the African average and has clearly been an important factor in maintaining investment levels.

The budget deficits are low, being 0.3% of GDP in the latest year for which figures are available, 1979, and this is typical of the budgetary restraint placed on members of the franc zone. Low budget deficits have contributed to the modest inflation rate of 8.2% a year from 1980-87, being half the inflation rate of Africa generally.

Overall, Benin's economic performance over the past decade has been good, particularly in view of the heavy dependence on agriculture, and a moderate stability record. The recent low levels of investment must cause some concern for the future. The reliance on agriculture will continue to make Benin vulnerable to year to year fluctuations in GDP as a result of climate variation, though less so than the Sahelian countries.

Recent Economic Developments

Benin's border with Nigeria, closed since April 1984, was re-opened in March 1986. This event was expected to improve Benin's balance of payments position as traditionally Benin exports significant quantities of food stuffs to Nigeria. The demand generated, however, affected the domestic price structure in Benin, with rice prices in particular rising several-fold, and increasing general inflationary pressures.

Expansion of petroleum production has been hampered by the withdrawal of the Swiss-based Pan Ocean Oil Company (PANOCO) after disagreements over payments. PANOCO has originally hoped to raise oil liftings from around 10,000 barrels per day to 25,000 barrels per day. This would enable a net export of three-quarters of oil production. In 1988 the US company Ashland Exploration took over development. Output in 1989 was estimated to be under 4,000 barrels per day.

At the beginning of 1987, IDA, the concessional lending arm of the World Bank, agreed to lend $US 15m for the reform of the parastatal sector, with the prospect of some enterprises being liquidated and others privatised. The major concerns under review are state-owned palm-oil and groundnut producers, oil seed and cocoa processing enterprises, the government monopoly of the importation and distribution of food, three shipping and transport firms, a textile factory and two cement producers.

Benin signed its first Structural Adjustment Programme agreement with the IMF in June 1989 and a $US 27.2m structural adjustment facility credit was offered for a 3 year recovery programme, and this was followed by a further credit facility of $US 41.8m from the World Bank. The government agreed to take measures to reduce fiscal and balance of payment deficits and the private sector is to play a greater role in the economy. GDP growth during the recovery programme is aimed at 3% with the current account deficit being reduced to around 2% of GDP.

World Bank estimates for 1988 put Benin's total external debt at $US 908m, nearly 70% of GNP. Meanwhile, the country's debt service ratio continues to grow and was expected to more than double in 1989.

Despite being a former French colony and a menber of the Franc Zone, West Germany is the largest aid donor, with $US 30.9m in 1987, while French aid was $25.8m. Multilateral aid included $US 21.5m from the IDA and $12m from the European Community. Following the 1989 IMF agreement in 1989 further aid was forthcoming, including a $US 17m loan from the European Development Fund.

In the 1989 budget, proposed expenditure was set at CFA 104.5b ($US 328.7m), while revenue was estimated at CFA 75.7b ($US 238.1m), leading to a deficit of CFA 28.7b ($US 903.8m). Cuts in personnel in government and the armed forces are planned as well as reductions in their allowances and bonuses of 50%.

Economic Outlook and Forecasts

Benin's economic prospects depend on prices of cocoa and palm oil, and success in obtaining agreement with an oil company to take over the PANOCO operation. Cocoa prices are expected to fall in real terms over the medium term. Palm oil prices are expected to fall too, but more slowly. Benin's oil operation is offshore, and in the absence of a sustained improvement in oil prices, it may be a while before an oil company can be persuaded to take on the Benin operation. Prospects have improved with the structural adjustment agreement with the IMF, and this will help to encourage foreign involvement, and improve recent low investment rates, particularly if the reform programme is able to press on with privatisation of the parastatal sector.

The economic forecasts assume no improvement in Benin's stability, but indicate modest increases in GDP and GDP per head into the next decade, despite relatively poor prospects for agricultural prices. Good growth in the industrial and services sectors will be important if these forecasts are to be realised. Export volumes are expected to show positive growth, but import volumes will grow more rapidly as a result of higher inflows of aid. Price stability will be good at under 10% annual inflation.

Benin: Economic Forecasts
(average annual percentage change)

	Actual	Forecast Base	Forecast High
	1980-87	1990-95	1990-95
GDP	2.8	3.0	4.0
GDP per Head	-0.4	0.1	1.1
Inflation Rate	8.2	8.0	5.0
Export Volumes	-0.1	3.0	4.5
Import Volumes	0.4	2.5	3.5

BENIN: Comparative Data

	BENIN	WEST AFRICA	AFRICA	INDUSTRIAL COUNTRIES
POPULATION & LAND				
Population, mid year, millions, 1989	4.6	12.3	10.2	40.0
Urban Population, %, 1985	35	28.7	30	75
Population Growth Rate, % per year, 1980-86	3.2	3.2	3.1	0.8
Land Area, thou. sq. kilom.	113	384	486	1,628
Population Density, persons per sq kilom., 1988	39.8	31.1	20.4	24.3
ECONOMY: PRODUCTION & INCOME				
GDP, $US millions, 1986	1,320	4,876	3,561	550,099
GNP per head, $US, 1986	270	510	389	12,960
ECONOMY: SUPPLY STRUCTURE				
Agriculture, % of GDP, 1986	49	40	35	3
Industry, % of GDP, 1986	13	26	27	35
Services, % of GDP, 1986	37	34	38	61
ECONOMY: DEMAND STRUCTURE				
Private Consumption, % of GDP, 1986	90	77	73	62
Gross Domestic Investment, % of GDP, 1986	13	13	16	21
Government Consumption, % of GDP, 1986	9	13	14	17
Exports, % of GDP, 1986	14	18	23	17
Imports, % of GDP, 1986	26	20	26	17
ECONOMY: PERFORMANCE				
GDP growth, % per year, 1980-86	3.6	-2.03	-0.6	2.5
GDP per head growth, % per year, 1980-86	0.4	-4.7	-3.7	1.7
Agriculture growth, % per year, 1980-86	3.0	1.3	0.0	2.5
Industry growth, % per year, 1980-86	10.2	-3.6	-1.0	2.5
Services growth, % per year, 1980-86	1.8	-2.5	-0.5	2.6
Exports growth, % per year, 1980-86	-3.5	-4.0	-1.9	3.3
Imports growth, % per year, 1980-86	-1.2	-12.9	-6.9	4.3
ECONOMY: OTHER				
Inflation Rate, % per year, 1980-86	6.6	13.0	16.7	5.3
Aid, net inflow, % of GDP, 1985	10.0	3.7	6.3	-
Debt Service, % of Exports, 1986	28.8	21.4	20.6	-
Budget Surplus (+), Deficit (-), % of GDP, 1979	-0.3	-3.4	-2.8	-5.1
EDUCATION				
Primary, % of 6-11 group, 1985	65	75	76	102
Secondary, % of 12-17 group, 1985	20	24	22	93
Higher, % of 20-24 group, 1985	2	2.6	1.9	39
Adult Literacy Rate, %, 1980	28	30	39	99
HEALTH & NUTRITION				
Life Expectancy, years, 1986	50	50	50	76
Calorie Supply, daily per head, 1985	2,248	2,105	2,096	3,357
Population per doctor, 1981	17,010	16,199	24,185	550

Notes: 'Southern Africa' and 'Africa' exclude South Africa. Dates are for the country in question, and do not always correspond with the Regional, African and Industrial averages.

BENIN: Leading Indicators

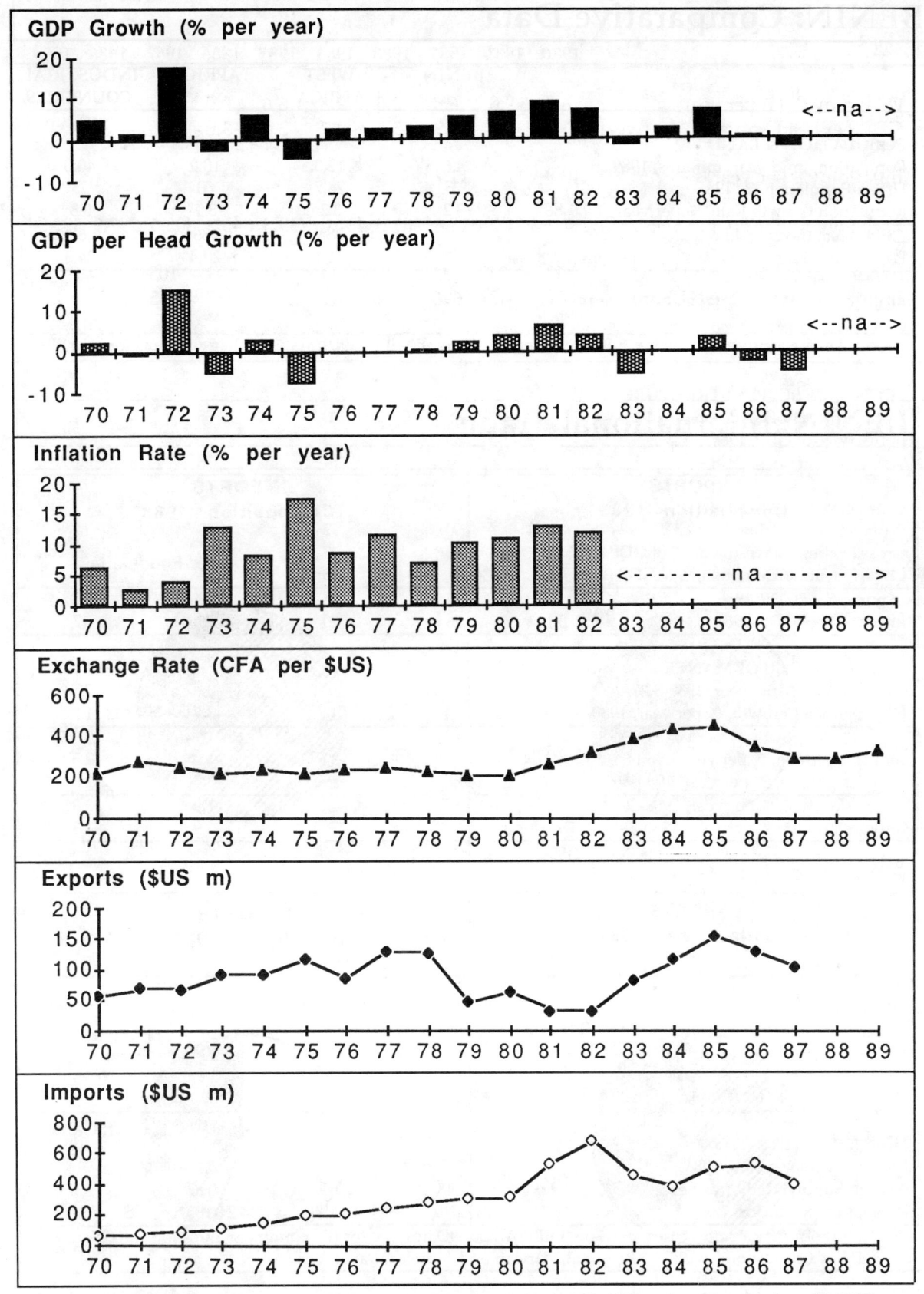

GDP Growth (% per year) — 1970–1987 data shown; 1988–89 na

GDP per Head Growth (% per year) — 1970–1987 data shown; 1988–89 na

Inflation Rate (% per year) — 1970–1982 data shown; 1983–89 na

Exchange Rate (CFA per $US) — 1970–1989

Exports ($US m) — 1970–1987

Imports ($US m) — 1970–1987

Note: Leading indicator data for 1989 are based on the first half of 1989. 1989 exchange rate is for mid-1989.

BENIN: Leading Indicator Data

	1980	1981	1982	1983	1984	1985	1986	1987	1988	1989
GDP growth (% per year)	6.5	9.0	6.8	-2.0	2.3	6.6	0.0	-2.5		
GDP per head growth (% per year)	3.7	6.1	3.9	-5.9	-0.7	3.6	-3.0	-5.6		
Inflation (% per year)	10.9	12.9	11.7							
Exchange rate (CFA per $US)	211.3	271.7	328.6	381.0	437.0	449.3	346.3	300.5	297.9	327.3
Exports, merchandise ($US m)	63	34	33	84	114	153	129	105		
Imports, merchandise ($US m)	331	543	696	463	395	516	537	415		

Note: Leading indicator data for 1989 are based on the first half of 1989. 1989 exchange rate is for mid-1989.

BENIN: International Trade

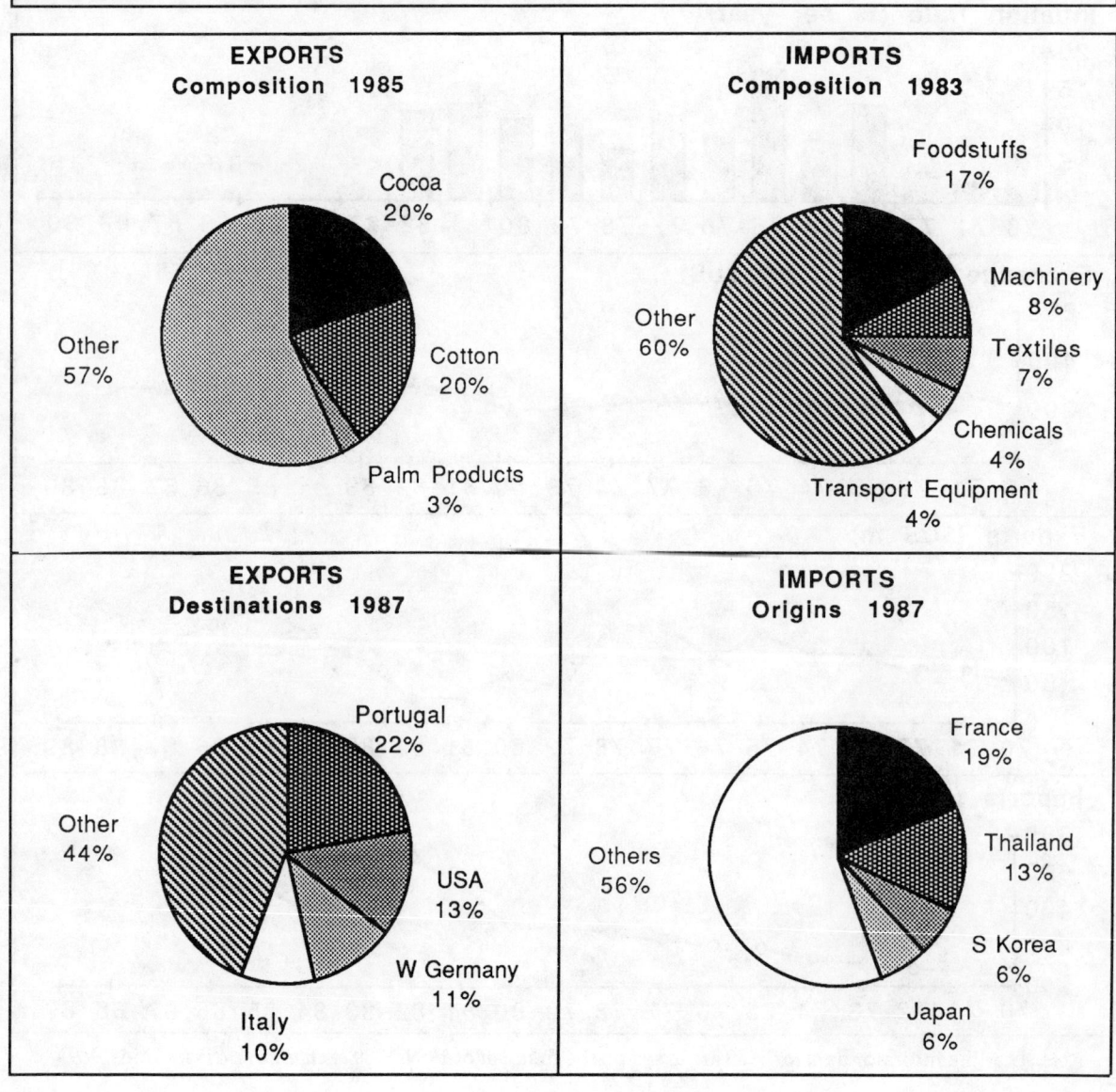

EXPORTS Composition 1985
- Cocoa 20%
- Cotton 20%
- Palm Products 3%
- Other 57%

IMPORTS Composition 1983
- Foodstuffs 17%
- Machinery 8%
- Textiles 7%
- Chemicals 4%
- Transport Equipment 4%
- Other 60%

EXPORTS Destinations 1987
- Portugal 22%
- USA 13%
- W Germany 11%
- Italy 10%
- Other 44%

IMPORTS Origins 1987
- France 19%
- Thailand 13%
- S Korea 6%
- Japan 6%
- Others 56%

3 BOTSWANA

Physical Geography and Climate

Botswana, in the Southern African region, is a small land-locked country bordered by Namibia, South Africa, Zambia and Zimbabwe. It is characterised by a plateau at 900m with hills to the east. The major river system is the Okavango and its swamp, with the Kalahari Desert to the south and west. Temperatures range from 18°-24°C. Rainfall ranges from 60cm per annum in the north to 38-50cm in the east to 13cm in the southwest to negligible amounts in the Kalahari Desert. Vegetation patterns follow the rainfall distribution from the thorny semi-desert of the Kalahari to the dry woodland savannah of the east.

Mineral deposits include diamonds, coal, copper, nickel, soda ash, salt, manganese, potash and sodium sulphate.

Population

Estimates for 1989 suggest a population of 1.2 million most of whom live in the east, including Europeans and Asians and over 80% of the Batswana tribes. The majority of the population follow traditional beliefs, with 15% Christians and small numbers of Moslems and followers of Bahá'i. The majority of people are Tswanas, though there are also some minority groups such as Kalangas and Basarwa (Bushmen) as well as several thousand Asians and Europeans. The official languages are Setswana and English. The density of population is very low at 2 per square kilometre compared to the regional average of 12.7 per square kilometre, though 80% of the population live in the east of the country. Urbanisation is near to the regional average at 23% and the rate of population growth has fallen from a high figure for 1970-82 of 5.0% to a figure just below the regional average of 3.4% for 1980-1987.

History

Bechuanaland came under British influence in 1885. In 1895 agreements were made whereby each chief should rule his people under the protection of the British Sovereign represented by a resident commissioner.

Botswana became fully independent within the Commonwealth in 1966. Multi-party elections have continued to return the BDP (Botswana Democratic Party) to office. On the death of Sir Seretse Khama in 1980, the leadership passed to Dr Quett Masire.

The economy follows a conservative, but mixed strategy, but relies on South Africa for all its petroleum imports and supplies her with 19,000 miners. In June 1985 South African Defence Forces attack targets in Gaborone and twelve members of the ANC were killed and six injured. The attack was condemned by the UN. Following talks with President Botha in February 1986 the ANC refugees were expelled to other frontline states. Since then South Africa has continued to carry out cross border raids and bombings, mainly around the capital which is only 20 km from the border. Later in the year Zimbabwe accused Botswana of aiding dissidents resident in the Dukwe refugee camps near the Zimbabwian border. However, the dissident problem is now minimal in Zimbabwe and most of the Dukwe refugees have returned to Zimbabwe in the last year.

Stability

Since independence in 1966 Botswana has held

Botswana

regular multi-party elections, with the Botswana Democratic Party (BDP) returned to office at each election. Dr. Quett Masire succeeded smoothly to the leadership of the BDP when Sir Seretse Khama died in 1980. There are no less than nine political parties in the country. The only concerns surround the robustness of democratic institutions should an opposition party, such as the left-wing Botswana National Front (BNF), appear to be likely to win an election, and the security position along Botswana's borders.

Economic strategy has been consistent with a strong private sector and substantial foreign investment in the mining and industrial sectors.

Following the 1989 general election, President Masire brought Brig. Gen. Merhafe, the chief of the armed forces into the cabinet, implying that the Botswana Defence Force may have been presenting a threat to Botswana's democracy.

Overall, Botswana has had an excellent record of stability, and there are good reasons to expect this to continue, particularly now that South Africa is embarking on a programme of political reforms.

Economic Structure

Total GDP was estimated at $US 990m in 1984, making the size of the economy half the regional average. GNP per head is $US 960, and this makes Botswana a middle income country with an average living standard double the average regional level and higher than in any other Southern African country with the excepting of South Africa and Namibia.

The agriculture sector generates only 6% of GDP, as compared with the typical 20% in the region, and only South Africa in the region, with a substantially more developed economy, has a comparably small agriculture sector. The industrial sector, which includes mining activities is responsible for 45% of GDP, and this is half as large again as the regional norm. Services contribute 48% of GDP and this is a little higher than the Southern African average of 42%.

Exports are 61% of GDP, and only Namibia and Swaziland have higher export dependence in the region. 62% of expenditure goes on imports, and this is 50% higher than the average for Southern Africa, although certain small countries, notably Lesotho, Namibia and Swaziland do have higher reliance on imports. Export earnings in 1988 were $US 1,489m while imports were worth $US 1,140m in a country which has little restriction on the import of even luxury goods.

Exports are dominated by diamonds, which made up 73% of earnings, with copper and nickel 16%, and meat products 3%. Over half of expenditure on imports is on industrial products, with machinery comprising 29% of imports, manufactures 21% and chemicals 8%. Foodstuffs are 16% of imports, fuels 9% and ores and metals 9%.

Botswana's external debt service is 3.8% of export earnings, but it must be borne in mind that exports are a high proportion (48%) of GDP. Nevertheless, this does make Botswana's debt one of the more manageable in the region.

Educational provision is among the best in the region, with 96% of the relevant age group in primary school, 22% in secondary and 2% in higher. The last literacy survey was in 1976 when 35% of the adult population was recorded as literate, but high enrolments will have raised this.

Life expectancy is the highest in the region at 58 years, average calorie intake is 93% of minimum requirements. With one doctor for every 9,250 people and a nurse for every 700, health provisions are substantially better than is common in Southern Africa

Overall, Botswana's economic structure is one of heavy dependence on the diamond mining sector for both export earnings and generation of GDP, although the government is attempting to diversify the economy through investment in other sectors. This has led to a noticeably open economy with substantial reliance on world markets for sales of goods produced, and for expenditures. The small size of the agriculture sector is partly a result of middle income country production structure, but also specialisation in mining where Botswana clearly has natural advantages. High levels of income per head are responsible for good educational and health provision.

Economic Performance

GDP has grown at an astonishing 13% a year over the period 1980-87, and this has allowed average living standards to rise by 8.9%. This performance is easily the best in Africa, the next best performance being that of Cameroon, with an annual GDP growth rate of 7%.

Growth has been generated by the industrial sector (including most noticeably mining), which has expanded output at 19.2% a year, and this has led to growth of 9.5% a year in the services sector. Resources have shifted to these two sectors from agriculture, and agricultural output has declined -7.8%, partly due to drought, and Botswana has increased imports of foodstuffs as a result.

The output of the mining sector is predominantly sold overseas, and export volumes have risen by 18.3% a year, and this consequent rapid increase in export earnings has allowed import of goods to increase at 10.4% a year in the period 1980-87.

Investment was running at 21% of GDP in 1984, and this is 30% higher than the average for the region. Aid was 11.6% of GDP in 1984, and this is particularly high in view of Botswana's middle-income GDP levels.

The budget deficit was 11.5% of GDP in 1983, and only Zambia in the region ran a larger deficit. However, the deficit, which leads to an increase in the money supply, needs to be judged in relation to the rate of growth of output in gauging the effect on the price level. Botswana's exceptional 10.7% annual growth rate has meant that despite the size of the budget deficit, the inflation rate has only averaged 8.4% over the period 1980-87, among only three countries with single-figure rates of price increase in Southern Africa.

Overall, Botswana has had an outstanding record of economic performance, and this has extended to all the leading indicators, GDP, GDP per head, inflation and the balance of payments. Over the period since 1965, this record has been not only the best in Africa, but the best GDP growth performance in the world.

Recent Economic Developments

Botswana continues to pursue economic policies that capitalise on exploitation of Botswana's mineral resources with the co-operation of foreign investment, predominantly from South Africa. The main concern is to diversify the economy away from dependence on the diamond sector which has so successfully allowed Botswana to have the world's fastest growing GDP for the past two decades. At present Botswana is estimated to have over $US 2.5b in foreign exchange reserves.

GDP growth shows every indication of being maintained at more than 8% annually, which allows a close to 5% annual rise in GDP per head. While there are doubts as to whether the diamond sector can maintain its phenomenal expansion rate, the copper and nickel mines have experienced a resurgence with the rise in metal prices at the end of 1988. The six years of drought that have severely constrained the agricultural sector, particularly beef production, appeared to have ended in early 1988 giving rise to hopes that the agricultural sector will begin to show positive growth.

The current inflation rate is around 10%, but the high level of inflation in South Africa at present is likely to affect Botswana too. The budget surplus of recent years is necessary if the balance of payments surplus is not to generate an acceleration in the rate of increase of prices through expansion of the money supply.

The balance of trade continues to be healthy, with the value of merchandise exports running at twice the level of imports in 1987. Exports were almost their 1986 value, but this is unlikely to be maintained as 1987 saw the sale of a substantial diamond stock pile worth approximately $US 500 m.

External indebtedness is modest as phenomenal growth of GDP and of export earnings have avoided the need to increase borrowings. Total external debt was estimated at $US 518m in 1987, and taking up only 4.3% of export earnings in debt service.

The exchange rate has undergone periodic adjustment, which has attempted to strike a balance between the need to encourage non-diamond exports, and the pressure to appreciate the Pula which comes from the current trade surplus. Botswana's long-run concern to diversify export earnings led to an 8.2% depreciation of the currency in 1988.

The 1988/9 budget showed modest rises in real expenditure in comparison with the actual

Botswana

outcome for 1987/8, which was higher than projected. Total expenditures are projected 5% in real terms to reach BP 1,497m ($US 894m), with revenue at BP 1,777m ($US 1,061m), also a 5% real rise in actual 1987/8 revenues. Investment expenditures rose from BP 44m ($US 268m) in the 1987/8 budget to BP 710m ($US 424), while recurrent expenditures rose from BP 660 m ($US 378m) to BP 755m ($US 451m). Emphasis is therefore on investment spending rather than current consumption, and this is reflected in the decision to raise Civil Service pay by 7%, below the rate of inflation. A budget surplus of BP 280m ($US 167) is projected, which is approximately 10% of GDP.

Botswana being a middle income country, attracts relatively little concessionary aid. In 1988 the Kuwait Fund for Arab Economic Development contributed $US 13m for road works, and Norway $US 9m for health.

Despite a vigorous campaign to encourage foreign investment, Botswana has been disappointed in the results. However, toward the end of 1988 there was a substantial investment in the Sua Pan soda dish project, in which South Africa has taken up 52% of the equity in a $US 380m scheme. Heinz (US) took over Kgalaghadi Soap Industries for $US 5.8m in April 1988.

Economic Outlook and Forecasts

Botswana's prospects depend heavily in the medium-term on the international price of diamonds. Improving stability in South Africa may improve world supply, although careful control of the marketing of diamonds should ensure their real price is maintained.

The reduction of confrontation by the countries in the region with South Africa will ease problems, particularly in the areas of transport and communications. Botswana's main trade routes are through South Africa, and although diamond exports are shipped by air, 80% of imports come from South Africa, including essential equipment, fuel and materials for the mining sector.

There is no reason to expect Botswana's political and economic stability to deteriorate, particularly now South Africa's de-stabilising activities are likely to be reduced.

The forecasts asume no deterioration in Botswana's stability, stable real diamond prices, and improvements in trade links.

On the above basis, the forecasts show continuing excellent GDP and GDP per head growth, together with very rapid expansion of export volumes. Import volumes are forecast to increase more rapidly than in the 1970s, while inflation will be just above single figures.

Botswana: Economic Forecasts
(average annual percentage change)

	Actual	Forecast Base	High
	1980-87	1990-95	1990-95
GDP	13.0	8.0	10.0
GDP per Head	9.6	5.7	7.7
Inflation Rate	8.4	12.0	9.5
Export Volumes	18.3	10.5	15.5
Import Volumes	10.4	9.5	12.5

BOTSWANA: Comparative Data

	BOTSWANA	SOUTHERN AFRICA	AFRICA	INDUSTRIAL COUNTRIES
POPULATION & LAND				
Population, mid year, millions, 1989	1.2	6.2	10.2	40.0
Urban Population, %, 1985	20	23.8	30	75
Population Growth Rate, % per year, 1980-86	3.5	3.1	3.1	0.8
Land Area, thou. sq. kilom.	600	531	486	1,628
Population Density, persons per sq kilom., 1988	2	11.3	20.4	24.3
ECONOMY: PRODUCTION & INCOME				
GDP, $US millions, 1986	1,150	2,121	3,561	550,099
GNP per head, $US, 1986	840	383	389	12,960
ECONOMY: SUPPLY STRUCTURE				
Agriculture, % of GDP, 1986	4	25	35	3
Industry, % of GDP, 1986	58	33	27	35
Services, % of GDP, 1986	38	42	38	61
ECONOMY: DEMAND STRUCTURE				
Private Consumption, % of GDP, 1985	47	67	73	62
Gross Domestic Investment, % of GDP, 1985	26	14	16	21
Government Consumption, % of GDP, 1985	28	21	14	17
Exports, % of GDP, 1985	63	34	23	17
Imports, % of GDP, 1985	64	36	26	17
ECONOMY: PERFORMANCE				
GDP growth, % per year, 1980-86	11.9	-3.7	-0.6	2.5
GDP per head growth, % per year, 1980-86	8.1	-6.6	-3.7	1.7
Agriculture growth, % per year, 1980-86	-9.8	-6.8	0.0	2.5
Industry growth, % per year, 1980-86	19.1	-3.8	-1.0	2.5
Services growth, % per year, 1980-86	7.6	-0.9	-0.5	2.6
Exports growth, % per year, 1970-81	18.3	-5.1	-1.9	3.3
Imports growth, % per year, 1970-81	10.4	-2.5	-6.9	4.3
ECONOMY: OTHER				
Inflation Rate, % per year, 1980-86	7.6	18.1	16.7	5.3
Aid, net inflow, % of GDP, 1986	10.4	8.6	6.3	-
Debt Service, & of Exports, 1986	4.3	7.8	20.6	-
Budget Surplus (+), Deficit (-), % of GDP, 1986	31.8	-4.1	-2.8	-5.1
EDUCATION				
Primary, % of 6-11 group, 1985	104	95	76	102
Secondary, % of 12-17 group, 1985	29	18	22	93
Higher, % of 20-24 group, 1985	1	1.3	1.9	39
Adult Literacy Rate, %, 1976	35	47	39	99
HEALTH & NUTRITION				
Life Expectancy, years, 1986	59	50	50	76
Calorie Supply, daily per head, 1985	2,159	1,998	2,096	3,357
Population per doctor, 1981	7,400	26,118	24,185	550

Notes: 'Southern Africa' and 'Africa' exclude South Africa. Dates are for the country in question, and do not always correspond with the Regional, African and Industrial averages.

BOTSWANA: Leading Indicators

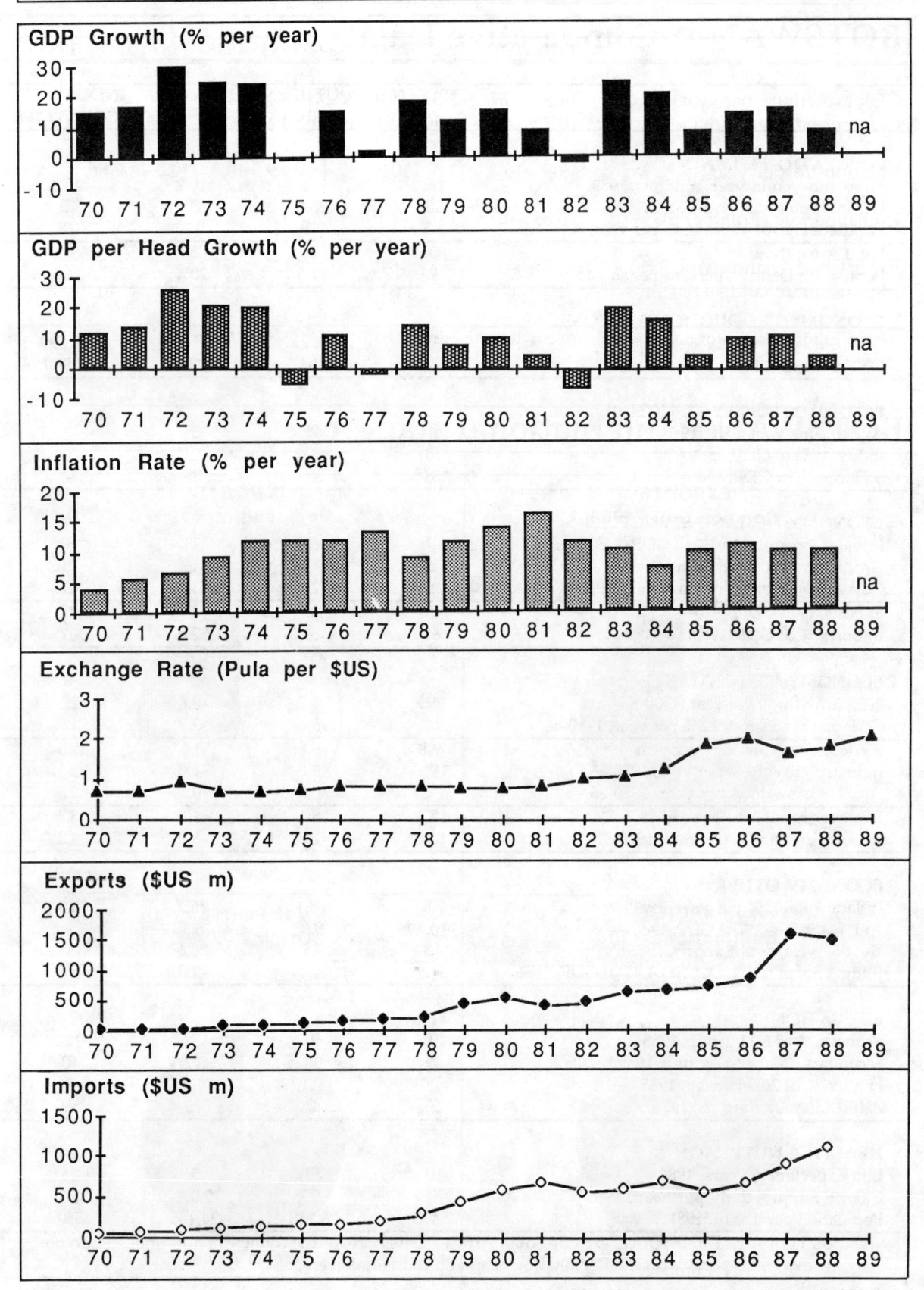

GDP Growth (% per year)

GDP per Head Growth (% per year)

Inflation Rate (% per year)

Exchange Rate (Pula per $US)

Exports ($US m)

Imports ($US m)

Note: Leading indicator data for 1989 are based on the first half of 1989. 1989 exchange rate is for mid-1989.

BOTSWANA: Leading Indicator Data

	1980	1981	1982	1983	1984	1985	1986	1987	1988	1989
GDP growth (% per year)	14.9	8.7	-2.3	24.0	20.0	8.1	14.0	14.7	8.1	
GDP per head growth (% per year)	10.7	4.6	-6.3	20.1	16.2	4.4	10.4	11.1	4.5	
Inflation (% per year)	13.9	16.2	11.5	10.3	7.4	9.9	10.8	10.0	9.8	
Exchange rate (Pula per $US)	0.83	0.78	1.02	1.10	1.28	1.89	2.04	1.68	1.82	2.08
Exports, merchandise ($US n.)	545	401	461	640	678	727	853	1582	1489	
Imports, merchandise ($US n.)	603	687	580	615	707	583	684	904	1140	

Note: Leading indicator data for 1989 are based on the first half of 1989. 1989 exchange rate is for mid-1989.

BOTSWANA: International Trade

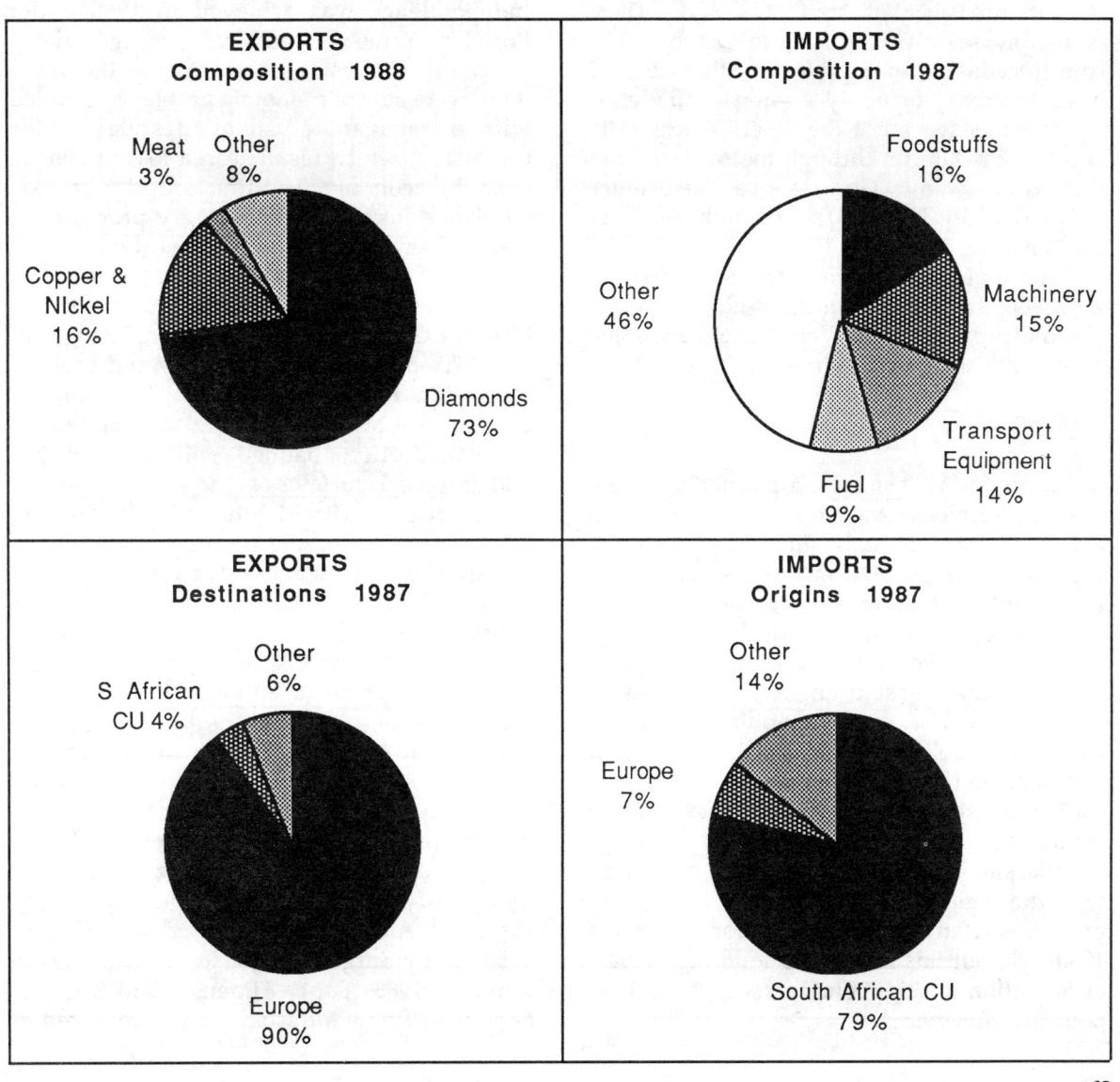

EXPORTS Composition 1988
- Meat 3%
- Other 8%
- Copper & Nickel 16%
- Diamonds 73%

IMPORTS Composition 1987
- Foodstuffs 16%
- Machinery 15%
- Transport Equipment 14%
- Fuel 9%
- Other 46%

EXPORTS Destinations 1987
- S African CU 4%
- Other 6%
- Europe 90%

IMPORTS Origins 1987
- Other 14%
- Europe 7%
- South African CU 79%

4 BURKINA FASO

Physical Geography and Climate

Burkina Faso, in the West Africa region, is a small, landlocked country, bounded to the north by Mali, Niger, and to the south by Benin, Togo, Ghana and the Côte d'Ivoire. The temperature is hot but dry, 25°C to 32°C. There is a rainy season from June to October. The country consists of a plateau with extensive infertile areas of mainly savanna and desert where water is scarce. The Rivers White Volta and Black Volta run through the country and there is the swampy Gourma area. The country suffered badly in the Sahel drought of 1969-1974 and also in the late 1970s.

Hydro-electric schemes are projected for Kompienga and Bagre on the White Volta and Noumbiel on the Black Volta. There are some mineral resources such as iron ore.

Population

Estimates for 1989 suggest a population of 8.7 million. Ethnically the population is Bobo in the south west; nomadic Fulani and Mossi in the north and Gourma in the east. Generally the country is seriously overpopulated in relation to fertile land, and there is much emigration to the Côte d'Ivoire and Ghana. It is one of the oldest traditional kingdoms in Africa with 55% of the population having traditional beliefs, although in the north there has been Islamic expansion.

The density of population is near average for the region at 31 per thousand and the urban population is low at 8% for 1987 compared with the regional average of 23%. Population growth is relatively low at 2.5% for the period 1980-1986 but this is partly due to high levels of migration to neighbouring states, as well as poor health services.

History

Burkina Faso, formerly known as Upper Volta, emerged following boundary changes within AOF, the French West African Federation, between the 1920s and World War II. Independence was achieved in 1960 under President Yameogo and a 75 member National Assembly was established. During the 1960s there were severe economic problems, coupled with extravagant personal lifestyles of the leadership, which resulted in a loss of control over the economy. The first coup by Lt. Col. Lamizana instituted an austerity programme, state of emergency and dissolution of the National Assembly. In 1970 a civilian government was installed, led by Prime Minister Ouedraogo. The Sahalian Drought of the 1970s disrupted the economy and brought starvation to the rural areas. The military coup of 1974 by President Lamizana suspended the constitution, banned political activities and imposed curfews. Later in the year the border dispute with Mali turned into war with Mali troops occupying an area in the north, rich in oil and manganese. The war was settled by the OAU in Burkina Faso's favour. In February 1976 there was a return to civilian government followed by a general strike and a promise to return to an elected government. Elections took place in 1979, but the turn-out was low at 40% and party leaders were not allowed to stand for the Presidency. General Lamizana was returned with Dr Conombo as Prime Minister. Following teacher's strikes in 1980 there was a bloodless coup by Col. Zerbo and a Military Recovery Committee was instituted. Another coup followed in November 1982 led by Surgeon-Major Ouedraogo and in January 1983 Capt. Thomas Sankara was appointed Prime Minister. He was removed in

Burkina Faso

May 1983 following a purge of left wing elements; the country returned to civilian government and Western aid was restored. In 1986 FAO declared the country to be seriously affected by the famine. Sankara was restored as Head of State. There was a dispute with the Côte d'Ivoire following attacks on Burkinabe nationals and further fighting in the long running dispute with Mali resulting in troop withdrawal and public reconciliation of the two Presidents in early 1986.

Despite his popularity within the country Sankara was killed in a further military coup in 1987 led by his one time friend, Capt. Blaise Compaore. Relations with Ghana became strained (Rawlings and Sankara had been close friends) and Côte d'Ivoire and France were accused of being behind the coup. Compaore is seen as a less radical figure than Sankara and has begun to steer the country away from hard line socialism, encouraging the growth of the private sector in the economy and inviting more foreign participation.

Stability

There have been six coups in Burkina Faso, and there have been four periods of civilian rule.

The political stance of the Sankara regime was originally revolutionary, but this did not appear to have been expressed in a radical change in economic policies. French influence via the France zone has been continuous, and Sankara committed the regime to support for the private sector with no nationalisations. There were, however, frequent exhortations against exploitation of the masses, and arbitrary economic interventions such as a decree that there should be no rent payments except by foreign firms.

The government of President Compaore has shown less concern to regulate and control the economy.

Overall, Burkina Faso's stability record is very poor. There can be little confidence that Compaore's regime will prove enduring. The moderate stance of the government on economic issues gives some encouragement for the future.

Economic Structure

GDP in 1987 was estimated at $US 1,650m, and although this is one third of the West African average size, it would be placed about half way in a size ranking of economies in the region. GNP per head in 1987 was $US 190, and this placed Burkina Faso in the low income category, and only Guinea-Bissau in the region has a lower average income.

Agriculture provides 38% of GDP, and this is typical of the region, but is characteristic of economies generating low levels of income per head (80% of the population depend on agriculture at subsistance level). Industry provides 25% of GDP and although low is perhaps high when viewed in the light of Burkina Faso's general development level. The services sector generates 38% of GDP, and this is fairly representative of a country with such a low average income per head.

Private consumption in 1987 took 74% of GDP, with only five countries in the region being lower, reflecting the low level of incomes and austerity of the Sankara period. Gross domestic investment of 24% of GDP during this period was bettered only by Cape Verde and Gambia, again a reflection of Sankara's policies. Government consumption was also high, almost twice the regional average, though attempts were made to reduce this. Imports comprise 40% of expenditure and this is double the import dependence, on average, of the region, and is financed by substantial inflows of aid and remittances from migrant workers in neighbouring Côte d'Ivoire. Attempts to reduce imports have been frustrated by drought years requiring food imports during the last decade.

Exports, which represent 17% of GDP, are predominantly from the agriculture sector. Cotton contributed 43% of exports earnings in 1987, with gold 25%, live animals 5% and hides 4%. Over half of imports are machinery, manufacturers and chemicals, with the other main categories being foodstuffs at 19% of the total, and fuels at 8%.

Debt service took up only 14.8% of export earnings in 1986, about two thirds the regional average, and this is a reflection of Burkina Faso's low income status which has lead to

Burkina Faso

more borrowing being made on highly concessional terms. In addition, Burkina Faso's poor stability record, low income and few exploitable mineral deposits inhibited commercial lending in the 1970s. Adult literacy, from the most recent survey in 1980, was estimated at 5%. Not only is this the lowest rate in West Africa, it is probably the lowest in Africa. The enrolment in primary education in 1986 was 35% of the relevant age group and this is one of the lowest in Africa, and compares unfavourably with the 75% primary enrolment in the region. Secondary enrolment is 6%, only one quarter of the regional average and this is one of the lowest in Africa, although the 1% in higher education is only bettered by seven of the West Africa group.

Life expectancy is 47 years, and this is below the regional average, although six other countries in the region have lower life expectancies. Provision of doctors is the joint lowest with Guinea in the region with one per 57 thousand of the population and provision of nurses at one per 1.7 thousand persons is well below average but fairly typical of many countries in the region.

Overall, the economic structure is one of a low level of mainly subsistence production, with heavy reliance on the agriculture sector for GDP and exports, making the country susceptible to drought. Provision of education and health services is a reflection of low levels of income per head.

Economic Performance

The growth of GDP over the period 1980-87 has been at an average rate of 5.6% a year, substantially better than the WA average of 1.3%. The low rate of population growth has allowed a rise in GNP per head at the rate of 1.6% per year, and this is markedly better than for the region where GDP per head fell at -4.7% a year. Growth in agriculture was 6.1% a year, better than the rate of population growth, despite drought, and this compares well with the rest of West Africa where agricultural sector growth was only 1.3%. There was impressive performance in the industrial sector, bearing in mind the low general level of development, which expanded at 3.9% a year, noticeably better than the region, where industrial output contracted. The expansion of the services sector at 5.8% a year was about the same as for the region.

There has been a positive increase in export volumes at 4.9% a year, 1980-87, which is generally good and must be viewed in the context of the region where export volumes have contracted at -4.0% a year. Import volumes have increased modestly at 2.0% a year, whereas in the region they have contracted by -12.9% a year.

Investment was 24% of GDP in 1987, well above the 13% regional average, and mostly made possible by aid inflows at 16.2% of GNP.

There was a budget surplus in 1987 at 1.6% of GDP, a third of the size of deficit typically run in West Africa. Budgetary restraint has led to a low inflation rate of 4.4%, about one-third the general regional rate of annual price increase.

Overall, economic performance has been good. A positive increase in living standards is a notable achievement when set against the adverse weather conditions and oil price rises of the early part of the period. Positive export expansion has been the norm in Franc zone countries, where the exchange rate regime has not resulted in over-valued currencies as elsewhere in Africa. Franc zone membership has enforced low budget deficits and superior price stability.

Recent Economic Developments

A five year development plan was launched in August 1986 in a bid to increase the levels of foreign investment and aid flows. Two major dams are envisaged in the plan, one to be built jointly with Ghana in the South on a tributary of the Volta. The second is a $US 108m project on the River Taroa on the border with Benin. Both dams will provide hydro-electric power, drinking water and irrigation. Finance for the Taroa dam appears to be in place, from a consortium of banks, but most of the rest of the projects in the plan require funding.

The drought in the Sahel made Burkina Faso dependent on Ghana for grain imports, and in 1986 the country needed to rely on international food aid.

Burkina has not as yet made any agreement with the IMF or World Bank on a structural adjustment programme though this is expected in 1990. The 1989 budget forecast revenues of CFA 100.6b ($US 316m) with expenditure set at CFA 107.2b ($US 337m), giving a budget deficit of around 1% of GDP.

Economic Outlook and Forecasts

Burkina Faso's immediate economic prospects depend on the international price of cotton, climatic conditions and continuing inflow of international aid. Cotton prices are expected to improve in real terms to something approaching their level in the mid 1960s. There is optimism that the drought conditions of the early 1980s will not recur, and expectation that high levels of aid inflows will continue.

Growth GDP is expected to accelerate, and to allow improvements in GDP per head into the next decade. Inflation is expected to be modest. Export volumes are expected to increase, with import volumes increasing more rapidly as a result of greater aid flows.

Burkina Faso: Economic Forecasts
(average annual percentage change)

	Actual	Forecast Base	Forecast High
	1980-87	1990-95	1990-95
GDP	5.6	4.5	6.0
GDP per Head	3.0	1.6	3.1
Inflation Rate	4.4	5.0	3.5
Export Volumes	4.9	4.5	5.5
Import Volumes	2.0	4.0	5.0

BURKINA FASO: Comparative Data

	BURKINA FASO	WEST AFRICA	AFRICA	INDUSTRIAL COUNTRIES
POPULATION & LAND				
Population, mid year, millions, 1989	8.7	12.3	10.2	40.0
Urban Population, %, 1985	8	28.7	30	75
Population Growth Rate, % per year, 1980-86	2.5	3.2	3.1	0.8
Land Area, thou. sq. kilom.	274	384	486	1,628
Population Density, persons per sq kilom.,1988	31.0	31.1	20.4	24.3
ECONOMY: PRODUCTION & INCOME				
GDP, $US millions, 1985	930	4,876	3,561	550,099
GNP per head, $US, 1986	150	510	389	12,960
ECONOMY: SUPPLY STRUCTURE				
Agriculture, % of GDP, 1985	45	40	35	3
Industry, % of GDP, 1985	22	26	27	35
Services, % of GDP, 1985	33	34	38	61
ECONOMY: DEMAND STRUCTURE				
Private Consumption, % of GDP, 1985	91	77	73	62
Gross Domestic Investment, % of GDP, 1985	20	13	16	21
Government Consumption, % of GDP, 1985	15	13	14	17
Exports, % of GDP, 1985	16	18	23	17
Imports, % of GDP, 1985	42	20	26	17
ECONOMY: PERFORMANCE				
GDP growth, % per year, 1980-85	2.5	-2.03	-0.6	2.5
GDP per head growth, % per year, 1980-85	0.0	-4.7	-3.7	1.7
Agriculture growth, % per year, 1980-85	2.7	1.3	0.0	2.5
Industry growth, % per year, 1980-85	2.1	-3.6	-1.0	2.5
Services growth, % per year, 1980-85	2.4	-2.5	-0.5	2.6
Exports growth, % per year, 1980-86	1.6	-4.0	-1.9	3.3
Imports growth, % per year, 1980-86	-0.9	-12.9	-6.9	4.3
ECONOMY: OTHER				
Inflation Rate, % per year, 1980-86	6.3	13.0	16.7	5.3
Aid, net inflow, % of GDP, 1986	19.3	3.7	6.3	-
Debt Service, % of Exports, 1986	148	21.4	20.6	-
Budget Surplus (+), Deficit (-), % of GDP, 1985	1.6	-3.4	-2.8	-5.1
EDUCATION				
Primary, % of 6-11 group, 1985	32	75	76	102
Secondary, % of 12-17 group, 1985	5	24	22	93
Higher, % of 20-24 group, 1984	1	2.6	1.9	39
Adult Literacy Rate, %, 1980	5	30	39	99
HEALTH & NUTRITION				
Life Expectancy, years, 1986	47	50	50	76
Calorie Supply, daily per head, 1985	2,003	2,105	2,096	3,357
Population per doctor, 1981	55,760	16,199	24,185	550

Notes: 'Southern Africa' and 'Africa' exclude South Africa. Dates are for the country in question, and do not always correspond with the Regional, African and Industrial averages.

BURKINA FASO: Leading Indicators

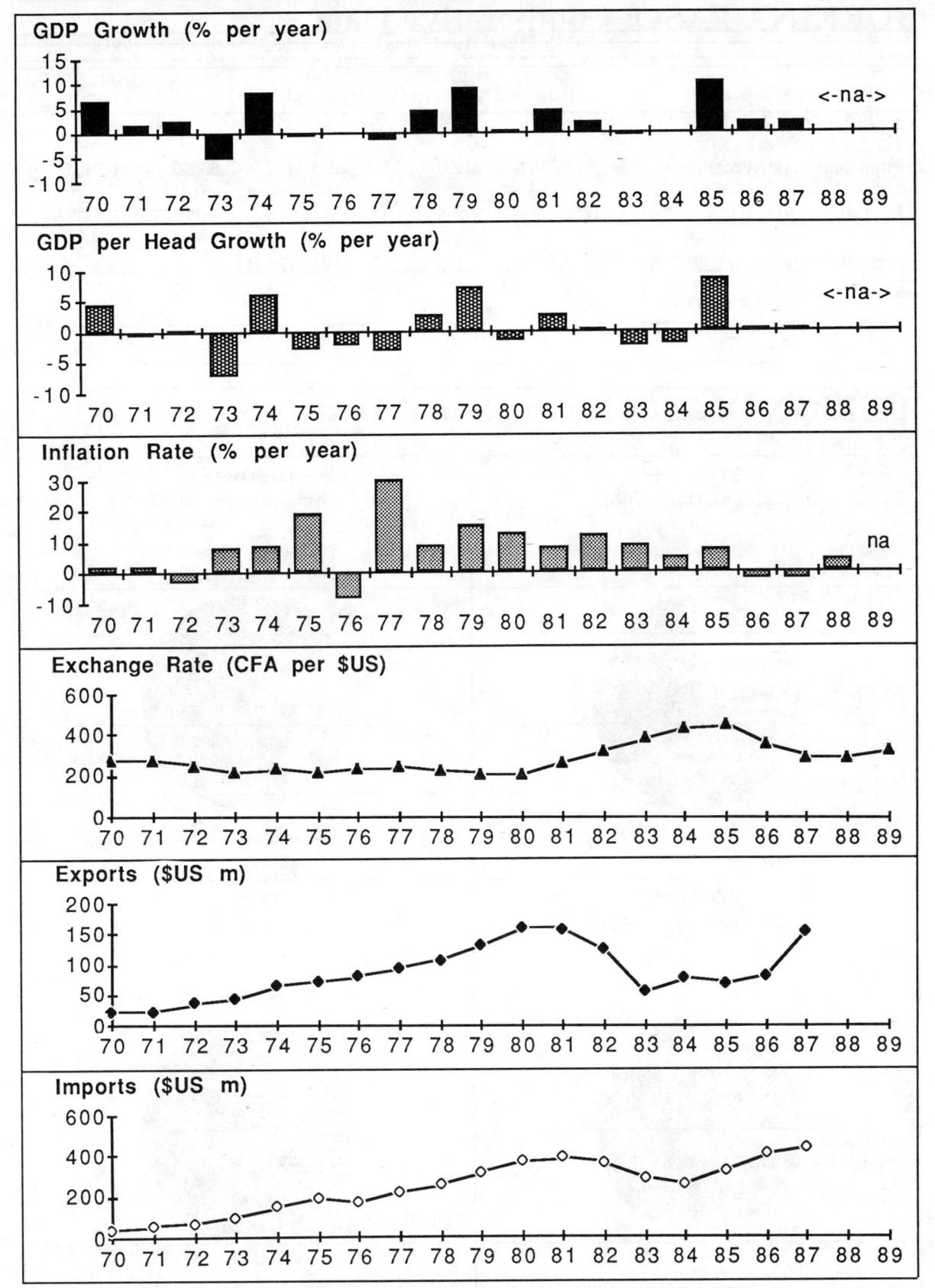

GDP Growth (% per year)

GDP per Head Growth (% per year)

Inflation Rate (% per year)

Exchange Rate (CFA per $US)

Exports ($US m)

Imports ($US m)

Note: Leading indicator data for 1989 are based on the first half of 1989. 1989 exchange rate is for mid-1989.

BURKINA FASO: Leading Indicator Data

	1980	1981	1982	1983	1984	1985	1986	1987	1988	1989
GDP growth (% per year)	0.0	4.3	2.0	-0.8	-0.5	10.2	2.1	2.1		
GDP per head growth (% per year)	-1.8	2.5	0.1	-2.7	-2.4	8.3	0.2	0.1		
Inflation (% per year)	12.3	7.6	12.0	8.3	4.8	6.9	-2.6	-2.9	4.2	
Exchange rate (CFA per $US)	211.3	271.7	328.6	381.1	437.0	449.3	346.3	300.5	297.9	327.3
Exports, merchandise ($US m)	161	159	126	57	79	70	88	155		
Imports, merchandise ($US m)	368	384	360	287	255	326	403	434		

Note: Leading indicator data for 1989 are based on the first half of 1989. 1989 exchange rate is for mid-1989.

BURKINA FASO: International Trade

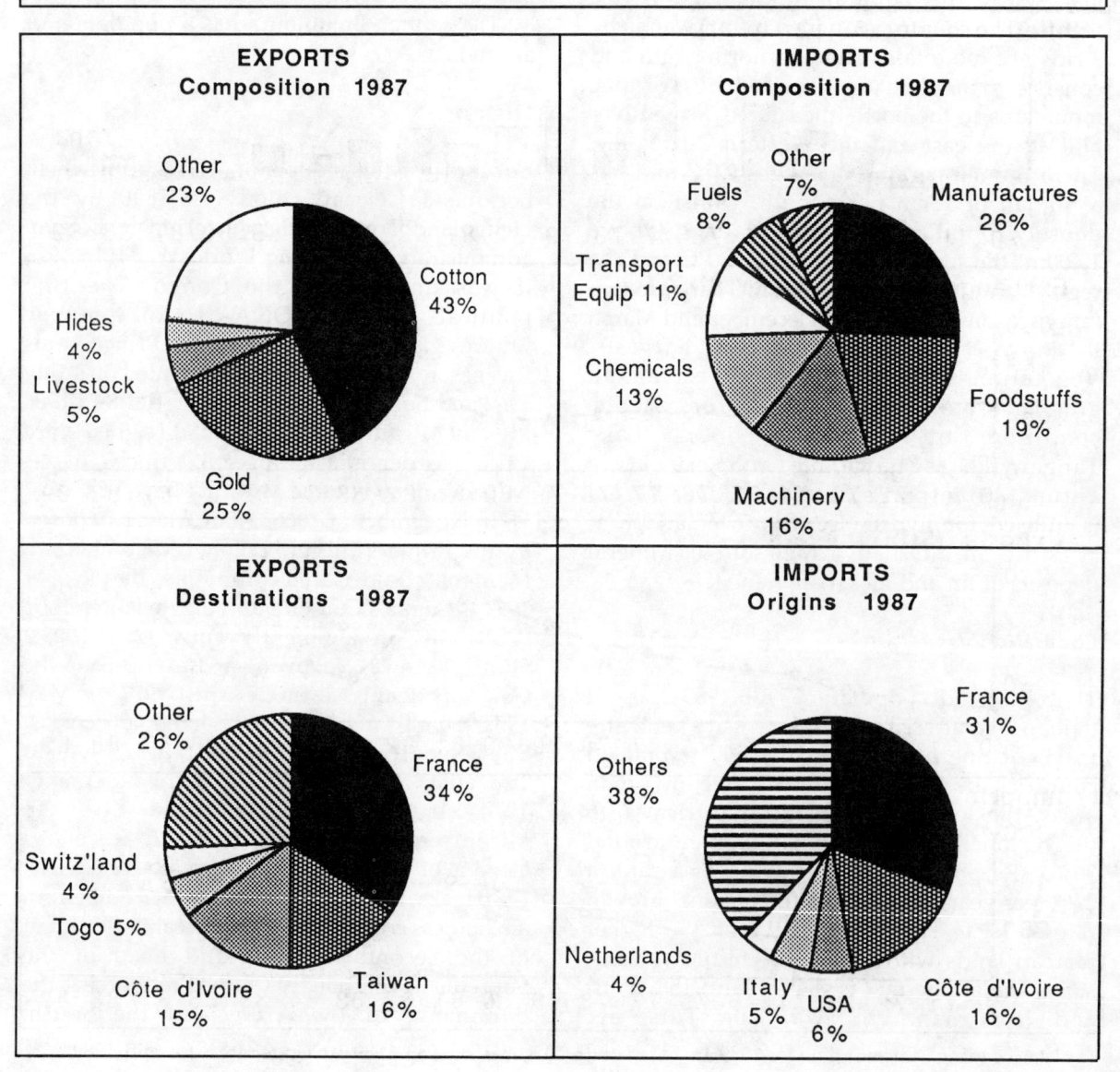

EXPORTS Composition 1987
- Cotton 43%
- Gold 25%
- Other 23%
- Livestock 5%
- Hides 4%

IMPORTS Composition 1987
- Manufactures 26%
- Foodstuffs 19%
- Machinery 16%
- Chemicals 13%
- Transport Equip 11%
- Fuels 8%
- Other 7%

EXPORTS Destinations 1987
- France 34%
- Other 26%
- Taiwan 16%
- Côte d'Ivoire 15%
- Togo 5%
- Switz'land 4%

IMPORTS Origins 1987
- Others 38%
- France 31%
- Côte d'Ivoire 16%
- USA 6%
- Italy 5%
- Netherlands 4%

5 BURUNDI

Physical Geography and Climate

Burundi is one of Africa's smallest, least urbanised and inaccessible states, bounded by Zaïre, Uganda, Tanzania and Rwanda lying just below the Equator in the East Africa region. The country is divided from Rwanda by a ridge of mountains running north-south and consists of a plateau bounded by volcanic mountains to the north, the source of the River Nile to the east and the Western Rift valley. There the temperature averages 23°C and has a rainfall of 75cm per annum, whilst on the densely populated higher plateau (above 1,400m) the temperature reaches 20°C and the regular annual rainfall is 120cm. There are two rainy seasons - October to December and March to May - which enables two harvests. Vegetation is typically high savannah grassland and forest with hot dry steppe conditions on the valley floors. Lake Tanganyika is navigable and crucial to Burundi's transport system. The River Kagera is utilised for hydro-electric power and peat bogs are an alternative fuel source. Mineral deposits of tin and nickel are mined.

Population

The estimated population for 1989 was 5.2 million and at 182 persons per square kilometre makes it one of the most densely populated countries in Africa. Only 2% live in urban areas, the lowest percentage in Africa, with the regional and continental average around 30%. The population is growing at a rate of 2.7% per year, slower than the regional growth rate of 3.1%. The majority live on the higher plateau lands with their more fertile volcanic soils. The ethnic mix is 84% Hutu; 15% Tutsi and less than 1% Twa. The Tutsi are economically and politically dominant. Internal violence in the 1970s and a shortage of cultivatable land have led to Hutu migration to neighbouring states. The official languages are French and Kirundi; 95% speak Ikirundi and 10% Kiswahili.

The capital, Bujumbura has a population of around 200,000.

History

Burundi was an independent kingdom which became a German colony in 1890 by the Heligoland Treaty and transferred to Belgian administration following World War I. In 1925 it was united with the Congo. The first political party UPRONA (Union pour le Progrès National) was formed by Prince Louis Rwagasore. Independence was granted on 1 July 1962 and Burundi remained a kingdom. In 1966 Mwami Mwambatsa IV, who had reigned since 1915, was deposed by his son, Prince Charles Ndozeye, who became Mwami Ntare V.

In November of 1966, Ntare V was deposed by his Prime Minister, Colonel Micombero, a Mantan, who declared a republic. By October 1971 the junta's Conseil Suprême de la République (CSR) was predominently Tutsi: (23 Tutsi, 2 Hutu, 2 Ganwa). Following a Hutu uprising the CSR was dismissed on 29 April 1972. In May 1972 the all-Tutsi army massacred 450 Hutus. Subsequently over 250,000 Hutus fled to Tanzania, Zaïre and Rwanda. In November 1974 elections were held and Lt. Gen. Micombero became President and Secretary-General of UPRONA. On 1 November 1976 Col. J-B Bagaza, a Rutovu, overthrew Micombero in a bloodless coup. Col. Bagaza became President of the Second Republic and head of the Supreme Revolutionary Council reinforcing the dominance of the Army and the South.

Burundi

Agrarian reforms were introduced whereby the Tutsis ceded land to the Hutu peasants. In 1978 the first five year plan was introduced. The first National Congress of UPRONA took place in 1979 and elected a Central Committee to take over the functions of the Supreme Revolutionary Council. In January 1980 President Bagaza was endorsed as President and Head of the Central Committee for a five year period. In 1981 a new constitution was adopted and elections followed in October 1982. President Bagaza, the sole candidate, was confirmed in power. Bagaza was deposed in a further coup in September 1987 by Major Pierre Buyoya because of allegations of corruption. In 1988 there was a large scale massacre of BaHutu by the security forces, which consist of mainly Tutsi troops. Huge numbers of BaHutu fled into neighbouring Rwanda. The government, concerned about its international image, has since begun to involve BaHutu much more in government and has tried to encourage the refugees to return by guaranteeing their safety.

Stability

After a period of civilian rule following independence, there have been three military coups, an attempted coup, and two episodes of violent reprisals against the majority Hutu group. There has been a substantial disruption of society with the Hutu leaving together with an influx of members of the ruling Tutsi group from nearby Ruanda. There is still tremendous distrust between the two communities.

A development strategy based on a mixed economy has remained throughout various changes of government.

Overall, Burundi's stability record is very poor. Major Buyoya has only been in power for two years, and the underlying basis for insecurity, the tension between the minority ruling Tutsi and the majority Hutu, persists. Instability in Uganda has contributed to disruption of economic life in Burundi, but with the advent of the Musevini regime in early 1986, this source of disruption will perhaps lessen.

Economic Structure

GDP, estimated at $US 1,150m in 1987 is under half the regional average, and there are only three smaller economies in the region, Comoros, Djibouti and Seychelles. Income per head in 1987 was estimated at $US 250, which is about the regional average, and which places Burundi in the low-income category.

Agriculture generates 57% of GDP, and only Uganda in the region which has seen a reversion to subsistence production in the face of political instability, has greater reliance on farming. Industry accounts for 14% of GDP, close to the East African average, and services account for 27%. The small size of the service sector, which on average in EA provides 42% of GDP is a reflection of low levels of income per head, and of the low degree of urbanisation.

Private consumption was about average for the region, taking about 76% of GDP while government consumption of 17% was 2% higher than average. Investment of 20% was reasonably high compared to regional and African averages of 16%.

Exports generate 9% of GDP in 1987 and this is well below the regional average of 16%. Only Rwanda and Sudan in the region have lower export dependence. Imports make up 21% of expenditure, which is just below the regional average.

Exports are mostly coffee, with 68% of earnings in 1987. There is a little cotton and tea, which represent 4% and 6% respectively. Imports are mostly industrial goods, with chemicals, manufactures and machinery making up 62% of the total. 14% of imports by value are fuels and 7% is food.

Debt service took up 38.5% of export earnings in 1987, and this is well above the regional average.

In 1981, 25% of the adult population were literate, and this is well below the East African average of 41%. In 1986, 59% of the relevant age group were in primary education, which is similar to the regional average. There was a 4% enrolment in secondary education, a sixth of the regional average, and only Rwanda and Tanzania had lower enrolments at this level, although secondary education in Uganda has been severely disrupted in the last decade and figures are not available. 1% were in higher education, and this is in line with the average for the region.

Life expectancy was 49 in 1987, almost the same as the East African average. Nutritional levels were on average better than elsewhere in the region, with the minimum calorie requirement just being met. Provision of doctors is better than the regional average with one practitioner for every 21,000 people, though only Ethiopia, Rwanda and Uganda have a worse ratio. There is one nurse for every 3,000 of the population.

Overall, Burundi's economic structure is one of a densely populated, small, low-income country, with a predominantly rural population relying heavily on agriculture to provide income, and on coffee to provide foreign exchange for manufactured goods. Educational and health provisions are poor, and reflect the low average standard of living.

Economic Performance

Burundi's GDP grew at 2.6% a year over the period 1980-87, and this was a growth rate over 60% faster than the average for the region. GDP per head fell at 0.2% a year over this period, and this was similar to the rest of the region.

From 1980-87 the agriculture sector showed a rate of growth of 1.7% a year, which is below the rate of increase of population. The industrial sector grew at 4.9% a year, and this was the fastest rate in mainland East Africa, though better growth was recorded in Mauritius and the Seychelles. The services sector grew at 3.5% and this was faster than the regional average.

Export volumes increased at an impressive 8.3% a year, and this compared well with the region. Imports expanded at 2.4% a year, and this was above the East African average. Aid inflows in 1987 were 15.3% of GDP, and this was above the level for the region. Aid contributed substantially to the level of investment achieved.

The budget deficit was 0.9% of GDP in 1983, and this was a third the size of the budget deficit on average in the region. Inflation averaged only 7.5% a year in the period 1980-87, and this was lower by about two thirds than was experienced in the region.

Overall, Burundi's economic performance has been good with substantial aid inflows enabling high investment levels, a steady expansion of all sectors, and improving living standards. Budgetary policy has been cautious, and rate of inflation modest, with falls to under 5% inflation in some recent years.

Recent Economic Developments

Burundi received a $US 25m stand-by loan and a $US24m structural adjustment loan from the IMF in August 1986. In addition there were funds from the IDA $US 15m and from the World Banks Special Facility for Sub-Saharan Africa $US 17m in support of a structural adjustment programme. The programme involves a 28% increase in coffee producer prices, a reduction in public sector spending, reduction in tariffs and rises in interest rates. The Burundi franc has been steadily devalued.

1986 saw boom conditions in the economy with the increase in coffee prices, which, for part of the year, saw a doubling of prices. Coffee performance has been good, with a 20% increase in production in 1985, and a further improvement of 17% in 1987 with the high prices discouraging inter-cropping and leading to intensified harvesting.

Oil has been discovered under Lake Tanganyika, and Amoco of the US is conducting test drilling. The current low price of oil may prevent immediate exploitation as costs of lifting will be high.

Government expenditure emphasises infrastructural development, and major projects in hand include extending and up-grading the road network, installing new hydro-electric capacity and extending electrification.

There has been some worsening of relations between the Catholic Church and the government, with a series of measures that have involved expelling missionaries, arresting priests, and banning Catholic Parish Councils, lay preaching and weekday mass.

Economic Outlook and Forecasts

Burundi's economic prospects depend on the continued inflow of aid, the world price of

Burundi

coffee, and the ability of the Buyoya regime to estabilish stability. Burundi can expect, by virtue of her low level of income per head, and the willingness of the governement to undertake the structural reforms instigated by the IMF and World Bank adjustment programmes, to continue to receive substantial donor support. World coffee prices are expected to deteriorate sharply. Over the longer-term, Burundi faces serious problems posed by land shortage exacerbated by the likelihood of faster rates of population growth. Prospects will depend heavily on the success of the structural adjustment programme in diversifying exports away from dependence on coffee, and continuing the expansion of the industrial sector.

The forecasts assume no significant improvement in the stability record of Burundi over the forecast period, but that the government persists with the structural adjustment programme. The growth rate of GDP is forecast to accelerate, which will allow improvements in the growth of GDP per head in the more optimistic scenario.

Inflation is expected to accelerate slightly with more government expenditure restrictions being offset by the effects of regular adjustment of the exchange rate. The structural adjustment programme, and, in particular, regular adjustment of the exchange rate, and improvements in produce prices are expected to lead to continued expansion of exports. Import volumes are projected to increase faster than export volumes as a result of increased inflows of aid.

Burundi: Economic Forecasts
(average annual percentage change)

	Actual 1980-87	Forecast Base 1990-95	Forecast High 1990-95
GDP	2.6	3.0	3.5
GDP per Head	-0.2	-0.2	0.3
Inflation Rate	7.5	9.5	8.5
Export Volumes	8.3	6.5	7.0
Import Volumes	2.4	4.5	5.5

BURUNDI: Comparative Data

	BURUNDI	EAST AFRICA	AFRICA	INDUSTRIAL COUNTRIES
POPULATION & LAND				
Population, mid year, millions, 1989	5.2	12.2	10.2	40.0
Urban Population, %, 1985	2	30.5	30	75
Population Growth Rate, % per year, 1980-86	2.7	3.1	3.1	0.8
Land Area, thou. sq. kilom.	28	486	486	1,628
Population Density, persons per sq kilom., 1988	182	24.2	20.4	24.3
ECONOMY: PRODUCTION & INCOME				
GDP, $US millions, 1986	1,090	2,650	3,561	550,099
GNP per head, $US, 1986	240	250	389	12,960
ECONOMY: SUPPLY STRUCTURE				
Agriculture, % of GDP, 1986	58	43	35	3
Industry, % of GDP, 1986	17	15	27	35
Services, % of GDP, 1986	25	42	38	61
ECONOMY: DEMAND STRUCTURE				
Private Consumption, % of GDP, 1986	79	77	73	62
Gross Domestic Investment, % of GDP, 1986	17	16	16	21
Government Consumption, % of GDP, 1986	12	15	14	17
Exports, % of GDP, 1986	12	16	23	17
Imports, % of GDP, 1986	20	24	26	17
ECONOMY: PERFORMANCE				
GDP growth, % per year, 1980-86	2.3	1.6	-0.6	2.5
GDP per head growth, % per year, 1980-86	-0.4	-1.7	-3.7	1.7
Agriculture growth, % per year, 1980-86	1.3	1.1	0.0	2.5
Industry growth, % per year, 1980-86	4.9	1.1	-1.0	2.5
Services growth, % per year, 1980-86	3.2	2.5	-0.5	2.6
Exports growth, % per year, 1980-86	11.6	0.7	-1.9	3.3
Imports growth, % per year, 1980-86	3.6	0.2	-6.9	4.3
ECONOMY: OTHER				
Inflation Rate, % per year, 1980-86	6.4	23.6	16.7	5.3
Aid, net inflow, % of GDP, 1986	15.7	11.5	6.3	-
Debt Service, % of Exports, 1986	19	18	20.6	-
Budget Surplus (+), Deficit (-), % of GDP, 1983	-0.9	-3.0	-2.8	-5.1
EDUCATION				
Primary, % of 6-11 group, 1985	53	62	76	102
Secondary, % of 12-17 group, 1985	4	15	22	93
Higher, % of 20-24 group, 1985	1	1.2	1.9	39
Adult Literacy Rate, %, 1981	25	41	39	99
HEALTH & NUTRITION				
Life Expectancy, years, 1986	48	50	50	76
Calorie Supply, daily per head, 1985	2,233	2,111	2,096	3,357
Population per doctor, 1980	40,020	35,986	24,185	550

Notes: 'Southern Africa' and 'Africa' exclude South Africa. Dates are for the country in question, and do not always correspond with the Regional, African and Industrial averages.

BURUNDI: Leading Indicators

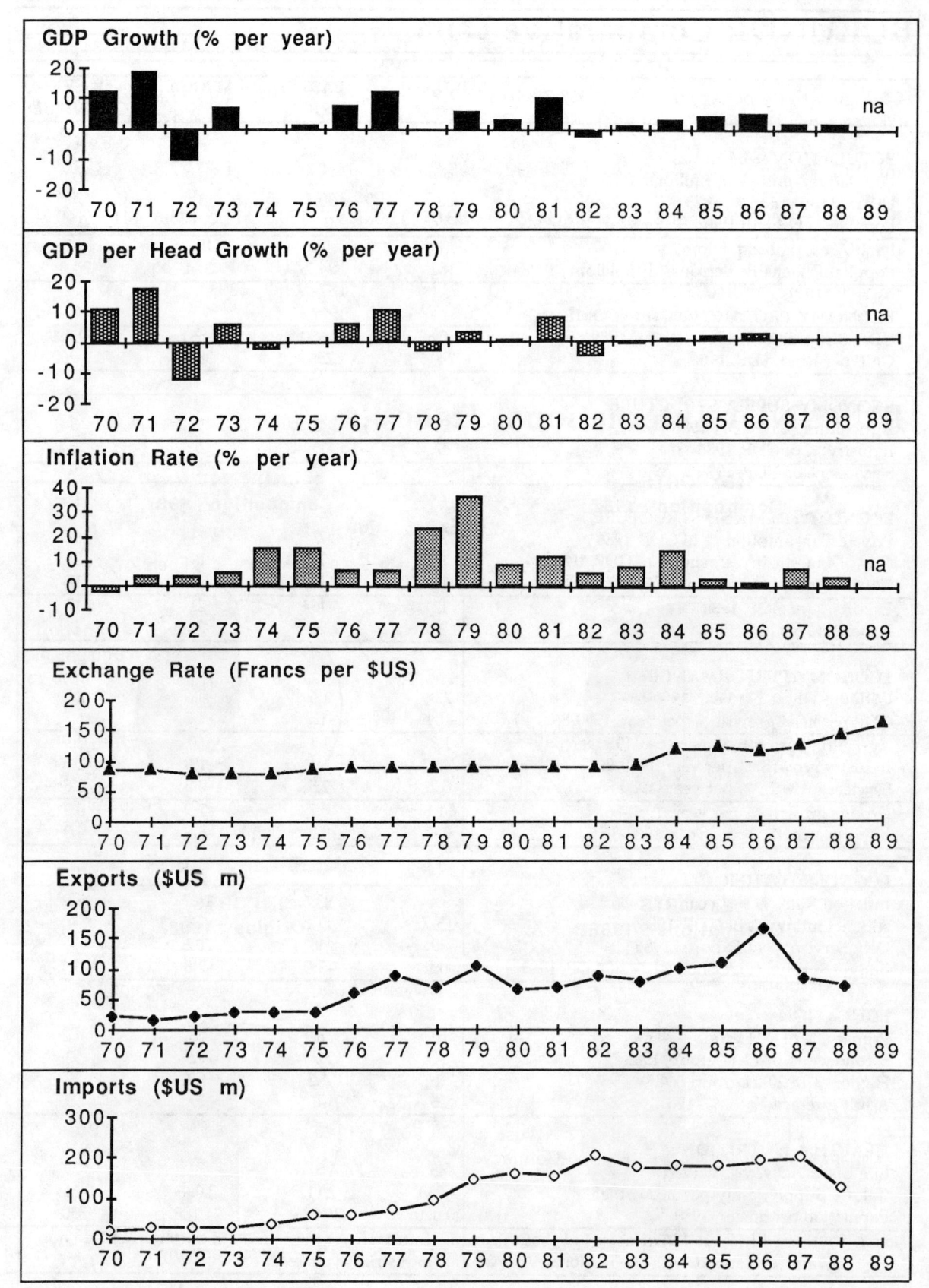

GDP Growth (% per year)

GDP per Head Growth (% per year)

Inflation Rate (% per year)

Exchange Rate (Francs per $US)

Exports ($US m)

Imports ($US m)

Note: Leading indicator data for 1989 are based on the first half of 1989. 1989 exchange rate is for mid-1989.

BURUNDI: Leading Indicator Data

	1980	1981	1982	1983	1984	1985	1986	1987	1988	1989
GDP growth (% per year)	3.3	10.3	-2.7	1.0	3.1	4.1	4.9	1.8	2.0	
GDP per head growth (% per year)	0.7	7.8	-5.3	-1.7	-0.3	1.2	1.9	-1.2	-1.0	
Inflation (% per year)	9.4	12.0	5.8	8.4	14.4	3.6	1.8	7.3	4.3	
Exchange rate (Francs per $US)	90.0	90.0	90.0	93.0	119.7	120.7	114.2	123.6	141.7	160.3
Exports, merchandise ($US m)	65	71	88	80	103	112	169	85	73	
Imports, merchandise ($US m)	168	161	214	183	187	189	203	212	137	

Note: Leading indicator data for 1989 are based on the first half of 1989. 1989 exchange rate is for mid-1989.

BURUNDI: International Trade

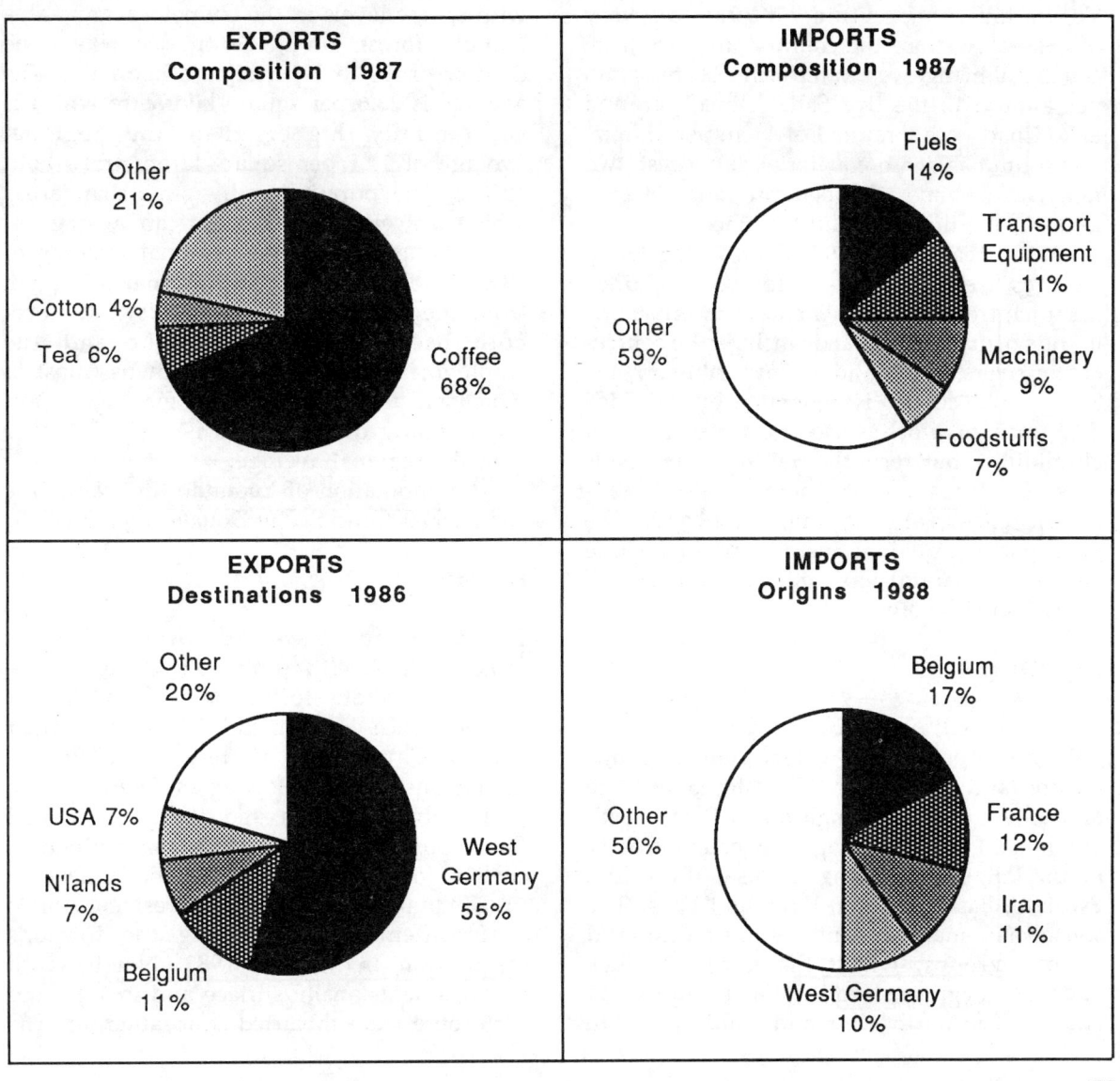

EXPORTS Composition 1987
- Coffee 68%
- Other 21%
- Tea 6%
- Cotton 4%

IMPORTS Composition 1987
- Other 59%
- Fuels 14%
- Transport Equipment 11%
- Machinery 9%
- Foodstuffs 7%

EXPORTS Destinations 1986
- West Germany 55%
- Other 20%
- Belgium 11%
- USA 7%
- N'lands 7%

IMPORTS Origins 1988
- Other 50%
- Belgium 17%
- France 12%
- Iran 11%
- West Germany 10%

6 CAMEROON

Physical Geography and Climate

Cameroon is a tropical middle-sized coastal state in the Central African region, bounded by Nigeria, Chad, the Central African Republic, the Congo, Gabon and Equatorial Guinea. Vegetation, temperature and rainfall vary considerably, since the country stretches from equatorial mangrove swamp with 500cm of rain per annum to the dry Sahel climate around Lake Chad with a rainfall of 61cm per annum. The south has dense equatorial rain forest, two rainy seasons and a temperature range of 25°C to 33°C, modified by altitude since the south central plateau rises to 670 metres. Vegetation changes from dense rain forest to Guinea savanna; from Sudan savanna to thorn steppe in the north. The highlands include the sources of the rivers Benue and a Zaïre tributary, the Sangha. Electricity is generated by the Edea dam, most of which is used for the smelting of aluminium, but recently falling river levels have hampered output. There are also hydro-electric schemes at Song Loulou and Lagdo. The main port is Douala. Mineral deposits include an oil and natural gas field and deposits of bauxite and iron ore.

Population

Estimates for 1989 suggest a population of 11.6 million composed of a wide variety of ethnic groups and more than 250 languages. The official languages are English and French with around 20% speaking Pidgin English; the other main languages being Hausa, Fulfulde, Ewondo, Basaa, Bamun, Bulu, and Fang. The south has mainly Bantu, semi-Bantu and Pygmie groups, whilst the north includes Sudanese negroes, Hamitic Fulani and Choa Arabs. The north west and south west are anglophone, whilst the east is francophone. The highest population densities are on the fertile soils of the western mountains, where densities can be as high as 135 persons to the square kilometre. The west, south central and the Sudan savanna area are the most populous, with sparse levels in the Guinea savanna area and the forests of the south east where the density drops to 2 per square kilometre. The average is 23.6 per square kilometre which is substantially higher than the regional average of 11.1 per square kilometre. About 42% of the population live in urban areas which is well above the African average of 30%, though nearer the regional average of 38.6% and there has been internal migration from the west central highlands and north both because of the drought and the employment prospects of the towns (most of which are in the south) and forestry sector. Population growth at 3.2% is slightly higher than the regional average.

The population of Yaounde (the capital) is about 850 000 and that of Douala about 580 000.

History

The Cameroons were a former German protectorate which was placed under an Anglo-French mandate following World War I. Independence was gained in 1960. The former British Cameroons parted from French Cameroons in 1961 following a referendum, the north joined Nigeria and the south joined Cameroon. Cameroon has maintained a civilian government since 1960 and achieved a smooth transfer of power from President Ahidjo to President Biya following the former's resignation in August 1983. North/south tensions occasionally surface and in 1984 and 1985 there were thwarted coup attempts. The visit of Pope John Paul II allowed the return from exile in Canada of the former Bishop of Nkongsamba, who had assisted in a plot to overthrow President Ahidjo. There were allegations of repression of the anglophone minority in January 1986, coupled with clandestine opposition which has led to tighter internal security and the need to placate northern interests. There is a mixed economy with a vigorous private sector.

Stability

Cameroon has been ruled by civilian governments since Independence, with first President Ahidjo in power for 22 years before resigning in favour of the present President Biya. There are tensions between the North and South of the country, and there was a thwarted coup in 1983, and a further coup attempt in 1984 was only suppressed after heavy fighting.

On the economic front, Cameroon had persevered with a consistent development strategy which has relied upon a strong private sector, and continuous membership of the Franc Zone. Despite the 1984 coup attempt, Cameroon's stability has remained good. The events of 1984 have caused the government to take more care of security matters, and there are signs of more political tolerance. The important feature from the stability standpoint was that the coup attempts were unsuccessful.

Overall, Cameroon's stability record is very good, and there appears to be no reason why it should not continue.

Economic Structure

GDP was estimated at $US 12,660m in 1987, making it the biggest economy in the region, and only Nigeria and South Africa in Africa have larger economies. GNP per head was estimated at $US 970 in 1987, which is high by African standards and only Gabon has a higher GNP per head in the region.

Agriculture generates 24% of GDP, and employs about 70% of workers, and this is above the Central African average, although it is a region with wide variation, between Gabon with 11% and São Tomé with 51%. The industry sector, which includes oil production and employs only 8% of the working population, provides 31% of GDP, and this is high by the standards of Africa, but reflects, in part, Cameroon's middle-income level of development. Manufacturing is limited to processing raw materials and some assembly. The service sector, which has increased in importance, contributes 45% of GDP, and this is above the regional average of 43%.

Private consumption was 74% of GDP in 1987, about the regional average, while government consumption was only 11% compared to 21% for the region. Investment was generally low at 18%.

16% of GDP is exported, about the regional average, while 20% of expenditure goes on imports, about half the percentage for Central Africa generally. Exports are dominated in value by petroleum, which in 1987 made up 46% of foreign exchange earnings. Coffee provided 10% of export revenue and cocoa 13%. There are also significant exports of cotton, timber, aluminium, bananas, rubber and tobacco. Imports are predominantly industrial goods, with machinery making up 22% of the total, chemicals 13% and foodstuffs 14%.

Debt service payments took up 27.9% of export earnings in 1987, and this was well above the average for the region, although it must be borne in mind that the debt service ratio for Cameroon is heavily dependent on the world oil price.

Adult literacy was estimated at 50% in 1983, about typical for the region, which has better literacy than for Africa generally. The primary enrolment rate is 107% of the 6-11 age group, (i.e. some people above that age range receive primary education) while the enrolment rate is 23% in secondary education and 2% in higher education. These enrolment rates are fairly typical of the region, which has greater participation in education than Africa generally, though the secondary enrolment is somewhat lower. Education levels are lower in Muslim areas. Life expectancy is 56 years, and this is slightly better than for most countries in the region, where the average is 54 years. Calorific intake was 88% of minimum requirement in 1983, and given Cameroon's middle-income status, is a reflection of disparities in income levels between the urban population and some of the

Cameroon

rural areas. Provision of doctors was one per 13,990 of population in 1983, which is about average for the region.

Overall, Cameroon has the economic structure of a medium-sized country with middle-income levels of GNP per head, with income generated by oil production in the export sector, but with the rest of the export structure well diversified. Education provision is generally good, and reflects Cameroon's middle-income status.

Economic Performance

Over the period 1980-87, GDP expanded at the rate of 7% per year, the fastest growth rate in the region. Cameroon's economy deteriorated after 1986, however, due to falling oil and agricultural commodity prices. This allowed GNP per head to rise by 3.8% per year. Only Botswana and Seychelles in Africa have had comparable improvements in output levels and productivity.

Agriculture growth was fairly good at 2.4% a year, although this did not keep pace with population growth at 3.2% a year. Prices for most commodities are presently low and will therefore restrict growth for the time being. The industry sector grew at 11% per year, and this was bettered only by Botswana in Africa, and is a reflection of the rapid expansion of Cameroon's oil sector since the mid-1970s. There is considerable scope for future growth of this sector if mineral deposits such as bauxite, uranium and tin are developed. Manufacturing is limited to small-scale enterprises. This has led to an expansion of the services sector at 6.9% a year, again faster than any other country in the region.

Export volumes have expanded at 9.7% a year in the 1980-87 period, by far the best performance in the region reflecting greater oil output to compensate for falling prices. Expanding export volumes have enabled import volumes to expand at 3.4% a year in the 1980s.

The budget deficit was -3.7% of GNP in 1987, and tighter budgetary control has contributed to better price stability than the regional average.

Overall, Cameroon's economic performance has been very good, with rising average living standards, expanding exports and a modest rate of inflation.

Recent Economic Developments

The fall in the international price of oil from around $US 30 a barrel in 1985 to under $US 12 a barrel in mid 1986 has caused the government severe financing problems, particularly with regard to funding government expenditure and servicing the external debt. Oil contributed some 44% of export earnings in 1984, and government policy has been directed toward diversifying away from oil dependency. Oil output has been fairly constant since 1985.

Cameroon's external debt was around $US 3.5b in 1989, and debt servicing requirements are put at $US 350m for 1989. Cameroon has run a trade surplus throughout the 1980s, (except for a small deficit in 1987) but still had to reschedule its debts with the Paris Club in 1989.

The government is committed to a privatisation programme that will sell off the assets of 62 enterprises held by the state-owned Societe Nationale d'Investissement (SNI). Despite monopoly positions in local markets, and considerable tariff protection, SNI has had a poor financial record accumulating $US 11.7m in losses in 1985. The privatisation which is scheduled to take place over 7 years, has to date been slow, and involves pulp and paper, aluminium, sugar, plywood, chemicals, tannery textiles, forest products and tobacco enterprises.

Cameroon's Sixth Development Plan for 1986-91 anticipates continuing growth of GDP at an average of 6.7% a year. Rural development receives special emphasis in the plan with 26% of total government spending. In particular, Cameroon will seek to diversify away from export dependency on oil, and encourage production of cocoa, coffee, cotton, timber, bananas, rubber, tobacco in agriculture as well as aluminium in the industrial sector.

The 1989/90 budget of CFA 600b ($US 1,887m) introduced spending cuts of 23% overall, with recurrent expenditure reduced by 15% to CFA 425b ($US 1,337m) and investment expenditure cut by 36%. These reductions will only be met by public sector job losses, and this will be widely unpopular. Planned reduction of civil service allowances are considerable but it has been difficult to keep expenditure as low as planned. There will be pressure to accelerate the privatisation programme and the reduction in investment spending will affect future growth prospects and slow down the diversification away from oil. A balanced budget is projected, and this will continue to constrain inflation, after reported reduction to 2% in 1988.

In 1989 a World Bank structural adjustment loan of $US 150m was agreed. The African Development Bank arranged a $US 90.8m loan for road construction.

Economic Outlook and Forecasts

Cameroon's immediate economic outlook depends on the international price of oil, which has experienced a 20% rise in 1989. Unfortunately this was somewhat offset by poor prospects for cocoa and coffee prices. Cameroon has a well diversified export structure, and has excellent prospects of expanding non-oil exports. The major immediate problem is that budget cuts have reduced the amount of infrastructure investment, and this will curtail the rate at which non-oil exports can be expanded. In the immediate future, Cameroon will find it difficult to maintain recent growth rates of around 5.0%. In the longer term overall prospects are good, and Cameroon will hope to expand GDP at around 7% a year, matching the performance of the 1970s.

The forecasts assume that Cameroon is able to maintain her good record of stability, and that the cautious budgetary policy and reform programme for the economy are both maintained.

It is not expected that the GDP growth rates of the early 1980s will be maintained, but rates will be higher than projected population growth, and will allow steady improvements in GDP per head.

Inflation is projected to be modest reflecting the budgetary discipline imposed by membership of the Franc Zone.

It is not expected that export volumes will continue to expand as rapidly as in the first half of the 1980s. Import volumes are projected to rise, but not as fast as export volumes.

Cameroon: Economic Forecasts
(average annual percentage change)

	Actual	Forecast Base	High
	1980-87	1990-95	1990-95
GDP	7.0	4.5	5.5
GDP per Head	3.8	1.3	2.3
Inflation Rate	8.1	7.5	4.5
Export Volumes	9.1	5.5	7.5
Import Volumes	3.4	4.5	5.5

CAMEROON: Comparative Data

	CAMEROON	CENTRAL AFRICA	AFRICA	INDUSTRIAL COUNTRIES
POPULATION & LAND				
Population, mid year, millions, 1989	11.6	7.3	10.2	40.0
Urban Population, %, 1985	42	38.6	30	75
Population Growth Rate, % per year, 1980-86	3.2	3.0	3.1	0.8
Land Area, thou. sq. kilom.	475	638	486	1,628
Population Density, persons per sq kilom., 1988	236	11.1	20.4	24.3
ECONOMY: PRODUCTION & INCOME				
GDP, $US millions, 1986	11,280	4,146	3,561	550,099
GNP per head, $US, 1986	910	395	389	12,960
ECONOMY: SUPPLY STRUCTURE				
Agriculture, % of GDP, 1986	22	18	35	3
Industry, % of GDP, 1986	35	41	27	35
Services, % of GDP, 1986	43	41	38	61
ECONOMY: DEMAND STRUCTURE				
Private Consumption, % of GDP, 1986	62	62	73	62
Gross Domestic Investment, % of GDP, 1986	25	25	16	21
Government Consumption, % of GDP, 1986	3	14	14	17
Exports, % of GDP, 1986	23	35	23	17
Imports, % of GDP, 1986	19	35	26	17
ECONOMY: PERFORMANCE				
GDP growth, % per year, 1980-86	8.2	4.2	-0.6	2.5
GDP per head growth, % per year, 1980-86	4.8	-0.8	-3.7	1.7
Agriculture growth, % per year, 1980-86	2.0	0.5	0.0	2.5
Industry growth, % per year, 1980-86	15.9	7.9	-1.0	2.5
Services growth, % per year, 1980-86	7.0	3.2	-0.5	2.6
Exports growth, % per year, 1973-86	13.8	4.5	-1.9	3.3
Imports growth, % per year, 1973-86	-0.5	0.4	-6.9	4.3
ECONOMY: OTHER				
Inflation Rate, % per year, 1980-86	11.0	17	16.7	5.3
Aid, net inflow, % of GDP, 1986	2.1	5.4	6.3	-
Debt Service, % of Exports, 1986	22.8	28	20.6	-
Budget Surplus (+), Deficit (-), % of GDP, 1985	0.8	-1.4	-2.8	-5.1
EDUCATION				
Primary, % of 6-11 group, 1984	107	93	76	102
Secondary, % of 12-17 group, 1984	23	42	22	93
Higher, % of 20-24 group, 1984	2	1.9	1.9	39
Adult Literacy Rate, %, 1980	50	52	39	99
HEALTH & NUTRITION				
Life Expectancy, years, 1986	56	52	50	76
Calorie Supply, daily per head, 1985	2,080	2,115	2,096	3,357
Population per doctor, 1980	13,990	13,835	24,185	550

Notes: 'Southern Africa' and 'Africa' exclude South Africa. Dates are for the country in question, and do not always correspond with the Regional, African and Industrial averages.

CAMEROON: Leading Indicators

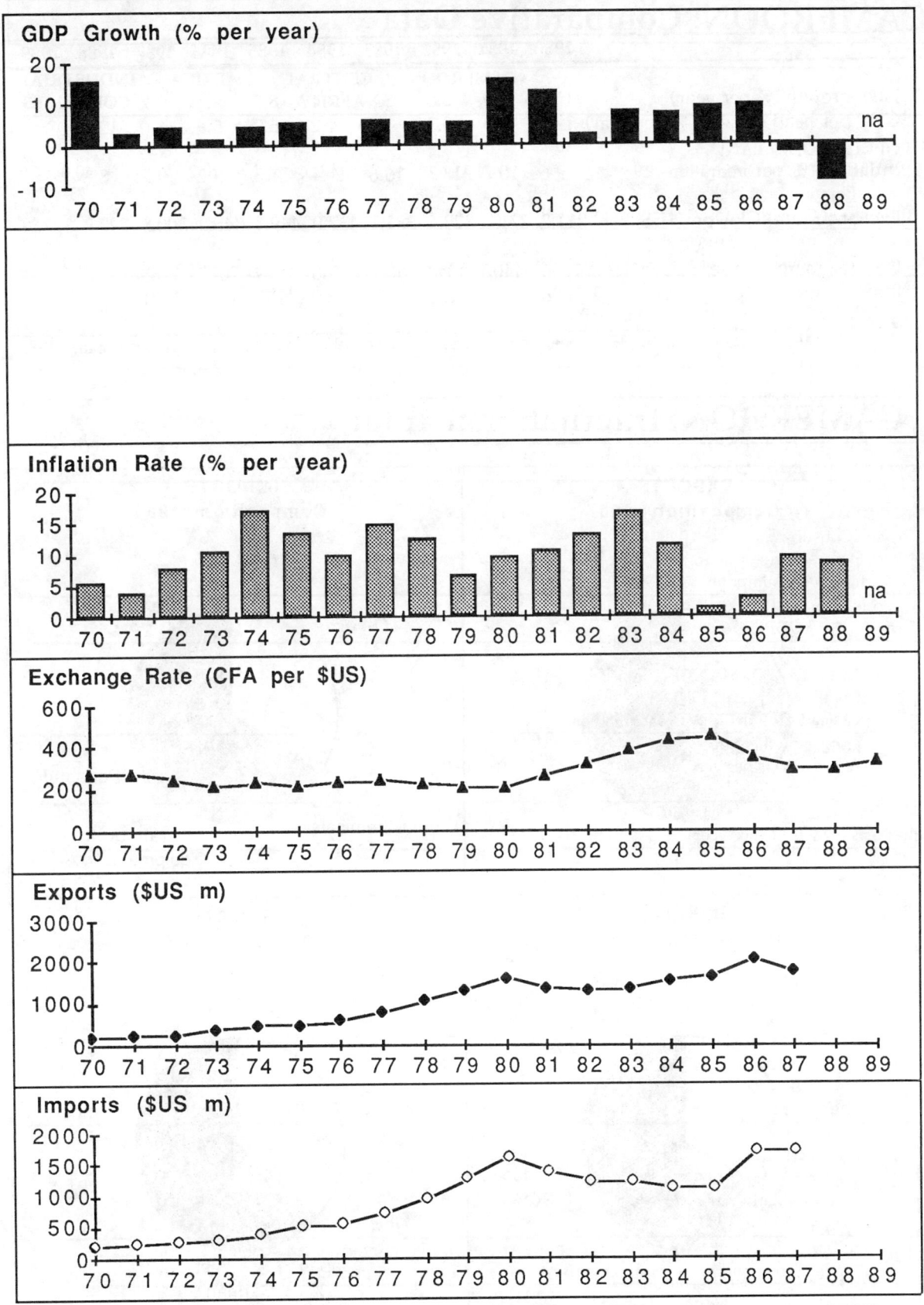

GDP Growth (% per year)

Inflation Rate (% per year)

Exchange Rate (CFA per $US)

Exports ($US m)

Imports ($US m)

Note: Leading indicator data for 1989 are based on the first half of 1989. 1989 exchange rate is for mid-1989.

CAMEROON: Leading Indicator Data

	1980	1981	1982	1983	1984	1985	1986	1987	1988	1989
GDP growth (% per year)	15.6	12.9	2.6	7.8	7.4	8.9	9.1	-2.7	-9.3	
GDP per head growth (% per year)	12.5	9.8	-0.5	4.6	4.2	5.7	5.9	-6.0	-12.6	
Inflation (% per year)		9.6	10.7	13.3	16.6	11.4	1.3	3.2	9.7	8.6
Exchange rate (CFA per $US)		211.3	271.7	328.6	381.1	437.0	449.3	346.3	300.5	297.9	327.3
Exports, merchandise ($US m)	1656	1407	1348	1364	1568	1679	2077	1768		
Imports, merchandise ($US m)	1608	1368	1220	1223	1109	1132	1705	1721		

Note: Leading indicator data for 1989 are based on the first half of 1989. 1989 exchange rate is for mid-1989.

CAMEROON: International Trade

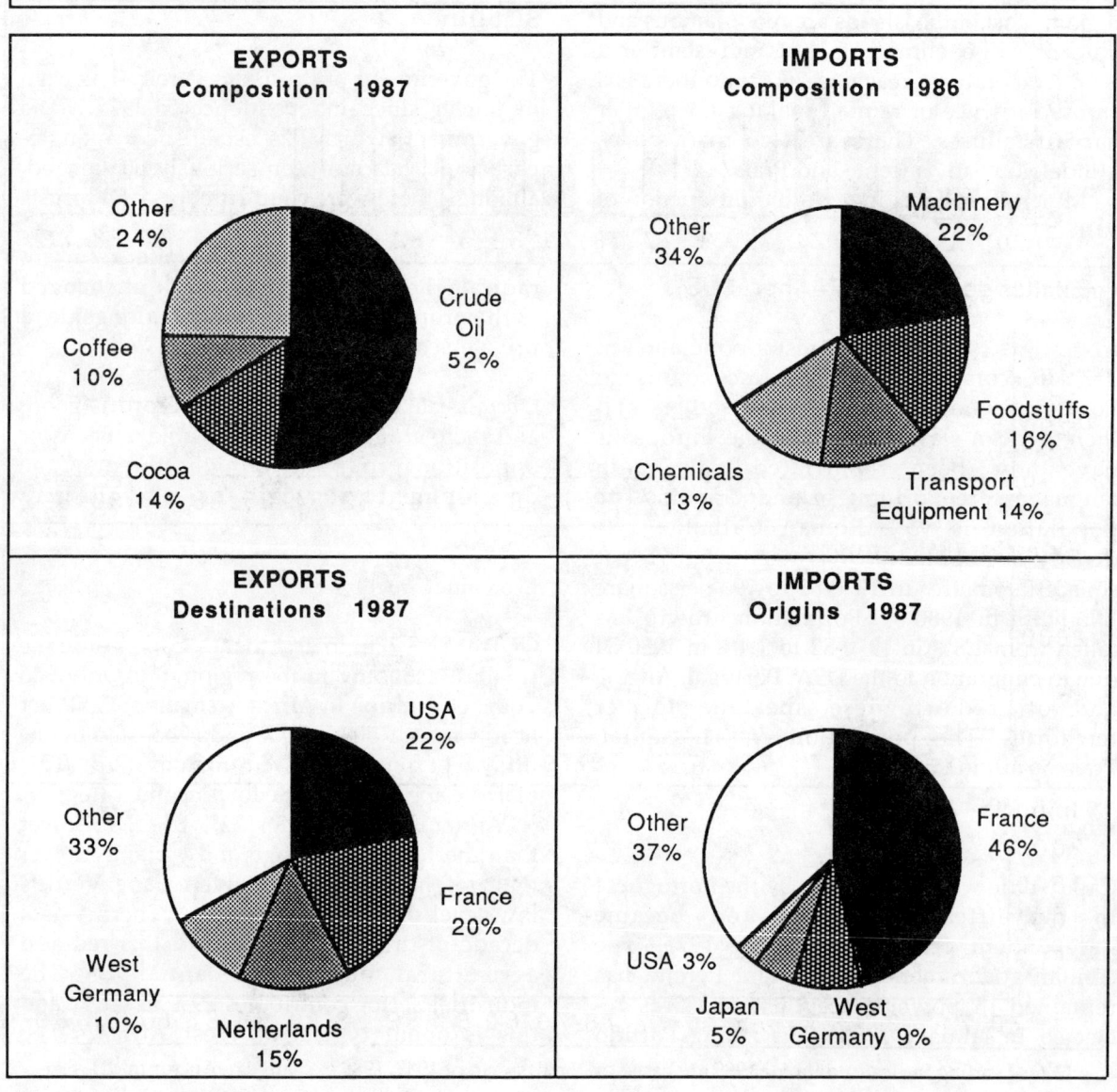

EXPORTS Composition 1987
- Crude Oil 52%
- Other 24%
- Cocoa 14%
- Coffee 10%

IMPORTS Composition 1986
- Machinery 22%
- Foodstuffs 16%
- Transport Equipment 14%
- Chemicals 13%
- Other 34%

EXPORTS Destinations 1987
- USA 22%
- France 20%
- Netherlands 15%
- West Germany 10%
- Other 33%

IMPORTS Origins 1987
- France 46%
- West Germany 9%
- Japan 5%
- USA 3%
- Other 37%

7 CAPE VERDE

Physical Geography and Climate

The Cape Verde Islands are a group of nine inhabited islands, five islets and an uninhabited island in the Atlantic Ocean, west of Senegal and Mauritania in the West African region. Most of the islands are mountainous and volcanic. The climate is hot and semi-arid with temperatures reaching 26°C and there is a chronic absence of rainfall making for regular harvest failures. There are two main ports, Mindelo on São Vincente and Praia.

Mineral products include the production of salt.

Population

UN estimates for 1984 suggest a population of 317,000, composed of the descendants of Portuguese Catholic and Jewish settlers and their African slaves. Most of the Europeans have now been repatriated. The main languages are Portuguese and Crioulo. The population is 95% Roman Catholic. The density of population has risen from 75 per square kilometre in 1970-82 to 79 per square kilometre in 1980-84. Population growth has fallen from 1.8% in 1970-82 to 1.4% in 1980-84 due to emigration to the USA, Portugal, Angola and other Portuguese speaking former territories. The population of the capital, Praia, is around 58,000.

History

Cape Verde was first visited by the Portuguese in the fifteenth century and became independent from Portugal in 1975. The civilian government of President Pereira has remained in power since then and was re-elected in January 1986. His ruling Partido Africano da Indepencia do Guine e Cabo Verde (PAIGC) finally dropped the idea of any form of unity with Guinea-Bissau and the party became PAICV. In 1986 FAO declared that the country was seriously affected by famine.

Stability

The government of President Pereira has been in power since independence in 1975. The government, which has established a single-party rule, has not been seriously challenged, although there were coup rumours and arrests in 1977.

The economic strategy, despite a certain radical rhetoric, has remained unchanged with strong state involvement alongside a private sector.

Overall, Cape Verde's stability record is good, and there are no major discernable tensions or opposition groups that might threaten the development strategy or the Pereira government.

Economic Structure

GDP at $US 77m in 1984 makes Cape Verde the smallest economy in the region, and only São Tomé et Príncipe in Africa is smaller. GNP per head was $US 500 in 1987, and is boosted by the inflow of remittances from migrants. This places Cape Verde in the low-income category.

Agriculture generates 22% of GDP, lower than the regional average and certainly lower than might be expected given Cape Verde's low level of income per head. There was a decade of drought, 1968-78, and this reduced agricultural output substantially. The industrial sector contributes 29% of GDP, and this is about typical of West Africa. The

Cape Verde

services sector generates over half of GDP, higher than general in the region by a fifth, and is accounted for by the expansion of air and sea transport facilities and tourism.

Exports were 36% of GDP in 1982, twice the export reliance observed in West Africa generally, and is mostly a function of Cape Verde's small size, where the domestic market is not large enough to allow a range of manufacturing enterprises at a size where they would be efficient. Imports comprised 58% of GDP in 1987. This heavy reliance is partly the effect of remittances from migrants, and partly from the substantial inflow of aid.

Merchandise exports are predominantly fish and fish products, making up 73% of all exports in 1984. Mining products are the next largest category, at 21%, and are mostly volcanic ash used in hydraulic cement, and salt. Fruit makes up 5% of exports.

Imports are mostly foodstuffs, at 27% of the total in 1984, and this is a reflection of the high population density, and the small size of the agriculture sector. Otherwise imports are mostly manufactures of consumer and capital goods, with 44% of imports, fuels make up 9% of the total, and chemicals 7% in 1984.

Debt service takes up 19% of export earnings, and although this is below the regional average of 23%, the average is dominated by the debt service problems of Nigeria. Only six countries of the West Africa group have high debt services to export earnings ratios.

There was an estimated 37% adult literacy rate in 1970 and this is above the regional average of 28%. Adult literacy will have improved considerably since the 1970 survey, as primary enrolments at 110% of the primary age group, and this is the best record in West Africa. The enrolment rate is 11% in secondary education, two-thirds the general enrolment rate in Africa, and there is negligible higher education. Most pupils undertaking post-secondary education receive it in Portugal, and this contributes to the high rate of migration.

Life expectancy is high, at 64, the best in West Africa, and bettered in Africa only in Mauritius and Seychelles, both middle-income countries with good provision of medical services, and equalled by São Tomé et Príncipe. High nutritional levels contribute to longevity in Cape Verde, where calorie intake is a third above the minimum on average, and this is the best in Africa. Provision of medical facilities is better than would be expected, given Cape Verde's low income per head with one doctor per 5000 people, and one nurse per thousand, and these are overall the best provision rates for medical staff in West Africa, and are a further factor in Cape Verde's high level of life expectancy.

Overall, Cape Verde's economic structure is one of a small, densely populated island with a small economy and low level of income per head. Cape Verde has a relatively small agriculture sector, is not self-sufficient in food, and has a large service sector. Exports are fish and minerals, and there is heavy dependence, particularly for food, financed by migrant remittances and aid. Basic educational provision is excellent. Nutrition and medical facilities are very good, resulting in a high life expectancy.

Economic Performance

GDP has grown at 7.5% a year over the period 1980-87, and this has allowed GDP per head to rise at 5.5% a year. Only Niger in West Africa has managed to raise living standards at this rate, and in general in West Africa living standards have fallen at -1.8% a year.

Agriculture has grown at 4.8% a year, but it must be borne in mind that some of this growth represents recovery from the contraction in agriculture caused by the droughts beginning in 1968. The industry sector has grown at 2.5% a year, and this compares very favourably with the region where industrial output has fallen at an average rate of -0.4% a year. The services sector has grown at 2.7% a year, and this is below the regional average, and reflects the caution shown by the government in expanding the size of the public sector.

Gross domestic investment was 83% of GDP in 1982. This is unrepresentative, however, as the $US 32m expansion of the harbour, together with a new shipyard and airport improvements made a massive one-off contribution to investment in Cape Verde's small economy.

The budget deficit was almost 14% of GDP in 1977, the latest year for which figures are available. Where such a large deficit is met

Cape Verde

by bank borrowing, this would lead to a rapid rate of inflation. However, much of the gap is filled by foreign aid, and this has enabled Cape Verde to maintain a modest rate of inflation at around 12% a year, slightly better than the general West African rate.

Overall, economic performance has been good, with increases in living standards being sustained over the past decade, rises in export volumes and moderate price inflation. Cape Verde is particularly reliant, however, on a sustained flow of remittances from migrants and steady receipts of aid in order to maintain her substantial deficit in merchandise trade.

Recent Economic Developments

Half of Cape Verde's foreign exchange earnings of around $US 7m come from landing fees paid by South African Airways (SAA). The halting of the four weekly flights by SAA which used Cape Verde en route to the US have been halted because the Americans refused SAA landing rights. The US is considering increasing aid to Cape Verde to compensate for the loss of revenue. Cape Verde agreed to stop allowing SAA landing rights in 1988 but their aircraft are still flying there. However, a number of other international airlines have now started flying to Cape Verde.

Remittances from the 350,000 Cape Verdeans living abroad, mostly in Portugal and the US, provide 56% of GNP, but remittances have declined in the 1980s as a result of world recession.

Brazil plan to use Cape Verde as a distribution centre for the export of Brazilian goods to the rest of Africa. The project is at present suspended as the Brazilian government is unwilling to commit funds given her constrained financial position. Brazil has signed an agreement to train Cape Verde's army and police.

Aid continues to be of considerable importance, and there is currently a $US 4m African Development Bank project to rehabilitate the agriculture sector. West Germany has provided $US 2m for cargo boats and a butane gas plant, the Netherlands is providing $US 11m for rural development, and Switzerland has committed up to $US 8m for adult literacy and rural development.

Economic Outlook and Forecasts

Cape Verde's economic prospects depend to a large part on the continuation of remittances from migrants, sustained high levels of aid inflows and the one or two key contracts for use of airport and port facilities. Migrant remittances are now expected to stabilise at their current level after falling in the early 1980s. The agriculture sector is expected to continue its recovery from the drought which ended in 1978, and to show good growth performance at above the overall rate for the whole economy.

The forecasts assume that Cape Verde maintains its good stability record, that climatic conditions continue to be favourable, and that aid and remittances are sustained at their current levels.

GDP is forecast to rise at between 4.1% and 4.9%, which represents better performance than achieved in the 1973-82 period. This will give GDP per head rises of between 3.1% and 3.9% a year. Export volumes are projected to grow at between 4.6% and 5.5%, and this is faster than previously achieved. Import growth is expected to be constrained by lower levels of remittances, and to expand at between 5.6% and 7.5% a year. Price stability is set to improve, if aid inflows are sustained, with inflation falling to between 10.9% and 12.5% annually.

Cape Verde: Economic Forecasts
(average annual percentage change)

	Actual 1980-87	Forecast Base 1990-95	Forecast High 1990-95
GDP	3.2	4.1	4.9
GDP per Head	2.2	3.1	3.9
Inflation Rate	16.0	12.5	10.9
Export Volumes	1.5[a]	4.6	5.5
Import Volumes	12.0[a]	5.6	7.5

Note: [a] 1970-81.

CAPE VERDE: Comparative Data

	CAPE VERDE	WEST AFRICA	AFRICA	INDUSTRIAL COUNTRIES
POPULATION & LAND				
Population, mid year, millions, 1989	0.349	12.3	10.2	40.0
Urban Population, %, 1981	25	28.7	30	75
Population Growth Rate, % per year, 1970-81	1.0	3.2	3.1	0.8
Land Area, thou. sq. kilom.	4	384	486	1,628
Population Density, persons per sq kilom., 1988	85.5	31.1	20.4	24.3
ECONOMY: PRODUCTION & INCOME				
GDP, $US millions, 1981	77	4,876	3,561	550,099
GNP per head, $US, 1986	460	510	389	12,960
ECONOMY: SUPPLY STRUCTURE				
Agriculture, % of GDP, 1982	22	40	35	3
Industry, % of GDP, 1982	29	26	27	35
Services, % of GDP, 1982	51	34	38	61
ECONOMY: DEMAND STRUCTURE				
Private Consumption, % of GDP, 1982	103	77	73	62
Gross Domestic Investment, % of GDP, 1982	83	13	16	21
Government Consumption, % of GDP, 1982	34	13	14	17
Exports, % of GDP, 1982	36	18	23	17
Imports, % of GDP, 1982	157	20	26	17
ECONOMY: PERFORMANCE				
GDP growth, % per year, 1973-82	3.2	-2.03	-0.6	2.5
GDP per head growth, % per year, 1973-82	2.2	-4.7	-3.7	1.7
Agriculture growth, % per year, 1973-82	4.8	1.3	0.0	2.5
Industry growth, % per year, 1973-82	2.6	-3.6	-1.0	2.5
Services growth, % per year, 1973-82	2.7	-2.5	-0.5	2.6
Exports growth, % per year, 1970-81	1.5	-4.0	-1.9	3.3
Imports growth, % per year, 1970-81	12.0	-12.9	-6.9	4.3
ECONOMY: OTHER				
Inflation Rate, % per year, 1980-86	16.0	13.0	16.7	5.3
Aid, net inflow, % of GDP, 1984	9.4	3.7	6.3	-
Debt Service, % of Exports, 1984	18.8	21.4	20.6	-
Budget Surplus (+), Deficit (-), % of GDP, 1977	-13.9	-3.4	-2.8	-5.1
EDUCATION				
Primary, % of 6-11 group, 1983	110	75	76	102
Secondary, % of 12-17 group, 1983	11	24	22	93
Higher, % of 20-24 group, 1983	0.0	2.6	1.9	39
Adult Literacy Rate, %, 1970	37	30	39	99
HEALTH & NUTRITION				
Life Expectancy, years, 1986	66	50	50	76
Calorie Supply, daily per head, 1980	2,660	2,105	2,096	3,357
Population per doctor, 1979	5,510	16,199	24,185	550

Notes: 'Southern Africa' and 'Africa' exclude South Africa. Dates are for the country in question, and do not always correspond with the Regional, African and Industrial averages.

CAPE VERDE: Leading Indicators

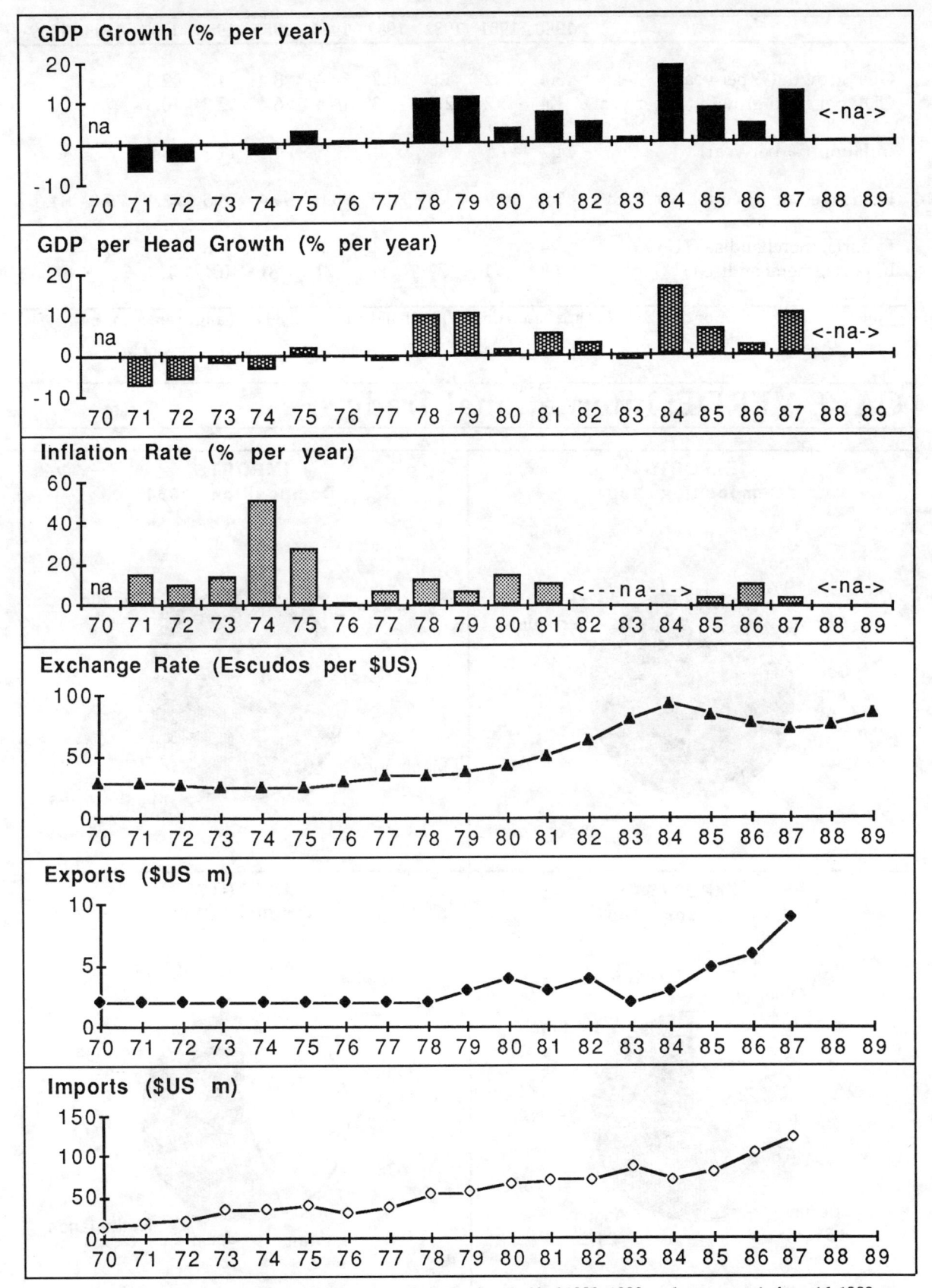

GDP Growth (% per year)

GDP per Head Growth (% per year)

Inflation Rate (% per year)

Exchange Rate (Escudos per $US)

Exports ($US m)

Imports ($US m)

Note: Leading indicator data for 1989 are based on the first half of 1989. 1989 exchange rate is for mid-1989.

CAPE VERDE: Leading Indicator Data

	1980	1981	1982	1983	1984	1985	1986	1987	1988	1989
GDP growth (% per year)	3.4	7.2	4.8	0.7	18.4	8.4	4.3	12.3		
GDP per head growth (% per year)	1.4	5.2	2.8	-1.3	16.4	6.4	2.3	10.3		
Inflation (% per year)	14.7	11.5				4.5	11.1	3.8		
Exchange rate (Escudos per $US)	42.5	50.9	63.0	80.0	93.0	84.4	76.5	72.4	75.0	84.6
Exports, merchandise ($US m)	4	3	4	2	3	5	6	9		
Imports, merchandise ($US m)	68	71	72	86	71	81	104	124		

Note: Leading indicator data for 1989 are based on the first half of 1989. 1989 exchange rate is for mid-1989.

CAPE VERDE: International Trade

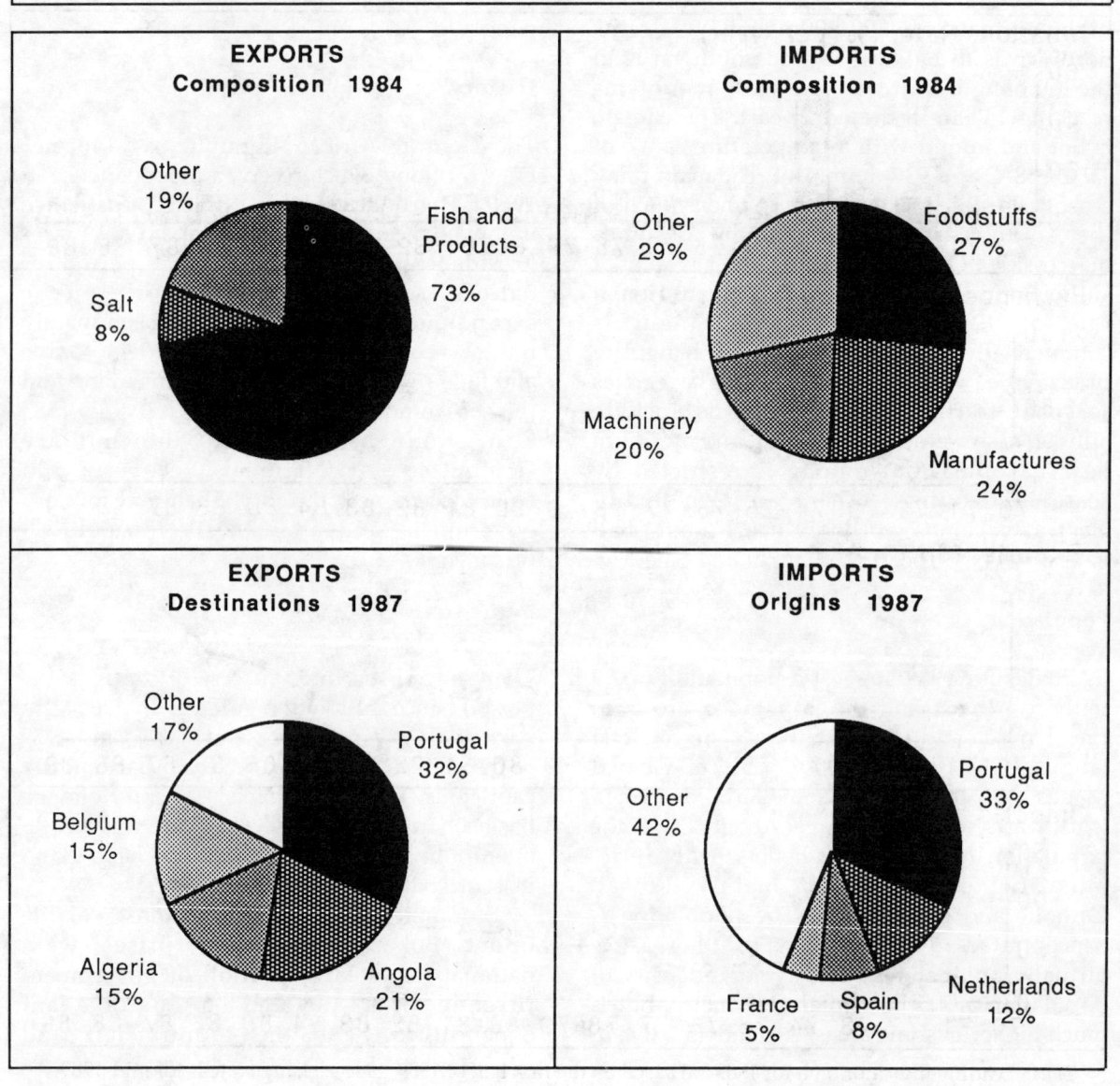

EXPORTS Composition 1984
- Fish and Products 73%
- Other 19%
- Salt 8%

IMPORTS Composition 1984
- Foodstuffs 27%
- Manufactures 24%
- Machinery 20%
- Other 29%

EXPORTS Destinations 1987
- Portugal 32%
- Angola 21%
- Algeria 15%
- Belgium 15%
- Other 17%

IMPORTS Origins 1987
- Portugal 33%
- Netherlands 12%
- Spain 8%
- France 5%
- Other 42%

8 CENTRAL AFRICAN REPUBLIC

Physical Geography and Climate

The Central African Republic is a large landlocked country lying north of the equator, bounded by Cameroon, Chad, Sudan and Zaïre in the Central African region. It consists of a high plateau at 600-900m which drains northwards to Lake Chad and southwards to the Oubangui-Zaïre basin, with mountains reaching 1400m in the north east. The climate is hot and humid with a temperature range of 14°C - 38°C and 93% humidity. The main rainy season is from July to October, when much of the south east of the country becomes inaccessible. Only in the south west is rainfall sufficient for forest vegetation at an annual rate of 125cm. The River Obangui is commercially navigable only below Bangui to Brazzaville, whilst the River Sangha carries seasonal traffic. Mineral deposits include alluvial diamonds in the west and uranium near Bakouma, but mining is restricted by seasonal flooding. There are two hydro-electric schemes at Boali near the M'Bali Falls. The major ports are Bangui and Salo.

Population

Estimates for 1989 suggest a population of 2.9 million. Almost half the population are from the Banda and Baya tribes. The official language is French, 39% speaking Gbaya-Manza, 34% Banda, with some Fulfulde. Sango is used as a second language. About 60% of the population retain animist beliefs, with a 35% minority of Christians, mainly Roman Catholic and 5% Moslem. The population is concentrated in the west, whilst the east is virtually uninhabited. The average density of population is 4.5 per square kilometre, but is much higher in some areas. Around 45% live in urban areas, which is above the regional average and is one of the highest rates in Africa. The rate of population growth is 2.5% for 1980-86, below the regional average of 3.0%.

The capital, Bangui, has about 474,000 people.

History

The Central African Republic is a former French colony which received independence in 1960. There have been three coups since independence. Between 1965 and 1979 under President Bokassa there was a period of nationalization and extravagant public expenditure. Since then there has been mainly military rule. Attempts were made to leave and then rejoin the regional customs union and this has hindered economic activity. In 1985 Gen. Kolingba dissolved the military Committee for National Recovery and announced a new mixed civilian and military government and an amnesty for political prisoners.

Stability

There have been three successful coups in the period since full independence in 1960. In particular, the period between 1965 and 1979 under Bokassa was marked by particularly capricious changes in policy. General Kolingba has been in power since 1981. There was a coup attempt in 1982 and events in Chad have led to insecurity along the northern border.

Economic policy was at its most volatile under Bokassa when enterprises were nationalised, a large portion of government revenue was devoted to extravagant expenditures, and the country left and

Central African Republic

thenrejoined the regional customs union, UDEAC.

Overall, stability remains poor in the Central African Republic despite continued membership of the Franc Zone and support from the French. This instability will persist while there are organised dissidents in exile and lack of security in Chad.

Economic Structure

GDP in 1987 was estimated at $US 1,010m, a quarter the average economy size in the region, and only Chad, Equatorial Guinea and São Tomé et Príncipe had smaller economies. GNP per head was $US 330 in 1987, and this places Central African Republic in the low income category, with below average income per head when compared with the region and Africa.

The agricultural sector generates 41% of GDP and this is roughly in line with what would be expected in a country with the low, income status, and is almost twice the size of the average agriculture sector in the region. Industry provides 13% of GDP and contains the diamond mining sector, which is an important source of exports. Services supply 46% of GDP, and this is roughly typical of the region and of Africa generally.

Private consumption of 89% of GDP was third highest in the region after Chad and Equatorial Guinea, while government consumption was a low 13%. Investment was also low, only 14% of GDP.

Exports are 17% of GDP, and this is below the general export dependency for Central Africa, and has a larger orientation towards exports than Africa as a whole as the result of the presence of three middle income oil exporters in the region. Imports are 33% of GDP and the trade deficit is financed by substantial aid inflows, mostly from France.

Diamonds are the most important export, making up 38% of total export earnings, and this is followed by timber with 15% and coffee at 13%. Cotton provides 5%, and with five categories each contributing over 5% of total earnings, the export structure is well diversified.

Imports are predominantly industrial goods, with machinery, manufactures and chemicals comprising over 70% of the total. Foodstuffs represent a further 13%, and only 2% of imports are fuels. The low imports of fuels are partly explained by chronic shortages resulting from poor accessibility to supplies, and the availability of hydro-electric power.

Debt service was 12.1% of the value of exports in 1987, and this represents a modest degree of debt burden compared with the 15% level for the region generally. The period of President Bokassa's rule between 1965 and 1979 did not lead to confidence among donor countries or commercial banks, and this limited the accumulation of external debt.

Adult literacy rate was estimated at 33% in 1980, and this is below the 48% average in the region, but more or less in line with the literacy level that might be expected in a low-income country. There is a 66% enrolment in primary education, 13% in secondary and 1% in higher. All these are lower than the regional averages, but are again in line with educational provision in a low income country, and are perhaps better than might be expected in a country with such a poor stability record.

Life expectancy is 50 years, and this is about average for the region and Africa generally. Nutrition level, as indicated by 91% of minimum calorie requirements being met, is better than the regional average, although this latter is heavily influenced by the poor calorie provision in the most populous country, Zaïre. There is one doctor per 23,000 people, and one nurse per 2,000, and these are average for the region, and good considering Central African Republic's low income status.

Overall, Central African Republic's economic structure is one of small population and economic size, quite highly urbanised and sparsely populated, and generating low average incomes. Agriculture and services are the largest sectors, with agricultural products providing the bulk of exports, although there are important diamond exports from the industrial sector. Debt burden is modest, educational attainments and enrolments are in line with, and health services are slightly better than, general low income country provision.

Economic Performance

GDP has grown at 2.0% per year, 1980-87, and

Central African Republic

this is below the regional average and well below the general record for Franc Zone countries. Living standards have fallen at -0.3% a year, about average for the region. The agriculture sector has grown at 2.4% per year, and this has just about kept pace with the 2.5% annual rate of population growth. Although agriculture growth has been better than in Central Africa generally, it is in the context of a region with three oil exporters who have concentrated resources in their industrial sectors at the expense of agriculture. Industry has grown at 2.2% a year, while the services sector has experienced 1.6% growth.

Growth in export volumes has been at 1.0% a year, less than a quarter of the rate for the region. Import volumes have contracted at -1.8% a year.

Inflation ran at a reasonable 7.9% over the period 1980-87, about half the regional average rate.

Overall, the economic performance of the Central African Republic has been poor, and is mostly attributable to the lack of stability experienced during the Bokassa period. In recent subsequent years there have been hopeful indications of rising GDP per head, lower rates of inflation and improving export performance.

Recent Economic Developments

Central African Republic received a $US 30m structural adjustment loan from the World Bank in 1986 to support a programme of economic reform. The main features of the programme are a concentration of investment on agriculture. Producer coffee prices have doubled and cotton fertiliser subsidies reduced. There is to be a new code for foreign investment. The private sector is to be encouraged and public sector spending reduced. Import restrictions are to be eased. The IMF confirmed a $US 28.5m credit in June 1987 to continue support for the adjustment programme.

French aid of $US 11.6m is involved in rationalising cotton production in response to poor world prices by restricting production to the most productive areas. Three of the present seven ginneries are to be closed, and the marketing system reformed. Further French aid totalling $US 120m is committed to road maintenance, education and training of administrators, and to cover losses by the main agricultural marketing board.

Economic Outlook and Forecasts

The economic prospects for the Central African Republic depend upon the ability of the Kolingba regime to consolidate the improvements in stability that have been achieved since 1981. The structural adjustment programme will provide a test for the government's resolve particularly with regard to implementing cuts in public expenditure. The commitment of the regime to liberalising reforms, together with the budget and exchange rate disciplines imposed by membership of the Franc Zone, lead to optimism that economic performance can be significantly improved. The price prospects for diamonds are considered to be good, but commodity prices for coffee, timber, cotton and animal products are expected to fall relative to world prices in the medium term. Overall, Central African Republic will be looking to good growth rates for agricultural production to compensate for any deterioration in the terms of trade.

The forecasts assume a significant consolidation of stability and improved performance generated by the liberalising reforms of the World Bank and IMF structural adjustment programme.

GDP is forecast to expand at a rate that will permit small increases in GDP per head. Inflation is expected to remain modest. Export volumes are forecast to expand more rapidly, and import volumes to expand, but not as fast as export volumes.

Central A. R. : Economic Forecasts
(average annual percentage change)

	Actual	Forecast	
		Base	High
	1980-87	1990-95	1990-95
GDP	2.0	2.8	3.6
GDP per Head	-0.6	0.2	1.0
Inflation Rate	7.9	7.6	6.3
Export Volumes	1.0	4.4	5.4
Import Volumes	-1.8	2.3	4.4

CENTRAL AFRICAN REPUBLIC: Comparative Data

	CENTRAL A R	CENTRAL AFRICA	AFRICA	INDUSTRIAL COUNTRIES
POPULATION & LAND				
Population, mid year, millions, 1989	2.9	7.3	10.2	40.0
Urban Population, %, 1985	45	38.6	30	75
Population Growth Rate, % per year, 1980-86	2.5	3.0	3.1	0.8
Land Area, thou. sq. kilom.	623	638	486	1,628
Population Density, persons per sq kilom., 1988	4.5	11.1	20.4	24.3
ECONOMY: PRODUCTION & INCOME				
GDP, $US millions, 1986	900	4,146	3,561	550,099
GNP per head, $US, 1986	290	395	389	12,960
ECONOMY: SUPPLY STRUCTURE				
Agriculture, % of GDP, 1986	41	18	35	3
Industry, % of GDP, 1986	12	41	27	35
Services, % of GDP, 1986	47	41	38	61
ECONOMY: DEMAND STRUCTURE				
Private Consumption, % of GDP, 1986	88	62	73	62
Gross Domestic Investment, % of GDP, 1986	16	25	16	21
Government Consumption, % of GDP, 1986	9	14	14	17
Exports, % of GDP, 1986	20	35	23	17
Imports, % of GDP, 1986	33	35	26	17
ECONOMY: PERFORMANCE				
GDP growth, % per year, 1980-86	1.1	4.2	-0.6	2.5
GDP per head growth, % per year, 1980-86	-1.3	-0.8	-3.7	1.7
Agriculture growth, % per year, 1980-86	2.5	0.5	0.0	2.5
Industry growth, % per year, 1980-86	1.7	7.9	-1.0	2.5
Services growth, % per year, 1980-86	-0.5	3.2	-0.5	2.6
Exports growth, % per year, 1980-86	2.0	4.5	-1.9	3.3
Imports growth, % per year, 1980-86	-2.7	0.4	-6.9	4.3
ECONOMY: OTHER				
Inflation Rate, % per year, 1980-86	11.5	17	16.7	5.3
Aid, net inflow, % of GDP, 1986	14.8	5.4	6.3	-
Debt Service, % of Exports, 1986	9.6	28	20.6	-
Budget Surplus (+), Deficit (-), % of GDP, 1981	-3.5	-1.4	-2.8	-5.1
EDUCATION				
Primary, % of 6-11 group, 1985	73	93	76	102
Secondary, % of 12-17 group, 1985	13	42	22	93
Higher, % of 20-24 group, 1985	1	1.9	1.9	39
Adult Literacy Rate, %, 1980	33	52	39	99
HEALTH & NUTRITION				
Life Expectancy, years, 1986	50	52	50	76
Calorie Supply, daily per head, 1985	2,059	2,115	2,096	3,357
Population per doctor, 1981	22,530	13,835	24,185	550

Notes: 'Southern Africa' and 'Africa' exclude South Africa. Dates are for the country in question, and do not always correspond with the Regional, African and Industrial averages.

CENTRAL AFRICAN REPUBLIC: Leading Indicators

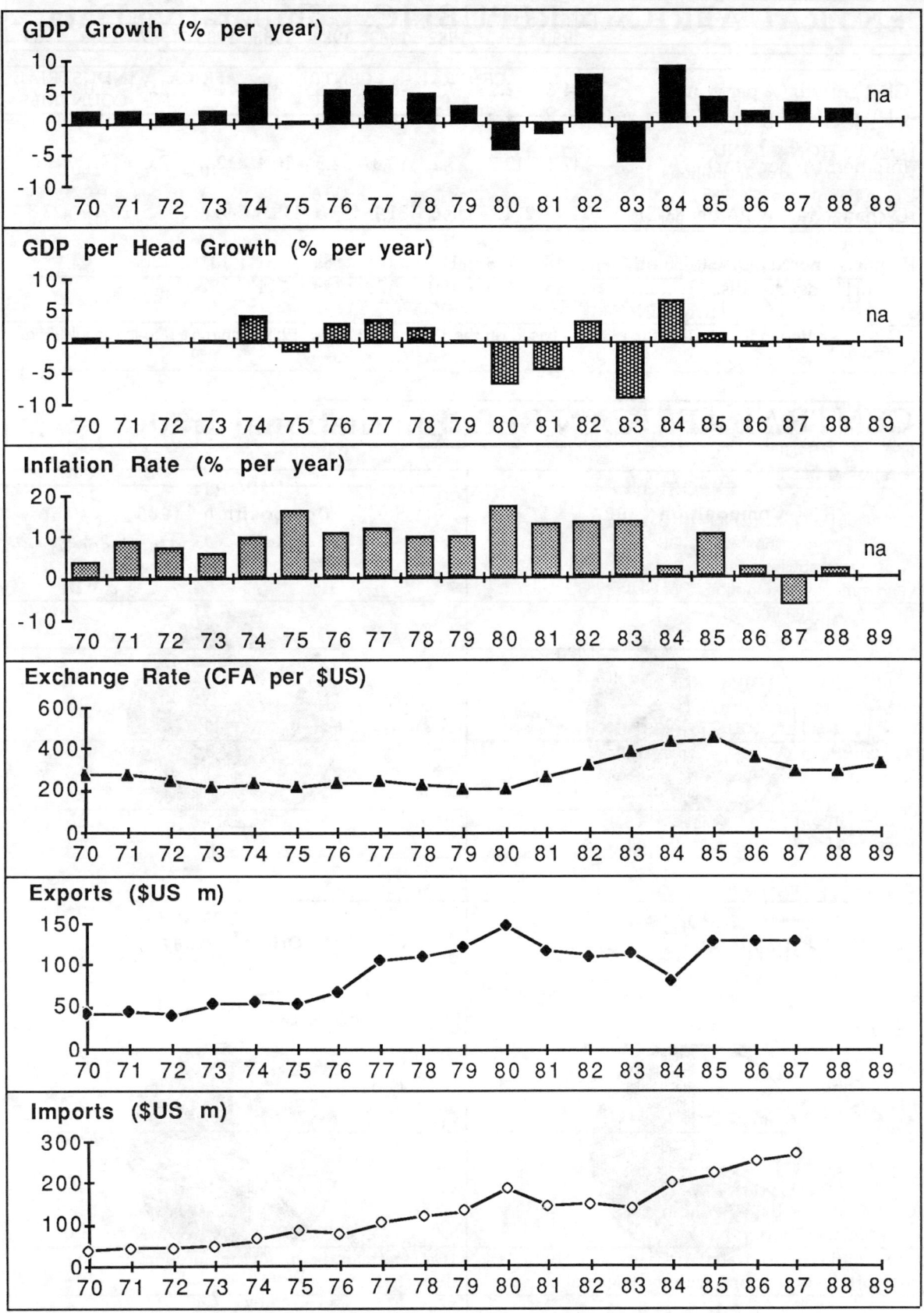

GDP Growth (% per year)

GDP per Head Growth (% per year)

Inflation Rate (% per year)

Exchange Rate (CFA per $US)

Exports ($US m)

Imports ($US m)

Note: Leading indicator data for 1989 are based on the first half of 1989. 1989 exchange rate is for mid-1989.

CENTRAL AFRICAN REP.: Leading Indicator Data

	1980	1981	1982	1983	1984	1985	1986	1987	1988	1989
GDP growth (% per year)	-4.6	-2.2	7.5	-6.7	8.8	3.8	1.4	2.7	2.0	
GDP per head growth (% per year)	-6.9	-4.6	3.1	-9.2	6.3	1.3	-1.2	-0.1	-0.7	
Inflation (% per year)	17.1	12.7	13.2	13.3	2.5	10.4	2.2	-7.0	2.1	
Exchange rate (CFA per $US)	211.3	271.7	328.6	381.1	437.0	449.3	346.3	300.5	297.9	327.3
Exports, merchandise ($US m)	147	118	112	115	85	131	131	130		
Imports, merchandise ($US m)	185	145	150	138	199	221	252	270		

Note: Leading indicator data for 1989 are based on the first half of 1989. 1989 exchange rate is for mid-1989.

CENTRAL AFRICAN REP.: International Trade

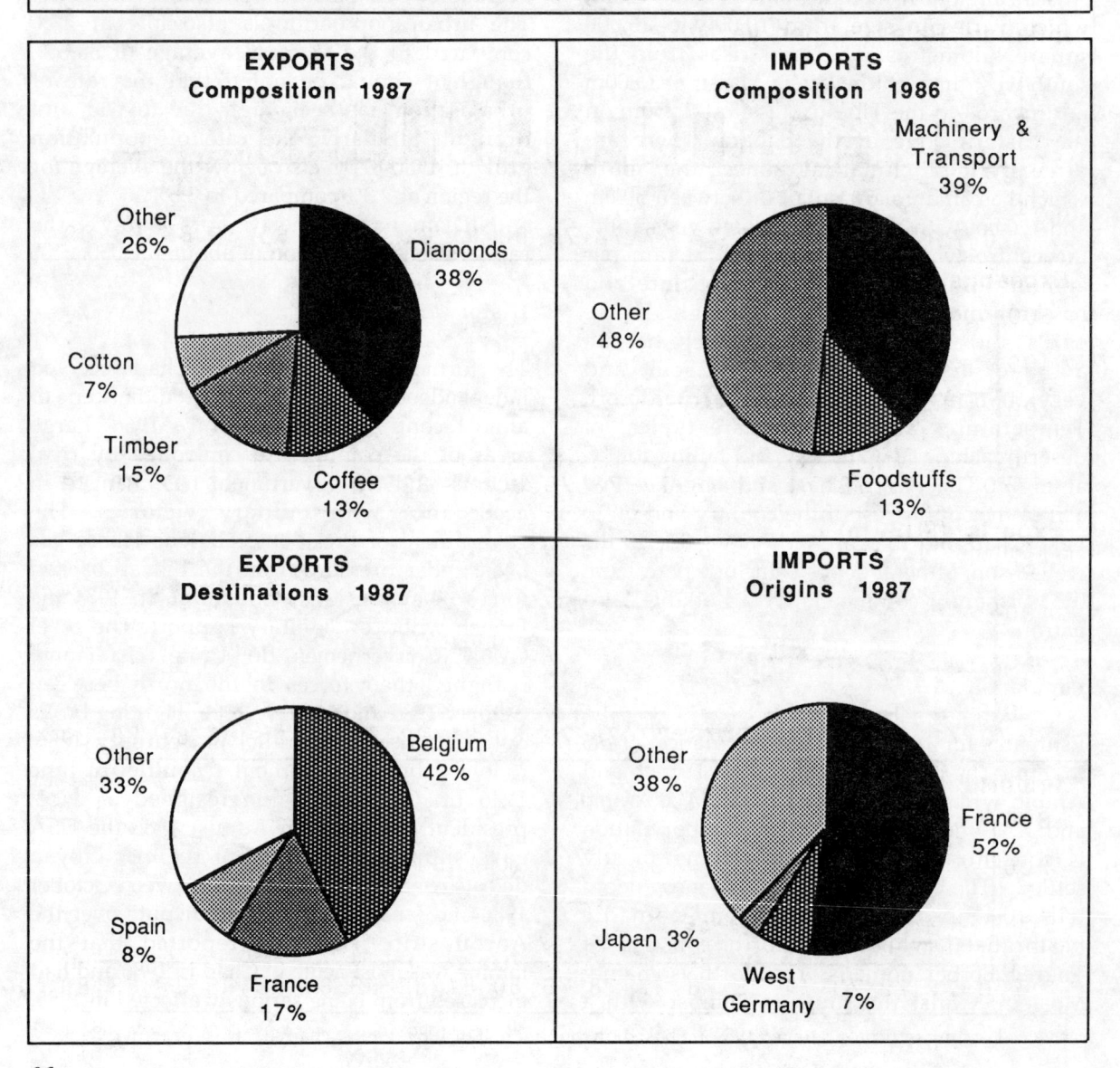

EXPORTS Composition 1987
- Diamonds 38%
- Other 26%
- Timber 15%
- Coffee 13%
- Cotton 7%

IMPORTS Composition 1986
- Machinery & Transport 39%
- Other 48%
- Foodstuffs 13%

EXPORTS Destinations 1987
- Belgium 42%
- Other 33%
- France 17%
- Spain 8%

IMPORTS Origins 1987
- France 52%
- Other 38%
- West Germany 7%
- Japan 3%

9 CHAD

Physical Geography and Climate

Chad is a very large landlocked state, bounded by Niger, Libya, Sudan, the Central African Republic, Cameroon and Nigeria in the Central African region. It includes part of Lake Chad, which varies in size from 10,000 to 26,000 square kilometres. The land rises from the south west through the Guera Massif at 1,800m northwards to the Tibesti Massif at 3,350m on the eastern edge of the Sahara. There are three defined climatical zones: the south which has an annual rainfall of between 50 cm and 120 cm with savanna woodland vegetation; the centre which receives 25-50 cm rain per annum with marginal Sahelian pasture and has been most affected by the droughts of the 1970s and 1980s; the north which has an average annual rainfall of 2.5cm and vegetation ranging from scrub to true desert. Temperatures at N'Djamena are typical of desert reaching 31°C by day and falling to 0°C at night. The rivers Chari and Logone flow across the south west of the country and on to Lake Chad, but they are only navigable in the wet season. Mineral deposits include petroleum (not presently commercially exploitable) and natron.

Population

Estimates for 1989 suggest a population of 5.5 million. The official languages are French and Arabic with 19% speaking Sara, 14% Arabic and 12% Buramabang. Most of the population is concentrated in the south being mostly settled. The centre consists of Sudanese negro tribes who are mainly semi-nomadic pastoralists, whilst the north consists of Tuareg Berber nomads. The north is mainly Moslem, whilst the south is animist with a small percentage of Christians. Population density is low at 4.1 persons per square kilometre (though this varies from less than 0.5 in the north to concentrations of over 30 per sq km in the south), compared to the regional average of 11.1 persons per square kilometre. The urban population is also low at 27% compared to the regional average of 38.6%, though it should be noted that the rate of urbanisation has been high due to war and drought. Similarly, the rate of population growth, 1980-87, is also below the average for the region at 2.3% compared to 3.0%.

N'Djamena (formerly Fort Lamy), the capital, has a population of about 402,000.

History

The former French colony of Chad received independence in 1960 and has been the scene of almost continual fighting since 1966. Large areas of the country are controlled by rival factions and the government has changed in accordance with military victories. The government of President Hissène Habré has been under pressure from the Libyan backed forces of ex-President Goukouni. In 1984 the French withdrew military support. The rebel GUNT (Gouvernement de l'Unité Tchadienne) realigned their forces in the north between October 1985 and January 1986. Fighting broke out near the 16th parallel in February 1986, followed by a government reshuffle. In June 1986 Lt. Col. Kamougue resigned as vice-president of GUNT. In August 1986 the USA gave support to the Habré regime. Libyan defeats continued in the north between October 1986 and January 1987 in the dispute over the Aozou strip. The FAO reported that the famine was most acute in Chad in 1986 and had increased from being seriously affected in 1985.

Chad

By 1989 all Libyan troops were supposed to have withdrawn from Chad. An attempted coup was reported in 1989, and ethnic divisions in the government appear as divisive as ever.

Stability

In recent years Chad has been one of the most unstable countries on the continent with civil war and conflicts with neighbouring Libya and Sudan. In 1980 the rebel forces of Goukouni Oueddei removed the government of Hissène Habré from N'Djamena, but by 1982 Habré was back in power. Libyan intervention led to Oueddei effectively controlling the north while Libya itself laid claim to the Aozou strip along the border between the two countries. However, Habré's government now has control over the whole country, except for the Aozou strip over which Libya has now agreed to international arbitration.

Overall, Chad has a very poor stability record and is heavily dependent on France for defence.

Economic Structure

GDP in 1987 was $US 980m which is only one quarter of the regional average but it must be remembered that three of the other eight countries in the region are oil producers. GNP per head of $US 150 is the lowest in Africa with the exception of Ethiopia.

The agricultural sector generates 43% of GDP and this is typical for a low-income economy in Africa. The industrial sector generates a further 18%, with little manufacturing and few minerals to develop this sector. Services account for 39% of GDP. It should be noted that Chad has a strong informal sector of the economy with up to 40% of GDP coming from such activities.

On the demand side, private consumption comprises 104% of GDP due to the large inflow of foreign aid. Government consumption of 8% is one of the lowest in Africa. Investment comprised a further 18%, which is low for the region.

Exports represented 17% of GDP and were composed of 43% cotton and a further 39% livestock, which means Chad is highly dependent on climatic conditions. Spending on imports was equal to 48% of GDP, and was comprised of food produce 19%, fuels 17%, chemicals 12%, and machinery 19% in 1983. Unofficial cross-border trade is considerable with a large range of commodities involved, including petroleum, mostly from Nigeria.

In education, 43% of primary-aged children attend school, and only 6% of the relevant age group are in secondary education, one of the lowest rates in Africa, with a considerable regional imbalance, most schools being in the south.

Life expectancy is 46 years. Provision of health is poor, with only one doctor for every 38,000 people, three times the regional level, though basic health care is better with one nurse to every 3,400 people and this is twice as good as the regional average.

Overall, Chad is a low-income country, heavily dependent on an agricultural sector and very vulnerable to climatic factors. Educational and health provision is poor.

Economic Performance

The economy has shown impressive growth, especially when the poor stability record is considered, with GDP growth rate of 5.1% from 1980-87, and this is better than average for the region with only Cameroon having a faster growth rate.

GDP per head growth for 1980-87 at 2.8% a year is better than regional and African performance.

Agricultural sector growth at 2.6% a year for 1980-87 has been better elsewhere in Africa, as have the growth rates for industry at 10.0%, and services at 6.3%.

Export performance in the 1980-87 period has been very impressive, with an 11.7% annual rate of expansion recorded, and this is substantially better than the average for the rest of the continent.

Inflation at 5.3% a year in the 1980-87 period was significantly lower than the region and Africa generally.

Overall, Chad has had a good economic performance in the 1980s, although to a degree this reflects recovery brought about through the ending of severe drought, greater aid flows and a marginal improvement in stability.

Chad

Recent Economic Developments

In 1988 the government began efforts to rebuild the economy after the war. Massive investment was needed in agriculture and infrastructure, especially road improvements which are needed for transport of cotton and other agricultural commodities on which the economy is dependent. A total of about $US 45m credit was made available by the World Bank, plus $US 38m from the African Development Bank. Other sources include $US 20m from the French and $US 4.5m from Saudi Arabia.

The IMF approved a three year Structural Adjustment Facility of $US 27.5m in 1987. World Bank credits of $US 154m were made available in 1989.

A $US 56.1m credit facility was made available by the IDA in 1989 for road improvements (particularly important as the country has no railway). French military support is being reduced, requiring Chad to find military assistance elsewhere. Around 90% of the investment budget, which totalled CFA 80b in 1988 ($US 268.6m), comes from foreign loans and grants.

Economic growth of 4% was forecast in the 1989 budget and government estimates for 1988 suggest 2.3% growth in the agricultural sector, 9.2% growth in the industrial sector and 5.6% in the services.

Chad has had increased budget deficits over the last few years. Recent estimates suggest reductions in current expenditure, though large capital expenditure increases are expected. The government is trying to reduce expenditure on personnel. Considerable infrastructural investment is needed following the war. Total external debt was $US 350m in 1989.

Economic Outlook and Forecasts

The stability of Chad appears as fragile as ever, but this uncertainty does not appear to have reduced the willingness of the donor community to commit aid. Cotton prices are expected to increase through 1990 and this will enhance benefits from higher cotton output.

The forecasts assume no significant improvement in stability, but that efforts to liberalise the economy under the structural adjustment programme continue.

GDP growth is forecast to give modest improvements in GDP per head, but export volumes are forecast to expand at lower rates than experienced in the 1980-87 period. The expansion of import volumes is forecast to be substantially slower than in the early 1980s. Inflation will be modest as slow monetary growth continues to be imposed by Franc Zone membership.

Chad: Economic Forecasts
(average annual percentage change)

	Actual 1980-87	Forecast Base 1990-95	Forecast High 1990-95
GDP	5.1	3.5	5.5
GDP per Head	2.8	0.9	2.9
Inflation Rate	5.3	47.4	4.5
Export Volumes	11.7	4.5	8.1
Import Volumes	32.0	6.5	8.5

CHAD: Comparative Data

	CHAD	CENTRAL AFRICA	AFRICA	INDUSTRIAL COUNTRIES
POPULATION & LAND				
Population, mid year, millions, 1989	5.5	7.3	10.2	40.0
Urban Population, %, 1985	27	38.6	30	75
Population Growth Rate, % per year, 1980-86	2.3	3.0	3.1	0.8
Land Area, thou. sq. kilom.	1,284	638	486	1,628
Population Density, persons per sq kilom., 1988	4.1	11.1	20.4	24.3
ECONOMY: PRODUCTION & INCOME				
GDP, $US millions, 1982	400	4,146	3,561	550,099
GNP per head, $US, 1982	80	395	389	12,960
ECONOMY: SUPPLY STRUCTURE				
Agriculture, % of GDP, 1984	50	18	35	3
Industry, % of GDP, 1984	8	41	27	35
Services, % of GDP, 1984	41	41	38	61
ECONOMY: DEMAND STRUCTURE				
Private Consumption, % of GDP, 1981	101	62	73	62
Gross Domestic Investment, % of GDP, 1981	9	25	16	21
Government Consumption, % of GDP, 1981	22	14	14	17
Exports, % of GDP, 1981	38	35	23	17
Imports, % of GDP, 1981	69	35	26	17
ECONOMY: PERFORMANCE				
GDP growth, % per year, 1970-81	-2.6	4.2	-0.6	2.5
GDP per head growth, % per year, 1970-81	-4.6	-0.8	-3.7	1.7
Agriculture growth, % per year, 1970-81	-1.0	0.5	0.0	2.5
Industry growth, % per year, 1970-81	-2.0	7.9	-1.0	2.5
Services growth, % per year, 1970-81	-1.1	3.2	-0.5	2.6
Exports growth, % per year, 1973-84	-2.9	4.5	-1.9	3.3
Imports growth, % per year, 1973-84	-7.7	0.4	-6.9	4.3
ECONOMY: OTHER				
Inflation Rate, % per year, 1970-82	7.8	17	16.7	5.3
Aid, net inflow, % of GDP, 1986	41	5.4	6.3	-
Debt Service, % of Exports, 1986	2.2	28	20.6	-
Budget Surplus (+), Deficit (-), % of GDP, 1976	-0.6	-1.4	-2.8	-5.1
EDUCATION				
Primary, % of 6-11 group, 1984	38	93	76	102
Secondary, % of 12-17 group, 1984	6	42	22	93
Higher, % of 20-24 group, 1984	0	1.9	1.9	39
Adult Literacy Rate, %, 1975	15	52	39	99
HEALTH & NUTRITION				
Life Expectancy, years, 1986	45	52	50	76
Calorie Supply, daily per head, 1985	1,733	2,115	2,096	3,357
Population per doctor, 1977	47,530	13,835	24,185	550

Notes: 'Southern Africa' and 'Africa' exclude South Africa. Dates are for the country in question, and do not always correspond with the Regional, African and Industrial averages.

CHAD: Leading Indicators

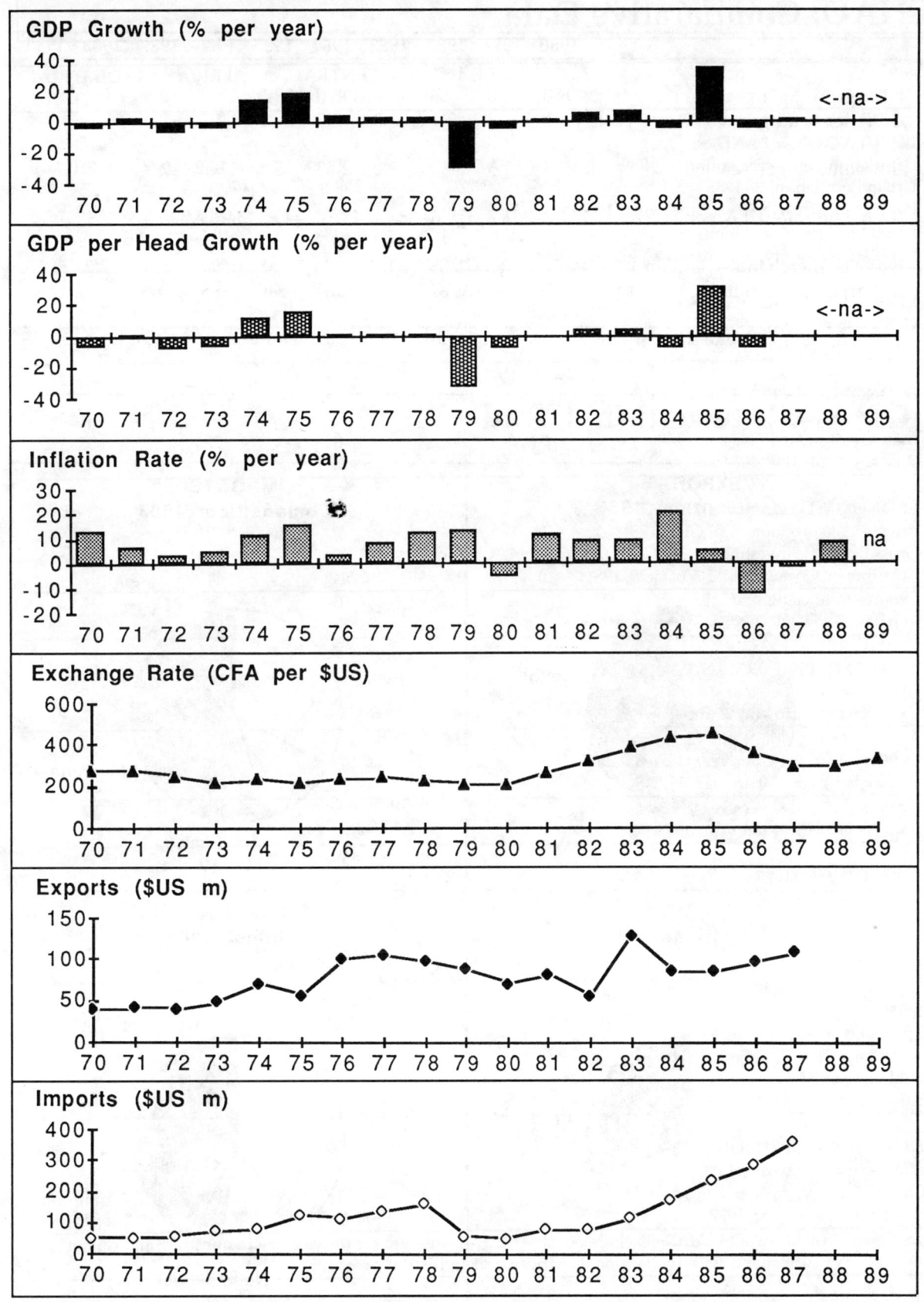

GDP Growth (% per year)

GDP per Head Growth (% per year)

Inflation Rate (% per year)

Exchange Rate (CFA per $US)

Exports ($US m)

Imports ($US m)

Note: Leading indicator data for 1989 are based on the first half of 1989. 1989 exchange rate is for mid-1989.

CHAD: Leading Indicator Data

	1980	1981	1982	1983	1984	1985	1986	1987	1988	1989
GDP growth (% per year)	-6.0	1.0	5.3	5.6	-6.0	33.0	-5.3	0.5		
GDP per head growth (% per year)	-8.2	-1.2	3.1	3.3	-8.3	30.7	-7.7	-1.9		
Inflation (% per year)	-5.5	11.5	9.0	9.0	20.3	5.1	-13.0	-2.7	8.2	
Exchange rate (CFA per $US)	211.3	271.7	328.6	381.1	437.0	449.3	346.3	300.5	297.9	327.3
Exports, merchandise ($US m)	71	83	58	131	88	88	99	111		
Imports, merchandise ($US m)	55	81	82	117	181	240	288	366		

Note: Leading indicator data for 1989 are based on the first half of 1989. 1989 exchange rate is for mid-1989.

CHAD: International Trade

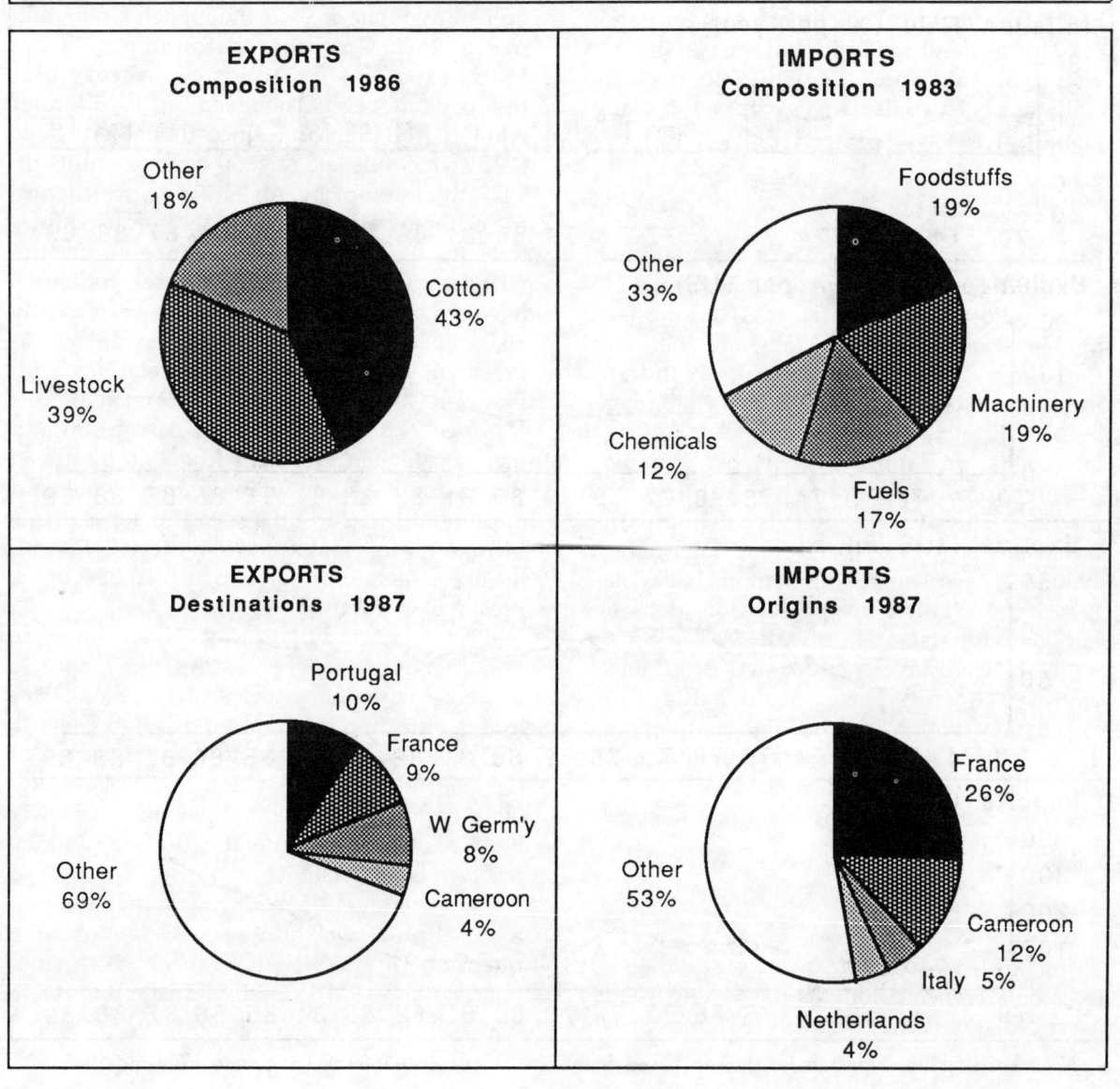

EXPORTS Composition 1986
- Cotton 43%
- Livestock 39%
- Other 18%

IMPORTS Composition 1983
- Foodstuffs 19%
- Machinery 19%
- Fuels 17%
- Chemicals 12%
- Other 33%

EXPORTS Destinations 1987
- Portugal 10%
- France 9%
- W Germ'y 8%
- Cameroon 4%
- Other 69%

IMPORTS Origins 1987
- France 26%
- Cameroon 12%
- Italy 5%
- Netherlands 4%
- Other 53%

10 COMOROS

Physical Geography and Climate

The Comoros are a group of four small volcanic islands covering a total land area of 2,236 sq km lying in the Mozambique Channel between northern Mozambique and Madagascar in the East Africa region. The climate and vegetation vary from island to island, but are classified as wet tropical and are suitable for the cultivation of vanilla, cloves, copra and ylang-ylang.

Population

Estimates for 1989 indicate a population of 453,000. The islands were originally settled by Melano-Polynesians, Africans, Indonesians, Madagascans, Persians, Arabs and then later Portuguese, Dutch, French and Indians. Swahili and Arabic were the main languages before French colonisation. The official languages are French and Arabic, but most people speak Comoran, a Swahili-Arabic mixture. The avarage density of population has risen from 182.5 per square kilometre in 1980 to 217 per square kilometre in 1988. There is great variation in population densities between the islands, Mwali averaging 66, whilst Nzwani averages 349 and is increasing rapidly, despite large-scale emigration. Population growth remains high; in the period 1973-82 it was 3.1%. There has been some recent immigration from Mozambique.

The capital, Moroni, has 17,000 inhabitants.

History

Prior to 1841 Comoros was a sultanate. It became a French colony as the various islands were ceded to France between 1841 and 1886. In 1909 the Comoros became a dependency of Madagascar and remained isolated in a pre-colonial state as a backwater of a culturally different island. In 1940 the islands were occupied by Britain following the declaration of the Vichy regime in France. In 1946 the 4th French Republic created the French Union and made the Comoros autonomous from Madagascar. Following referenda, independence was achieved in 1975 under Ahmed Abdallah and since then there have been two coups and several coup attempts. In 1977 the attempt by Ali Soilih to restructure the economy along socialist decentralised lines led to a break with France and the withdrawal of French aid. These measures have not been successful and have been mostly abandoned. Following the 1978 coup by ex-President Abdallah diplomatic relations with France have been restored and French aid has resumed. After allegations of corruption and mismanagement President Abdallah dissolved the Federal Assembly in January 1982 and appointed Ali Mroudjae as the new Prime Minister. In March 1985 there was an unsuccessful attempted coup by members of the Presidential Guard.

President Abdallah was assassinated in 1989 after moves, under pressure from France, to dispense with the services of the mercenary leader of the presidential guard, Bob Denard. As the islands appeared to have come under the control of the mercenaries, France and other donors suspended aid payments. Exactly how a confused political situation will be resolved is uncertain.

Voting in the 1975 independence referendum the island of Mayotte was in favour of a retention of links with France. A further referendum in 1976 voted unofficially to retain overseas territory status, but in December

Comoros

France decided that the island should become a *collectivité territoriale* with an intermediate status. The UN has passed several resolutions reaffirming sovereignty by the Comoros over the island and urging France to come to an agreement.

Stability

Since gaining independence from France in 1975, there have been two successful coups and several other attempts to overthrow the government before the 1989 assassination. In 1977, under Ali Soilih, there were radical attempts to reorganise society on a mixture of Maoist and Islamic lines.

This period introduced a major change in development strategy when the regime began moves to restructure the economy into decentralised socialist units. All French property was nationalised and French officials expelled. The socialist experiment has been ended for the most part, and Comoros formally joined the Franc Zone in 1976 and, with the restoration of Abdallah in 1978, French aid was re-established to be suspended with his death.

Overall, the stability record of Comoros has been very poor. The future course of government is unclear. Although economic policy has been more practical over the past nine years, the dislocation introduced by the Soilih era has had an enduring effect in damaging confidence.

Economic Structure

Comoros had a population of under half a million in 1989, making it one of the smallest countries in the African region. In 1982, 12.3% of the population were estimated to live in urban areas, well below average in both the East African region and Africa generally, which is close to 30%. Population is growing at 4.1% each year, and is a reflection of reasonably good medical care reducing death rates, with little influence of urbanisation and wider opportunites for women contributing to reductions in the birth rate.

The islands only comprise 2,236 square kilometres, and the population density is consequently very high at 217.5 persons a square kilometre, and only Rwanda and Mauritius in Africa have a higher density. The population density is nine times greater than the average for Africa.

GDP in 1982 was $US 99m. Income per head was estimated at $US 320, and this is above the average of $US 250 for East Africa, but below the African average of $US 389.

The most important sector is agriculture which generates 45% of GDP, which is representative of the East African region, industry generating 18%, and services 38%. East Africa has a rather lower level of industrialisation than Africa generally, where 27% of GDP comes from industry.

On the demand side, private consumption was equivalent to 66% of GDP in 1982, and this is lower than elsewhere in the region. Investment was 28% of GDP, and this represents a substantial development effort, being almost twice the 16% investment ratio in Africa generally. Government consumption was 22% of GDP, and this is 56% greater than is typical for Africa.

Exports generate 22% of GDP and this is a rather higher dependence on outside earnings than in the East Africa region. Expenditure on imports was equivalent to 40% of GDP, and was made possible by substantial aid flows which were equivalent to 16% of GDP.

In 1980, Comoros had an adult literacy ratio of 48%, better than in East Africa, or in Africa generally. 93% of the relevant age-group were in primary education, an achievement 50% better than in the East African region. Secondary enrolments were 50% better than the average for the East African region at 24% of the relevant age group. There were negligible enrolments in higher education.

Life expectancy was estimated at 56 years in 1986, and this is better than the regional and African average of 50. Daily calorie supply averaged 1,920, which is little below the regional and African averages. The most recent figures for availability of physicians suggests good provision of medical services at one doctor for every 15,000 inhabitants in 1973, whereas in East Africa the average was one doctor for every 35,000 and in Africa generally, one doctor for every 24,000; these latter figures being for 1981.

Overall, Comoros is a small economy with a low-income status, but among the better-off low income countries. It has a small population, but high population density. There is heavy dependence on agriculture, high investment and levels of government consumption. Aid flows allow a high proportion of total expenditure to be on imports. Educational provision is good, as is health provision.

Economic Performance

GDP growth was estimated at 0.3% annually for 1973-81 by the Economic Commission for Africa, but this is affected by two particularly poor years in 1975 and 1976. GDP growth was below the rate of population growth, and GDP per head fell at 3.4% in this period. GDP is estimated to have risen by 1.7% a year, 1980-87, and this indicates a fall in GDP per head of -1.4% a year.

Agricultural growth has been at 2.4% a year in the period 1973-81, but industrial output fell at -7.7% in the same period. Services grew at 5%, the only sector to have grown faster than the rate of expansion of the population.

Export volumes grew by 3.5% annually 1973-81, helped by membership of the Franc Zone which maintains high prices for exports. Import volumes declined in the period by -0.6% annually, although it needs to be observed that there is substantial year-to-year variation, with the $US value of imports halving between 1974 and 1976 and then almost tripling by 1980.

The 1982 investment ratio of 28% is high, but subsequent growth between 1983 and 1985 at 3.5% annual average implies poor capital productivity, perhaps reflecting a rather wasteful public sector investment programme.

Inflation has been modest at 11% a year for the period 1973-84, and this is a result of the fiscal discipline imposed by membership of the Franc Zone.

Overall, Comoros has had relatively slow GDP growth and, until recently, falling GDP per head, although performance has been less bad than in many African countries. Industry has performed poorly, especially considering the high investment rate. Growth in export volumes has been good, and inflation performance has been significantly better than that elsewhere in Africa.

Recent Economic Developments

Vanilla, which provides 65% of export revenue in 1985 is subject to considerable variation in annual output. The 1982 export total of 259 tonnes was ten times the volume exported in 1984 of 25 tonnes. Similarly with cloves which make up 20% of export revenue. 1981 saw 500 tonnes exported, but this grew to 1,200 tonnes in 1983. Ylang-ylang made up 10% of export revenue in 1985, and 29 tonnes were exported in 1984, and 60 tonnes in 1985. These factors cause substantial overall variations in export revenue, with this more than doubling in 1985 to $US 16m from the low of $US 7m in 1984. Production prices for cloves were reduced by 67% between 1983 and 1985, and this will have an effect in limiting the expansion of clove production from 1987 on, as clove trees take four years before new plantings produce crops.

Overall external debt totals $US 285m, and debt-service is estimated at $US 9.9m in 1987 rising to $US 10.7m in 1988. Debt service payments will comprise 50% of export earnings even in years of good export performance, and Comoros continues to depend on aid, particularly from France to meet the deficit on the current account on the balance of payments.

The International Development Association of the World Bank committed $US 7.9m in soft-loan funds to improve the educational system, and to provide vocational training. In May 1987, the OPEC Fund for International Development made a $US 1m loan on soft terms to improve petroleum storage facilities. A South African Consortium is to build four new hotels at a total cost of $US 13m.

Economic Outlook and Forecasts

The forecasts take into account the adverse impact of the unresolved political situation that has resulted from the 1989 assassination.

Prospects are poor, and the forecast rates of GDP growth indicate GDP per head declining. Import volumes are expected to rise as the resolution of the political situation will lead to restoration of aid flows. Provided Comoros remains a member of the Franc Zone, inflation is expected to be modest.

Comoros

Comoros: Economic Forecasts
(average annual percentage change)

	Actual	Forecast Base	Forecast High
	1980-87	1990-95	1990-95
GDP	1.7	0.0	2.0
GDP per Head	-1.4	-3.1	-1.1
Inflation Rate	7.7	8.5	5.5
Export Volumes	0.1	0.1	0.2
Import Volumes	6.3	4.0	6.5

COMOROS: Comparative Data

	COMOROS	EAST AFRICA	AFRICA	INDUSTRIAL COUNTRIES
POPULATION & LAND				
Population, mid year, millions, 1989	0.453	12.2	10.2	40.0
Urban Population, %, 1982	12.3	30.5	30	75
Population Growth Rate, % per year, 1973-82	4.1	3.1	3.1	0.8
Land Area, thou. sq. kilom.	2	486	486	1,628
Population Density, persons per sq kilom., 1988	217.5	24.2	20.4	24.3
ECONOMY: PRODUCTION & INCOME				
GDP, $US millions, 1982	99	2,650	3,561	550,099
GNP per head, $US, 1986	320	250	389	12,960
ECONOMY: SUPPLY STRUCTURE				
Agriculture, % of GDP, 1982	45	43	35	3
Industry, % of GDP, 1982	18	15	27	35
Services, % of GDP, 1982	38	42	38	61
ECONOMY: DEMAND STRUCTURE				
Private Consumption, % of GDP, 1982	66	77	73	62
Gross Domestic Investment, % of GDP, 1982	28	16	16	21
Government Consumption, % of GDP, 1982	22	15	14	17
Exports, % of GDP, 1982	22	16	23	17
Imports, % of GDP, 1982	40	24	26	17
ECONOMY: PERFORMANCE				
GDP growth, % per year, 1973-81	0.3	1.6	-0.6	2.5
GDP per head growth, % per year, 1978-81	-3.8	-1.7	-3.7	1.7
Agriculture growth, % per year, 1978-81	2.4	1.1	0.0	2.5
Industry growth, % per year, 1978-81	-7.7	1.1	-1.0	2.5
Services growth, % per year, 1978-81	5.0	2.5	-0.5	2.6
Exports growth, % per year, 1978-81	3.5	0.7	-1.9	3.3
Imports growth, % per year, 1978-81	-0.6	0.2	-6.9	4.3
ECONOMY: OTHER				
Inflation Rate, % per year, 1973-81	11.3	23.6	16.7	5.3
Aid, net inflow, % of GDP, 1983	16.6	11.5	6.3	-
Debt Service, % of Exports, 1984	24.6	18	20.6	-
Budget Surplus (+), Deficit (-), % of GDP, 1979	-1.0	-3.0	-2.8	-5.1
EDUCATION				
Primary, % of 6-11 group, 1980	93	62	76	102
Secondary, % of 12-17 group, 1980	24	15	22	93
Higher, % of 20-24 group, 1980	..	1.2	1.9	39
Adult Literacy Rate, %, 1980	48	41	39	99
HEALTH & NUTRITION				
Life Expectancy, years, 1986	56	50	50	76
Calorie Supply, daily per head, 1983	1920	2,111	2,096	3.357
Population per doctor, 1973	15,315	35,986	24,185	550

Notes: 'Southern Africa' and 'Africa' exclude South Africa. Dates are for the country in question, and do not always correspond with the Regional, African and Industrial averages.

COMOROS: Leading Indicators

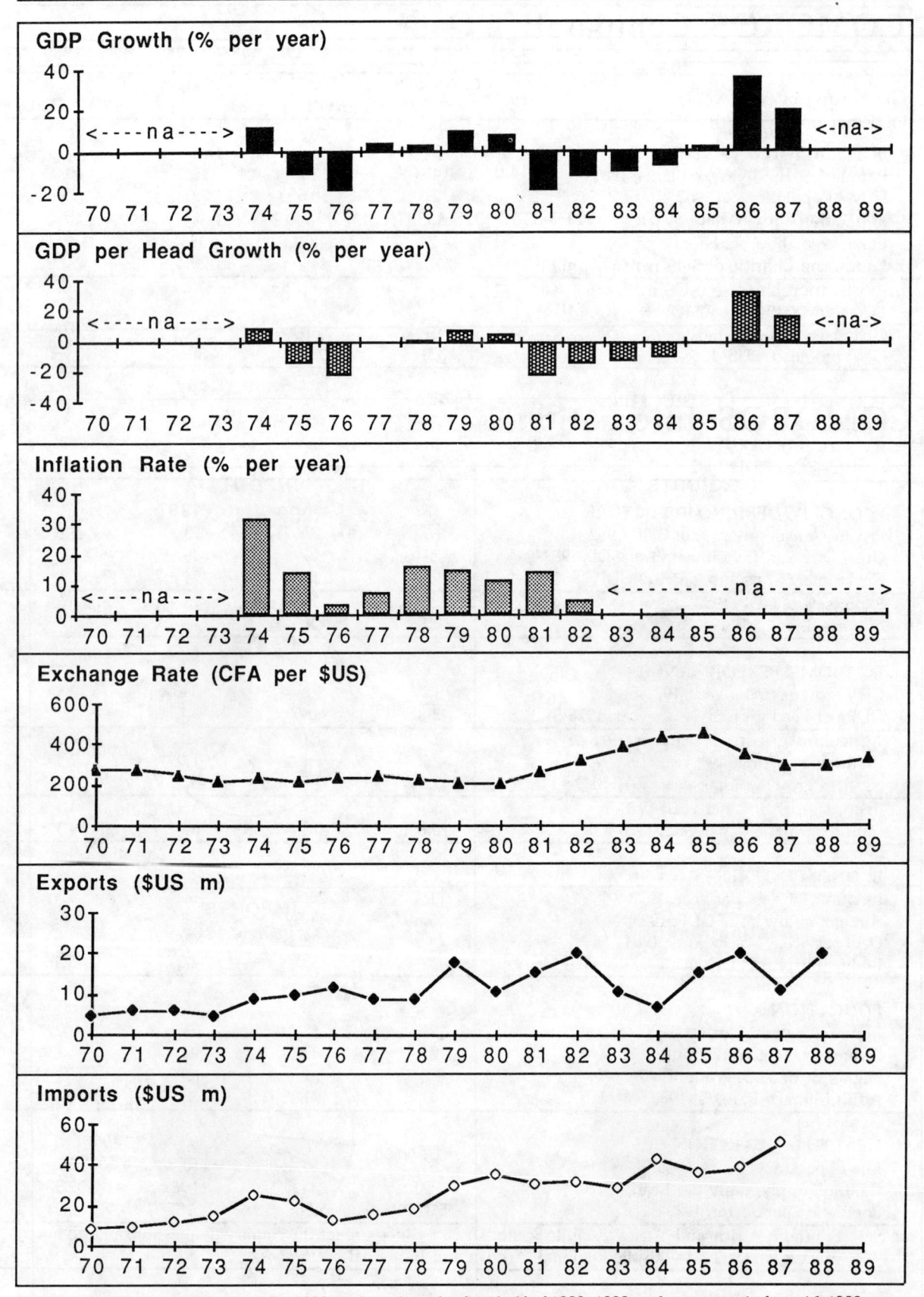

GDP Growth (% per year)

GDP per Head Growth (% per year)

Inflation Rate (% per year)

Exchange Rate (CFA per $US)

Exports ($US m)

Imports ($US m)

Note: Leading indicator data for 1989 are based on the first half of 1989. 1989 exchange rate is for mid-1989.

COMOROS: Leading Indicator Data

	1980	1981	1982	1983	1984	1985	1986	1987	1988	1989
GDP growth (% per year)	8.0	-19.4	-12.4	-10.6	-8.4	1.5	35.2	19.9		
GDP per head growth (% per year)	5.4	-22.1	-15.3	-13.6	-11.5	-1.6	32.1	16.8		
Inflation (% per year)	11.2	14.0	5.0							
Exchange rate (CFA per $US)	211.3	271.7	328.6	381.1	437.0	449.3	346.3	300.5	297.9	327.3
Exports, merchandise ($US m)	11	16	20	11	7	16	20	12	20	
Imports, merchandise ($US m)	36	32	33	30	43	37	39	52		

Note: Leading indicator data for 1989 are based on the first half of 1989. 1989 exchange rate is for mid-1989.

COMOROS: International Trade

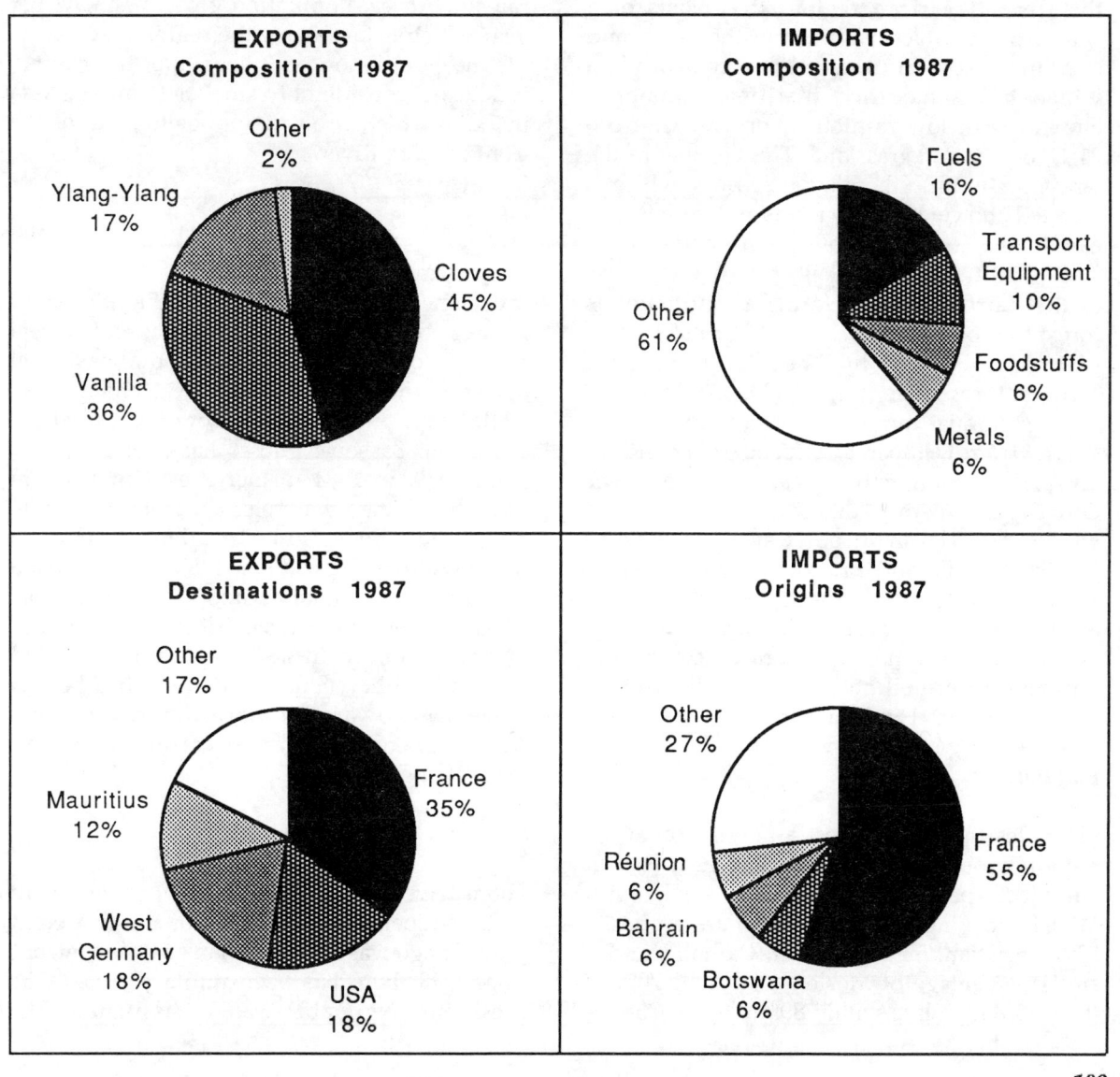

EXPORTS Composition 1987
- Cloves 45%
- Vanilla 36%
- Ylang-Ylang 17%
- Other 2%

IMPORTS Composition 1987
- Fuels 16%
- Transport Equipment 10%
- Foodstuffs 6%
- Metals 6%
- Other 61%

EXPORTS Destinations 1987
- France 35%
- USA 18%
- West Germany 18%
- Mauritius 12%
- Other 17%

IMPORTS Origins 1987
- France 55%
- Botswana 6%
- Bahrain 6%
- Réunion 6%
- Other 27%

11 CONGO

Physical Geography and Climate

The Congo is a medium-sized coastal state which straddles the Equator and is bounded by Gabon, Cameroon, the Central African Republic, Zaïre and the enclave of Cabinda in the Central African region. It consists of a coastal zone which is sandy in the north and swampy in the south, a narrow coastal plain which has a modified maritime equatorial climate with low rainfall, temperatures from 21°C to 25°C and grassland. This changes as the land rises to form the Mayombé ridge which is forested and cut by many rivers and gorges. The east of the country is lower with better soils. The Chaillu massif is the western watershed of the Zaïre basin and forms a mountainous forest region. The north east forms a drier plateau, bisected by the River Sangha. Temperatures range from 21°C to 27°C with an average annual rainfall of 142cm and a rainy season from October to December. The River Zaïre, its seven tributaries, and the River Oubangui form 2300km of navigable waterways. The main ports are Pointe-Noire on the Atlantic and Brazzaville on the River Zaïre. Hydro-electric capacity exists at Djoule on the Kouilou, Moukoukoulou Dam and Bouenza at Loubomo. Mineral resources include a large offshore oil and gas field, deposits of copper, gold, lead zinc, tin and uranium.

Population

Estimates for 1989 suggest a population of 2.2 million people. The official language is French with 50% speaking Kikonga, 20% Kiteke with Mbosi and Lingala. Around 50% are animists, 30% Christian, mainly Roman Catholic, and a small percentage of Moslems. Around 70% of the population live within 80km of the Congo-Ocean Railway which runs fron Brazzaville to Pointe Noire. The density of population is below the regional average of 11.1 persons per square kilometre at 6.1 per square kilometre. Urbanisation is at 46% compared to the regional average of 36%, one of the highest rates in Africa. Population growth is 3.3% per year which is just above the regional average.

The population of Brazzaville, the capital, is 302,000, considerably smaller than Kinshasa in Zaïre which is on the opposite bank of the Congo/Zaïre River.

History

The former French colony of Congo was proclaimed independent in 1960. Eight years of uneasy civilian rule were experienced before a military coup took place. In the past two decades there have been two changes of leadership, one resulting from assassination. Col. Denis Sassou-Nguesso has been in power since 1979. In 1984 former President Joachim Yhombi Opango was released from detention. There has been some student unrest in Brazzaville in November 1985 which closed primary and secondary schools for a short time.

There was an expansion of the public sector of the economy in 1964 which interrupted economic policy. Since 1981 there has been a movement towards a more mixed economy, which has encouraged investment and foreign aid.

Stability

There have been five changes of government since independence in 1960. For the first eight years Congo was under civilian government, but since 1968 there has been military rule. One President, Ngouabi, was assassinated, the other changes have occurred by means of bloodless coups. Col. Dennis Sassou-Nguesso has established some degree of continuity since 1979.

In the mid 1960s there were moves toward a socialist development strategy, with a large expansion of the public sector, but under Sassou-Nguesso there have been a series of liberal reforms. Congo has continued to be a member of the Franc Zone.

Overall, despite two distinctly uncertain decades since 1960, Congo's present stability under Sassou-Nguesso might be described as moderate. There is still tension between northern and southern groups, but economic policy is showing increasing signs of realism.

Economic Structure

Substantial oil income gives GDP in 1987 at $US 2,150m. GNP per head in 1987 was $US 870, placing it behind Cameroon, Botswana, Mauritius, South Africa and Gabon in average income.

Supply structure reflects the high level of oil earnings, with only 12% of GDP being generated by agriculture, and 33% from industry which includes the oil sector. Manufacturing, a sub-sector of industry, was limited to 6.5% of GDP. 55% of GDP comes from the services sector, and this is higher than most of Africa.

The structure of demand shows comparatively low allocation of income to private consumption which absorbed 58% of the value of GDP in 1987. 24% of GDP went to investment, and 21% of expenditure was accounted for by government consumption, both these two being high by the general structure of African economies.

Exports are almost half of all domestic product at 43% of GDP, and expenditure on imports is a similarly high proportion of GDP at 45%. Oil accounts for 72% of all exports, with timber a further 15%. Machinery and transport equipment made up 33% of imports with food accounting for a further 19%

The most recent estimates of adult literacy rates are for 1975, at 50%. There is expected to be a substantial improvement in these rates as in the late seventies Congo had enrolments in primary education equivalent to 156% of the 6-11 age group. Secondary school enrolments were also high, at 69% of the 12-17 age group. This is over three times the general African provision of secondary education, as is higher education provision at 6% of the 20-24 age group.

Life expectancy is high at 58 years, well over the African average of 50. Daily calorie supply averages 2,500. In 1984, there was one doctor for every 8,000 inhabitants, which is better than the level of physician provision in Africa generally.

Overall, Congo is a small, low population density country with comparatively high-income levels generated by the oil-sector. Agriculture has been neglected, and oil has led to a heavy external dependence. Educational and health provision are good, and reflect Congo's middle-income status.

Economic Performance

GDP grew at 5.5% annually over the period 1980-87, with the main expansions of the economy, generated by high oil prices, occurring in the early 1980s. There was a sharp fall in the rate of expansion of GDP after the collapse in the world oil price in 1986. GNP per head rose at 4.2% annually in the first half of the 1980s, in contrast to the average for Central Africa generally where GNP per head contracted at -0.8% annually. The agriculture sector, as a result of the emphasis on oil, grew at only 1.5% in the years 1980-87, while industry, which contain oil, expanded at 10.9% a year. The services sector contracted by -1.9%.

Export volumes, dominated by oil (with timber also important), grew at 3.9% a year up to 1987 this decade, while import volumes actually fell marginally by -0.7%.

The development effort, as represented by the investment ratio, which stood at 24% of GDP in 1987 was considerable. There was heavy concentration on infrastructure spending, much of which has been cut back after the collapse in oil prices after 1986.

Inflation has been moderate, averaging 1.8% a year 1980-87, one of the lowest rates in Africa and a reflection of the fiscal discipline imposed by Congo's membership of the Franc Zone.

Congo

Overall, Congo had good GDP growth performance in the early 1980s, but this was sharply reduced by the oil price fall in 1986. Investment allocations were high and inflation low, but the extent to which the 1986 oil price fall adversely affects economic performance in the late 1980s remains to be seen.

Recent Economic Developments

Congo accepted IMF supervision of the economy in May of 1986 and a three-year $US 12m IMF stand-by was agreed in July, though half of this was never drawn. This was Congo's first credit from the fund since 1979. The programme agreed with the Fund included a commitment to reduce government spending (following overambitious development spending in the early 1980s), a reduction in the budget deficit, and a schedule of reform and privatisation for public sector enterprises. Congo was expected to sign a new agreement by the end of 1989 which would also clear the way for rescheduling of debts with the Paris and London Clubs. The State monopoly of food crop marketing was ended late in 1986 and privatisation and foreign investment are now accepted as necessary for economic growth, though progress has been slow so far.

In July 1986, the Paris Club of creditor governments duly rescheduled arrears of debt service payments from 1985, and payment due in 1986 and 1987 over a ten-year period, with a five-year grace period. Further rescheduling is expected.

The agriculture sector suffered mixed fortunes in 1986, with timber sales rising 50%, a rise in coffee production, but falls in cocoa and foodstuffs output. Congo has good prospects for increasing exports of timber, coffee, cocoa, sugar, rubber, and tobacco, as well as becoming self-sufficient in food. Development of these sectors depends on the provision of transport infrastructure, but the budget cuts after 1986 in investment spending signified a set-back for prospects of diversifying away from oil dependency.

Total external debt in 1989 reached $US 4,800m according to the World Bank. As Congo is classified as a middle-income economy, it does not qualify for the debt relief given to the poorer countries, such as France's writing off of debts.

Aid commitments have helped to sustain parts of the government's investment programme. Caisse Centrale de Co-operation Economique of France approved a loan of $US 44m to support the balance of payments and the structural adjustment programme. West Germany is committed to increase its aid to Congo, which since 1979 has averaged $US 30m, and been directed mostly towards infrastructure. France is the largest bilateral donor with aid in 1989 totalling $US 118m in 1989.

Economic Outlook and Forecasts

The forecasts assume that Congo continues to restructure its economy under IMF and World Bank supervision, and that stability shows steady consolidation. The world oil price is clearly the major influence on Congo's prospects, and the assumption is that this remains unchanged in real terms until 1995.

GDP is projected to expand at a rate that will allow improvements in GDP per head in favourable world conditions. Inflation is expected to be faster than in the early 1980s, but to be at 5.0% or lower. Export volumes are forecast to expand more rapidly in the period up to 1995, and import volumes are expected to expand, but at a slower rate than exports.

Congo: Economic Forecasts
(average annual percentage change)

	Actual	Forecast Base	High
	1980-87	1990-95	1990-95
GDP	5.5	3.6	5.5
GDP per Head	2.2	0.0	1.9
Inflation Rate	1.8	5.0	3.0
Export Volumes	3.9	4.5	6.5
Import Volumes	-0.7	2.2	4.7

CONGO: Comparative Data

	CONGO	CENTRAL AFRICA	AFRICA	INDUSTRIAL COUNTRIES
POPULATION & LAND				
Population, mid year, millions, 1989	2.2	7.3	10.2	40.0
Urban Population, %, 1985	40	38.6	30	75
Population Growth Rate, % per year, 1980-86	3.3	3.0	3.1	0.8
Land Area, thou. sq. kilom.	342	638	486	1,628
Population Density, persons per sq kilom., 1988	6.1	11.1	20.4	24.3
ECONOMY: PRODUCTION & INCOME				
GDP, $US millions, 1986	2,000	4,146	3,561	550,099
GNP per head, $US, 1986	990	395	389	12,960
ECONOMY: SUPPLY STRUCTURE				
Agriculture, % of GDP, 1985	8	18	35	3
Industry, % of GDP, 1985	54	41	27	35
Services, % of GDP, 1985	38	41	38	61
ECONOMY: DEMAND STRUCTURE				
Private Consumption, % of GDP, 1986	50	62	73	62
Gross Domestic Investment, % of GDP, 1986	29	25	16	21
Government Consumption, % of GDP, 1986	20	14	14	17
Exports, % of GDP, 1986	47	35	23	17
Imports, % of GDP, 1986	46	35	26	17
ECONOMY: PERFORMANCE				
GDP growth, % per year, 1980-86	5.1	4.2	-0.6	2.5
GDP per head growth, % per year, 1980-86	1.7	-0.8	-3.7	1.7
Agriculture growth, % per year, 1980-86	-0.6	0.5	0.0	2.5
Industry growth, % per year, 1980-86	8.4	7.9	-1.0	2.5
Services growth, % per year, 1980-86	3.7	3.2	-0.5	2.6
Exports growth, % per year, 1980-86	5.4	4.5	-1.9	3.3
Imports growth, % per year, 1980-86	2.0	0.4	-6.9	4.3
ECONOMY: OTHER				
Inflation Rate, % per year, 1980-86	7.5	17	16.7	5.3
Aid, net inflow, % of GDP, 1986	5.9	5.4	6.3	-
Debt Service, % of Exports, 1986	39.8	28	20.6	-
Budget Surplus (+), Deficit (-), % of GDP, 1983	-3.0	-1.4	-2.8	-5.1
EDUCATION				
Primary, % of 6-11 group, 1979	156	93	76	102
Secondary, % of 12-17 group, 1979	69	42	22	93
Higher, % of 20-24 group, 1983	6	1.9	1.9	39
Adult Literacy Rate, %, 1975	50	52	39	99
HEALTH & NUTRITION				
Life Expectancy, years, 1986	58	52	50	76
Calorie Supply, daily per head, 1985	2,511	2,115	2,096	3,357
Population per doctor, 1979	5,150	13,835	24,185	550

Notes: 'Southern Africa' and 'Africa' exclude South Africa. Dates are for the country in question, and do not always correspond with the Regional, African and Industrial averages.

CONGO: Leading Indicators

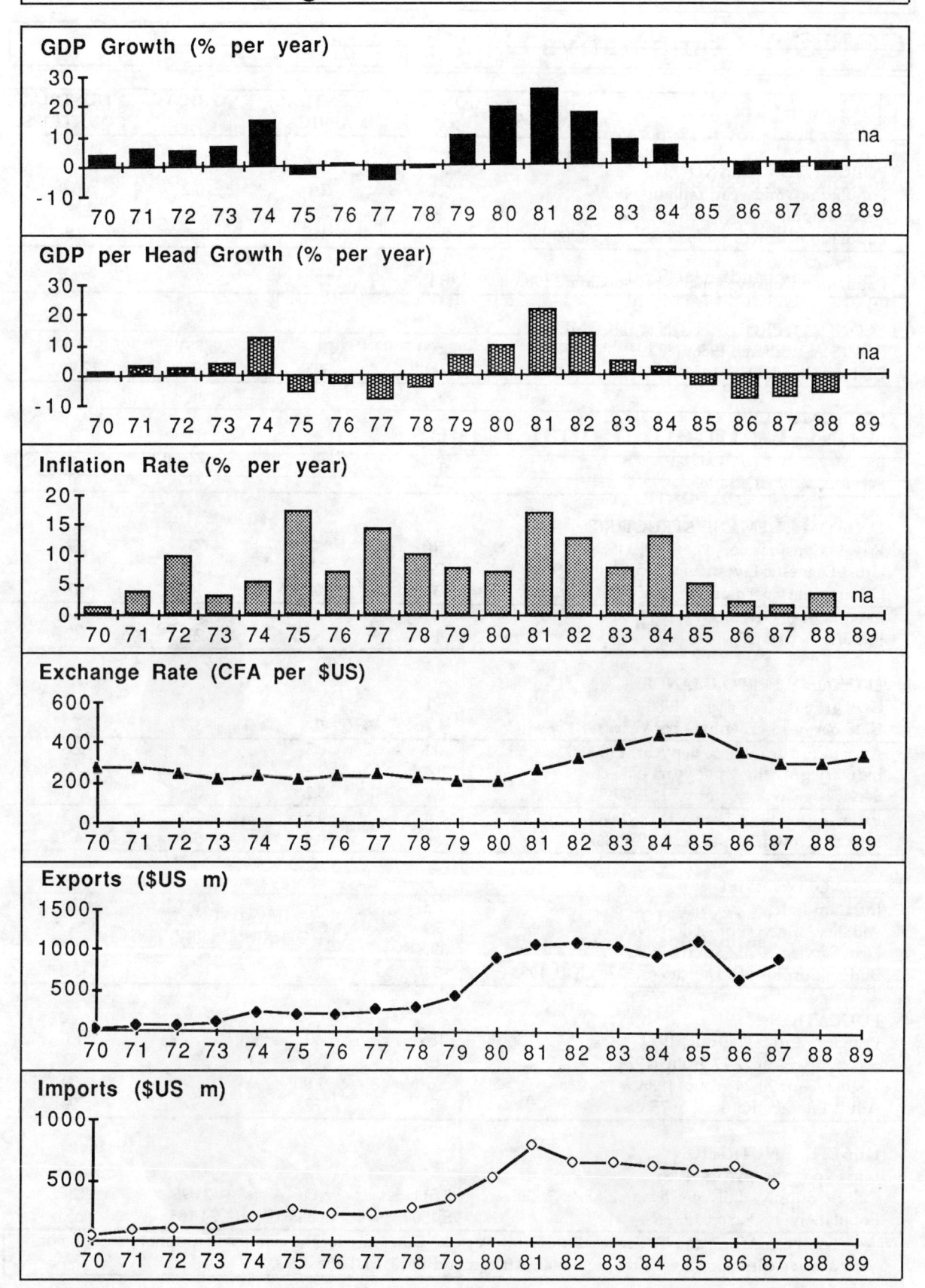

Note: Leading indicator data for 1989 are based on the first half of 1989. 1989 exchange rate is for mid-1989.

CONGO: Leading Indicator Data

	1980	1981	1982	1983	1984	1985	1986	1987	1988	1989
GDP growth (% per year)	18.8	24.9	17.1	7.7	6.2	-0.7	-4.8	-4.2	-3.0	
GDP per head growth (% per year)	9.6	21.7	13.9	4.4	2.9	-4.1	-8.2	-7.7	-6.6	
Inflation (% per year)	7.3	17.0	17.8	7.8	13.2	5.3	2.4	1.7	3.7	
Exchange rate (CFA per $US)	211.3	271.7	328.6	381.1	437.0	449.3	346.3	300.5	297.9	327.3
Exports, merchandise ($US m)	911	1073	1109	1066	930	1145	673	912		
Imports, merchandise ($US m)	545	804	664	650	618	598	627	495		

Note: Leading indicator data for 1989 are based on the first half of 1989. 1989 exchange rate is for mid-1989.

CONGO: International Trade

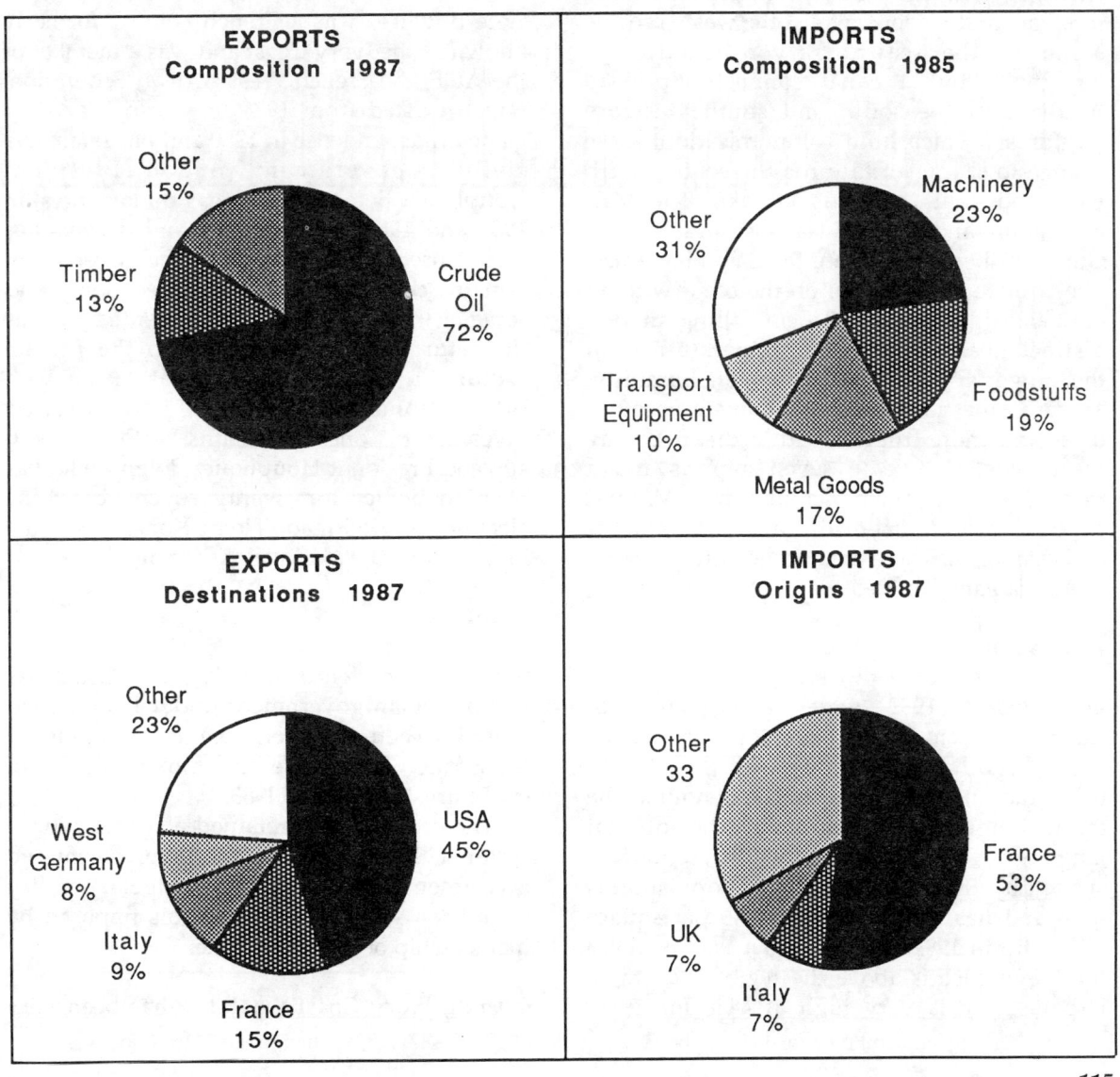

EXPORTS Composition 1987
- Crude Oil 72%
- Other 15%
- Timber 13%

IMPORTS Composition 1985
- Machinery 23%
- Foodstuffs 19%
- Metal Goods 17%
- Transport Equipment 10%
- Other 31%

EXPORTS Destinations 1987
- USA 45%
- Other 23%
- France 15%
- Italy 9%
- West Germany 8%

IMPORTS Origins 1987
- France 53%
- Other 33
- UK 7%
- Italy 7%

12 COTE D'IVOIRE

Physical Geography and Climate

Côte d'Ivoire is one of the geographically smaller of the Francophone West African coastal states, bounded by Liberia, Guinea, Mali, Burkina Faso and Ghana in the West Africa region. The eastern coastline is mainly sandbars and lagoons, whilst the west is rocky. None of the seaward river estuaries is navigable, but a canal connects Ebrié to Abidjan. In the south and south west are rainforests which limit communication; these change to savannah towards the centre, north west and south towards Bouaké. The Man mountains and the Guinea highlands on the Guinea/Liberia border are the only high areas. The climate is equatorial on the coast with an annual rainfall of 125-240cm falling in two distinct peaks, whilst the temperature is in the range 25°C to 30°C with high humidity. Towards the north west the climate becomes drier and more tropical with a distinct rainy season lasting five to seven months and a rainfall of 125-150cm per annum. Mineral deposits include haematite iron ore, oilfields and diamonds. There are two deep water ports at Abidjan and San Pedro.

Population

Estimates for 1989 suggest a population of 12.1m. The main groups being the Agri and the Boulé, who have an affinity with the Ashanti of Ghana and, in the north, the Senoufo. The main languages are French, the official language, Akan, spoken by 27%, and Kru spoken by 18%. The density of population is high and has increased from 27.6 per square kilometre in 1982 to 37.5 per square kilometre for 1989 which is above the regional average. Urbanisation is very high at 44% for 1987, compared with the regional average of 28.7%. The growth of population was 3.3% over the period 1980-87, and remains very high compared to the regional average of 3.2%.

History

Côte d'Ivoire was a French colony, formerly known as the Ivory Coast and was a member of the AOF, the French West African Federation, which existed from 1920 to World War II. It gained independence in 1960 and has retained a civilian government under President Houphouët-Boigny. There was minor unrest in 1981 and 1983 when schools and universities were closed for a short time. The government remains committed to a mixed economy and some changes have been implemented such as the return of state functions to the private sector, a move which has increased international confidence and private investment. Concern remains as to who will succeed President Houphouët-Boigny who has been in power for twenty seven years; the elections of 1985 made Henri Konan Bédié the President of the National Assembly.

Stability

Since independence in 1960, Ivory Coast has had a civilian government under Houphouët-Boigny. There has been no effective challenge to his rule. There have been minor episodes of civil unrest in 1981 and 1983.

The economy has retained a strong private sector with continued French involvement and investment. Monetary and budgetary policy have been subject to the controls imposed by membership of the Franc Zone.

Overall, Ivory Coast's stability has been very

Cote d'Ivoire

good. There is some concern over the transfer of power after Houphouët-Boigny, but no real reason to expect Ivory Coast's stability record not to continue.

Economic Structure

GDP in 1987 was $US 7,650m, almost double the average economy size in Africa. GNP per head was $US 740, and this gives a level of income that puts Côte d'Ivoire in the middle-income grouping, with GNP per head roughly twice the Africa average.

Agriculture generates 36% of GDP, industry 25% and services 39%, and this is fairly representative of the supply structure for African economies.

The demand structure allocates 65% of GDP to private consumption, 13% to investment and 17% to government consumption. These figures are not too different from the averages for Africa, but the level of investment in 1987 had fallen to a level below the average for the rest of the continent.

Exports were equivalent to 34% of GDP, and this indicates a degree of dependence on trade that is twice the regional average for West Africa. Imports are equivalent to 28% of GDP, 50% higher than the average for West Africa, and implying a capital outflow equivalent to around 6% of GDP.

The most recent estimate of the adult literacy rate was for 1978, when it was put at 41%. Although the overall rate of literacy has improved in the ensuing decade, it is still low for a country with middle-income status. 78% of the relevant age group were in primary education in 1984, and this is fairly typical of Africa. The secondary school enrolment ratio is lower than the Africa average by a couple of percentage points at 20%. The higher education enrolment rate is 3% of the 20-24 age group, and this is 50% higher than the African average which is close to 2%.

Life expectancy is 52, fairly close to the African average of 51. Food supply in calorie terms averages 2,300 a day, well up to the Africa average of 2,000. Provision of physicians is one for every 21,000 people, and this is below the West African average of one doctor per 16,000 people, but better than the African average of one for every 24,000 people.

Overall, Côte d'Ivoire is a medium-sized, well urbanised, middle-income economy. Industry and agriculture are both important in generating GDP, but export dependence for an economy without significant oil or mineral deposits is high. On the demand side, investment has fallen to comparatively low levels. Educational provision, with the exception of higher education, is lower than might be expected in a middle-income country as is the provision of medical services.

Economic Performance

GDP expanded at 2.2% a year in the period 1980-87, and with a comparatively high rate of population growth, GDP per head contracted at -2.0%. This is fairly representative of the general African experience in this period, but a marked down-turn from 6.7% annual expansion of GDP achieved by Côte d'Ivoire in the 1970s. Outflows due to debt servicing requirements were 21% of export earnings in 1989.

Agricultural growth has remained positive at 1.6% in the period 1980-87 but the industry sector has contracted at -2.4% a year, while services have recovered at 4.2% a year. Although these sectoral performances are not too dissimilar from those experienced on average elsewhere in Africa, they are a major disappointment in an economy with Côte d'Ivoire's impressive record in the 1970s.

Volumes of exports have grown at 3.4% a year, indicating that much of Côte d'Ivoire's problems have not arisen from poor production performance, but have come from reduced world prices for commodities exported. Import volumes have contracted at -3.1% a year and this reflects falling purchasing power of export earnings and the increasing diversion of foreign exchange receipts to debt-servicing.

Inflation at 4.4% a year for the period 1980-87 is well below the African average. This satisfactory performance in price stability stems from Côte d'Ivoire's membership of the Franc Zone and the fiscal discipline imposed.

Investment has fallen markedly to 13% of GDP in 1987, and at this rate the existing level of the capital stock is just being maintained. There must be some concern that future growth prospects will be harmed if investment remains at this level.

Overall, the performance of the economy has been disappointing, with a much lower rate of growth of GDP than in the 1970s and falling GDP per head. Export performance has been good, however, particularly in the context of generally poor export performance in Africa. Price stability has been impressive.

Recent Economic Developments

Ivory Coast's economic strategy is to continue to rely on world markets for primary products for her export earnings, and to encourage the private sector to take over loss-making state enterprises. After two poor years in 1983 and 1984 when GDP fell, there has been a recovery in 1985 and 1986 with rises in GDP of 5.9% and 5.4% respectively. However initial estimates indicate, with the return of drought conditions in some areas, falls in GDP for each of the three years 1987-89.

Agreement was reached with the IMF in June 1986 which released a structural adjustment loan of $US 250m. However, Côte d'Ivoire is reported to have failed to meet performance criteria in the first quarter of 1987, which has led to disbursements being suspended. A new four-year recovery programme was initiated in 1989 with World Bank and IMF guidance which gives priority to reducing the public sector deficit.

Oil production reached a peak of 1.25m tonnes in 1983, but is now at 1.0m tonnes annually, below domestic need which runs at 1.5m tonnes a year.

Structural changes in the economy have been cautious, but these changes have been moved to increase private sector participation, notably in urban water supply, urban bus transport, cocoa bean processing and involvement of a US firm in grain production.

The government now has plans to sell state holding in 103 state companies. Petroleum, airline, shipping, sugar, palm oil and commodity marketing enterprises are among the main parastatals being sold.

Inflation in 1989 was 6.5%, and this is an improvement on the 8.2% in 1988. Budgetary problems are likely to make the inflation performance difficult to sustain. The balance of trade has typically been in substantial surplus, with a $US 1.2b excess of merchandise exports over imports in 1985, and a $US 1.4b surplus in 1986. Falls in cocoa and coffee prices, together with drought affecting production levels, has led to a fall of $US 1b in export revenues since 1985, creating difficulties in meeting import, debt servicing and budgetary needs. Côte d'Ivoire's external debts are now estimated at $US 15.5b in 1989. A London Club rescheduling of commercial debts is expected in 1990.

Côte d'Ivoire is a member of the Franc Zone, and as the CFA Franc is linked to the French Franc, the currency has appreciated between 1985-88 against the dollar. There was some depreciation during 1989, but by the end of the year, the currency had appreciated further. This has added to Côte d'Ivoire's balance of payments problems by reducing domestic prices of export crops, and leading to severe financial difficulties for the main crop maketing organisation. There has been talk of Cote d'Ivoire breaking away from the Franc Zone to allow a depreciation of her currency, but such a development is extremely unlikely. It is quite possible, however, that the CFA Franc might be devalued against the French Franc.

Major aid projects include a World Bank funded urban project totalling $US 304m which embraces housing, land servicing and urban transport. Caisse Centrale de Cooperation Economique (CCCE), the World Bank and the Commonwealth Development Corporation are involved in coffee, rubber and livestock development schemes totalling $US 60.9m. The World Bank is contributing $US 26m to modernisation of port facilities at Abidjan, and the EEC and the European Investment Bank are contributing $US 185m to rehabilitation of the palm oil sector.

Economic Outlook and Forecasts

Côte d'Ivoire's outlook depends critically on the price of its agricultural exports and the prospect of obtaining favourable debt reschedulings. It is assumed that the world prices of Côte d'Ivoire's main exports, cocoa and coffee, continue to fall in real terms, and that Côte d'Ivoire is able to continue rescheduling her external debt to maintain servicing payments at their current levels.

It is difficult to be optimistic about Côte d'Ivoire's prospects up to 1995. GDP is expected, at best, to show zero growth which

Cote d'Ivoire

will imply falls in GDP per head. Export volumes are forecast to show slower expansion rates than in the 1980s. Import volumes are forecast, at best, to show zero growth. Inflation is expected to accelerate compared with the early 1980s, but to remain under 10%.

Côte d'Ivoire: Economic Forecasts
(average annual percentage change)

	Actual 1980-87	Forecast Base 1990-95	Forecast High 1990-95
GDP	2.2	-1.0	0.0
GDP per Head	-2.0	-4.6	-3.6
Inflation Rate	6.5	9.0	7.0
Export Volumes	3.4	1.0	3.0
Import Volumes	-3.1	-1.0	0.0

COTE D'IVOIRE: Comparative Data

	COTE D'IVOIRE	WEST AFRICA	AFRICA	INDUSTRIAL COUNTRIES
POPULATION & LAND				
Population, mid year, millions, 1989	12.1	12.3	10.2	40.0
Urban Population, %, 1985	40	28.7	30	75
Population Growth Rate, % per year, 1980-86	4.2	3.2	3.1	0.8
Land Area, thou. sq. kilom.	323	384	486	1,628
Population Density, persons per sq kilom.,1988	35.9	31.1	20.4	24.3
ECONOMY: PRODUCTION & INCOME				
GDP, $US millions, 1986	7,320	4,876	3,561	550,099
GNP per head, $US, 1986	730	510	389	12,960
ECONOMY: SUPPLY STRUCTURE				
Agriculture, % of GDP, 1986	36	40	35	3
Industry, % of GDP, 1986	24	26	27	36
Services, % of GDP, 1986	40	34	38	61
ECONOMY: DEMAND STRUCTURE				
Private Consumption, % of GDP, 1986	62	77	73	62
Gross Domestic Investment, % of GDP, 1986	12	13	16	21
Government Consumption, % of GDP, 1986	15	13	14	17
Exports, % of GDP, 1986	40	18	23	17
Imports, % of GDP, 1986	29	20	26	17
ECONOMY: PERFORMANCE				
GDP growth, % per year, 1980-86	-0.3	-2.03	-0.6	2.5
GDP per head growth, % per year, 1980-86	-4.5	-4.7	-3.7	1.7
Agriculture growth, % per year, 1980-86	0.9	1.3	0.0	2.5
Industry growth, % per year, 1980-86	-1.9	-3.6	-1.0	2.5
Services growth, % per year, 1980-86	-0.5	-2.5	-0.5	2.6
Exports growth, % per year, 1980-86	3.5	-4.0	-1.9	3.3
Imports growth, % per year, 1980-86	-5.4	-12.9	-6.9	4.3
ECONOMY: OTHER				
Inflation Rate, % per year, 1980-86	8.3	13.0	16.7	5.3
Aid, net inflow, % of GDP, 1986	2.1	3.7	6.3	-
Debt Service, % of Exports, 1984	19.7	21.4	20.6	-
Budget Surplus (+), Deficit (-), % of GDP, 1985	-3.1	-3.4	-2.8	-5.1
EDUCATION				
Primary, % of 6-11 group, 1984	78	75	76	102
Secondary, % of 12-17 group, 1984	20	24	22	93
Higher, % of 20-24 group, 1984	3	2.6	1.9	39
Adult Literacy Rate, %, 1976	41	30	39	99
HEALTH & NUTRITION				
Life Expectancy, years, 1986	52	50	50	76
Calorie Supply, daily per head, 1985	2,308	2,105	2,096	3,357
Population per doctor, 1980	21,040	16,199	24,185	550

Notes: 'Southern Africa' and 'Africa' exclude South Africa. Dates are for the country in question, and do not always correspond with the Regional, African and Industrial averages.

COTE D'IVOIRE: Leading Indicators

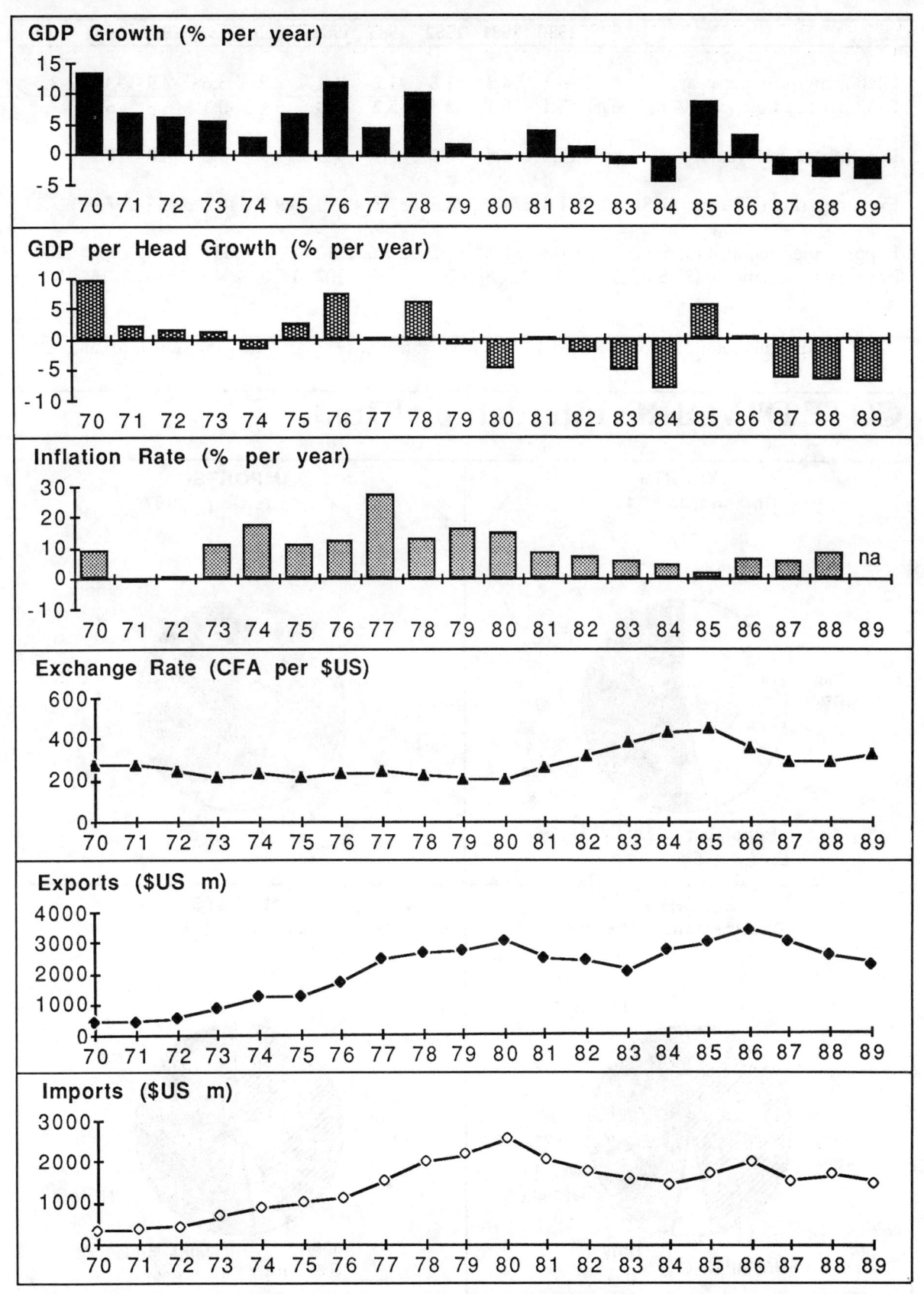

Note: Leading indicator data for 1989 are based on the first half of 1989. 1989 exchange rate is for mid-1989.

COTE D'IVOIRE: Leading Indicator Data

	1980	1981	1982	1983	1984	1985	1986	1987	1988	1989
GDP growth (% per year)	-0.8	4.3	1.6	-1.2	-4.4	9.2	3.6	-2.9	-3.2	-3.5
GDP per head growth (% per year)	-5.1	0.1	-2.5	-5.2	-8.3	5.4	-0.1	-6.6	-6.9	-7.2
Inflation (% per year)	14.7	8.8	7.3	5.9	4.3	1.9	6.6	6.0	8.2	
Exchange rate (CFA per $US)	211.3	271.7	328.6	381.1	437.0	449.3	346.3	300.5	297.9	327.3
Exports, merchandise ($US m)	3013	2435	2347	2066	2710	2934	3351	2965	2485	2250
Imports, merchandise ($US m)	2614	2068	1790	1635	1507	1723	2047	1582	1716	1550

Note: Leading indicator data for 1989 are based on the first half of 1989. 1989 exchange rate is for mid-1989.

COTE D'IVOIRE: International Trade

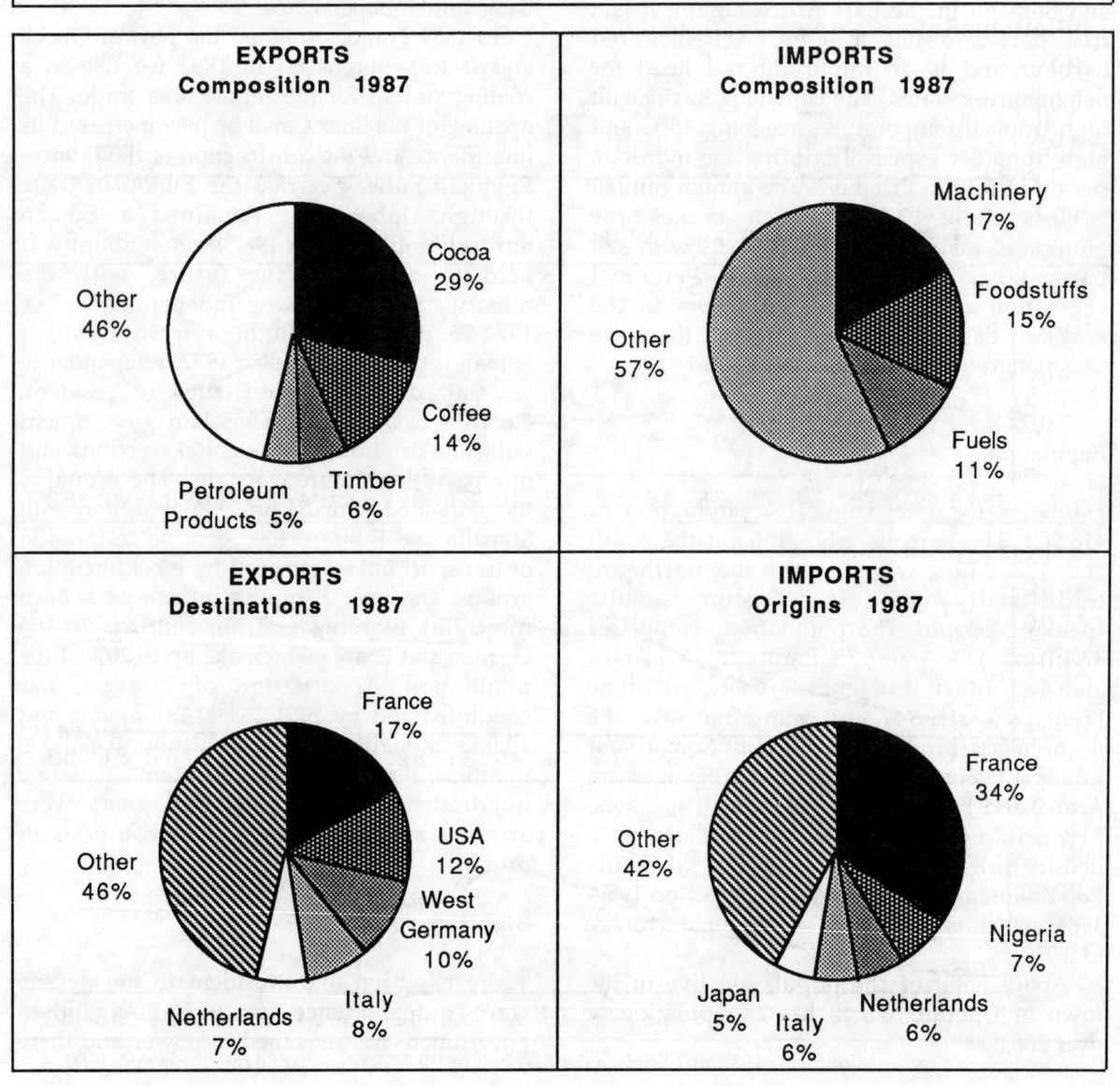

EXPORTS Composition 1987
- Cocoa 29%
- Coffee 14%
- Timber 6%
- Petroleum Products 5%
- Other 46%

IMPORTS Composition 1987
- Machinery 17%
- Foodstuffs 15%
- Fuels 11%
- Other 57%

EXPORTS Destinations 1987
- France 17%
- USA 12%
- West Germany 10%
- Italy 8%
- Netherlands 7%
- Other 46%

IMPORTS Origins 1987
- France 34%
- Nigeria 7%
- Netherlands 6%
- Italy 6%
- Japan 5%
- Other 42%

13 DJIBOUTI

Physical Geography and Climate

Djibouti is a small coastal enclave facing the Yemeni Peninsular and is bounded by Ethiopia and Somalia in the East Africa region. It is a free port and has a large, well-sheltered harbour and is an important rail head for neighbouring states. The climate is torrid with high tropical temperatures reaching 45°C and high humidity especially during the monsoon season in July to October. The annual rainfall is less then 12.5cm which makes the hinterland mainly arid desert areas with salt pans, lakes and douth palms. Perennial vegetation exists only above 1200m in the northern basalt mountains, though there are also mangrove swamps along the coast.

Population

Estimates for 1989 suggest a population of 418,000. The Somalis who inhabit the south and the Afars who live in the north are traditionally nomadic, Muslim Cushtic speaking people. The population comprises 37% Issas, 32% Afars, 7% Europeans, 6% Arabs and 18% other 'foreigners'. There are some French expatriates and more recently the drought has brought an influx of Somali and Ethiopian refugees. The official languages are Arabic and French, with 36% Somali speakers, 31% Eafaraf and 6% Arabic. There is a density of 18.5 persons to the square kilometre. Population growth is 3.4% for the period 1980-1984, which is just above the regional average of 3.1%.

About half of the population live in the town of Djibouti which has a population of over 200,000.

History

The country was known until 1967 as French Somaliland, then the Territory of Afars and Issas, until independence.

In 1859 France occupied the port of Obock and it was purchased in 1862 for use as a coaling station for the Indo-China trade. The opening of the Suez Canal in 1869 increased its importance and the construction of the Franco-Egyptian railway carried the Ethiopian trade through Djibouti. It remained a French territory until 1967. In 1967 a referendum was held to determine the future, with the minority Somalis seeking independence. The 1974-75 drought brought a fresh wave of Somali immigration and in 1977 independence was gained with Hassan Gouled as President. Djibouti has retained a civilian government with one candidate presidential elections and one party legislature elections. The economy has remained a mixed one. Co-operation with Somalia and Ethiopia has seen the restoration of transport links destroyed by rebel liberation groups. The most disruptive influence has been the influx of refugees from conflicts in the Ogaden and Eritrea who make up to 20% of the population. Repatriation of refugees has continued and by 1984 25,000 Ethiopians and 15,000 Somalis had left, although 20,000 unofficial refugees remain. In March 1986 five hundred and twenty nine foreigners were arrested and deported to their countries of origin.

Stability

There has been one President in the decade since independence in 1977. A civilian government has remained in power and there

Djibouti

was a single-party election in 1982 with no alternative candidates. Economic strategy has retained continuity, with commitment to a mixed economy.

In 1986 there was a bomb blast in the party headquarters, following which more than 1,000 people were arrested and 300-600 people expelled from Djibouti as foreigners.

Overall, Djibouti's stability would be judged good. There is some disruption caused by the influx of refugees from neighbouring countries. President Gouled has been in poor health, and there is concern over this transfer of power to his successor.

Economic Structure

GDP was estimated at $US 327 m in 1982, and only Comoros and Seychelles in East Africa have smaller economies. GNP per head in 1982 was estimated at $US 796 and this compares very favourably, with only Mauritius and Seychelles having higher average incomes per head. This places Djibouti in the lower-middle-income country grouping.

The economy relies heavily on the port facilities it provides, and this is reflected in the supply-side structure of the economy with agriculture providing only 4% of GDP, industry 19% and services 75%. Only Seychelles has a comparable supply structure, with services the leading growth sector, in the East African grouping.

The demand structure of the economy for 1982, the latest year for which figures are available, indicates that private consumption represents only 28% of GDP. This figure must be viewed with some suspicion as all the other East African countries have consumption at over 60% of GDP. Gross domestic investment absorbed 25% of GDP, and this is a high investment ratio, exceeded only by Comoros in East Africa, while government consumption was 38% of GDP, again high, but this may be inaccurate.

The openness of the economy is reflected in 35% of GDP being generated by exports, and only Mauritius and Seychelles in East Africa have higher dependence on exports. Expenditure on imports is equivalent to 68% of GDP, and the gap is mainly covered by foreign assistance with France the main benefactor.

The nominal figure for adult literacy in 1979 was 10%, the lowest in East Africa. Literacy has certainly improved in the ensuing decade, although with 59% of the relevant age group in primary education, only Ethiopia, Somalia, Sudan and Uganda have lower enrolments. Secondary enrolments by contrast are high, with 27% of the relevant age group in secondary education, and only Mauritius and Seychelles with middle-income status have higher enrolments at this level. There are negligible enrolments in higher education. The pattern of enrolments and the low overall literacy indicate a marked contrast between the urban areas where there is good provision going through to the end of secondary education, and the rural areas where educational provision is poor.

Life expectancy in 1986 was projected at 49 years, low for a middle-income country, with Mauritius, for example, with a comparable level of income per head in the region, having a 66-year life expectancy. Food supply compares well at an average of 2,400 calories a day, whereas the East African average is 2,100. The number of doctors in 1977 was one for every 1,700 persons, and this level of provision is usually only found in upper-middle income countries. The low life expectancy but good medical provision are reflections of the disparity between the urban areas and the rural parts of the country. Good medical provision in the urban areas influences the overall figures for doctors per head, but medical provision is poor in the rural areas where there is a life expectancy that is so low it drags down the overall figure.

Overall, Djibouti has a small economy with high urbanisation, moderate population density and a rapid increase in population due to the inflow of refugees. Income levels are in the lower-middle-income range. The economy depends heavily on the supply of port services, levels of government expenditure appear high, and there is a substantial deficit on the current account of the balance of payments.. Educational and health provision are uneven, good in the urban areas and poor elsewhere.

Djibouti

Economic Performance

GDP is estimated to have grown at 0.8% a year in the period 1973-82, and this implies a fall in GDP per head at an annual rate of -5.2%. Indications are that falls in GDP have been experienced in the years 1983-85. The 1973-82 record for GDP per head growth is the worst in East Africa, and this is mostly due to the influx of refugees leading to such a large increase in the population.

At the sectoral level, agriculture achieved a growth rate of 5.6% in 1973-84, almost keeping pace with population growth. The industry sector contracted at -2.1% a year, while the service sector grew at 2.5% a year.

Export volumes declined at -0.4% a year, although merchandise exports are very modest, being only $US 9m in 1982. Merchandise import volumes expanded at 5.1% a year, and this is rather more significant as merchandise imports were $US 225m in 1982.

Inflation 1970-81 ran at 12.6% a year. Evidence indicates that it has been even lower, at around 7.0% a year for 1980-86. This is considerably better than the East African average of 23.6% a year for the early 1980s, and has been aided by the openness of the economy.

Overall, Djibouti's growth performance has been poor, and inflow of refugees has led to sharply falling GDP per head. Sectorally, the industrial sector has performed worst, with the agriculture sector showing a creditable rate of expansion. Growth in import volumes has been steady, and price stability has been good.

Recent Economic Developments

The completion of a new container terminal in February 1984, at a cost of $US 142m, is expected to be the country's major source of revenue up to the end of the 1990s. Customs-free port facilities are offered for storage and freight in transit. Traffic with Ethiopia fell by 70% after the 1977-78 Ogaden war, and Djibouti hoped to increase trans-shipment traffic to other regional ports to compensate for this loss. However, results have been disappointing, with port revenue falling in 1986 by 25%.

Parastatal organisations have accumulated arrears, with the airport $US 2.8m behind with payments to creditors on a $US 22m upgrading programme. Air Djibouti has an accumulated deficit of $US 4m.

External debt-servicing has risen to $US 7m per year, and is projected to double by 1991. Debt servicing is expected to be a considerable burden on Djibouti's balance of payments. The IMF is pressing for freezes on wage rises and cuts in government expenditure, bolstered by tighter credit restrictions on the private sector.

Two geothermal drillings of a four-well $US 16.6m drilling programme have not yielded water of high enough temperature to make them commercially viable. It had earlier been hoped that the wells could meet Djibouti's electricity needs, and the programme is being funded by the International Development Association of the World Bank, Italy, the African Development Bank, the OPEC Fund for International Development and the UN Development Programme.

French financial support is expected to continue for strategic reasons, but aid from Middle-eastern countries is projected to fall as a result of the fall in the world oil price. Aid from France for 1987 was set at $US 13.6m, a rise of 33% over the 1986 level of $US 9.6m. Japan has committed $US 1.3m in April 1987 to buy rice to relieve food shortages arising from crop failures and the needs of Ethiopian refugees. China is to build a $US 5m sports complex. The Abu Dhabi Fund for Arab Development approved $US 5m in June 1987 to continue the up-grading of the airport.

Economic Outlook and Forecasts

It is assumed that Djibouti's stability will remain good. Djibouti's outlook is particularly sensitive to the strength of world economic growth. Performance in the period up to 1995 is projected to be better than in the early 1980s with GDP expanding at between 1.5% and 4.3% annually. It is difficult to estimate Djibouti's population growth, but it is assumed that refugee inflow continues at present rates giving falls in GDP per head of between -1.7% and -4.5% a year. Price stability is expected to remain good, with single-figure inflation, between 9.1% and 8.5% a year. Export volumes are projected to expand at between 1.5% a year and 2.1%, and import volumes to continue

Djibouti

steady growth at between 5.6% and 7.6% a year, although continuing foreign assistance will be necessary for these rates of import expansion to occur.

Djibouti: Economic Forecasts
(average annual percentage change)

	Actual 1973-82	Forecast Base 1990-95	Forecast High 1990-95
GDP	0.8	1.5	2.5
GDP per Head	-5.2	-4.5	-3.5
Inflation Rate	12.0[a]	9.1	8.5
Export Volumes	-0.4	1.5	2.1
Import Volumes	5.1	5.6	7.6

Note: [a] 1970-81.

DJIBOUTI: Comparative Data

	DJIBOUTI	EAST AFRICA	AFRICA	INDUSTRIAL COUNTRIES
POPULATION & LAND				
Population, mid year, millions, 1989	0.418	12.2	10.2	40.0
Urban Population, %, 1982	53.4	30.5	30	75
Population Growth Rate, % per year, 1970-82	6.3	3.1	3.1	0.8
Land Area, thou. sq. kilom.	22	486	486	1,628
Population Density, persons per sq kilom., 1986	18.5	24.2	20.4	24.3
ECONOMY: PRODUCTION & INCOME				
GDP, $US millions, 1982	327	2,650	3,561	550,099
GNP per head, $US, 1982	796	250	389	12,960
ECONOMY: SUPPLY STRUCTURE				
Agriculture, % of GDP, 1982	4	43	35	3
Industry, % of GDP, 1982	19	15	27	35
Services, % of GDP, 1982	75	42	38	61
ECONOMY: DEMAND STRUCTURE				
Private Consumption, % of GDP, 1982	28	77	73	62
Gross Domestic Investment, % of GDP, 1982	25	16	16	21
Government Consumption, % of GDP, 1982	38	15	14	17
Exports, % of GDP, 1982	35	16	23	17
Imports, % of GDP, 1982	68	24	26	17
ECONOMY: PERFORMANCE				
GDP growth, % per year, 1973-82	0.8	1.6	-0.6	2.5
GDP per head growth, % per year, 1973-82	-5.2	-1.7	-3.7	1.7
Agriculture growth, % per year, 1973-82	5.6	1.1	0.0	2.5
Industry growth, % per year, 1973-82	-2.1	1.1	-1.0	2.5
Services growth, % per year, 1973-82	2.5	2.5	-0.5	2.6
Exports growth, % per year, 1973-82	-0.4	0.7	-1.9	3.3
Imports growth, % per year, 1973-82	5.1	0.2	-6.9	4.3
ECONOMY: OTHER				
Inflation Rate, % per year, 1970-81	12.6	23.6	16.7	5.3
Aid, net inflow, % of GDP, 1982	1.7	11.5	6.3	-
Debt Service, % of Exports, 1982	13.7	18	20.6	-
Budget Surplus (+), Deficit (-), % of GDP, 1982	0.0	-3.0	-2.8	-5.1
EDUCATION				
Primary, % of 6-11 group, 1980	59	62	76	102
Secondary, % of 12-17 group, 1980	27	15	22	93
Higher, % of 20-24 group, 1980	..	1.2	1.9	39
Adult Literacy Rate, %, 1979	10	41	39	99
HEALTH & NUTRITION				
Life Expectancy, years, 1986	49	50	50	76
Calorie Supply, daily per head, 1983	2,444	2,111	2,096	3,357
Population per doctor, 1973	1,734	35,986	24,185	550

Notes: 'Southern Africa' and 'Africa' exclude South Africa. Dates are for the country in question, and do not always correspond with the Regional, African and Industrial averages.

DJIBOUTI: Leading Indicators

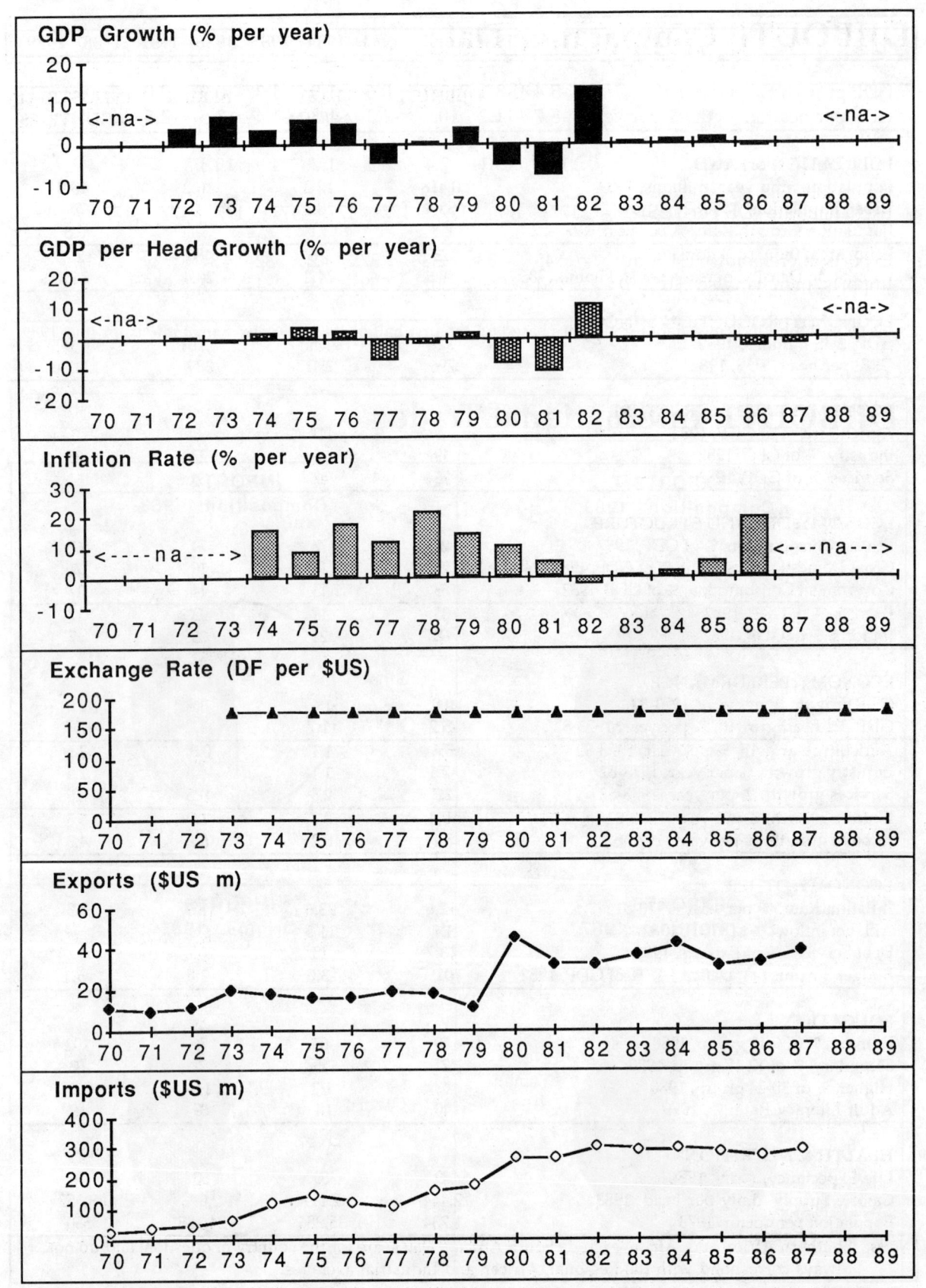

GDP Growth (% per year)

GDP per Head Growth (% per year)

Inflation Rate (% per year)

Exchange Rate (DF per $US)

Exports ($US m)

Imports ($US m)

Note: Leading indicator data for 1989 are based on the first half of 1989. 1989 exchange rate is for mid-1989.

DJIBOUTI: Leading Indicator Data

	1980	1981	1982	1983	1984	1985	1986	1987	1988	1989
GDP growth (% per year)	-5.4	-8.1	13.9	0.6	1.0	1.3	-0.9	0.0		
GDP per head growth (% per year)	-8.7	-11.2	10.9	-2.3	-1.8	-1.4	-3.5	-2.5		
Inflation (% per year)	10.6	5.1	-2.4	0.9	1.7	4.9	19.3			
Exchange rate (DF per $US)	177	177	177	177	177	177	177	177	177	180
Exports, merchandise ($US m)	45	33	33	37	42	32	34	39		
Imports, merchandise ($US m)	261	264	302	289	294	284	267	288		

Note: Leading indicator data for 1989 are based on the first half of 1989. 1989 exchange rate is for mid-1989.

DJIBOUTI: International Trade

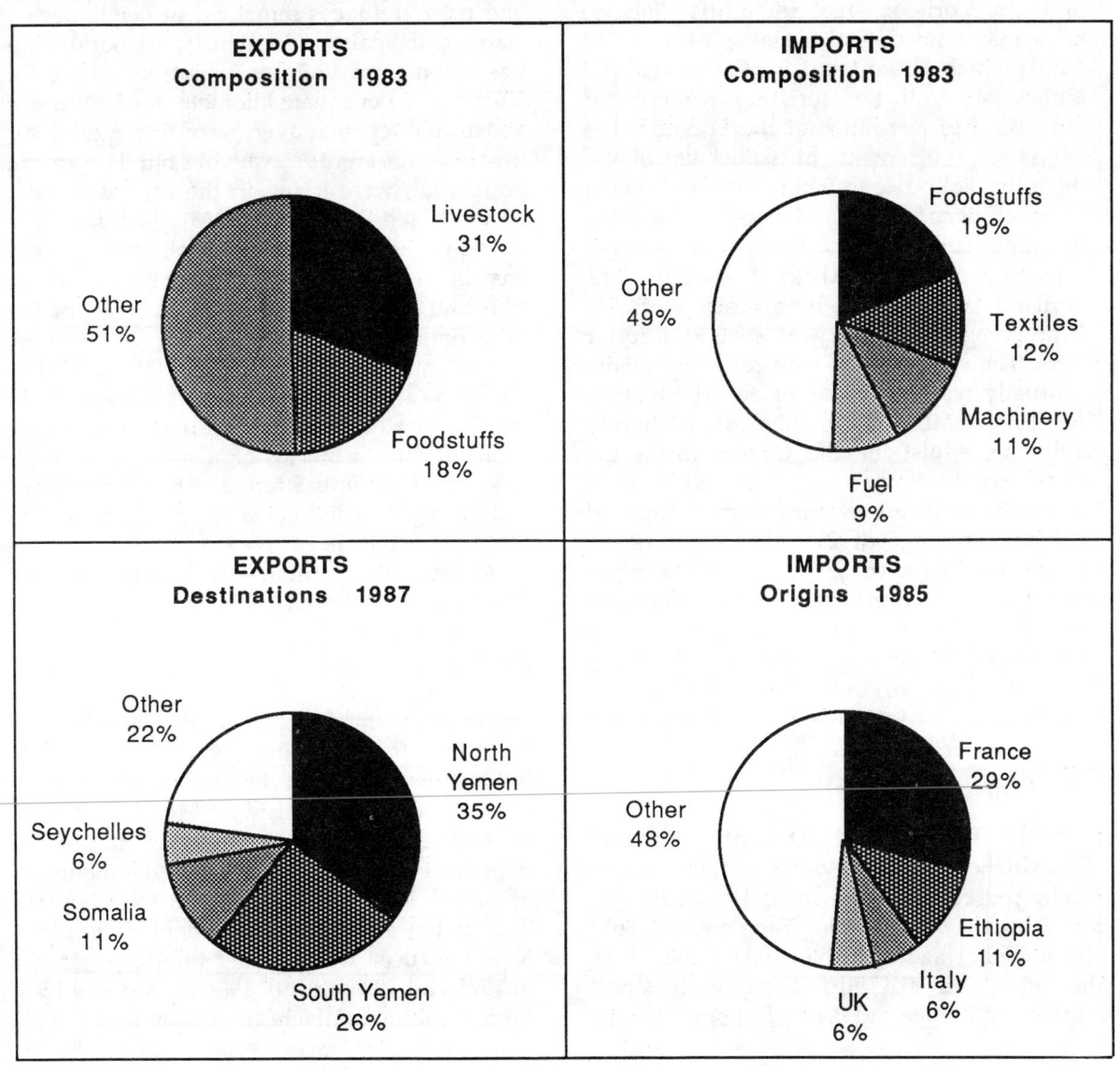

EXPORTS Composition 1983
- Livestock 31%
- Foodstuffs 18%
- Other 51%

IMPORTS Composition 1983
- Foodstuffs 19%
- Textiles 12%
- Machinery 11%
- Fuel 9%
- Other 49%

EXPORTS Destinations 1987
- North Yemen 35%
- South Yemen 26%
- Somalia 11%
- Seychelles 6%
- Other 22%

IMPORTS Origins 1985
- France 29%
- Ethiopia 11%
- Italy 6%
- UK 6%
- Other 48%

14 EQUATORIAL GUINEA

Physical Geography and Climate

Equatorial Guinea, in the central African region, consists of a group of five small scattered islands off the coasts of Cameroon and Gabon : Bioko (formerly Fernando Póo); Annobón; Corisco; Great and Little Elobeys) and a mainland coastal enclave, Mbini (Río Muni), which is bordered by Cameroon and Gabon. Most of the territory, except for Annobón, lies just north of the Equator. The islands are the remains of extinct volcanoes, which on Bioko rise to alpine height allowing the ranching of cattle and horses. There are high mountain lakes and fertile volcanic soil. A natural harbour exists at Malabo. The mainland consists of a long beach with Río Muni being the estuary of several interior rivers. The coastal plain rises towards Gabon, eventually reaching peaks up to 1200 metres. The River Mbini (Río Bonito) is barely navigable, whilst the Río Campo forms the border with Cameroon. The climate is equatorial with an average temperature of 25°C and a range of 21°C-34°C and heavy rainfall, with an average rainfall of 200cm per annum, heaviest from October to December and from January to May. Temperatures are modified on the islands by altitude. The major ports are Malabo, Luba, Bata, Mbini (Río Benito) and Kongo. Mineral deposits are poor.

Population

Estimates for 1989 suggest a population of 408,000 who were 80% Roman Catholic before recent persecution. The main languages are Spanish with 68% speaking Fang and 6% Bubi and some Portuguese Kriolo. The inhabitants of the mainland are 90% Fang with small numbers of Kombe and Bejeba Balongue. On the island of Bioko the Bubi were the original occupants, the Ferdinando were descendants of former colonists and slaves and then came immigrants from Sierra Leone, Ghana, Nigeria and Cameroon. Many Nigerian contract workers were repatriated from 1976 onwards and many refugees remained in neighbouring states and Spain. The density of population has fallen from 18.2 per square kilometre for 1970 to 14.6 per square kilometre for 1989 and is above the regional average of 9.6. Figures for urbanisation are not available but 11% of the population were living in the capital Malabo on the island of Bioko in 1965 which was 56% of the island's population. Population growth has fallen from 2.3% for 1970-82 to 2.1% for 1989 and is below the regional average for the same period of 2.9%.

It is notable that over 100,000 Equatorial Guineans are thought to be still living outside the country, following the mass exodus of refugees during Macías Nguema's regime, 20-30% of the population at the time. With continuing poor living standards, there is little incentive to return.

Malabo, the capital, has a population of 15,000, with 24,000 living in Bata.

History

Equatorial Guinea became a Spanish colony in the eighteenth century. It acheived independence in 1968. Following ten years of civilian rule under President Macías Nguema, it is currently under the military rule of his nephew President Teodoro Obiang Nguema Mbasogo. Relations are strained with Nigeria due to the alleged maltreatment of migrant Nigerian cocoa workers and their repatriation in 1975 and 1976. In July 1986 a coup plot by a former military attaché to Spain was foiled, and amidst army unrest many soldiers were transferred to Mbini (Río Muni) on the mainland.

The Nguema regime was characterised by haphazard intervention in the economy and the persecution of particular groups, combined with a policy of isolation and withdrawal from contact with the rest of the world, consequently dislocating economic activity. The Mbasogo regime has tried to alleviate these problems and in January 1985 entered the Franc Zone.

Stability

After independence in 1968, Equatorial Guinea entered a period of withdrawal from international contact and persecution of minority groups. In the 1979 military coup, there was some improvement with provisional plans for a return to civilian rule, but the regime has been plagued by coup attempts.

The 1970s were characterised by capricious intervention in the economy, the squandering of aid from Spain and pervasive corruption. In 1983 Equatorial Guinea joined the area's customs union, UDEAC, and in 1985 joined the Franc Zone, and these developments indicate a move toward more stable economic conditions.

Overall, the stability of Equatorial Guinea is poor, although there are signs of improvement while the present regime continues political reforms and adjusts to the disciplines imposed by the Franc Zone.

Economic Structure

GDP was estimated at $US 139m in 1987, making it small in economic size, with only São Tomé et Príncipe in Central Africa smaller. GDP per head in 1987 was estimated at $US 340, placing the country in the low-income world grouping, with Chad and Zaïre in the region being the only countries with lower GDP per head.

As is to be expected in a low-income country, agriculture generates a high proportion of GDP, with 41% in 1982, well above the Central African average of 18%. Industry generated 10% of GDP, and the services sector, 48%. This production structure is a reflection of the absence of minerals, and a reliance on agricultural crops for export.

On the demand side, private consumption stood at 119% of GDP in 1982, with investment at 12% and government expenditure at 50% of GDP. These levels of expenditure are only made possible by substantial foreign assistance, mostly from Spain, but increasingly latterly from France since 1985 when Equatorial Guinea joined the Franc Zone.

Exports, mostly cocoa and timber, ran at 69% of GDP in 1982 and imports, comprising food, fuel and machinery for the most part, were 151% of GDP. Aid is the key factor in enabling imports to be sustained at such high levels. It needs to be observed that there are large year-to-year fluctuations in both export revenues and import expenditure, and this results from climatic variations affecting agriculture and the impact of large items of investment expenditure in particular years.

Adult literacy was estimated at 37% in 1980, below the regional average of 52%. This figure can be expected to have increased in subsequent years as 108% of the relevant age-group were in primary education in 1983. 11% of the relevant age-group were in secondary education, and this is well below the regional average of 42%, which is heavily influenced by the reported 57% in secondary education in Zaïre, the most populous country. There are negligible enrolments in higher education.

Life expectancy was 46 years in 1987, below the regional average of 52, and a reflection of low-income status and poor medical provision. Food supply averaged 2,400 calories per person per day and compares well with the African average of 2,100. Fertile agricultural land, low population densities, adequate rainfall in most years and an emphasis on subsistence agriculture are responsible for good food supply. Medical provision in 1975 was one doctor per 62,000 inhabitants, the worst provision in a region which averaged one doctor for every 24,000 inhabitants.

Overall, Equatorial Guinea is a small, low-income country with considerable dependence on agriculture to generate GDP. There is a great reliance on foreign assistance to maintain import levels. Medical provision is poor, but food supply and primary education enrolments are considerably better than might be expected in a low-income country.

Equatorial Guinea

Economic performance

Economic performance has been very poor in the period for which figures are available, 1973-82. GDP contracted at -12.4% a year, and there was a massive disruption of the economy in the mid-1970s following the withdrawal of Nigerian cocoa workers and political instability. GDP per head growth contracted at -14.7% a year and over the ten-year period, GDP per head will have fallen, on these figures, to 16% of its level in 1973.

Contraction has been spread over all sectors, but the industry sector, which has contracted at -17.7% a year, has declined most rapidly. Agriculture has declined at -14.7% a year, and services, cushioned somewhat by external assistance, has contracted by -9.6% a year. These are quite the worst performance figures in all of Africa, with the next poorest performance being in Angola, where GDP per head declined at -7.3% a year in the same period.

Export volumes, given the disruptions in the cocoa sector, have contracted at -7.5% a year, and this has caused imports to decline at the rate of -3.8% a year, this being for 1973-83. In recent years, particularly 1985 and 1986, after Equatorial Guinea joined the Franc Zone, there appear to have been improvements in export performance.

Inflation in the period 1973-82 has been 14.5% annually, and this is about the level observed in Africa generally. Since joining the Franc Zone, Equatorial Guinea can be expected to retain and perhaps improve her record of price stability.

Investment at 12% of GDP in 1982 is barely enough to maintain the capital stock, and is caused by the general lack of stability in the economy, and in turn has contributed to the very poor economic performance.

Overall, Equatorial Guinea has had disastrous economic performance in the producing sectors. Inflation has been moderate, but development effort, as indicated by the investment ratio, has been very poor.

Recent Economic Developments

Equatorial Guinea joined the Franc Zone in January 1985, and since that date the economy has shown signs of recovery. Cocoa had fallen to 5,000 tonnes in 1979, having been at 38,000 tonnes in 1967. Improvements in producer prices, which are now the highest in the region, have resulted in improvements such that 8,000 tonnes are estimated to have been produced in 1988.

Coffee production had fallen from 8,450 tonnes in 1968 to 108 tonnes in 1980. Production has now risen to 482 tonnes in 1985, and there are hopes that 2,000 tonnes will be reached by 1990.

Timber exports have a potential of 300,000 cu metres a year with an effective re-planting policy. In 1989, it was estimated that 170,000 cu metres were exported and timber is now the major export.

Equatorial Guinea had no budgets between 1974 and 1979 as Macías Nguema exercised personal control over public finances. A budget deficit of CFA 747m ($US 2.5m, 1.9% of GDP) was projected for 1988.

The World Bank is providing $US 9.3m, the Arab Bank for Economic Development $US 2.8m, the OPEC Fund for International Development $US 1m, and the European Development Fund $US 1m for a cocoa rehabilitation scheme on Bioko island, begun in 1985. The World Bank is also funding, with UMDP and UNESCO, a $US 6.1m scheme approved in May 1987 to rehabilitate primary school education. France has figured more prominently in aid provision since Equatorial Guinea joined the Franc Zone. Caisse Centrale de Co-operation Economique (CCCE) committed $US 2m in 1987 to upgrade Bioko island airport, as well as $US 3m for telecommunications improvements and $US 19.5m for a hydro-electric scheme on Bioko island. The African Development Bank is financing a $US 3.7m water supply scheme, and the EEC is providing $US 13.6m for forestry development.

With the present government committed to a reducing government role in the economy, and good prospects of finding oil and other minerals, foreign assistance should be expected to increase.

An IMF standby credit of $US 10.2m was agreed in 1985 and this was followed by a $US 16.5m structural adjustment facility in 1988, following agreement by the government over improvements in the bureaucracy and moves

Equatorial Guinea

towards greater private sector involvement.

The World Bank estimated total external debt to be in the region of $US 210m in 1989. Equatorial Guinea will be one of the beneficiaries of France's decision to cancel loans to the poorer countries, with Spain also cancelling a third of the country's debt.

Economic Outlook and Forecasts

Long-term prospects for Equatorial Guinea have improved markedly with the decision to join the Franc Zone in 1985, and the inception of an economic reform programme under IMF and World Bank supervision. The forecasts assume no relapse to the political instability of the 1970s.

Growth in GDP is expected to be restored at between 3.5% and 4.8% annually up to 1995, which will allow some improvement in GDP per head at between 1.2% and 2.5% a year. Price stability is expected to improve with continued membership of the Franc Zone. Export volumes are projected to show strong recovery at between 6.5% and 8.8% annually as cocoa plantations are rehabilitated. Import volumes will show steady growth as a result of improved export earnings and external assistance.

Equatorial Guinea: Economic Forecasts
(average annual percentage change)

	Actual 1973-82	Forecast Base 1990-95	Forecast High 1990-95
GDP	-12.4	3.5	4.8
GDP per Head	-14.7	1.2	2.5
Inflation Rate	14.5	8.5	6.2
Export Volumes	-7.5[a]	6.5	8.8
Import Volumes	-3.8[a]	4.2	6.2

Note: [a] 1973 - 83.

EQUATORIAL GUINEA: Comparative Data

	EQUATORIAL GUINEA	CENTRAL AFRICA	AFRICA	INDUSTRIAL COUNTRIES
POPULATION & LAND				
Population, mid year, millions, 1989	0.408	7.3	10.2	40.0
Urban Population, %, 1982	6.9	38.6	30	75
Population Growth Rate, % per year, 1972-82	2.3	3.0	3.1	0.8
Land Area, thou. sq. kilom.	28	638	486	1,628
Population Density, persons per sq kilom., 1988	14.3	11.1	20.4	24.3
ECONOMY: PRODUCTION & INCOME				
GDP, $US millions, 1981	72	4,146	3,561	550,099
GNP per head, $US, 1981	180	395	389	12,960
ECONOMY: SUPPLY STRUCTURE				
Agriculture, % of GDP, 1982	41	18	35	3
Industry, % of GDP, 1982	10	41	27	35
Services, % of GDP, 1982	48	41	38	61
ECONOMY: DEMAND STRUCTURE				
Private Consumption, % of GDP, 1982	119	62	73	62
Gross Domestic Investment, % of GDP, 1982	12	25	16	21
Government Consumption, % of GDP, 1982	50	14	14	17
Exports, % of GDP, 1982	69	35	23	17
Imports, % of GDP, 1982	151	35	26	17
ECONOMY: PERFORMANCE				
GDP growth, % per year, 1973-82	-12.4	4.2	-0.6	2.5
GDP per head growth, % per year, 1973-82	-14.7	-0.8	-3.7	1.7
Agriculture growth, % per year, 1973-82	-15.6	0.5	0.0	2.5
Industry growth, % per year, 1973-82	-17.7	7.9	-1.0	2.5
Services growth, % per year, 1973-82	-9.6	3.2	-0.5	2.6
Exports growth, % per year, 1973-82	-7.5	4.5	-1.9	3.3
Imports growth, % per year, 1973-82	-3.8	0.4	-6.9	4.3
ECONOMY: OTHER				
Inflation Rate, % per year, 1973-82	14.5	17	16.7	5.3
Aid, net inflow, % of GDP, 1981	21.7	5.4	6.3	-
Debt Service, % of Exports, 1981	8.8	28	20.6	-
Budget Surplus (+), Deficit (-), % of GDP, 1982	-19.4	-1.4	-2.8	-5.1
EDUCATION				
Primary, % of 6-11 group, 1983	108	93	76	102
Secondary, % of 12-17 group, 1975	108	42	22	93
Higher, % of 20-24 group, 1983	108	1.9	1.9	39
Adult Literacy Rate, %, 1980	37	52	39	99
HEALTH & NUTRITION				
Life Expectancy, years, 1986	45	52	50	76
Calorie Supply, daily per head, 1984	2,444	2,115	2,096	3,357
Population per doctor, 1975	62,000	13,835	24,185	550

Notes: 'Southern Africa' and 'Africa' exclude South Africa. Dates are for the country in question, and do not always correspond with the Regional, African and Industrial averages.

EQUATORIAL GUINEA: Leading Indicators

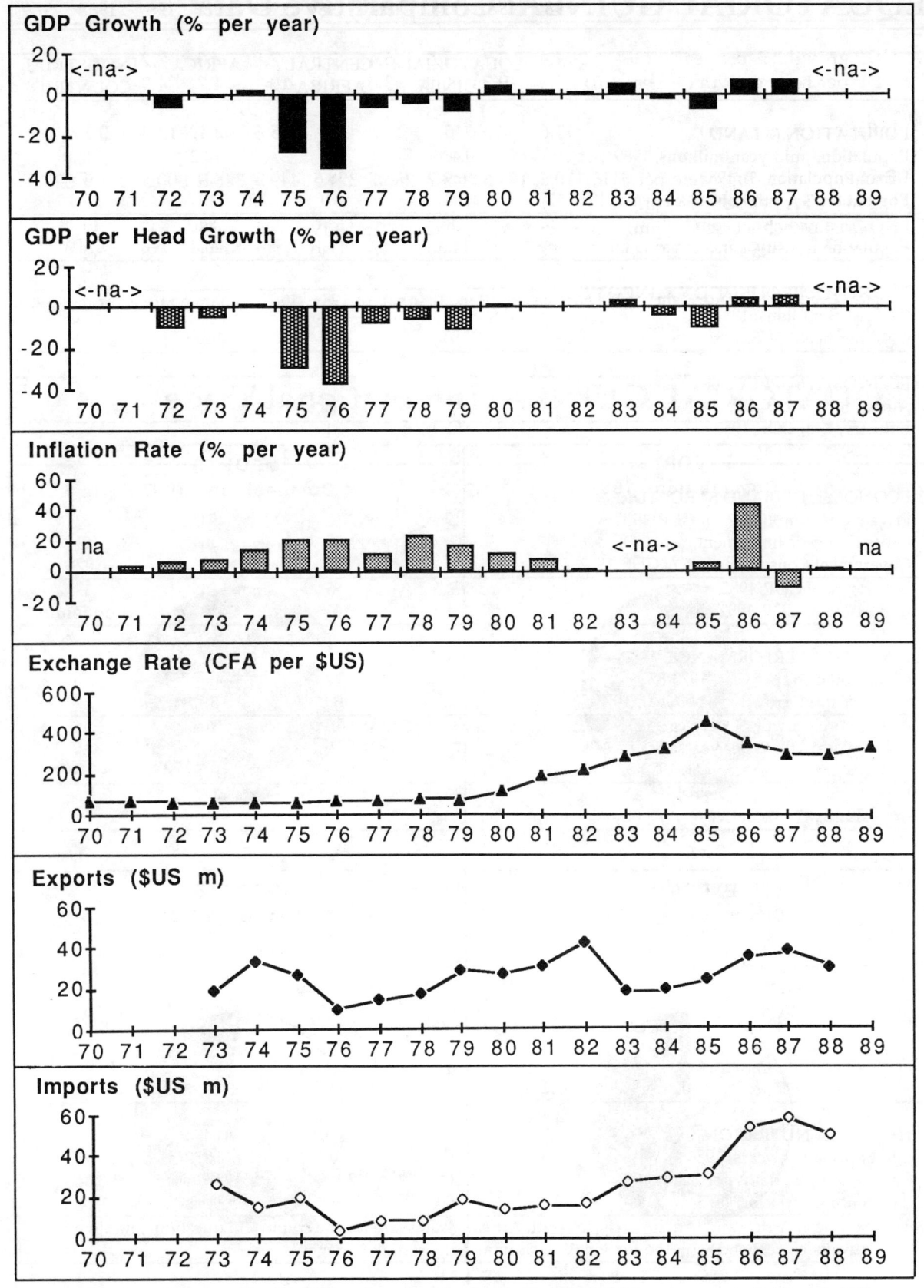

GDP Growth (% per year)

GDP per Head Growth (% per year)

Inflation Rate (% per year)

Exchange Rate (CFA per $US)

Exports ($US m)

Imports ($US m)

Note: Leading indicator data for 1989 are based on the first half of 1989. 1989 exchange rate is for mid-1989.

EQUATORIAL GUINEA: Leading Indicator Data

	1980	1981	1982	1983	1984	1985	1986	1987	1988	1989
GDP growth (% per year)	4.4	2.2	1.5	5.4	-2.5	-7.8	7.0	7.2		
GDP per head growth (% per year)	1.9	-0.3	-0.9	3.0	-4.9	-10.1	4.7	4.9		
Inflation (% per year)	11.6	7.6	1.1			5.6	43.1	-12.3	0.1	
Exchange rate (Bipkwele per $US to 1984. CFA per $US 1985 on)	110.6	184.6	219.7	286.7	321.5	449.3	346.3	300.5	297.9	327.3
Exports, merchandise ($US m)	26	31	42	18	19	23	35	37	30	
Imports, merchandise ($US m)	14	16	17	28	30	32	54	58	50	

Note: Leading indicator data for 1989 are based on the first half of 1989. 1989 exchange rate is for mid-1989.

EQUATORIAL GUINEA: International Trade

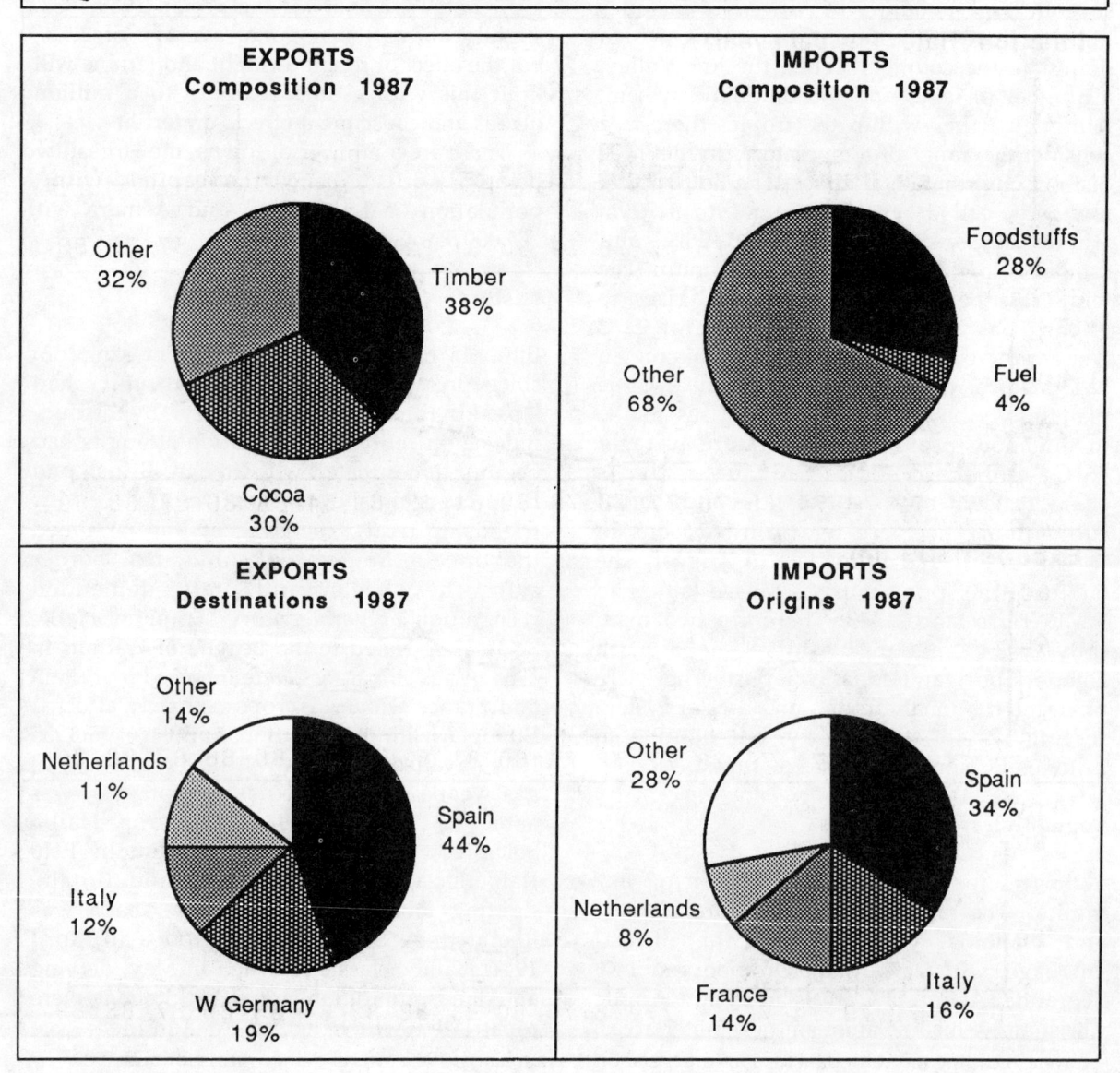

EXPORTS Composition 1987
- Timber 38%
- Cocoa 30%
- Other 32%

IMPORTS Composition 1987
- Foodstuffs 28%
- Fuel 4%
- Other 68%

EXPORTS Destinations 1987
- Spain 44%
- W Germany 19%
- Italy 12%
- Netherlands 11%
- Other 14%

IMPORTS Origins 1987
- Spain 34%
- Italy 16%
- France 14%
- Netherlands 8%
- Other 28%

15 ETHIOPIA

Physical Geography and Climate

Ethiopia, in the East African region, stretches westwards from the Red Sea and is bounded by Sudan, Kenya, Somalia and Djibouti. It is the seventh largest country in Africa and includes high plateau and mountain ranges which dominate the country's relief, the Rift Valley, a number of lakes and the Blue Nile system. Although it lies within the tropics there is a considerable range of temperatures in the three distinct environmental zones. The hot tropical lowlands called 'kwolla' extend to 1,700m where temperatures can exceed 26°C and rainfall is low, less than 50cm per annum. The temperate belt called 'woina dega' rises to 2,400m, has an average temperature of 22°C and a rainfall of 100cm per annum; this region has pasture, forest and Mediterranean type plant life. Above this the 'dega' rises to the mountainous peaks; the temperature averages 16°C and rainfall is adequate, it is characterised by pasturelands and cereal cultivation. There are two rainy seasons in spring (Belg) and from June to August. The failure of the spring rain has caused famine in Wallo-Tigre since 1973. There are two main railways connecting Addis Ababa with Djibouti port and Massawa port. There are three international airports and a road system serving 12% of the country, but fighting in Eritrea and Tigré has caused disruptions.

Population

Estimates for 1989 are a population of 46.7 million. The main ethnic groups are Hamitic and Danakil with Amharic the official language, with 27% Afaan Oromoo and 14% Tegranna. About 45% are Islamic, 40% Coptic Christian with some animism in remoter areas. Population density is well above average for East Africa at 37.3 persons per square kilometre, but densities fall dramatically in the desert and mountains. Urbanisation at 15% is about average for the region. Population grew at 2.4% a year for the period 1980-1987, slightly above the regional average of 2.9%, but the effect of recent drought and famine will limit this with estimates of up to a million deaths and consequent effects on fertility.

There are a number of towns, the largest two being Addis Ababa, the capital, with a population of 1.5 million, and Asmara with 275,000.

History

Ethiopia had been an independent kingdom since pre-classical times, but its unity had lapsed in the seventeenth century. By the mid-nineteenth century reunification attempts had begun but conflicted with Italian, British and Egyptian colonial expansion. Despite a 28-year friendship treaty, the King of Italy assumed the title of Emperor of Abyssinia, border skirmishes began and Italy demanded recognition of her territory. Emperor Hailie Selassie appealed to the League of Nations in 1934. Arms embargoes were imposed by Britain and France which disproportionately affected Ethiopia, with no munitions industry. In 1935 Italy invaded, but made little progress until gas weapons were used. The conquest was achieved in May 1936. Emperor Hailie Selassie went into exile in Britain. In 1940 Italy declared war on France and Britain, fighting took place and Addis Ababa was liberated by South African troops in April 1941. Hailie Selassie returned in May, defying an Allied prohibition and 40,000 Italians were repatriated which led to an administrative

Ethiopia

collapse. In 1942 Ethiopia was recognised as an independent state and became eligible for British aid. In 1952 a UN enquiry recommended that Eritrea be federated with Ethiopia. In 1963 Addis Ababa was chosen as headquarters of the OAU. Between 1972 and 1974 a famine in Wallo and Tigre left 200,000 dead. Following strikes and army mutinies, on 12 September 1974 Hailie Selassie was deposed; he died in detention eleven months later.

A Provisional Military Advisory Council was formed, known as the Dergue, headed by Brig. Gen. Banti and a programme of nationalisation was introduced. In 1977 General Andom was succeeded by Col. Mengistu. There were political conflicts between the All Ethiopia Socialist Movement (MEISON) who want a Soviet-style Communist Party, but accept a military government and the more popular Marxist-Leninist Party (EPRP) who want a people's civil government and support the Eritrean struggle. Purges and executions followed the MEISON victory and a new party was formed in 1978 called Commission for Organizing Party of Working People of Ethiopia, (COPWE) which has a 70% military membership. The civil war in Eritrea erupted again in 1972 and continues to defy resolution. Other groups such as the Somali's, Oromo's, Tigre's and Afar's have organized liberation movements on guerilla lines and in 1977-78 the conflict with Somalia, who supported the Ogaden liberation movement, paralysed economic activity and disrupted communications between Addis Ababa and the ports of Asmara and Djibouti. The famine in Tigre and Wallo reappeared in 1979-81 and 1983 and by February 1985 it affected 8.5 million people, with 6,000 per day dying in March 1985. North-east Ethiopia continues to be severely affected and resistance has been shown to the 'villagisation' resettlement of the population in the west.

Stability

The present regime has been in power since 1974, and Mengistu has been leader since 1977, but political life has been characterised by internal struggles and purges. There have been armed struggles with Eritrean, Somali, Oromo, Tigre and Afar communities, and war with Somalia over the Ogaden. None of these conflicts has been satisfactorily resolved, and they have disrupted economic life and communications in the regions concerned.

In 1974 the development strategy underwent a major change with an increase in public ownership and the nationalisation of all rural and urban land. Rural administration was re-oganised with emphasis on de-centralisation.

Overall, the stability of Ethiopia has been very poor, with the enduring qualities of the present regime more than offset by the continuing civil struggles and the socialist interventions in the economy.

Economic Structure

GDP in 1987 was $US 4,800m, and it is the third largest economy in the region after Sudan and Kenya. GNP per head was estimated at $US 130, and this made Ethiopia the poorest country in the world.

Agriculture is the mainstay of the economy, providing 42% of GDP in 1987, with industry providing 18% and services 40%. This supply structure is fairly representative of the East African region which has negligible mineral resources and relies on agriculture for export revenue.

On the demand side, private consumption was equivalent to 77% of GDP in 1987, gross domestic investment was 14% and government consumption was 19%. The high commitment to private consumption is a reflection of the low income levels generated and the low allocation to investment results from the need to meet immediate consumption needs as a priority in the short run.

Exports generated 11% of GDP in 1987, and this indicates that the economy is rather less open than the average in East Africa. Coffee is the main export, and provides 44% of export earnings, with hides and skins providing 13%. Imports comprise 22% of total expenditure and again this is a lower dependence than in the region or in Africa generally. The gap between imports and exports is filled by foreign assistance from a variety of sources including the World Bank, the EEC, Libya and China.

Literacy among adults was estimated at 15% in 1981, and this is among the lowest literacy levels in Africa, and well below the East African average of 40%. 36% of the relevant age group are in primary education, which is well under the regional average of 62%. 12% of the 12-17 age group were in secondary education, and this is more in line with averages elsewhere in Africa, as is the 1% in higher education. The poor adult literacy and inferior primary enrolments are reflections of the low income generated in the rural areas which contain most of the population.

Life expectancy in 1987 was 47 years, lower than the regional average of 52, food supply was 1,700 calories per person per day, and this is well below the 2,100 a day average in the rest of Africa. Medical provision at one doctor for every 88,000 inhabitants in 1981 was the worst in Africa.

Overall, Ethiopia is a large country in terms of population but with low urbanisation, poor medical and food provision, and very low income per head. It takes only limited advantage of the opportunities offered by world markets, and its economy is heavily dependent on agriculture.

Economic Performance

GDP has grown at 0.9% in the period 1980-87, and this is below the East Africa performance of 1.6% for the same period. With population growing at 2.4% a year, GDP per head contracted at -1.5% annually.

The performance in the major sectors in the early eighties was varied. Agriculture contracted at -2.1% a year, and was badly affected by drought and political instability. On the other hand, the industry sector grew at 3.8% and the services sector at 3.5%, and these two sectoral performances were better than the average elsewhere in the region and in Africa.

Export volumes contracted at -0.6% a year in 1980-87, and this was the result of policies which sought to reduce external dependence and a dogged refusal to depreciate the exchange rate in order to encourage exports. Import volumes grew at 7.6% a year in the same period as a result of the response of the international community to the stress experienced by the population under the impact of drought and armed conflicts.

Inflation figures for 1980-87 show a modest annual rate of price increase of 2.6%, and this is the best record of price stability in East Africa.

Development effort in respect of the ratio of gross domestic investment to GDP was low in that this ratio was 14% in 1987. At this level, there is grave danger of delapidation of the country's infrastructure and capital stock, making high growth rates in the future even more difficult to attain.

Overall, Ethiopia has had a poor economic performance in the early 1980s with GDP per head falling, and agriculture the sector most badly affected. Export volumes have fallen, investment ratios have been low, but prices have been fairly stable.

Recent Economic Developments

Ethiopia continues to shift rural populations, ostensibly from drought hit areas to more fertile zones, with 2.8 million people relocated into new villages. This programme is expected to have an adverse effect on agricultural output over the next few years. The new three-year plan introduced in mid-1987 contains proposals to encourage agricultural output by allowing farmers to sell 60% of their target output on the open market. Finally, the coffee sector, which provides 60% of Ethiopian foreign exchange earnings, and where 97% of putput is grown by small farmers, is to be re-organised. Coffee growing is to be transferred to state-owned collectives, and marketing is to be handled by the government's Ethiopian Coffee Marketing Corporation. Fourteen private sector coffee marketing firms are to be phased out. After the droughts in 1984 and 1985 there appear to have been good rains in 1986 and 1987, and agricultural output has improved considerably, with perhaps a 35% increase in production in 1986. Coffee output is estimated to have risen by 4% in 1988.

The 1986-89 plan is to be evaluated by the World Bank, and contains proposals for joint ventures between foreign investors and the Ethiopian government, with the latter maintaining majority shareholding. The plan calls for $US 402m in grants and $US 1,047m in loans. In view of donor misgivings over the population resettlement schemes, agricultural collectivisation and the state monopoly of

Ethiopia

coffee marketing, these aid levels, which would raise commitments from 7.7% of GDP to almost 11%, are considered optimistic.

Donor support continues to be steady, with EEC funding of $US 218m committed in 1986. Sweden, France, the OPEC group, the World Bank and China have agreed projects in 1987, with the emphasis on the agricultural sector and infrastructure.

An Italian firm is involved in a $US 18m project to rehabilitate the textile sector and Ethiopian airlines are to construct crop-spraying aircraft for the African market. International oil companies have shown little interest in taking up oil concessions, with only a single bid for one of 24 concessions on offer in January 1987.

Economic Outlook and Forecasts

The forecasts assume that there will be no improvement in the political stability of Ethiopia, and that there will be only limited moves to introduce economic reforms. Such reforms as have been introduced are expected to assist the export sector to recover.

The forecasts project growth rates for GDP between 1.1% and 1.3% a year, which will imply continuing falls in GDP per head. Inflation is expected to remain modest, although there will be a sharp acceleration if economic reforms include depreciation of the exchange rate. Exports are expected to show some recovery. Export recovery in conjunction with continuing external assistance in reponse to the stress from drought, locust plagues and armed conflicts, will allow import volumes to rise steadily.

Ethiopia: Economic Forecasts
(average annual percentage change)

	Actual	Forecast Base	Forecast High
	(1980-87)	1990-95	1990-95
GDP	0.9	1.1	1.3
GDP per Head	-1.5	-2.0	-1.8
Inflation Rate	2.6	6.0	4.0
Export Volumes	-0.6	4.0	5.0
Import Volumes	7.6	3.0	3.5

ETHIOPIA: Comparative Data

	ETHIOPIA	EAST AFRICA	AFRICA	INDUSTRIAL COUNTRIES
POPULATION & LAND				
Population, mid year, millions, 1989	46.7	12.2	10.2	40.0
Urban Population, %, 1985	15	30.5	30	75
Population Growth Rate, % per year, 1980-86	2.4	3.1	3.1	0.8
Land Area, thou. sq. kilom.	1,222	486	486	1,628
Population Density, persons per sq kilom., 1988	37.3	24.2	20.4	24.3
ECONOMY: PRODUCTION & INCOME				
GDP, $US millions, 1986	4,960	2,650	3,561	550,099
GNP per head, $US, 1986	120	250	389	12,960
ECONOMY: SUPPLY STRUCTURE				
Agriculture, % of GDP, 1986	48	43	35	3
Industry, % of GDP, 1986	15	15	27	35
Services, % of GDP, 1986	36	42	38	61
ECONOMY: DEMAND STRUCTURE				
Private Consumption, % of GDP, 1986	80	77	73	62
Gross Domestic Investment, % of GDP, 1986	9	16	16	21
Government Consumption, % of GDP, 1986	17	15	14	17
Exports, % of GDP, 1986	13	16	23	17
Imports, % of GDP, 1986	20	24	26	17
ECONOMY: PERFORMANCE				
GDP growth, % per year, 1980-86	0.8	1.6	-0.6	2.5
GDP per head growth, % per year, 1980-86	-1.6	-1.7	-3.7	1.7
Agriculture growth, % per year, 1980-86	-3.9	1.1	0.0	2.5
Industry growth, % per year, 1980-86	3.8	1.1	-1.0	2.5
Services growth, % per year, 1980-86	5.1	2.5	-0.5	2.6
Exports growth, % per year, 1980-86	-2.5	0.7	-1.9	3.3
Imports growth, % per year, 1980-86	10.7	0.2	-6.9	4.3
ECONOMY: OTHER				
Inflation Rate, % per year, 1980-86	3.4	23.6	16.7	5.3
Aid, net inflow, % of GDP, 1986	11.5	11.5	6.3	-
Debt Service, % of Exports, 1986	25.8	18	20.6	-
Budget Surplus (+), Deficit (-), % of GDP, 1980	-4.5	-3.0	-2.8	-5.1
EDUCATION				
Primary, % of 6-11 group, 1985	36	62	76	102
Secondary, % of 12-17 group, 1985	12	15	22	93
Higher, % of 20-24 group, 1985	1	1.2	1.9	39
Adult Literacy Rate, %, 1981	15	41	39	99
HEALTH & NUTRITION				
Life Expectancy, years, 1986	46	50	50	76
Calorie Supply, daily per head, 1985	1708	2,111	2,096	3,357
Population per doctor, 1981	88,150	35,986	24,185	550

Notes: 'Southern Africa' and 'Africa' exclude South Africa. Dates are for the country in question, and do not always correspond with the Regional, African and Industrial averages.

ETHIOPIA: Leading Indicators

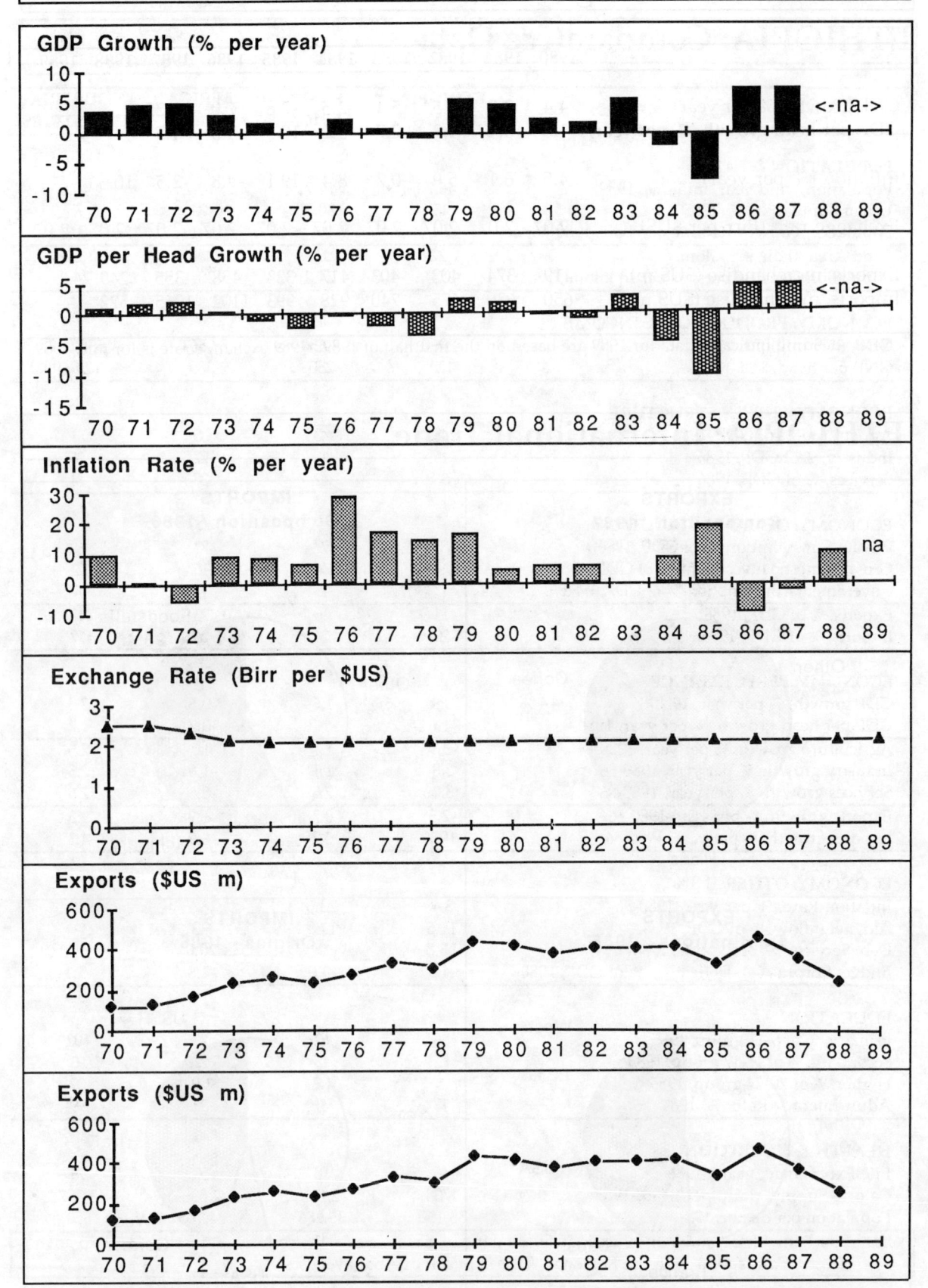

GDP Growth (% per year)

GDP per Head Growth (% per year)

Inflation Rate (% per year)

Exchange Rate (Birr per $US)

Exports ($US m)

Exports ($US m)

Note: Leading indicator data for 1989 are based on the first half of 1989. 1989 exchange rate is for mid-1989.

ETHIOPIA: Leading Indicator Data

	1980	1981	1982	1983	1984	1985	1986	1987	1988	1989
GDP growth (% per year)	4.4	2.2	1.5	5.4	-2.5	-7.8	7.0	7.25		
GDP per head growth (% per year)	1.6	-0.4	-1.2	2.7	-5.2	-10.5	4.3	4.5		
Inflation (% per year)	4.5	6.1	5.9	-0.7	8.4	19.1	-9.8	-2.5	10.66	
Exchange rate (Birr per $US)	2.07	2.07	2.07	2.07	2.07	2.07	2.07	2.07	2.07	2.07
Exports, merchandise ($US m)	419	374	403	403	417	333	458	355	249	
Imports, merchandise ($US m)	650	630	675	740	928	993	1102	1065	522	

Note: Leading indicator data for 1989 are based on the first half of 1989. 1989 exchange rate is for mid-1989.

ETHIOPIA: International Trade

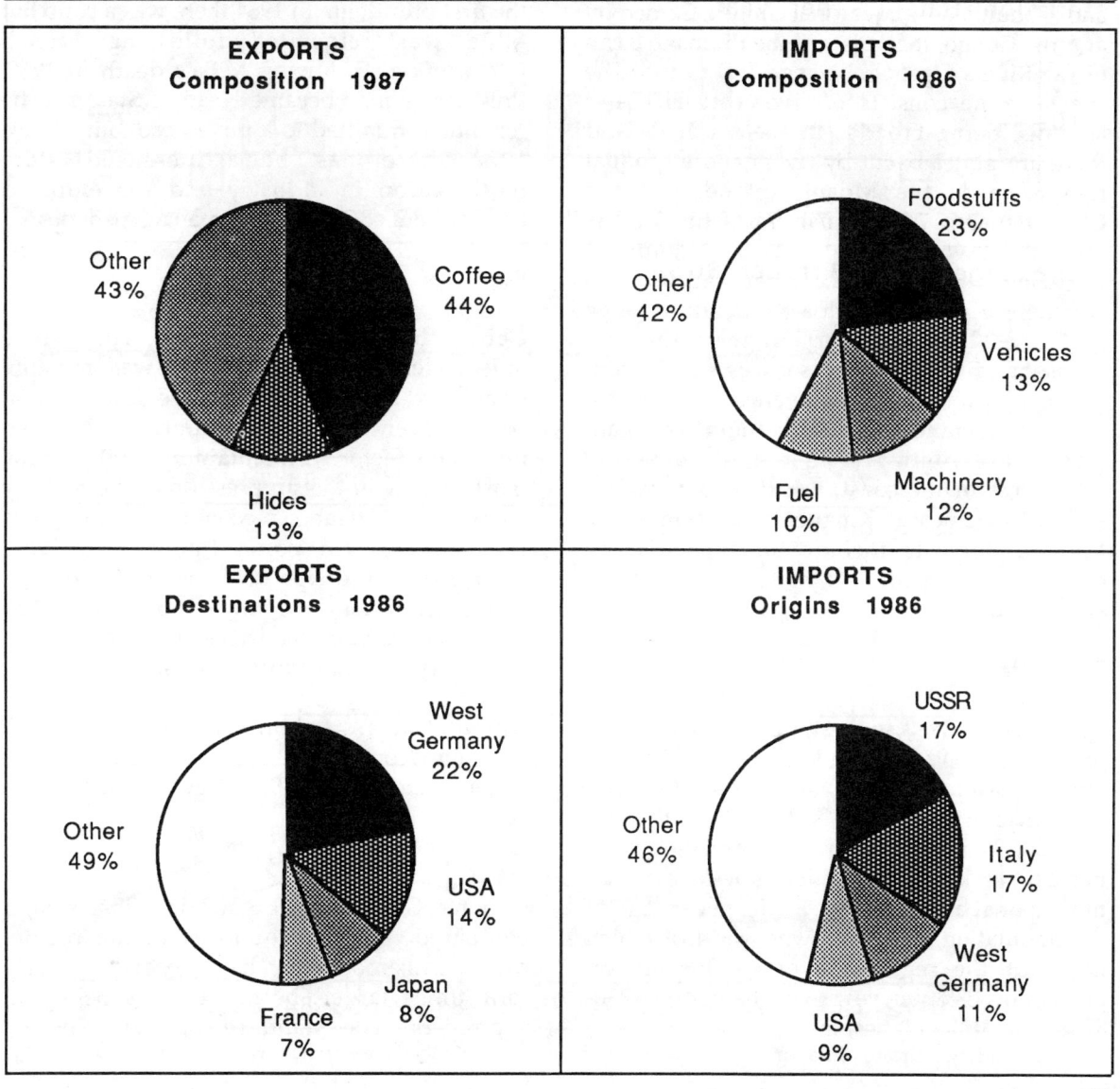

EXPORTS Composition 1987
- Coffee 44%
- Other 43%
- Hides 13%

IMPORTS Composition 1986
- Foodstuffs 23%
- Vehicles 13%
- Machinery 12%
- Fuel 10%
- Other 42%

EXPORTS Destinations 1986
- West Germany 22%
- USA 14%
- Japan 8%
- France 7%
- Other 49%

IMPORTS Origins 1986
- USSR 17%
- Italy 17%
- West Germany 11%
- USA 9%
- Other 46%

16 GABON

Physical Geography and Climate

Gabon is a small coastal state in the Central African region which straddles the Equator and is bounded by Equatorial Guinea, Cameroon and the Congo. It consists of the drainage basin of the Rivers Ogooué, Nyanga and Como. The coastal zone consists of river estuaries and lagoons, being broader in the south. Inland there are plateaux cut by rivers and a gradual rise towards the Moabi upland and the Chaillu massif. Three quarters of the land is rain forest with coastal grassland south of Port-Gentil and in some river valleys. The climate is equatorial with a temperature range of 21°C-32°C and very high humidity. Monsoons give two wet seasons, especially from October to mid-May and average rainfall can be from 150cm to 300cm per annum. The main rivers are only navigable within 150 kilometres of the coast, but there are hydro-electric schemes at Kinguélé and Tchimbélé. Mineral deposits include on and offshore oilfields, gold, diamonds, manganese, iron ore and uranium.

Population

Estimates for 1989 suggest a population of 1.1 million. Around 50% are Roman Catholic, 40% follow tradtional beliefs, and 1% are Moslem. The official language is French with 35% speaking Fang. The density of population is 4.1 per square kilometre, which is less than half the regional average. In 1987 the percentage of the population living in towns was 43% which is about the regional average. Population growth of 4.3% per year over the period 1980-87 is one of the highest in Africa.

History

Gabon was a former French colony which gained independence in 1960, with Leo M'Ba the first president. In 1964 there was a coup but M'Ba was reinstated following French intervention. Following M'Ba's death in 1967, President Bongo became Head of State and he remains committed to a mixed economy. Since 1972 there has been increased state participation in industry and the enforced employment of Gabonese in management posts.

Stability

Gabon has retained civilian rule since independence in 1960, and power was smoothly transferred to Omar Bongo on the death of Leon M'Ba. French military support has been an important factor in maintaining civilian rule, most notably in the suppression of the military coup in 1964. Gabon's economic strategy has involved a firm commitment to private enterprise and foreign investment, particularly in the oil sector. Membership of the Franc Zone has provided the background to cautious budgetary and monetary policies.

Overall, Gabon's stability has been very good, and minor coup and assassination alarms do not seriously threaten the Bongo government.

Economic Structure

Gabon's GDP of $US 3,500 in 1987 is high compared to other countries in the region, except Cameroon which has an economy nearly four times larger, both countries being oil producers. The standard of living is the highest in Africa with a GNP per head of $US 2,700 in 1987, and this makes Gabon an upper-middle-income economy according to the World Bank.

The agricultural sector is one of the smallest in Africa and accounts for only 11% of GDP while the industrial sector, which includes oil, generated 41%, one of the heaviest dependencies on this sector in Africa. The service sector is also large as a result of oil revenues and accounts for 48% of GDP.

There is a comparatively low proportion of income allocated to private consumption which absorbed only 43% of GDP, while investment took 31%, the second highest allocation in Africa, after Somalia. Government consumption accounted for a further 23%.

Exports are about 41% of GDP, while expenditure on imports is equal to 38% of GDP. The oil industry continues to be affected by the world market but in 1988 made up 62% of exports while the remainder consisted mainly of only three other exports; timber products 16%, manganese 10% and uranium 7%. Main imports are food and machinery, 22% each, although the government austerity measures aim to reduce reliance on food imports. The high proportion of machinery is needed for the oil and mining industry. The completion of the Transgabonaise railway should reduce the 16% of imports which comprise transport equipment.

Provision of education is good and has steadily improved since independence. The proportion of the primary age group in school is 126% (this being accounted for by adult education), while 27% of the secondary age group were in school. Education up to the age of 16 is supposed to be compulsory and free. A large 4% (the highest in Africa) of 20-24 year age group were in higher education as Gabon has its own university and polytechnic, with a considerable number in higher education abroad, with oil money available to pay for this. Gabon is trying to replace its large expatriate population with local employees.

Health provision is good as a result of considerable investment, with life expectancy of 52 years. There are about 2,800 people per doctor, one of the best ratios in Africa. The ratio of nurses per population is also very good with a ratio of one to every 270 people.

Overall, Gabon has a small population but has exceptional resources which have allowed the economy to develop with a fairly strong infrastructure. The structural adjustment programme may encourage diversification, for example food import substitution. A high level of investment has resulted in health and education bettered by few in sub-Saharan Africa.

Economic Performance

From 1980-87 GDP grew by only 0.6%, well below the regional average of 4.2%. This implies a fall in GDP per head of -3.7% a year. From 1973 to 1983 the agricultural sector contracted by -2.2%, while the industrial sector grew by 1.7%, with 2.5% growth in the service sector. Updated figures are not available, but the industrial sector which includes oil and mining, has developed substantially since 1983, despite falling oil prices, with services growth also growing as a result. The agricultural sector has expanded in an effort to reduce dependence on food imports.

Exports fell by -1.9% during the 1980-87 period while regional growth was 4.5%. Imports grew by 3.0% though in the region growth was limited to 0.4%.

The inflation rate in Gabon during this period was only 2.6% when regional and African inflation reached about 17%.

Overall, Gabon has found it difficult to maintain economic performance in the 1980s with GDP per head falling and slower trade growth, although the inflation record has been good.

Recent Economic Developments

In December of 1986 Gabon signed an agreement with the IMF which released a $US 126m standby credit for balance of payments support. The most recent IMF standby credit (September 1989) was for $US 55.1m in support of the structural adjustment programme. In August 1987 it was announced that the World Bank would provide a Structural Adjustment loan which would come into force in October and this was followed by a further $US 50m in 1988. The main measures to be implemented as

Gabon

part of the agreements are to diversify the economy away from dependence on oil, increase producer prices in the agriculture sector to encourage production, and to reform the parastatal sector.

Gabon has been severely hit by the fall in oil prices in 1986, and the resulting cuts in government investment spending compounding the declines in GDP. An effort has been made to compensate for lower oil prices by expanding output, and production of oil increased by 32% in 1989. In the longer term, the prospects for the economy have been improved by the discovery of new oil deposits totalling 80m tonnes, equal to the present level of proven reserves.

Inflation has been contained by austerity measures, and the annual rate of price increase in the three years 1987-1989 has fallen on average by -2.0% a year. In the four years 1980 to 1983, inflation averaged 12 per cent annually.

Gabon has always run a substantial surplus on the balance of trade, with imports being only a third of export earnings. With export earnings running at about $US 2bn up to the mid 1980s, a deficit on in visible earnings led to a current account surplus of around $US 100m in 1984. With oil prices halving in 1986 from their 1984 level, a substantial current account deficit has emerged which Gabon is financing partly from accumulated reserves, by debt rescheduling and by borrowing from the IMF and World Bank.

In September of 1986, Gabon suspended payment on its external debt, which is now estimated at $US 2.6b in 1989. Two-thirds of this debt is thought to be with commercial banks, accumulated during the 1970s when Gabon's oil revenues made it attractive to commercial lenders. In January of 1987, the Paris Club of creditor governments rescheduled $US 350m over a ten-year term, with a four-year grace period, and there was a further Paris Club re-scheduling in September 1989. In June 1987, the London Club of commercial creditors is reported to have rescheduled $US 70m of debt, on similar terms to the Paris Club.

As a member of the Franc Zone, Gabon has seen the currency appreciate steadily against the US dollar, from CFA 449.3 = $US 1 in 1985, to CFA 289.3 = $US 1 at the end of 1989. This currency appreciation reduces prices for non-oil exporters and works against Gabon's intentions to diversify the economy, and it is possible that the CFA franc will be devalued against the French franc.

Economic Outlook and Forecasts

The forecasts assume that Gabon's good stability record is maintained, and that the government presses on with the economic reform programme.

Expanded output in the oil sector is expected to stimulate GDP growth to above 5% a year up to 1995. This will allow increases in GDP per head of around 3.0% a year or better. Export volumes are expected to rise rapidly with the expansion of oil production, but import volume growth to be much slower. Inflation is expected to remain at below 5% a year.

Gabon : Economic Forecasts
(average annual percentage change)

	Actual	Forecast Base	High
	1980-87	1990-95	1990-95
GDP	0.6	5.5	6.4
GDP per Head	-3.7	2.9	3.8
Inflation Rate	2.6	4.5	3.7
Export Volumes	-1.9	15.0	17.0
Import Volumes	3.0	2.0	3.3

GABON: Comparative Data

	GABON	CENTRAL AFRICA	AFRICA	INDUSTRIAL COUNTRIES
POPULATION & LAND				
Population, mid year, millions, 1989	1.1	7.3	10.2	40.0
Urban Population, %, 1985	38.8	38.6	30	75
Population Growth Rate, % per year, 1980-86	4.4	3.0	3.1	0.8
Land Area, thou. sq. kilom.	262	638	486	1,628
Population Density, persons per sq kilom., 1988	4.1	11.1	20.4	24.3
ECONOMY: PRODUCTION & INCOME				
GDP, $US millions, 1986	3,190	4,146	3,561	550,099
GNP per head, $US, 1986	3,080	395	389	12,960
ECONOMY: SUPPLY STRUCTURE				
Agriculture, % of GDP, 1986	10	18	35	3
Industry, % of GDP, 1986	35	41	27	35
Services, % of GDP, 1986	55	41	38	61
ECONOMY: DEMAND STRUCTURE				
Private Consumption, % of GDP, 1986	55	62	73	62
Gross Domestic Investment, % of GDP, 1986	37	25	16	21
Government Consumption, % of GDP, 1986	26	14	14	17
Exports, % of GDP, 1986	37	35	23	17
Imports, % of GDP, 1986	55	35	26	17
ECONOMY: PERFORMANCE				
GDP growth, % per year, 1980-86	1.5	4.2	-0.6	2.5
GDP per head growth, % per year, 1980-86	-2.8	-0.8	-3.7	1.7
Agriculture growth, % per year, 1973-82	-2.2	0.5	0.0	2.5
Industry growth, % per year, 1973-82	1.7	7.9	-1.0	2.5
Services growth, % per year, 1973-82	2.5	3.2	-0.5	2.6
Exports growth, % per year, 1980-86	-0.6	4.5	-1.9	3.3
Imports growth, % per year, 1980-86	3.1	0.4	-6.9	4.3
ECONOMY: OTHER				
Inflation Rate, % per year, 1980-86	4.8	17	16.7	5.3
Aid, net inflow, % of GDP, 1986	2.7	5.4	6.3	-
Debt Service, % of Exports, 1986	17.5	28	20.6	-
Budget Surplus (+), Deficit (-), % of GDP, 1985	0.1	-1.4	-2.8	-5.1
EDUCATION				
Primary, % of 6-11 group, 1984	123	93	76	102
Secondary, % of 12-17 group, 1984	25	42	22	93
Higher, % of 20-24 group, 1984	4	1.9	1.9	39
Adult Literacy Rate, %, 1974	12	52	39	99
HEALTH & NUTRITION				
Life Expectancy, years, 1986	52	52	50	76
Calorie Supply, daily per head, 1985	2,448	2,115	2,096	3,357
Population per doctor, 1981	2,550	13,835	24,185	550

Notes: 'Southern Africa' and 'Africa' exclude South Africa. Dates are for the country in question, and do not always correspond with the Regional, African and Industrial averages.

GABON: Leading Indicators

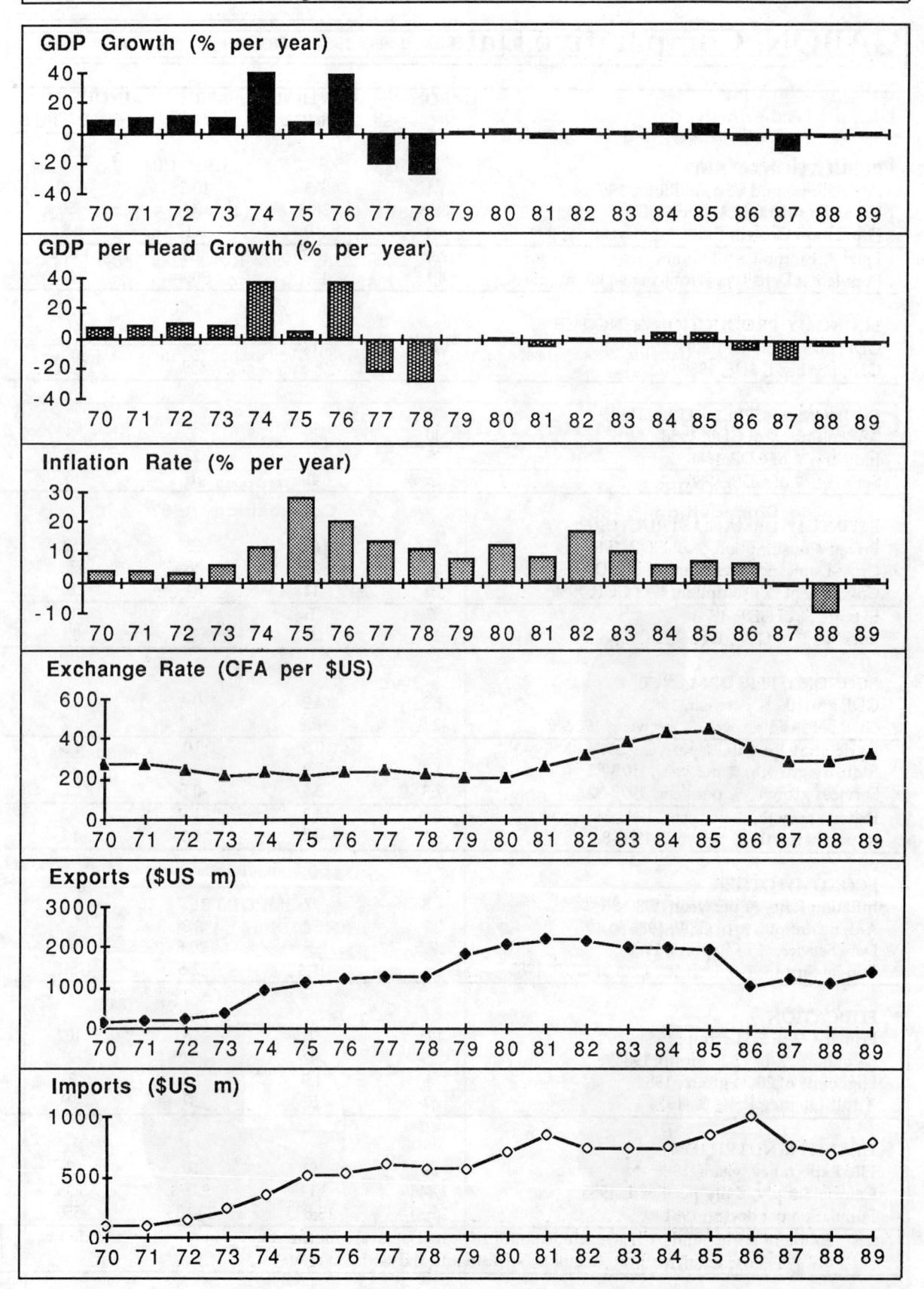

Note: Leading indicator data for 1989 are based on the first half of 1989. 1989 exchange rate is for mid-1989.

GABON: Leading Indicator Data

	1980	1981	1982	1983	1984	1985	1986	1987	1988	1989
GDP growth (% per year)	2.5	-4.0	2.7	0.9	6.4	6.3	-5.6	-12.0	-3.0	-0.5
GDP per head growth (% per year)	0.7	-5.8	0.9	-0.9	4.6	4.5	-7.4	-13.8	-4.9	-2.4
Inflation (% per year)	12.3	8.7	16.7	10.4	5.8	7.3	6.3	-1.0	-9.6	1.0
Exchange rate (CFA per $US)	211.3	271.7	328.6	381.1	437.0	449.3	346.3	300.5	297.9	327.3
Exports, merchandise ($US m)	2084	2200	2160	2000	2018	1952	1074	1286	1184	147
Imports, merchandise ($US m)	686	841	723	726	733	842	979	732	678	770

Note: Leading indicator data for 1989 are based on the first half of 1989. 1989 exchange rate is for mid-1989.

GABON: International Trade

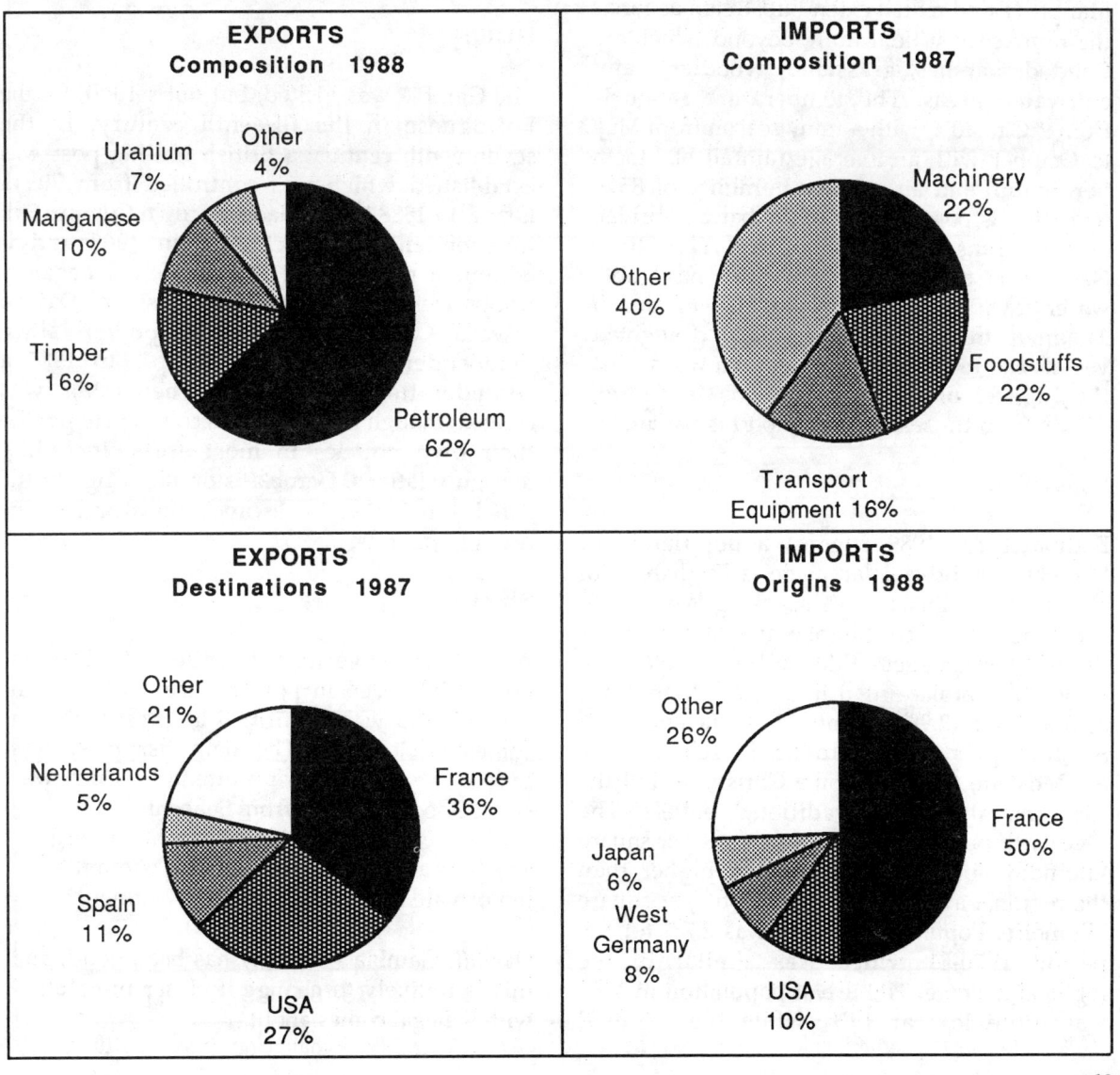

EXPORTS Composition 1988
- Petroleum 62%
- Timber 16%
- Manganese 10%
- Uranium 7%
- Other 4%

IMPORTS Composition 1987
- Machinery 22%
- Foodstuffs 22%
- Transport Equipment 16%
- Other 40%

EXPORTS Destinations 1987
- France 36%
- USA 27%
- Spain 11%
- Netherlands 5%
- Other 21%

IMPORTS Origins 1988
- France 50%
- USA 10%
- West Germany 8%
- Japan 6%
- Other 26%

17 GAMBIA

Physical Geography and Climate

Gambia, in the West African region, is a small enclave on the Atlantic coast, covering an area of 11,000 sq km, entirely surrounded by Senegal. It consists of the valley of the River Gambia. Mangrove and marsh extend upstream as far as the rainy season tidal limit, beyond which are flooded marshes, grassland, woodland and cultivated areas. The temperature range is from 9°C to 40°C with a rainy season from May to October with an average rainfall of 115cm per annum and an average humidity of 85%. Irrigation is possible due to the barrage-bridge of the Trans-Gambia Highway. The River Gambia is one of Africa's more navigable waterways; ocean-going vessels can reach 240km upstream to Kuntaur; smaller draughted vessels can reach 283km to Georgetown; whilst the highest of the wharf towns is Fatoto, 464km from the sea. The main port is Banjul.

Population

Estimates for 1989 suggest a population of 837,000. The official language is English, with 18% speaking Pulaar; 16% speaking Wolof; 10% speaking Joola with Sininke and Mandinka as minority languages. Ethnically in 1963 the population broke down into 40.8% Mandinka; 13.5% Fula; 12.9% Wolof; 7% Joola and 6.7% Serahuli, plus smaller groups. By religion 85% are Moslem, there are some Christians and the Joola mainly follow traditional beliefs. The density of population was 74 persons per square kilometre for 1988, considerably higher than the regional average of 31.1 persons per square kilometre. Population growth was 2.7% for the period 1970-81 which was similar to the regional average. The urban population in 1981 was quite low at 19%, while the regional average is now about 29%. However, with the development of the service sector, especially tourism, increased urban drift can be expected.

The population of Banjul, the capital, was 80,000 in 1973 at the time of the last census, but is probably close to 200,000 in 1990.

History

The Gambia was visited, but not settled, by the Portuguese in the fifteenth century. In the seventeenth century a British trading post was established which was controlled from Sierra Leone. In 1888 it was made a Crown Colony. Full independence was achieved in 1965 and it became a republic in 1970. In 1981 Senegalese troops were called in to help President Dawda Jawara, who has been in power since independence, suppress a coup. In 1982 a confederation known as Senegambia was created, though both countries continue to pursue their own policies in most areas, including foreign relations. Gambia is on the fringe of the Sahel, but has so far escaped the worst of the drought periods.

Stability

A civilian government under Sir Dawda Jawara has been in power since 1965, and in 1982 Jawara was confirmed as president in a contested election. The main disruption was caused by the 1981 coup attempt, which was suppressed with help from Senegal.

The economic strategy has not undergone any radical changes, and remains committed to the private sector and foreign investment.

Overall, Gambia's stability has been good, and this is unlikely to change if closer integration with Senegal comes about.

Economic Structure

Recent figures are not available for the Gambia for comparitive purposes. GDP in 1984 stood at $187m, and although this makes it one of the smallest economies, it is also one of the smallest countries in Africa. The standard of living is low with a GNP per capita of $US 220 in 1987 and 80% of the population are employed in the agricultural sector.

In 1982, only 32% of GDP was generated by the agricultural sector which is low by African standards. The industrial sector generated a further 15% which is also low as there is very little manufacturing or mining. The service sector however was about the largest in the region at 54% and has continued to grow as tourism and transport services have expanded in the last few years.

On the demand side, in 1981 private consumption was around 76% which is about average for the region, though government consumption at 28% was high. Investment was higher than in most other African countries at about 25%.

Exports were about 37% of GDP while imports took a large 66% in 1981. The main exports are agricultural products , especially groundnuts which make up 85% of merchandise exports. However, tourism is also a large foreign exchange earner. Food produce make up the largest proportion of imports (39%) followed by manufactured goods (23%) and machinery (21%).

Education levels are about average for the region with 73% of primary age children in school and 21% of the secondary age group. The adult literacy rate is low however, at only 20%.

Life expectancy is relatively low, at 43 years. The ratio of 11,470 people per doctor is better than the regional average, as is the ratio of 1,650 people per nurse.

Overall, with limited natural resources and a low level of industrialisation, Gambia needs to develop its agricultural sector toward food self-sufficiency and though it can continue to develop tourism, caution will be needed to prevent over-reliance on a changeable market. Confederation with Senegal has given more scope for a market for any industrial development.

Economic Performance

GDP grew at an estimated 3.1% a year in the period 1980-89, and this is rather better than the rest of the region and in Africa generally, where GDP declined. The period did exhibit considerable swings in growth performance, with negative GDP expansion in 1981, 1984 and 1985 being followed by years of good recovery. GDP per head has expanded modestly at 0.4% a year in the 1980s.

Sectoral growth rate data is only available for the 1970s, when the leading sector was industry with a 7.4% a year growth rate, followed by services with 4.4% a year and agriculture, at 3.2%, just keeping ahead of the expansion of the population. In the 1980s, it is expected that the services sector, with vigorous expansion of tourism, will have been the leading sector.

Export volumes expanded at 3.3% a year in the 1970s, and this was markedly better than the rest of Africa in the same period, when export volumes contracted. Import volumes, enabled by good earnings from invisible exports, expanded at 7.1% a year in the 1970s, and this was about twice the rate of expansion in Africa in the same period.

Inflation has averaged 17.2% a year in the 1980-89 period, and this is slightly above the Africa average. However, the figures are influenced by a sharp rise in prices of over 50% in 1986 when the currency was floated, since which time they have come more under control, averaging 14% for 1987-89.

Overall, Gambia has had a creditable growth performance in the 1980s when increases in GDP per head have been the exception rather than the rule. Industrial and service sector expansion have been above average, as has trade performance. Gambia's inflation record has been fairly typical of Africa, high by industrial country standards, bur manageable.

Recent Economic Developments

In 1986 Gambia concluded an agreement with the World Bank for a $US 37m Structural Adjustment loan. A wide-ranging set of economic reforms were instituted including raising producer prices for groundnuts, rice and cotton, and ending subsidies on agricultural

Gambia

inputs and fuels. The monopoly of the import of rice by the Gambia Produce Marketing Board was ended, and a programme to overhaul marketing boards, public utilities and ports was begun. Bank interest rates have risen 50 per cent, and equity in banks has been sold to private interests. The currency has been floated since the beginning of 1986.

The IMF granted an enhanced structural adjustment facility of $US 26m over three years in 1988. A second World Bank structural adjustment agreement was made in 1989, and the first instalment paid in April.

Economic performance in Gambia is heavily dependent on climatic factors which affect agricultural export volumes, particularly groundnuts. After increases in GDP and living standards in the 1970s, this performance has been maintained in the 1980s.

Gambia's merchandise trade was in balance until the mid-seventies, but at present inport values are 60% greater than exports. Despite improvements in tourism receipts, there is a deficit in invisibles, and the gap in the current account is financed by substantial aid flows. Gambia's low income status did not allow it to accumulate commercial debt in the 1970s, and debt servicing takes up around 8 per cent of export earnings.

The currency, the dalasi, was floated in January 1986, and the currency immediately depreciated by 44%. This depreciation has contributed to the inflationary pressure which drove up prices by over 50% in 1986.

Recent budgets have projected zero deficits, and this has led to the improving price stability.

Gambia's low-income status has led to considerable aid flows on concessionary terms, and lending to Gambia has increased with the adoption of the reform programme in 1986. France has provided $US 19m for a French firm to extend the telephone network, and the World Bank, Italy and the Netherlands are contributing $US 20.6m for health services. The World Bank, the African Development Fund and the European Development Fund are involved in a $US 19.4m water supply scheme, while Japan is donating $US3.5m for ferry services, and Saudi Arabia is contributing $US3.2m to general balance of payments support. The World Bank is providing $US 12m for loans to the private sector, USAID is spending $US 6m to improve economic and fiscal services, and the African Development Bank is involved in $US 2.4m of improvements in groundnut processing.

Economic Outlook and Forecasts

Gambia will always be vulnerable to climatic changes, particularly in agriculture. But assuming continuing good stability, the perceived success of the 1986 economic reforms should sustain GDP growth above the rate of population increase. Inflation performance is expected to improve and be close to single figures. Export volumes are forecast to continue to expand, with import volumes rising more rapidly as a result of tourist sector earnings and aid flows.

Gambia: Economic Forecasts
(average annual percentage change)

	Actual 1980-87	Forecast Base 1990-95	Forecast High 1990-95
GDP	3.1	4.2	4.5
GDP per Head	0.4	1.4	1.7
Inflation Rate	17.2	8.5	10.5
Export Volumes	3.3[a]	3.0	4.5
Import Volumes	7.1[a]	5.5	6.5

Note: [a] 1970 - 81.

GAMBIA: Comparative Data

	GAMBIA	WEST AFRICA	AFRICA	INDUSTRIAL COUNTRIES
POPULATION & LAND				
Population, mid year, millions, 1989	0.837	12.3	10.2	40.0
Urban Population, %, 1981	19	28.7	30	75
Population Growth Rate, % per year, 1970-81	2.7	3.2	3.1	0.8
Land Area, thou. sq. kilom.	11	384	486	1,628
Population Density, persons per sq kilom.,1988	74.1	31.1	20.4	24.3
ECONOMY: PRODUCTION & INCOME				
GDP, $US millions, 1984	187	4,876	3,561	550,099
GNP per head, $US, 1986	230	510	389	12,960
ECONOMY: SUPPLY STRUCTURE				
Agriculture, % of GDP, 1982	32	40	35	3
Industry, % of GDP, 1982	15	26	27	35
Services, % of GDP, 1982	54	34	38	61
ECONOMY: DEMAND STRUCTURE				
Private Consumption, % of GDP, 1981	76	77	73	62
Gross Domestic Investment, % of GDP, 1981	25	13	16	21
Government Consumption, % of GDP, 1981	28	13	14	17
Exports, % of GDP, 1981	37	18	23	17
Imports, % of GDP, 1981	66	20	26	17
ECONOMY: PERFORMANCE				
GDP growth, % per year, 1970-81	4.5	-2.03	-0.6	2.5
GDP per head growth, % per year, 1970-81	1.8	-4.7	-3.7	1.7
Agriculture growth, % per year, 1970-81	3.2	1.3	0.0	2.5
Industry growth, % per year, 1970-81	7.4	-3.6	-1.0	2.5
Services growth, % per year, 1970-81	4.4	-2.5	-0.5	2.6
Exports growth, % per year, 1970-81	3.3	-4.0	-1.9	3.3
Imports growth, % per year, 1970-81	7.1	-12.9	-6.9	4.3
ECONOMY: OTHER				
Inflation Rate, % per year, 1980-86	32.9	13.0	16.7	5.3
Aid, net inflow, % of GDP, 1984	5.3	3.7	6.3	-
Debt Service, % of Exports, 1983	8.3	21.4	20.6	-
Budget Surplus (+), Deficit (-), % of GDP, 1982	-7.0	-3.4	-2.8	-5.1
EDUCATION				
Primary, % of 6-11 group, 1984	73	75	76	102
Secondary, % of 12-17 group, 1984	21	24	22	93
Higher, % of 20-24 group, 1984	21	2.6	1.9	39
Adult Literacy Rate, %, 1982	20	30	39	99
HEALTH & NUTRITION				
Life Expectancy, years, 1986	39	50	50	76
Calorie Supply, daily per head, 1980	2,190	2,105	2,096	3,357
Population per doctor, 1978	11,470	16,199	24,185	550

Notes: 'Southern Africa' and 'Africa' exclude South Africa. Dates are for the country in question, and do not always correspond with the Regional, African and Industrial averages.

GAMBIA: Leading Indicators

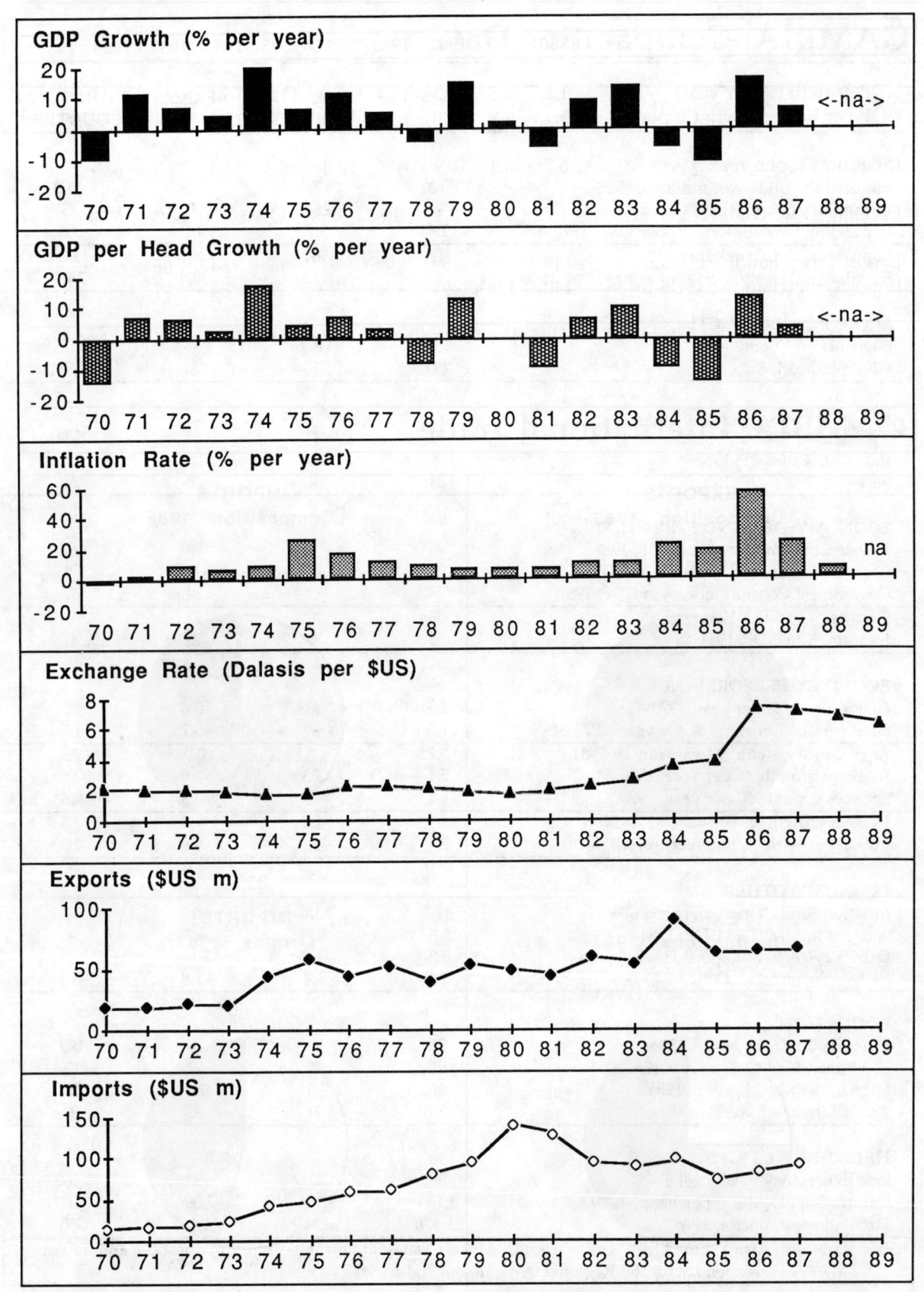

GDP Growth (% per year)

GDP per Head Growth (% per year)

Inflation Rate (% per year)

Exchange Rate (Dalasis per $US)

Exports ($US m)

Imports ($US m)

Note: Leading indicator data for 1989 are based on the first half of 1989. 1989 exchange rate is for mid-1989.

GAMBIA: Leading Indicator Data

	1980	1981	1982	1983	1984	1985	1986	1987	1988	1989
GDP growth (% per year)	1.8	-6.5	9.2	13.4	-6.9	-11.2	16.4	6.6		
GDP per head growth (% per year)	-1.1	-9.8	6.0	10.2	-10.1	-14.4	13.2	3.4		
Inflation (% per year)	6.7	6.1	10.9	10.6	22.1	18.3	56.3	23.5	6.3	
Exchange rate (Dalasis per $US)	1.72	1.97	2.29	2.63	3.58	3.86	7.30	7.06	6.70	6.23
Exports, merchandise ($US m)	49	45	59	55	91	63	64	66		
Imports, merchandise ($US m)	140	129	94	90	99	75	84	91		

Note: Leading indicator data for 1989 are based on the first half of 1989. 1989 exchange rate is for mid-1989.

GAMBIA: International Trade

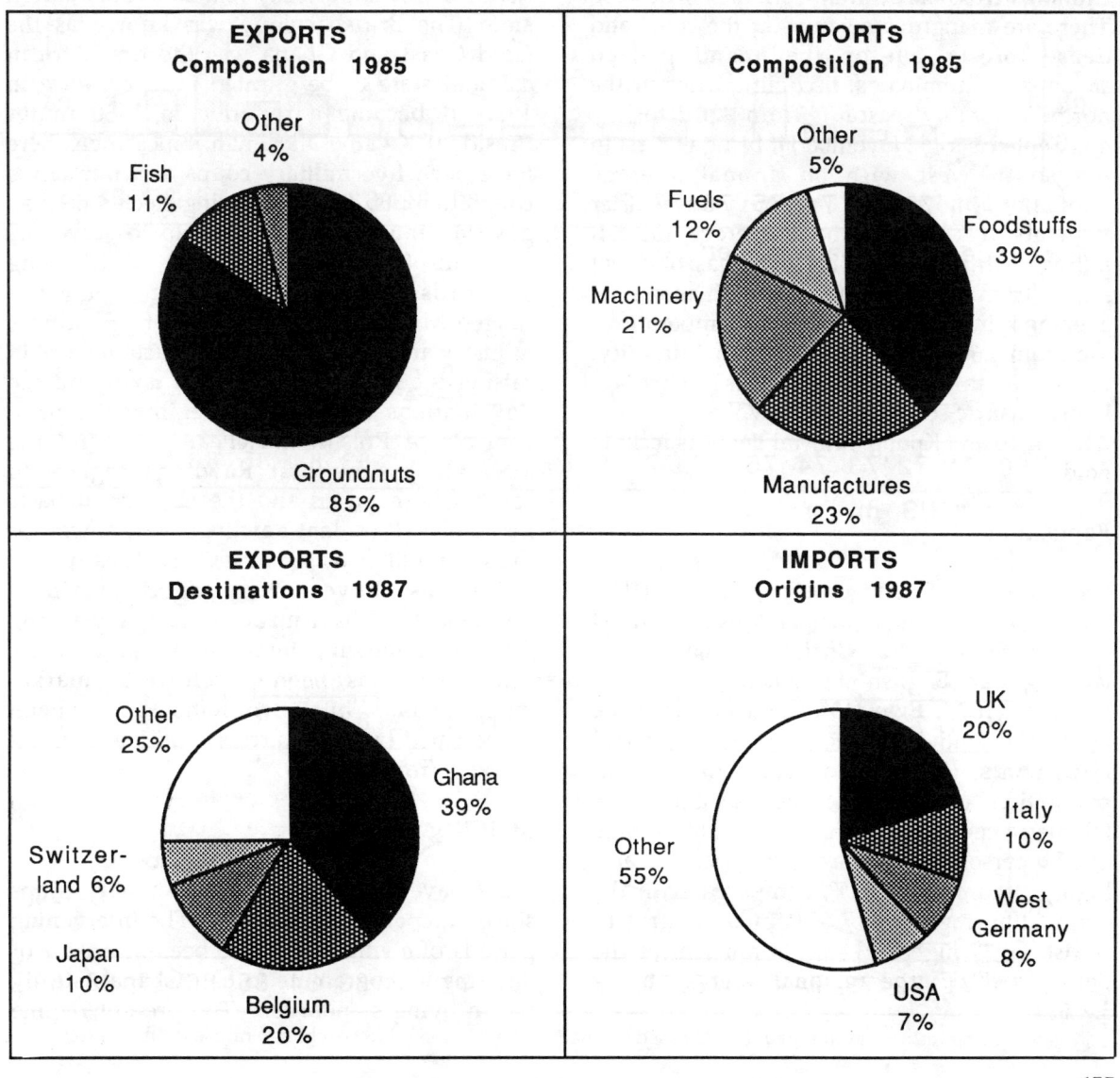

EXPORTS Composition 1985
- Other 4%
- Fish 11%
- Groundnuts 85%

IMPORTS Composition 1985
- Other 5%
- Fuels 12%
- Machinery 21%
- Foodstuffs 39%
- Manufactures 23%

EXPORTS Destinations 1987
- Other 25%
- Switzerland 6%
- Japan 10%
- Belgium 20%
- Ghana 39%

IMPORTS Origins 1987
- UK 20%
- Italy 10%
- West Germany 8%
- USA 7%
- Other 55%

18 GHANA

Physical Geography and Climate

Ghana, in the West African region, is a coastal state bounded to the east by the Côte d'Ivoire, Burkina Faso and Togo. It is mainly low lying land traversed by rivers, with the Black Volta running across the country from the north west. There are mangrove swamps at the coast and dense forests, but on the inland plateau savanna predominates, becoming drier in the north. The rainy season is from April to July and September to November, it being wettest in the south west with an annual average rainfall of up to 210 cm. The south east is drier with about 127 cm annual average rainfall falling to the north which only has one wet season from April to September with an annual average rainfall of 118 cm. Mean temperatures are from 26°C to 29°C with high humidity, especially in the Ashanti forests. There are hydro-electric schemes on the Volta system at Akosombo and Kpong. Mineral deposits include gold.

Population

Estimates for 1989 suggest a population of 14.6 million. The north is mainly Moslem, whilst the south is 42% Christian. The main languages are English and Akan (44%), Mole-Dagbani (16%), Ewe (13%) and Ga-Adangbe (8%). The south is the most densely populated region, the national average density of population being high at 59 persons per square kilometre compared with the regional average of 31.1 persons per square kilometre. It is also highly urbanised at 32%, compared with the regional average of 28.7%. Population growth is also fairly high at 3.4% per annum for the period 1980-87; the regional average being 3.2%.

The capital Accra is estimated to have a population of 1.9m in 1989.

History

The ancient medieval empire of the northern savanna was not directly related to the modern state. The British colony was known as the Gold Coast and Ghana was the first African colonial state to be granted independence in 1957. It became a republic in 1960 under President Kwame Nkrumah. Since then there have been five military coups and numerous coup attempts. The intervening elected civilian governments have been unable to solve the problem of Ghana's continuous fall in living standards. Intervention in the economy has ranged from the nationalisation of agriculture, industry and public sector construction work to raising the incidence of import taxes and the devaluations of the Busia regime; the price controls of President Acheampong and the attempts by the first Rawlings regime to combat high prices and the absence of basic necessities. President Rawlings took power for the second time in December 1981. These policy variations have discouraged business investment and have made production planning difficult. Following international pressure in 1983 there has been a return to market mechanisms, but problems have been exacerbated by the return of a million migrant workers from Nigeria.

Stability

There have been five successful military coups since independence in 1957. The intervening periods of civilian rule have been incapable of devising a programme to halt Ghana's steady fall in living standards. The present regime has been in place since 1981, and has been threatened by coup attempts and doubts concerning the loyalty of the armed forces.

In the 1960s the economic strategy involved a steady increase in the size of the public sector and, later, attempts to intervene in markets by fixing prices. Since 1983 there has been a return to more reliance on market forces in allocating resources and setting prices, and a more determined effort to encourage foreign investment.

Overall, Ghana's record has been very poor with political uncertainty compounded by two major changes in economic strategy. Recent liberalisation measures, the continued ability of the Rawlings regime to survive, and signs of a return to economic progress indicate that Ghana's stability record may be showing signs of gradual improvement.

Economic Structure

GDP in 1987 was $US 5,080m, the largest economy in the region after Nigeria and Côte d'Ivoire. The living standard was well below the regional average with GNP per capita at $US 390, though this is the same as the African average.

51% of GDP is generated by the agricultural sector, which employs about 55% of the labour force, and makes Ghana the most heavily dependent on agriculture in the region after Mali. The industrial sector generates only 16% of GDP, well below the regional average of 26%. The service sector is about average size for the region, generating 33% of GDP.

On the demand side, private consumption in 1987 was 87% of GDP (10% higher than the regional average), while government consumption was only 9%, as against a regional figure of 13%.

Exports were 20% of GDP while imports are 26%. In 1987 exports brought in earnings of $US 827m while $US 952m went out of the country for imports, a trade deficit of $US 125m. Cocoa is the major export, representing 56% of the total, while gold and timber represented a further 17% and 11% respectively. Fuels accounted for 19% (the largest proportion) of imports, with machinery a further 13%.

The adult literacy rate in 1975 was 30% which was then slightly better than the regional average. In 1986, 63% of primary-aged children were in school which is poor in comparison to the 75% regional average. However, 35% of the secondary age group are in education while the regional average was only 24%. About 2% of the 20-24 age group are in higher education.

Life expectancy in Ghana is 54 years which is better than the African and regional averages of 50%. In 1984 there was one doctor to every 14,900 of the population which was better than the regional average. However, there was one nurse to every 640 people, more than four times better than the regional average and the best level of basic health care other than Niger in West Africa.

Overall, Ghana has a low-income economy, heavily dependent on agriculture, and exports predominantly made up of cocoa. Educational provision at the primary level is poor, health provision is relatively good.

Economic Performance

GDP grew by 1.4% though most countries in the region grew faster, from 1980-87, and there was negative growth from 1981-83. Following several years of decline, GNP per head increased from 1984, after the Rawlings coup, though the average for the period 1965-87 was -1.6%.

Although no sectors showed actual decline, the agricultural and industrial sectors showed overall stagnation between 1980 and 1987 while the regional trend was for limited agricultural growth and reductions in the industrial sector. The service sector showed the greatest growth at 4.2% a year.

Inflation ran at a massive 48.6% over the period 1980-87, and was still 33% in 1988.

Overall, Ghana has had a poor GDP growth performance record in the 1980s, although this has improved in the second half of the decade. The annual rate of inflation has been three times the African average. Export volumes fell at -1.6% a year over the period 1980-87, worse than in Africa generally, where export volumes declined at -1.0% a year. Import volumes declined at -2.9% a year, compared with Africa generally, where they fell by -5.8% a year.

Recent Economic Developments

The Economic Recovery Programme (ERP) launched in November 1983 marked a break with the socialist development strategy that had been followed since Rawlings came to power for the second time in 1981. Agreements were made with the IMF and World Bank; Ghana has encouraged foreign involvement and aid flows, effectively abandoned any policies directed towards attaining self-sufficiency, and has begun the rehabilitation of the traditional cocoa, timber and gold export sectors. The currency has been successively devalued since September of 1986, the exchange rate has been determined by an auction system. The size of the government sector, and the level of public sector employment have been cut, and 180 parastatal organisations are being scrutinised with a view to privatisation.

Since the implementation of the ERP, economic performance has improved. Ghana's GDP grew at an estimated 3% in 1983, 5% in 1985 and 5.3% in 1986. This compares with an average annual decline at -2.1% for the first four years of the 1980s. With population growing at 3.3%, the growth rates achieved in 1985 and 1986 have meant improved average living standards. The rate of price inflation has varied widely from year to year in the 1980s, but the first four years of the decade saw prices rising at an average rate of 78% a year while for the three years 1984 to 1986 the yearly inflation rate averaged 24%. The variability of the inflation rate is underlined by the fact that it fell by three-quarters to around 10% in 1985, but then doubled in 1986. In 1989 inflation was estimated at 26%.

Ghana has invariably had a balance of merchandise trade which has shown no substantial surplus or deficit. The dollar value of exports fell by 60% between 1980 and 1984, and this severely constrained the ability to import. Exports totalled $US 1,104m in 1980, but fell to $US 440m in 1983. Since then, export receipts have gradually increased to reach $US 773m in 1986, allowing imports to rise to around the same level. The cocoa sector is undergoing extensive rehabilitation, with substantial increases in producer prices. These prices now represent 47% of the world price translated at the current exchange rate, and it is planned to increase cocoa farmers' receipts to 60% of the world price. The gold mining sector has benefited from foreign investment, particularly from Lonrho which owns 45% of the Ashanti Goldfields Corporation. Interest in Ghanaian gold mining has been enhanced by the prospect of increasing instability in South Africa, and expectations of substantial rises in gold prices as a result. Ashanti Goldfields raised its production by 7% in 1986. The timber sector produced 2-3m cubic metres in the 1960s, and half of this output was exported. In 1986 only 0.3 cubic metres was exported, but it is planned to double these export volumes by 1990.

Ghana's external debt was estimated at $US 2,100m at the beginning of 1987, and debt servicing was running at $US 433m which comprised 55% of merchandise export receipts. There has been a considerable rise in Ghana's external debt servicing commitments, as these were estimated at 14% of exports in 1984. However, increased output in the export and domestic manufacturing sectors is dependent on imports of parts, spares and inputs, and it will be hoped that increased exports will eventually ease the debt burden.

The exchange rate for the cedi has depreciated considerably with the implementation of the ERP. It stood at cedis 2.75 = $US 1 in 1982, but had reached cedi 306 = $US 1 at the end of 1989. When the foreign exchange auctions were introduced in 1986, there was a priority rate at cedis 90 = $US 1 which was used for certain government transactions. This was abolished early in 1987. The currency is expected to continue to depreciate in value at the rate of cedis 3 per month while Ghana's rate of domestic inflation continues to remains above world levels.

The 1989 budget projects a virtual balance of expenditures and revenues.

There has been extensive lending to Ghana, with heavy emphasis on infrastructure rehabilitation and extension, and general balance of payments support to facilitate the structural adjustments of the ERP. The main lending has been $US 210m from the World Bank, mostly for structural adjustment, but also for telecommunications, mining, oil exportation, and roads. Japan has provided $US 115m for roads, and the African

Development Bank has lent $US 100m for electricity, manufacturing and development bank credit. The EEC has put $US 70m mainly into rural development projects, and the European Investment Bank has committed $US 50m to gold mining and rural development; Italy, $US 4m to water supply and railways; the UK, $US 43m to forestry and agriculture; West Germany, $US 30m to transport and rural development, and France, $US 30m to telecommunications and agriculture.

Economic Outlook and Forecasts

The forecasts assume that Ghana's stability record continues to be consolidated under President Rawlings, and that the economic reforms proceed steadily. In real terms, cocoa prices are expected to fall by 20% up to 1995.

The projections show GDP expanding at a rate that allows increases in GDP per head. Inflation, based on the budget deficit projections, is expected to fall. Export volumes, boosted by expansion in the non-cocoa sector is forecast to show steady expansion, with import volumes slightly faster than exports.

Ghana: Economic Forecasts
(average annual percentage change)

	Actual 1980-87	Forecast Base 1990-95	Forecast High 1990-95
GDP	1.4	4.9	5.5
GDP per Head	-2.0	1.9	2.5
Inflation Rate	48.3	22.0	18.4
Export Volumes	-1.6	3.0	4.2
Import Volumes	-2.9	4.2	5.1

GHANA: Comparative Data

	GHANA	WEST AFRICA	AFRICA	INDUSTRIAL COUNTRIES
POPULATION & LAND				
Population, mid year, millions, 1989	14.6	12.3	10.2	40.0
Urban Population, %, 1985	32	28.7	30	75
Population Growth Rate, % per year, 1980-86	3.5	3.2	3.1	0.8
Land Area, thou. sq. kilom.	239	384	486	1,628
Population Density, persons per sq kilom., 1988	59.0	31.1	20.4	24.3
ECONOMY: PRODUCTION & INCOME				
GDP, $US millions, 1986	5,720	4,876	3,561	550,099
GNP per head, $US, 1986	239	510	389	12,960
ECONOMY: SUPPLY STRUCTURE				
Agriculture, % of GDP, 1985	45	40	35	3
Industry, % of GDP, 1985	17	26	27	35
Services, % of GDP, 1985	39	34	38	61
ECONOMY: DEMAND STRUCTURE				
Private Consumption, % of GDP, 1985	82	77	73	62
Gross Domestic Investment, % of GDP, 1985	10	13	16	21
Government Consumption, % of GDP, 1985	10	13	14	17
Exports, % of GDP, 1985	10	18	23	17
Imports, % of GDP, 1985	12	20	26	17
ECONOMY: PERFORMANCE				
GDP growth, % per year, 1980-86	0.7	-2.03	-0.6	2.5
GDP per head growth, % per year, 1980-86	-2.7	-4.7	-3.7	1.7
Agriculture growth, % per year, 1980-86	-0.2	1.3	0.0	2.5
Industry growth, % per year, 1980-86	-2.4	-3.6	-1.0	2.5
Services growth, % per year, 1980-86	3.3	-2.5	-0.5	2.6
Exports growth, % per year, 1980-86	-7.1	-4.0	-1.9	3.3
Imports growth, % per year, 1980-86	-4.6	-12.9	-6.9	4.3
ECONOMY: OTHER				
Inflation Rate, % per year, 1980-86	50.8	13.0	16.7	5.3
Aid, net inflow, % of GDP, 1986	6.6	3.7	6.3	-
Debt Service, % of Exports, 1986	10.8	21.4	20.6	-
Budget Surplus (+), Deficit (-), % of GDP, 1986	0.1	-3.4	-2.8	-5.1
EDUCATION				
Primary, % of 6-11 group, 1985	66	75	76	102
Secondary, % of 12-17 group, 1985	39	24	22	93
Higher, % of 20-24 group, 1984	2	2.6	1.9	39
Adult Literacy Rate, %, 1975	30	30	39	99
HEALTH & NUTRITION				
Life Expectancy, years, 1986	54	50	50	76
Calorie Supply, daily per head, 1985	1,785	2,105	2,096	3,357
Population per doctor, 1981	6,680	16,199	24,185	550

Notes: 'Southern Africa' and 'Africa' exclude South Africa. Dates are for the country in question, and do not always correspond with the Regional, African and Industrial averages.

GHANA: Leading Indicators

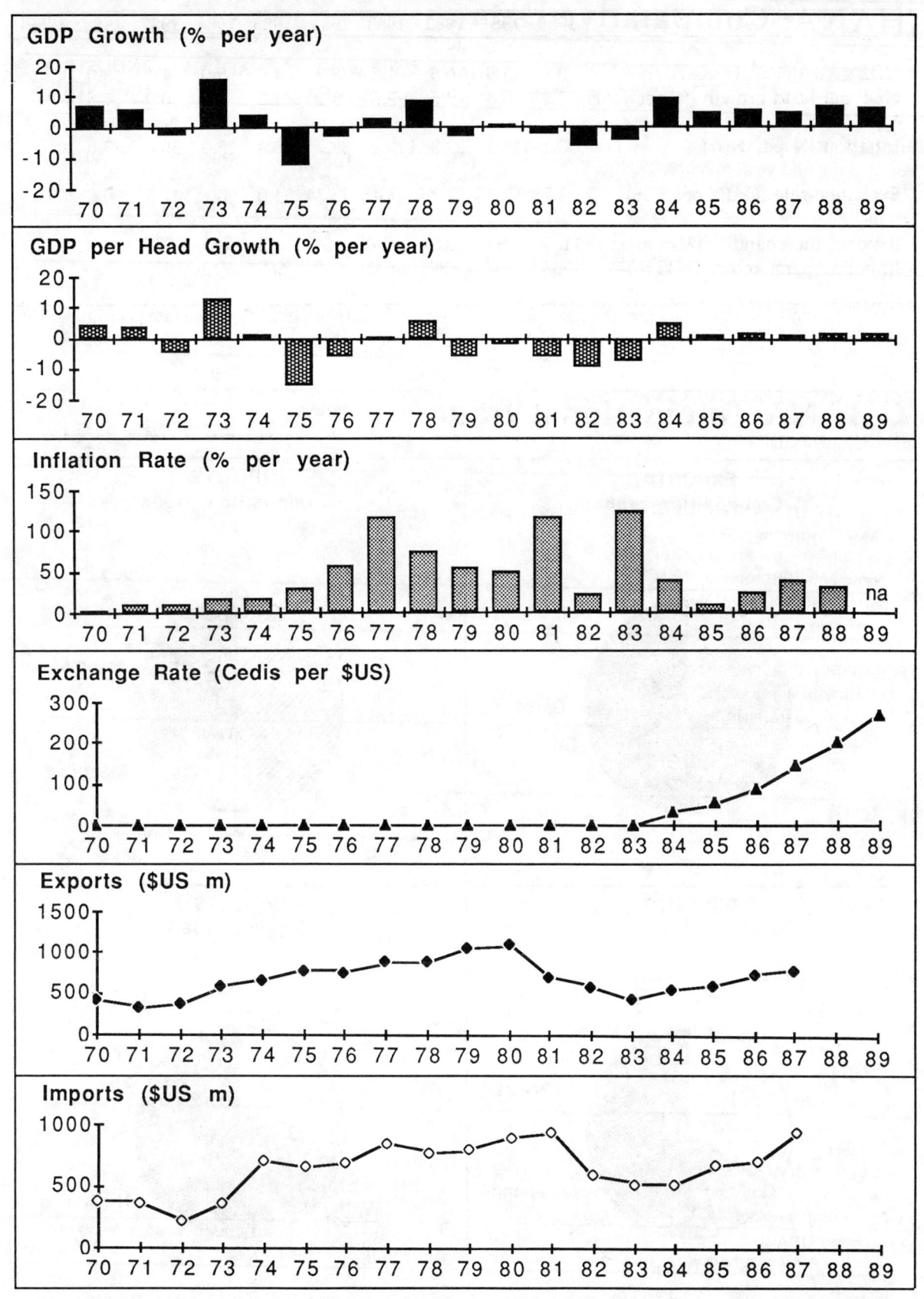

GDP Growth (% per year)

GDP per Head Growth (% per year)

Inflation Rate (% per year)

Exchange Rate (Cedis per $US)

Exports ($US m)

Imports ($US m)

Note: Leading indicator data for 1989 are based on the first half of 1989. 1989 exchange rate is for mid-1989.

GHANA: Leading Indicator Data

	1980	1981	1982	1983	1984	1985	1986	1987	1988	1989
GDP growth (% per year)	0.6	-2.9	-6.5	-4.5	8.7	4.5	5.0	4.4	6.0	
GDP per head growth (% per year)	-2.2	-5.8	-9.5	-7.6	5.2	1.2	1.6	0.9	1.5	2.0
Inflation (% per year)	50.1	116.5	22.3	122.9	39.7	10.3	25.0	39.0	33.0	
Exchange rate (Cedis per $US)	2.75	2.75	2.75	3.45	35.34	54.0	90.1	147.7	200.0	269.1
Exports, merchandise ($US m)	1104	711	607	440	566	632	773	827		
Imports, merchandise ($US m)	908	954	589	531	530	669	712	952		

Note: Leading indicator data for 1989 are based on the first half of 1989. 1989 exchange rate is for mid-1989.

GHANA: International Trade

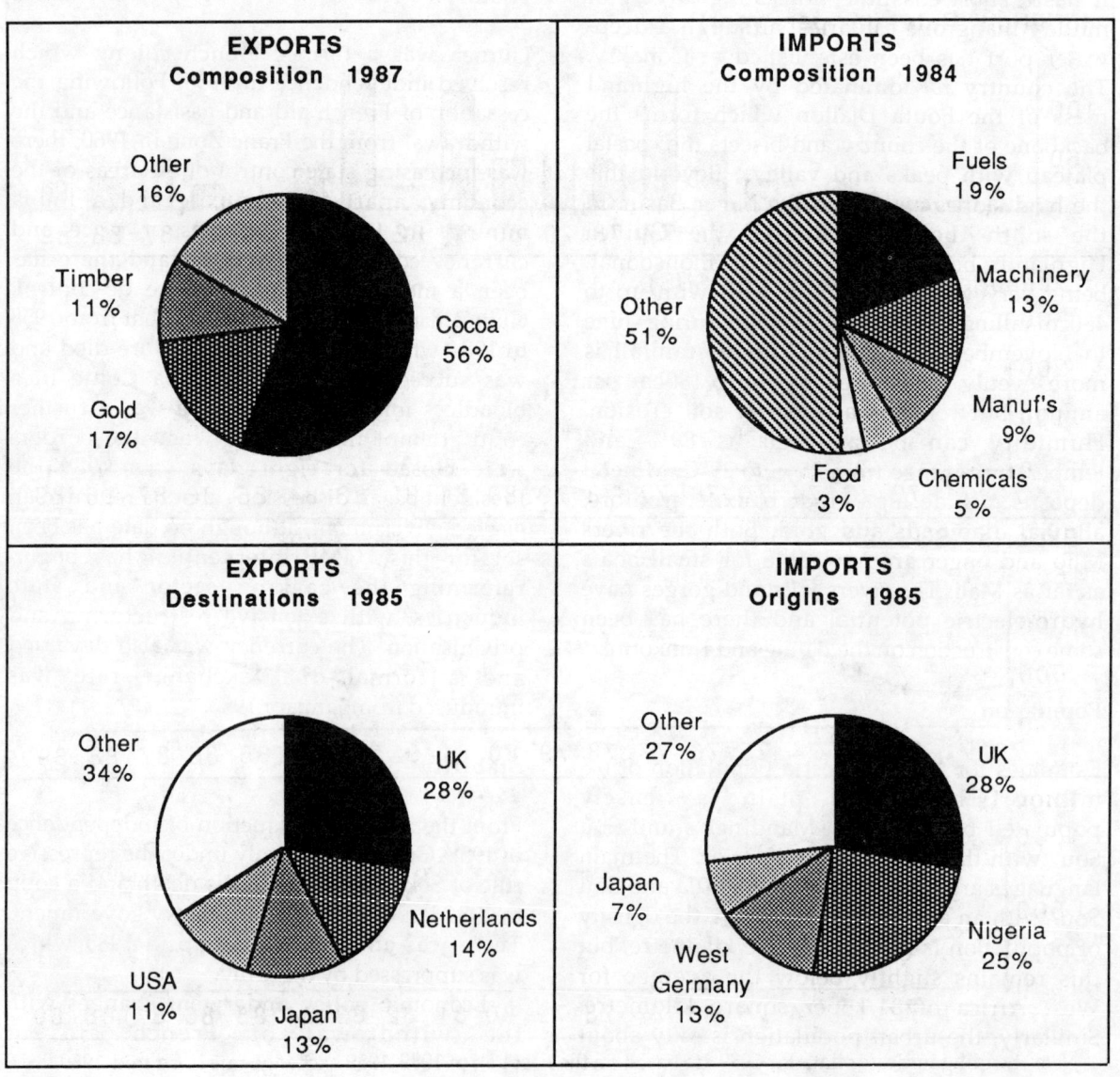

EXPORTS Composition 1987
- Cocoa 56%
- Gold 17%
- Timber 11%
- Other 16%

IMPORTS Composition 1984
- Fuels 19%
- Machinery 13%
- Manuf's 9%
- Chemicals 5%
- Food 3%
- Other 51%

EXPORTS Destinations 1985
- UK 28%
- Netherlands 14%
- Japan 13%
- USA 11%
- Other 34%

IMPORTS Origins 1985
- UK 28%
- Nigeria 25%
- West Germany 13%
- Japan 7%
- Other 27%

19 GUINEA

Physical Geography and Climate

Guinea in the West African region, is a coastal state bordered by Guinea Bissau, Senegal, Mali, Côte d'Ivoire, Liberia and Sierra Leone. It has a short coastline, consisting largely of muddy mangrove estuaries, although a deep-water port has been established at Conakry. The country is dominated by the highland mass of the Fouta Djallon which forms the backbone of the country and bisects the coastal plateau with peaks and valleys. Beyond this the land slopes away into the Niger Basin. In the south the mountains of the Guinea Highlands begin. The climate is monsoonal, being very wet in the south west with up to 430cm falling on the coastal region during June to November. At higher altitudes rainfall is more evenly distributed reaching 180cm per annum, but remains a cause of soil erosion. Humidity can be as high as 89% and temperatures range from 25°C to 29°C. Mineral deposits include high grade bauxite, iron ore, alluvial diamonds and gold. Both the rivers Milo and Niger are navigable for small boats as far as Mali. The rivers falls and gorges have hydro-electric potential and there has been some construction on the Bafing and Konkouré.

Population

Estimates for 1989 suggest a population of 6.8 million. The coastal plain is densely populated being mainly Mandingo's and Sou Sou with the Fulani in the interior. The main languages are French, Mandingo, Fula and Sou Sou. With an area of 246,000 sq km, the density of population is 26.8 per square kilometre, but this remains slightly below the average for West Africa of 31.1 per square kilometre. Similarly, the urban population is only about 22% while the regional average is much higher at 28.7%. Population growth is much lower than the regional average at 2.4% a year.

History

Guinea was a former French colony which received independence in 1958. Following the cessation of French aid and assistance and the withdrawal from the Franc Zone in 1960, there was increasing state control of all areas of the economy, apart from bauxite and alumina mining. In 1978 the policies of price and currency control were relaxed and there has been a movement towards price de-control. Guinea had continuity in government from 1958 to 1984 when President Sekou Touré died and was succeeded by Col Lansana Conté in a bloodless military coup. There was a further coup attempt in July 1985 when the borders were closed for eight days. In 1989 the President began discussions about a return to an elected government, although no date has been set for this. IMF intervention has begun reforming the banking sector and state industries, with extensive restructuring and privatisation. The currency was also devalued and a formal dual exchange rate was introduced from January 1986.

Stability

From the abrupt introduction of independence in 1958, Guinea was firmly under the repressive rule of Sekou Touré until his death and a coup in 1984 led to the present military government. There was an attempted coup in 1985 which was suppressed by the army.

Economic policy underwent changes with the withdrawal of French aid at

Guinea

independence, and withdrawal from the Franc Zone in 1960. With the exception of the mining sector, the government introduced substantial ownership and intervention in the 1960s and 1970s. The present government has begun a liberalisation programme.

Overall, Guinea's stability record has been poor. The enduring qualities of the Sekou Touré government were offset by the repressive nature of the regime, and the arbitrary interreactions of the economy. The present government shows signs of a more liberal approach.

Economic Structure

In 1985 GDP was estimated to be $US 1,980m, making it one of the smaller economies in the region. GNP per capita stood at $US 300, well below the regional average. The agricultural sector generated about 40% of GDP in 1985, about average for the region. Agriculture employs about 80% of the population. The industrial sector generates a further 22% of GDP though much of this is from mining (mainly bauxite and diamonds). There is limited manufacturing, and what there is was largely run down during the Sekou Touré period. The present government is selling off or closing two thirds of state enterprises. Guinea has about one quarter of the world's bauxite reserves, about 25% of the world total. The service sector is slightly larger than average at 38% of GDP.

On the demand side, private consumption was 73% of GDP in 1985, slightly below average for the region, while government consumption was a further 14%. Investment was one of the lowest in the region at around 9%.

While exports represented 25% of GDP, imports comprised 21%. In 1985 export values were around $US 559m with imports worth $US 448m, while in most African countries import costs exceeded export earnings. The main exports are bauxite 63%; aluminium 16% and diamonds and gold 15% (though much of the gold is not accounted for as it is smuggled into neighbouring Mali), a total of 94% being minerals, while in most other countries in Africa agricultural products make up a large proportion of exports. Iron ore deposits are now being developed for exploitation. Raw materials (partly for aluminium processing) account for 44% of imports, with machinery accounting for another 17%, fuels 14% and food produce 13%.

The level of education in Guinea is one of the lowest in Africa. Only 29% of primary children attend school, compared to regional and African averages of 75%, and only Mali in the region has a worse record. Only Burkina Faso, Mali and Niger have a lower proportion of secondary school aged children in school than Guinea's 9% and the regional average is 24%. Only 1% of the 20-24 age group are in higher education. The adult literacy rate of 28% is near to the regional average, however.

Life expectancy of 42 is lower only in Guinea Bissau and Sierra Leone in the region. The calorific intake level is one of the lowest in Africa. Guinea has a very high ratio of people per doctor of 57,000 to 1, compared with a regional average of about 16,000 to 1, about three and a half times worse. Provision of nurses is also poor with a ratio of 6,400 to 1, compared with a regional average of 2,700 to 1.

Overall, although Guinea is mineral rich and has a favourable climate for agriculture, there is still little industrialisation, poor infrastructure and poor provision of social services. The economy is heavily dependent on world bauxite prices, which are in decline.

Economic Performance

GDP grew by only a minimal 0.9% from 1980-85, though average regional growth was negative. Living standards fell by -1.5% a year over the same period. Growth in the agricultural sector was limited to 0.3% while industrial growth (including mining) was even smaller at 0.1%, though regionally this sector declined by -3.6% in comparison. Growth in the service sector was slightly better at 2.1% while services declined by -2.5% in the region.

Recent figures are not available for import and export growth but for the 1973-79 period export volume growth was an impressive 11.7% while imports only increased by 1%. There must be some doubt, however, over the accuracy of the export figures.

Inflation had been kept low and was only 4.5% over the period 1973-84. However, prices rose by 51% in 1986, falling to 43% the following year but seemed to be under control in 1989 when it had been brought down to around 30%.

Overall, figures for the first half of the decade are not encouraging, but GDP per head appears to have increased in the years 1987-89. Higher diamond sales, increased gold output and the opening of iron ore mines can only improve performance. The success of the structural adjustment programme may in the medium term be reflected in improved manufacturing output as the private sector becomes increasingly involved.

Recent Economic Developments

The military government which replaced the rule of Sekou Touré when the latter died waited until February 1986 before embracing an economic reform programme, which under IMF guidance reversed the intervention and state control of the Sekou Touré era. The currency (Syli) was replaced, at par, by the Guinea franc, and devalued one-sixteenth of its previous value, interest rates were raised, civil service jobs cut, and 81 state trading enterprises closed and 45 parastatals privatised. In July 1987 the IMF reaffirmed its support for the reform programme, with its commitment to a flexible exchange rate policy and further liberalisation. In 1988, the second phase was begun with continued efforts to make the public sector and administrative system more efficient, though the programme needs to be accelerated if overseas funding is to continue. The austerity measures are not popular domestically.

Inflation persisted in single figures for the first two years of the 1980s, but then began to accelerate, and ran at 52% in 1986. Some of this recent inflation is due to the substantial devaluation of the currency.

In the early 1980s, Guinea ran a surplus on the balance of merchandise trade, with exports a third higher than imports, and in 1982 the surplus was $US 150m. With the liberalisation of recent years this position has changed, and in 1989 Guinea is running a trade deficit of perhaps $US 100m. Bauxite is still expected to provide the bulk of export earnings, but gold earnings are expected to increase eightfold in the next three years, diamond earnings to rise by a third, and coffee and fish to provide $US 100m of earnings in total exports of $US 600m.

Guinea's external debt is estimated at $US 1,950m in 1989, and servicing takes up close to 20% of export earnings. Recent borrowing under the reform programme will have increased the debt service-to-exports ratio, but this will ease as the investments lead to increased export volumes.

The exchange rate allowed the currency to appreciate in the 1970s, and it remained virtually unchanged at around the syli 22 = $US 1 up to 1986 and the currency conversion and devaluation. In 1985, the black market exchange rate was ten to twelve times the official rate. The 1986 devaluation was large enough to have brought the official rate more or less into line with the market valuation, although with such a change it will be some time before the exchange rate shows stability. The Guinea franc had depreciated further to GFr 620 = $US 1 at the end of 1989.

In 1989 an austerity budget, in which the various ministries were made more accountable for their spending, was presented. Recent government measures since 1986 have reduced public sector employment by 10,000 out of a total government work-force of 88,000. The target is to lose 30,000 in total, although much of this will come through the winding-up or privatisation of state trading enterprises. However, attempts to reduce the government pay roll further have been slow and a job assessment scheme to remove inefficient workers has failed.

Donors have committed substantial aid, with the bulk of this directed toward infrastructure, particularly transport where the World Bank, the African Development Bank and West Germany are heavily involved. Japan has contributed $US 30.7m in general balance of payments support for structural adjustment, the European Development Fund is lending $US 9.4m to fishery development and the Islamic Development Bank is involved in $US 5.5m of forestry and water projects.

Private banks and the World Bank's commercial lending agency, the International Finance Corporation, are to co-ordinate the raising of $US 325m for a projected 12m tonnes per year iron-ore mining project at Mifergui-Nimba. Neighbouring Liberia will be

Guinea

involved in the project, and will mine 2m tonnes of the eventual total.

Economic Outlook and Forecasts

The forecasts assume Guinea remains stable despite the uncertainty concerning a return to an elected government. The liberalising reforms are expected to continue.

Bauxite prices are projected to improve in real terms by 12% up to 1995, and to maintain GDP growth and export performance to give steady improvements in GDP per head. Inflation is forecast to slow down steadily, but still be above 20% a year.

Guinea: Economic Forecasts
(average annual percentage change)

	Actual 1980-85	Forecast Base 1990-95	Forecast High 1990-95
GDP	0.9	4.7	5.2
GDP per Head	-1.5	2.3	2.8
Inflation Rate	36.0[a]	26.0	21.0
Export Volumes	11.7[b]	4.5	5.5
Import Volumes	1.0[b]	4.2	4.4

Note: [a] 1986-88 [b] 1973-79

GUINEA: Comparative Data

	GUINEA	WEST AFRICA	AFRICA	INDUSTRIAL COUNTRIES
POPULATION & LAND				
Population, mid year, millions, 1989	6.8	12.3	10.2	40.0
Urban Population, %, 1985	22	28.7	30	75
Population Growth Rate, % per year, 1980-86	2.4	3.2	3.1	0.8
Land Area, thou. sq. kilom.	246	384	486	1,628
Population Density, persons per sq kilom.,1988	26.8	31.1	20.4	24.3
ECONOMY: PRODUCTION & INCOME				
GDP, $US millions, 1985	1,980	4,876	3,561	550,099
GNP per head, $US, 1985	300	510	389	12,960
ECONOMY: SUPPLY STRUCTURE				
Agriculture, % of GDP, 1985	40	40	35	3
Industry, % of GDP, 1985	22	26	27	35
Services, % of GDP, 1985	38	34	38	61
ECONOMY: DEMAND STRUCTURE				
Private Consumption, % of GDP, 1985	73	77	73	62
Gross Domestic Investment, % of GDP, 1985	9	13	16	21
Government Consumption, % of GDP, 1985	14	13	14	17
Exports, % of GDP, 1985	25	18	23	17
Imports, % of GDP, 1985	21	20	26	17
ECONOMY: PERFORMANCE				
GDP growth, % per year, 1980-85	0.9	-2.03	-0.6	2.5
GDP per head growth, % per year, 1980-85	-1.5	-4.7	-3.7	1.7
Agriculture growth, % per year, 1980-85	0.3	1.3	0.0	2.5
Industry growth, % per year, 1980-85	0.1	-3.6	-1.0	2.5
Services growth, % per year, 1980-85	2.1	-2.5	-0.5	2.6
Exports growth, % per year, 1973-79	11.7	-4.0	-1.9	3.3
Imports growth, % per year, 1973-79	1.0	-12.9	-6.9	4.3
ECONOMY: OTHER				
Inflation Rate, % per year, 1973-84	4.5	13.0	16.7	5.3
Aid, net inflow, % of GDP, 1985	9.2	3.7	6.3	-
Debt Service, % of Exports, 1984	19.2	21.4	20.6	-
Budget Surplus (+), Deficit (-), % of GDP, 1983	19.2	-3.4	-2.8	-5.1
EDUCATION				
Primary, % of 6-11 group, 1985	30	75	76	102
Secondary, % of 12-17 group, 1985	12	24	22	93
Higher, % of 20-24 group, 1985	2	2.6	1.9	39
Adult Literacy Rate, %, 1976	20	30	39	99
HEALTH & NUTRITION				
Life Expectancy, years, 1986	42	50	50	76
Calorie Supply, daily per head, 1985	1,731	2,105	2,096	3,357
Population per doctor, 1981	56,170	16,199	24,185	550

Notes: 'Southern Africa' and 'Africa' exclude South Africa. Dates are for the country in question, and do not always correspond with the Regional, African and Industrial averages.

GUINEA: Leading Indicators

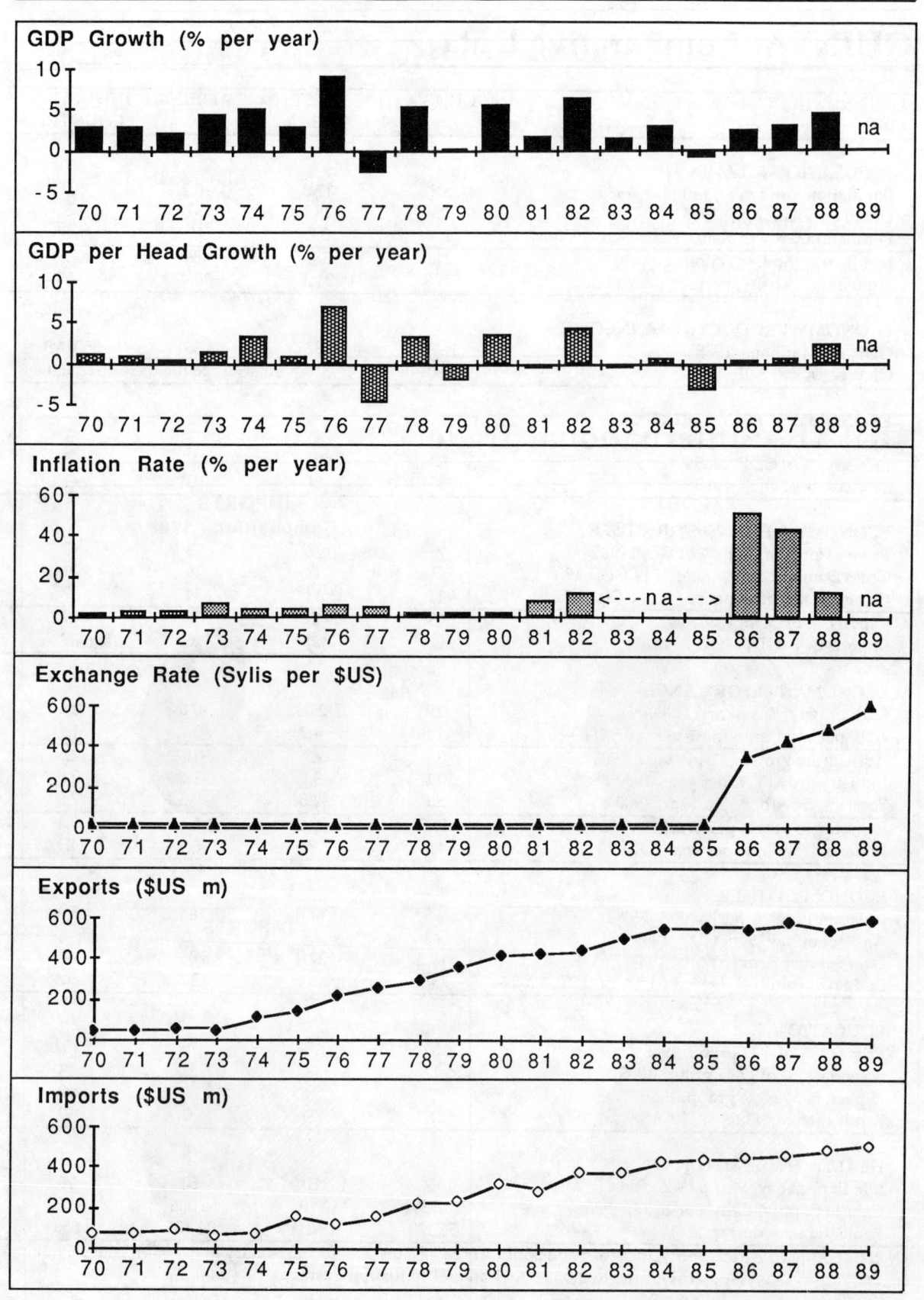

GDP Growth (% per year)

GDP per Head Growth (% per year)

Inflation Rate (% per year)

Exchange Rate (Sylis per $US)

Exports ($US m)

Imports ($US m)

Note: Leading indicator data for 1989 are based on the first half of 1989. 1989 exchange rate is for mid-1989.

GUINEA: Leading Indicator Data

	1980	1981	1982	1983	1984	1985	1986	1987	1988	1989
GDP growth (% per year)	5.6	1.6	6.3	1.4	2.8	-1.1	2.5	2.9	4.5	
GDP per head growth (% per year)	3.6	-0.4	4.3	-0.6	0.8	-3.1	0.5	0.8	2.4	
Inflation (% per year)	2.0	8.6	12.3				51.0	43.0	13.0	
Exchange rate (Sylis per $US)	19.0	20.9	22.4	23.1	24.1	22.5	347.0	424.0	485.0	590.3
Exports, merchandise ($US m)	418	429	444	503	553	559	555	584	548	592
Imports, merchandise ($US m)	326	290	380	380	438	448	451	468	491	513

Note: Leading indicator data for 1989 are based on the first half of 1989. 1989 exchange rate is for mid-1989.

GUINEA: International Trade

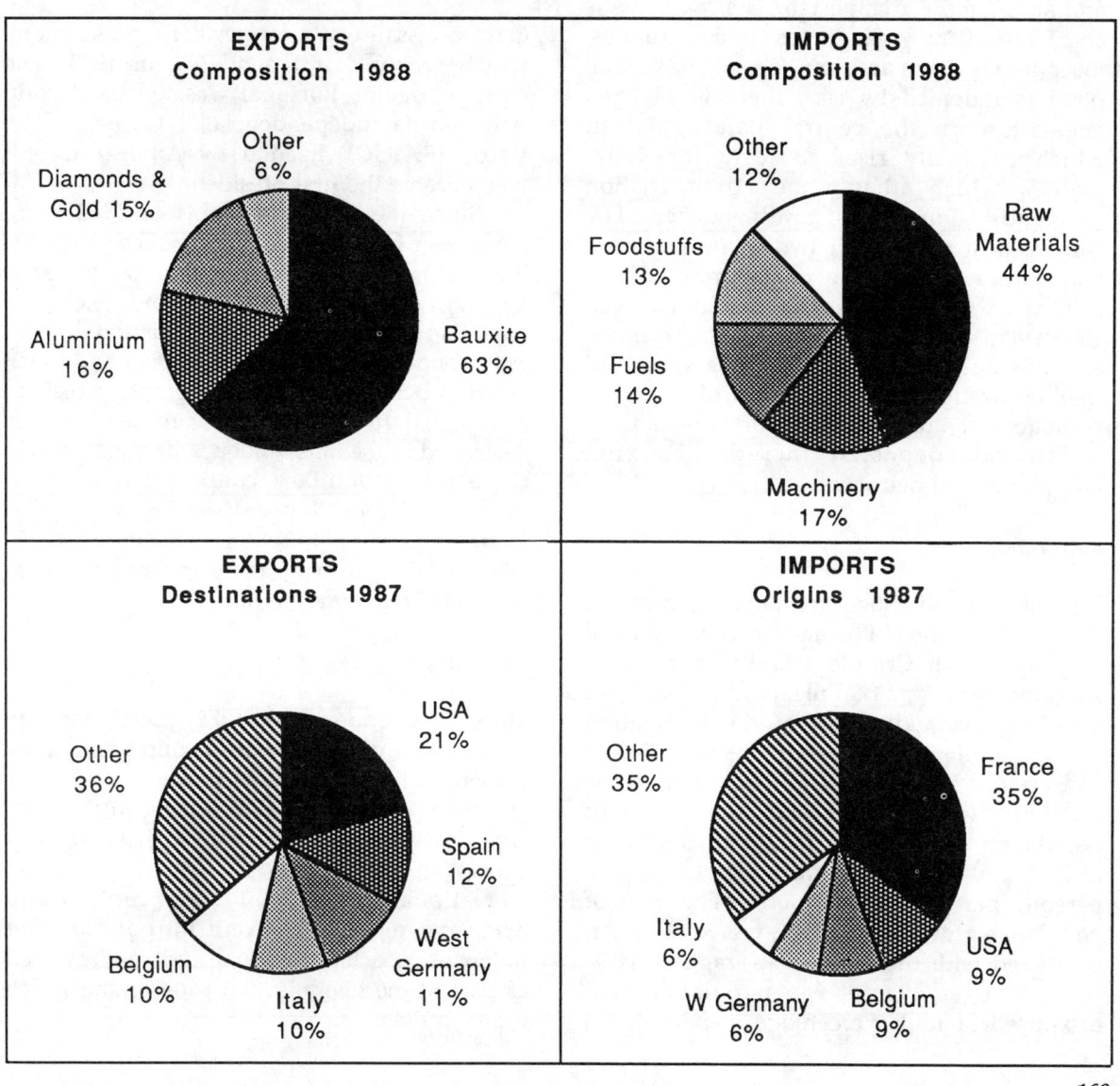

EXPORTS Composition 1988
- Bauxite 63%
- Aluminium 16%
- Diamonds & Gold 15%
- Other 6%

IMPORTS Composition 1988
- Raw Materials 44%
- Machinery 17%
- Fuels 14%
- Foodstuffs 13%
- Other 12%

EXPORTS Destinations 1987
- USA 21%
- Spain 12%
- West Germany 11%
- Italy 10%
- Belgium 10%
- Other 36%

IMPORTS Origins 1987
- France 35%
- USA 9%
- Belgium 9%
- W Germany 6%
- Italy 6%
- Other 35%

20 GUINEA-BISSAU

Physical Geography and Climate

Guinea-Bissau, in the West African region, is a small Atlantic coastal state bounded by Senegal and Guinea. It was a former Portuguese territory and includes several coastal islands and an off-shore archipelago. The coast line is characterised by mangroves, wide estuaries, meandering rivers and rain forest. The coastal plain is indented by rias; there is savanna vegetation on the central Bafatá plateau which gradually rises to form the Gabú plateau which adjoins the Fouta Djallon highlands along the Guinea border. The climate is tropical with two main seasons - hot, with a temperature range from 22°C to 28°C and wet, with an annual rainfall between 100cm and 200cm in the north and excessive amounts on the coast. Ocean-going vessels of shallow draught can reach most of the main population centres. The main port is Bissau.

Mineral deposits include bauxite, phosphates and petroleum.

Population

Estimates for 1989 suggest a population of just under one million. Portuguese is the official language, with Crioulo widely used; 26% speaking Balanta; 23% Pulaar; 12% Mandinka and 11% Manjuku. Most of the population follow traditional beliefs, but 35% are Moslem and around 5% Roman Catholic. Despite the small area (only 36,000 sq km) the density of population is lower than the regional average of 31.1 persons per square kilometre at 26.7 persons per square kilometre. The rate of population growth for 1973-82 was 3.0% a year, compared with the regional average of 3.2%. In 1982, over 39% of the population was urbanised, a high percentage, especially as there is almost no industry to attract rural migrants.

Bissau, the capital, has a population of about 109,000.

History

Guinea-Bissau was a former Portuguese colony which received independence in 1974. The struggle agains Portugal was led by Partido Africano da Indepêndencia de Guiné e Cabo Verde (PAIGC), headed by Amílcar Cabral who became the first president.

Since the nationalisations of 1975 there have been frequent economic policy reversals. Pressure from the IMF has reduced public sector holdings and relaxed price and exchange controls resulting in a 50% devaluation in 1983. The economic situation remains poor and much effort is being put into an attempt to reclaim abandoned rice fields that were neglected or destroyed in the independence struggle. In 1980 there was a military coup which deposed President Cabral, but not the Prime Minister João Bernardo Vieira who became President. In 1985 an alleged army coup attempt failed to displace President Vieira.

Stability

Since independence in 1974, there were six years of civilian rule before a coup brought the present military regime to power. There was an attempted coup in 1985 arising from ethnic rivalries and the government drive against corruption.

Nationalisations and state control and extensive regulations were introduced after independence, but these policies were reversed as part of the liberalisation programme which began in 1983.

Guinea-Bissau

Overall, Guinea-Bissau's stability record has been very poor with the turbulent events of the independence struggle with Portugal followed by the radical changes in economic strategy. There are some signs of improvement with introduction of the recent reform programme, although the attempted coup of 1985 indicates that the regime is not entirely secure.

Economic Structure

Recent reliable data on the economy of Guinea-Bissau are not available. GDP in 1984 was estimated at $US 130m which is small even when the size of the country is taken into account. The standard of living is very low; in 1987 GNP per head stood at only $US 160, giving it the poorest population in the region.

The agricultural sector, at 50%, is in relative terms one of the largest in the region and only Ghana and Mali have a larger dependence on this sector. The industrial sector, conversely, is by far the smallest in the region, reflecting the very low level of development At only 6% of GDP is more than four times smaller than the regional average. Extraction of the country's oil, bauxite and phosphate deposits has so far not been possible because of the lack of infrastructure and initial development costs. The service sector is comparatively large, generating 44% of GDP.

On the demand side, private consumption in 1982 was fairly high at 80%, with government consumption of 30% one of the highest in the region. Investment of 18% was above average for the region

While exports made up 11% of GDP, imports comprised 39%. Estimates for 1988 suggest a large balance of trade deficit, with imports worth $US 60m and only $US 22m exports. Main exports in 1985 were cashew nuts 41%; fish 28%; groundnuts 16% and palm kernals 9%, reflecting the country's almost total dependence on agriculture. Despite this, 25% of imports were still food produce, with petroleum products accounting for a further 13%.

In 1983 primary education was surprisingly high considering the low level of development, with 62% of the primary age group in schools. However, only 11% of the secondary age group were in full-time education. No figures were available for higher education. Adult literacy in 1979 was poor at only 20%.

Although the life expectancy of 39 years was about the lowest in Africa, the ratio of doctors to population was very good with one for every 6,750 in 1977, and one nurse to every 1,350 people better than is general in the region.

Overall, Guinea-Bissau is one of the poorest ten countries in the world. Infrastructure is poorly developed with almost no manufacturing industry. Although there is a heavy dependence on agriculture for both GDP and exports, the climate does not suffer from the unreliable rainfall that other countries in the region do.

Economic Performance

From 1970-81 GDP expanded by 3.0% a year. GDP per head from 1970-81 showed minimal growth of 0.1%. Africa's GDP per head rose by 1.6% a year in the same period.

Growth in the agricultural sector for the same period was limited to 0.5%, with industrial growth of 2.4% and an 8.4% growth rate in the service sector. Export growth was an encouraging 7.1% while imports fell by -5.4% a year.

The inflation rate over the period 1970-81 was 39.2%, whereas Africa in the 1970s had inflation at 10.3% a year.

Overall, economic performance appears worse than elsewhere in Africa, with GDP per head static and high inflation, although the data is not the most reliable.

Recent Economic Developments

In 1983, the government began a programme of economic reform which began to reverse the intervention, state ownership and control which followed independence in 1974. Early in 1987, a structural adjustment package was agreed with the IMF, the World Bank and other donors providing $US 44m of loans. The currency was to be devalued, agricultural prices decontrolled, marketing liberalised and a programme introduced to reduce public sector employment. Taxes were raised on beer and spirits, electricity tariffs quadrupled and fuel

Guinea-Bissau

prices raised 10 per cent. The reforms, which have raised urban prices are reported to have caused unrest, and the government has pledged to continue subsidising essential goods. There are no recent GDP figures, but the impression is that GDP has stagnated and average living standards have fallen in the 1980s. Inflation has accelerated to reach around 50 per cent in 1987.

The balance of merchandise trade shows a continuous deficit, with imports at around $US 60m, five times the level of exports. There are only modest prospects of improving this trade balance although the reform programme will aim to raise exports by raising producer prices. Debt service will increase with the borrowings under the reform programme, but it will be hoped that exports will be encouraged by the higher prices to producers. Portugal cancelled $5m of debt in 1987.

The exchange rate depreciated slightly in the 1980s up to the onset of the reform programme in 1983. In 1982 the currency stood at GBP 39.9 = $US 1, and steady substantial depreciations have occurred until with the substantial depreciations of 1986 and following years, the exchange rate stood at GBP 2,000 = $US 1 at the end of 1989. There is speculation that Guinea-Bissau is considering joining the Franc Zone, but it is likely that there will need to be an attainment of some stability in the exchange rate and positive response to the reform moves before this is seriously considered by France.

Guinea-Bissau is pledged to reducing government expenditure as part of the economic reform programme, and one-third of the 16,600 public sector jobs are to go at the rate of 1,500 a year. Determined efforts are being made to reduce the budget deficit by raising revenue with higher taxes on alcohol and certain exports. Subsidies were to be reduced, but the government is unwilling to risk unrest by decontrolling prices of basic foods.

Aid flows have increased with the agreement over the economic reforms, and in addition to the $US 44m of structural adjustment loans, Denmark is committing $US 5.5m to health projects, and the African Development Fund $US 6.4m to health and $US 5.6m to small farmer assistance. West Germany is lending $US 4.4m for agricultural and fishery projects, Sweden $US 24m for a set of projects covering all major sectors, and the UN Development programme $US 19.5m over four years to improve public enterprise performance.

Economic Outlook and Forecasts

The forecasts assume that the regime of President Vieira continues to consolidate its position, and that the liberalising reform remains in place.

It is not expected that the price of Guinea-Bissau's agricultural exports will change appreciably in real terms up to 1995. GDP growth is expected to allow GDP per head to be maintained, and perhaps expand slightly in favourable world conditions. Export volumes are expected to expand, with aid disbursements allowing import volumes to expand more rapidly. Inflation is forecast to slow down only slightly.

Guinea-Bissau : Economic Forecasts
(average annual percentage change)

	Actual 1970-81	Forecast Base 1990-95	Forecast High 1990-95
GDP	3.0	3.0	3.5
GDP per Head	0.1	0.0	0.5
Inflation Rate	32.9	30.0	27.0
Export Volumes	7.1	3.2	3.5
Import Volumes	-5.4	4.0	4.5

GUINEA-BISSAU: Comparative Data

	GUINEA-BISSAU	WEST AFRICA	AFRICA	INDUSTRIAL COUNTRIES
POPULATION & LAND				
Population, mid year, millions, 1989	0.989	12.3	10.2	40.0
Urban Population, %, 1982	39.2	28.7	30	75
Population Growth Rate, % per year, 1973-82	3.0	3.2	3.1	0.8
Land Area, thou. sq. kilom.	36	384	486	1,628
Population Density, persons per sq kilom., 1988	26.7	31.1	20.4	24.3
ECONOMY: PRODUCTION & INCOME				
GDP, $US millions, 1984	130	4,876	3,561	550,099
GNP per head, $US, 1986	170	510	389	12,960
ECONOMY: SUPPLY STRUCTURE				
Agriculture, % of GDP, 1982	50	40	35	3
Industry, % of GDP, 1982	6	26	27	35
Services, % of GDP, 1982	44	34	38	61
ECONOMY: DEMAND STRUCTURE				
Private Consumption, % of GDP, 1982	80	77	73	62
Gross Domestic Investment, % of GDP, 1982	18	13	16	21
Government Consumption, % of GDP, 1982	30	13	14	17
Exports, % of GDP, 1982	11	18	23	17
Imports, % of GDP, 1982	39	20	26	17
ECONOMY: PERFORMANCE				
GDP growth, % per year, 1970-81	3.1	-2.03	-0.6	2.5
GDP per head growth, % per year, 1970-81	0.1	-4.7	-3.7	1.7
Agriculture growth, % per year, 1970-81	0.5	1.3	0.0	2.5
Industry growth, % per year, 1970-81	2.4	-3.6	-1.0	2.5
Services growth, % per year, 1970-81	8.4	-2.5	-0.5	2.6
Exports growth, % per year, 1970-81	7.1	-4.0	-1.9	3.3
Imports growth, % per year, 1970-81	-5.4	-12.9	-6.9	4.3
ECONOMY: OTHER				
Inflation Rate, % per year, 1984-86	54.4	13.0	16.7	5.3
Aid, net inflow, % of GDP, 1984	24.9	3.7	6.3	-
Debt Service, % of Exports, 1984	21.3	21.4	20.6	-
Budget Surplus (+), Deficit (-), % of GDP, 1975	-8.0	-3.4	-2.8	-5.1
EDUCATION				
Primary, % of 6-11 group, 1983	62	75	76	102
Secondary, % of 12-17 group, 1983	11	24	22	93
Higher, % of 20-24 group, 1983	11	2.6	1.9	39
Adult Literacy Rate, %, 1979	20	30	39	99
HEALTH & NUTRITION				
Life Expectancy, years, 1986	39	50	50	76
Calorie Supply, daily per head, 1985	2,258	2,105	2,096	3,357
Population per doctor, 1977	6,750	16,199	24,185	550

Notes: 'Southern Africa' and 'Africa' exclude South Africa. Dates are for the country in question, and do not always correspond with the Regional, African and Industrial averages.

GUINEA-BISSAU: Leading Indicators

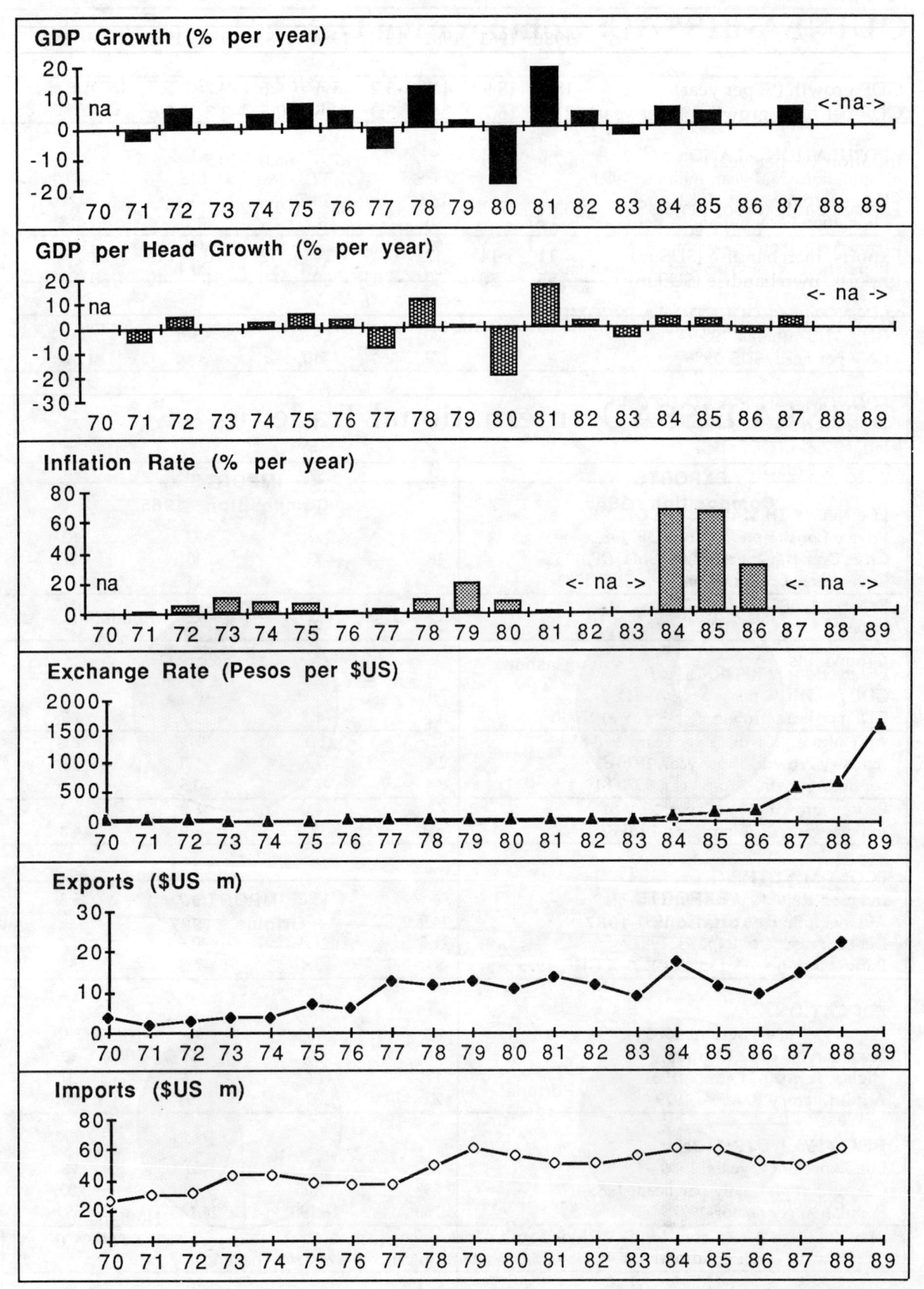

GDP Growth (% per year)

GDP per Head Growth (% per year)

Inflation Rate (% per year)

Exchange Rate (Pesos per $US)

Exports ($US m)

Imports ($US m)

Note: Leading indicator data for 1989 are based on the first half of 1989. 1989 exchange rate is for mid-1989.

GUINEA-BISSAU: Leading Indicator Data

	1980	1981	1982	1983	1984	1985	1986	1987	1988	1989
GDP growth (% per year)	-18.9	18.9	4.4	-3.2	5.6	4.6	-0.7	5.7		
GDP per head growth (% per year)	-21.1	16.7	2.2	-5.3	3.5	2.5	-3.8	3.6		
Inflation (% per year)	7.9	0.0				67.0	66.0	30.3		
Exchange rate (Pesos per $US)	33	37	40	42	105	159	204	559	650	1611
Exports, merchandise ($US m)	11	14	12	9	17	12	10	15	22	
Imports, merchandise ($US m)	55	50	50	55	60	60	51	49	60	

Note: Leading indicator data for 1989 are based on the first half of 1989. 1989 exchange rate is for mid-1989.

GUINEA-BISSAU: International Trade

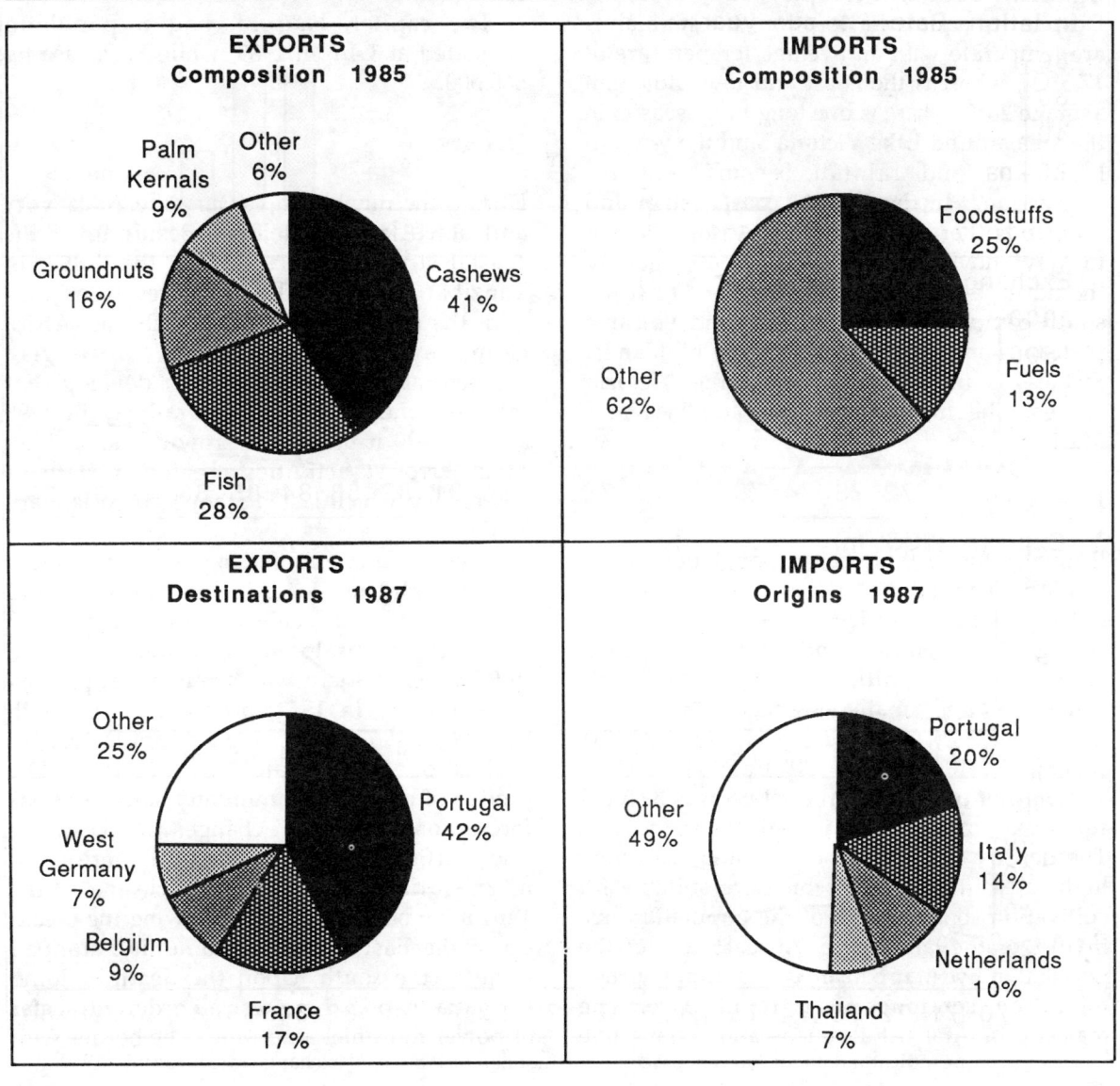

EXPORTS Composition 1985
- Cashews 41%
- Fish 28%
- Groundnuts 16%
- Palm Kernals 9%
- Other 6%

IMPORTS Composition 1985
- Foodstuffs 25%
- Fuels 13%
- Other 62%

EXPORTS Destinations 1987
- Portugal 42%
- France 17%
- Belgium 9%
- West Germany 7%
- Other 25%

IMPORTS Origins 1987
- Portugal 20%
- Italy 14%
- Netherlands 10%
- Thailand 7%
- Other 49%

21 KENYA

Physical Geography and Climate

Kenya, in the East African region, straddles the equator and includes part of Lake Victoria and Lake Turkana. It is bounded by Tanzania, Uganda, Sudan, Ethiopia, and Somalia. Despite its equatorial location, the highlands are temperate with an average temperature of 17.5°C, whereas the coast and lowlands can average 26°C. There is one long rainy season in the area around Lake Victoria and the western highlands and rainfall is sufficient for cultivation. In the eastern coastal strip and highlands there are two rainy seasons, March to May and September to October. Generally the north is arid and unproductive whilst the south comprises a coastal plain and volcanic plateau. Forests exist in the upper highlands. Fishery is important around Lake Victoria shores. The major ports are Mombasa and Malindi.

Population

Estimates for 1989 indicate a population of 23.9 million. The ethnic origin includes Bantu, Nilotic, Nilo-Hamitic and Cushtic with a strong Islamic cultural influence. The official language is Swahili, with English widely spoken. Other main languages spoken are Kikuyu (21%), Luhya (14%), Luo (14%), Kalenjin (11%), Kamba (11%) and Kisii (7%). The density of population for 1988 was 39.5 per square kilometre which is high for the region. The density is unevenly distributed with very high rates in the areas which are suitable for cultivation so about 75% of the population live in 10% of the land space. At least 20% of the population live in urban areas and rural to urban migration continues at a rapid pace. The majority of Arabs, Europeans and Asians live there. Population growth in Kenya over the period 1980-87 was one of the highest in Africa at 4.1% and is of serious concern to the government, especially with the increase in agricultural production of only 2.8% over the same period.

The capital, Nairobi, had a population estimated at 1.4m in 1989, while Mombasa had 660,000.

History

During the nineteenth century the Arab ivory and slave trade developed resulting in the coastal strip being claimed by the Sultan of Zanzibar. British influence dates from 1887 with the foundation of the British East Africa Company. Kenya became a British Protectorate in 1894; economic development followed the starting of the railway in 1895 and in 1920 it became a Crown Colony. Mau Mau terrorist activities created a state of emergency which lasted from 1952 to January 1960.

A constitution conferring self government was brought into force in June 1963 and full independence was achieved in December 1963 under President Jomo Kenyatta. Following independence there was a quickly supressed army mutiny in 1964; border fighting with Somalia in 1967 and a coup attempt in 1982. In 1965 a one-party state was declared. The exodus of the Asian community since 1968 has forced some structural changes to take place. The difficulties of the Amin period in neighbouring Uganda and the closing of the Tanzanian border in 1977, following the break-up of the East African Community, stopped trade to the south. Upon the death of Jomo Kenyatta in 1978 there was an orderly transfer of power to Daniel arap Moi. The border with

Tanzania reopened in 1983. 1986 saw student unrest in Nairobi, leading to the temporary closure of the University.

Moi celebrated ten years in power in 1988 and has shown no signs of giving up the leadership.

Stability

Kenya has retained a civilian government since independence in 1963, and the transfer of power from Kenyatta to Daniel arap Moi was fairly smooth. There was a supressed coup attempt in 1982, but security precautions have been intensified, and the survival of the coup has strengthened constitutional institutions. Economic policy has consistently supported a strong private sector with substantial foreign investment. Some disruption was caused by the exodus of part of the Asian community in 1968 and by the break-up of the East African Community in 1977, but the economy has weathered the changes.

Overall, Kenya's stability is very good. With improving conditions in Uganda and the reopening of the border with Tanzania in 1983, together with liberalising economic reforms in both Tanzania and Uganda, regional harmony should be enhanced. Concern over the transfer of power to Moi's successor has receded with the choice of George Saitoti as Vice-President.

Economic Structure

GDP in 1987 was estimated at $US 6,930m, making it one of the larger economies in Africa. GNP per head was $US 330, and this places Kenya firmly in the low-income group.

Agriculture generated about 31% of GDP in 1987, and this is below the regional and African averages. Industry generated 19% of GDP and services 50%, and these are above the regional averages.

On the demand side, private consumption accounted for 61% of GDP, below the regional and African averages. Investment was 25% of GDP, again high by African standards, where investment averaged about 16% of GDP. Government consumption was 19% of GDP, and again this is above the African average of 15%. Dependence on foreign trade, with exports comprising 21% of GDP in 1987 and imports 26% of GDP is fairly representative of Africa.

Adult literacy was estimated at 47% in 1981, better than the regional and African averages, but will improve with 97% primary school enrolments, one of the best rates of provision of basic education in Africa. Secondary school enrolments at 20% are better than the regional average, but not as good as in Africa generally. Higher education enrolments in 1985 were 1%, but there have been recent substantial expansions that are expected to place Kenya's 1989 enrolment ratio above this and close to the African average of 2%.

Life expectancy is 57 years and significantly above the African average of 50 years. Average daily calorie supply is above the African average, and provision of doctors and nurses per head of population is twice as good.

Overall, although Kenya is a low-income country, it has one of the larger economies, and better developed industrial and service sectors than the level of income per head alone might indicate. Educational and health provisions are above average for Africa.

Economic Performance

GDP grew at 3.8% a year in the period 1980-87, but with population growth at 4.1% a year, GDP per head declined at -0.3% a year. This is rather better than in the region and in Africa generally, where GDP per head fell more rapidly.

Growth in the 1980s has been reasonably evenly spread across the sectors. Agriculture has expanded at 3.4% a year, industry at 3.0% and services at 4.4%. These are all very much better than African averages, where agriculture grew at 1.2% a year, industry declined at -1.2% a year, and services expanded at 1.2% a year.

Export volumes, however, have declined by -0.6% a year, 1980-87. Import volumes affected by increasing debt service payments and terms of trade falling by 20% since 1980, have declined by -3.0% a year. Export and import volumes have declined more rapidly, however, in Africa generally.

Inflation has been around 10% a year, 1980-87, and this is a slight improvement compared with 11% in the 1970s, and much better than

Kenya

the rest of Africa, where inflation was 15% a year in the 1980-87 period.

Overall, although Kenya's economic performance has involved small annual falls in GDP per head and trade volumes in the 1980s, it has been better than elsewhere in Africa. Inflation has been moderate, at just above single figure annual price rises.

Recent Economic Developments

Kenya's economic strategy maintains reliance on a strong private sector in manufacturing and services as well as in the small-farm sector. Foreign investment is encouraged, although the regulations have recently been uncertain, and there are periodic efforts to increase local participation in foreign-owned enterprises. Kenya is in receipt of structural adjustment loans from the World Bank and an enhanced structural adjustment arrangement was agreed with the IMF for $US 301m over three years in May 1989. Policy changes involve gradual changes to bring domestic prices in line with world prices, although this stops short of floating the exchange rate. Moves to end marketing board monopolies and to undertake widespread privatisation of parastatal enterprises have been resisted.

The growth of GDP has recovered well from the three years 1982 to 1984 when GDP growth was positive but slower than population growth. GDP grew at 5.7% in 1986, and in 1987 and 1988 averaged 5%, compared with a population growth rate of 4.0%. Inflation performance has been good, with the annual rate of price increase following a declining trend in the 1980s, with single-figure inflation achieved in 1986 and 1987, but faltering slightly in 1988 and 1989 to above 10%.

The financial sector has been subject to a series of failures by privately-owned domestic institutions. There have been collapses of five financial groups, where three have shown evidence of irregularities, and two have suffered from the ensuing lack of confidence. Banking regulations have been tightened, and one of the stricken banks has reopened, and there are plans to re-open another, with creditors taking up equity. Banks with foreign ownership and control, namely Barclays and Standard Chartered, have increased their share of banking business, and have realised higher profits, as has the state-owned Kenya Commercial Bank.

Efforts are being made to reform and improve the performance of the parastatal sector with changes in management personnel. The grain purchasing body, the National Cereals and Produce Board (NCPB), provides a continuing problem as maize is bought at well above the world price, and in recent years of good harvests, the Board is accumulating stocks and runs at a continual loss. Kenya seems inclined to solve problems in the parastatal sector by reforms rather than privatisation, although there are moves to end the monopoly of the NCPB by making it a purchaser of last resort.

Kenya's trade balance has been affected by world commodity prices in the 1980s, particularly the prices for coffee and tea. As export revenues fell between 1980 and 1983, imports were compressed to half their 1980 level. The boom in coffee prices in 1986 allowed imports to expand, but they still stand at only two-thirds of the level achieved in 1980, when $US 2.3b of goods were imported.

Some $US 5.7b of external debt was estimated outstanding in 1989. Debt service takes up just over a quarter of export earnings at around $US 250m a year. At present this is within Kenya's ability to service providing export revenues can be maintained.

Kenya's exchange rate policy involves periodic adjustments such that the official rate responds to the market rate. The black market in foreign exchange is not particularly vigorous, but there is evidence that the currency is overvalued by perhaps 10-15%.

Kenya continues to be a major aid recipient with USAID adding $US 5m to the $US 25m already committed to housing, and lending a further $US 15m to rural enterprise schemes. Finland has allocated $US 74m over four years to dairy, water, health and electrification projects. The UK has allocated $US 9.4m to telecommunications, the Netherlands $US 7m to sugar rehabilitation, the African Development Fund and OPEC $US 29m to rice schemes, and the World Bank $US 28m to railway improvements. Japan is contributing $US 20m to balance of payments assistance to support the structural adjustment programme and a further $US 20m for projects.

Kenya

Foreign investment has been steady despite uncertainty regarding local ownership provisions, and the UK's Mirror Group is investing $US 43m in publishing, printing and newspapers. Phillips is expanding its involvement in assembly of electrical goods, and the World Bank's International Finance Corporation has lent up to $US 29m to private sector paper production.

Economic Outlook and Forecasts

The forecasts assume that Kenya maintains her good stability record and that liberalising reforms continue at the same cautious rate of implementation.

Coffee prices are expected to fall by 8% in real terms up to 1995 while tea prices rise by 15%.

GDP is forecast to expand at a rate that will allow modest increases in GDP per head. Inflation is expected to show a marginal acceleration, while export and import volumes show positive growth.

Kenya: Economic Forecasts
(average annual percentage change)

	Actual 1980-87	Forecast Base 1990-95	Forecast High 1990-95
GDP	3.8	4.2	4.7
GDP per Head	-0.3	0.3	0.8
Inflation Rate	10.3	13.5	10.5
Export Volumes	-0.6	3.2	3.9
Import Volumes	-3.0	4.3	5.2

KENYA: Comparative Data

	KENYA	EAST AFRICA	AFRICA	INDUSTRIAL COUNTRIES
POPULATION & LAND				
Population, mid year, millions, 1989	23.9	12.2	10.2	40.0
Urban Population, %, 1985	20	30.5	30	75
Population Growth Rate, % per year, 1980-86	4.1	3.1	3.1	0.8
Land Area, thou. sq. kilom.	583	486	486	1,628
Population Density, persons per sq kilom.,1988	39.5	24.2	20.4	24.3
ECONOMY: PRODUCTION & INCOME				
GDP, $US millions, 1986	5,960	2,650	3,561	550,099
GNP per head, $US, 1986	300	250	389	12,960
ECONOMY: SUPPLY STRUCTURE				
Agriculture, % of GDP, 1986	30	43	35	3
Industry, % of GDP, 1986	20	15	27	35
Services, % of GDP, 1986	50	42	38	61
ECONOMY: DEMAND STRUCTURE				
Private Consumption, % of GDP, 1986	55	77	73	62
Gross Domestic Investment, % of GDP, 1986	26	16	16	21
Government Consumption, % of GDP, 1986	19	15	14	17
Exports, % of GDP, 1986	27	16	23	17
Imports, % of GDP, 1986	27	24	26	17
ECONOMY: PERFORMANCE				
GDP growth, % per year, 1980-86	3.4	1.6	-0.6	2.5
GDP per head growth, % per year, 1980-86	-0.6	-1.7	-3.7	1.7
Agriculture growth, % per year, 1980-86	2.8	1.1	0.0	2.5
Industry growth, % per year, 1980-86	2.7	1.1	-1.0	2.5
Services growth, % per year, 1980-86	4.2	2.5	-0.5	2.6
Exports growth, % per year, 1980-86	-0.9	0.7	-1.9	3.3
Imports growth, % per year, 1980-86	-5.2	0.2	-6.9	4.3
ECONOMY: OTHER				
Inflation Rate, % per year, 1980-86	9.9	23.6	16.7	5.3
Aid, net inflow, % of GDP, 1986	6.9	11.5	6.3	-
Debt Service, % of Exports, 1984	21.5	18	20.6	-
Budget Surplus (+), Deficit (-), % of GDP, 1985	-6.7	-3.0	-2.8	-5.1
EDUCATION				
Primary, % of 6-11 group, 1985	94	62	76	102
Secondary, % of 12-17 group, 1985	20	15	22	93
Higher, % of 20-24 group, 1985	1	1.2	1.9	39
Adult Literacy Rate, %, 1981	47	41	39	99
HEALTH & NUTRITION				
Life Expectancy, years, 1986	57	50	50	76
Calorie Supply, daily per head, 1985	2,214	2,111	2,096	3,357
Population per doctor, 1981	10,120	35,986	24,185	550

Notes: 'Southern Africa' and 'Africa' exclude South Africa. Dates are for the country in question, and do not always correspond with the Regional, African and Industrial averages.

KENYA: Leading Indicators

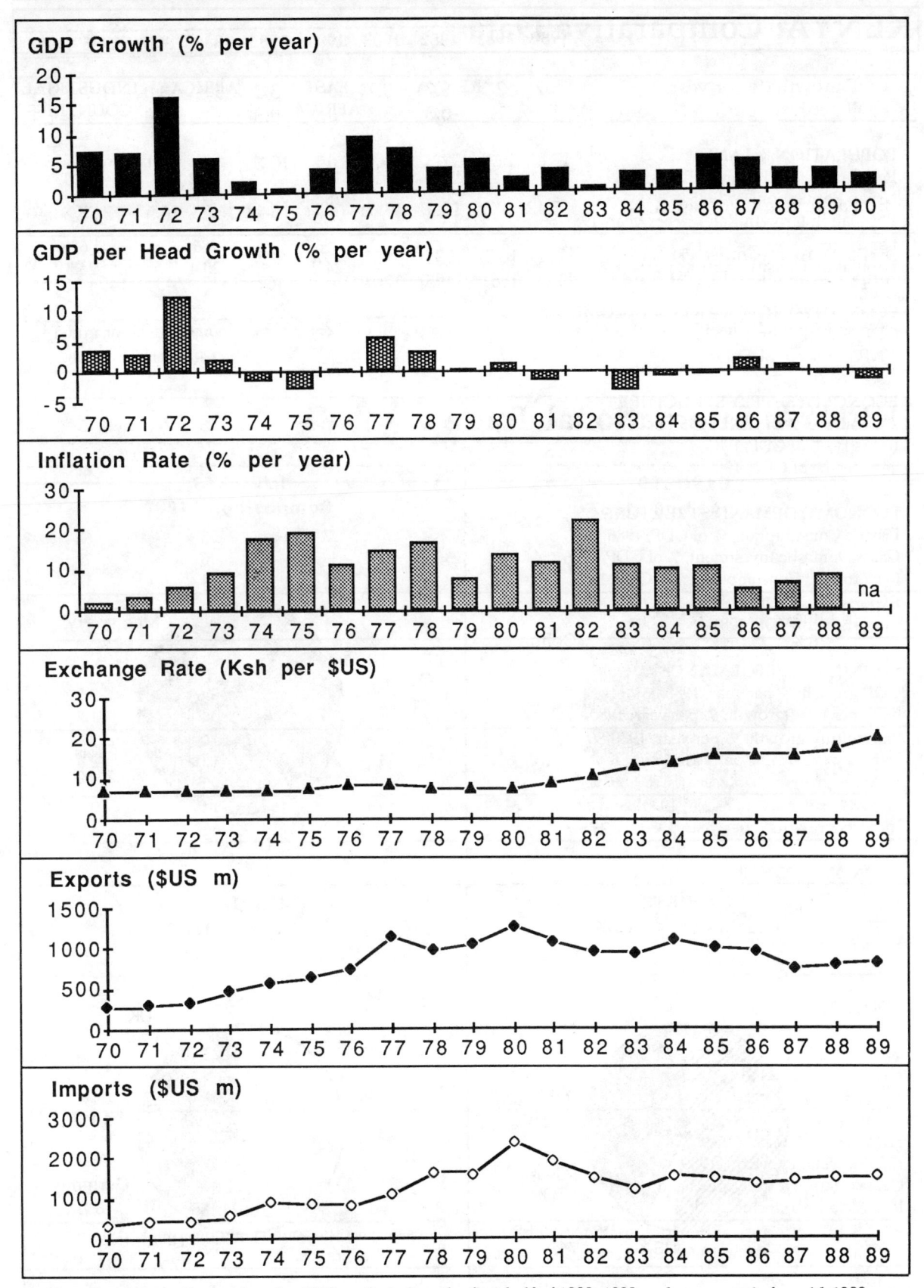

GDP Growth (% per year)

GDP per Head Growth (% per year)

Inflation Rate (% per year)

Exchange Rate (Ksh per $US)

Exports ($US m)

Imports ($US m)

Note: Leading indicator data for 1989 are based on the first half of 1989. 1989 exchange rate is for mid-1989.

KENYA: Leading Indicator Data

	1980	1981	1982	1983	1984	1985	1986	1987	1988	1989
GDP growth (% per year)	5.2	2.3	3.7	0.7	3.1	3.2	5.8	5.0	3.5	2.5
GDP per head growth (% per year)	1.2	-1.7	-0.3	-3.3	-0.9	-0.8	1.8	1.0	-0.5	-1.5
Inflation (% per year)	13.8	11.8	22.33	11.5	10.2	10.7	5.7	7.1	9.0	
Exchange rate (Ksh per $US)	7.42	9.05	10.92	13.31	14.41	16.43	16.23	16.5	17.75	20.9
Exports, merchandise ($US m)	1261	1072	934	925	1078	977	949	748	785	816
Imports, merchandise ($US m)	2345	1881	1495	1204	1552	1462	1337	1431	1495	1520

Note: Leading indicator data for 1989 are based on the first half of 1989. 1989 exchange rate is for mid-1989.

KENYA: International Trade

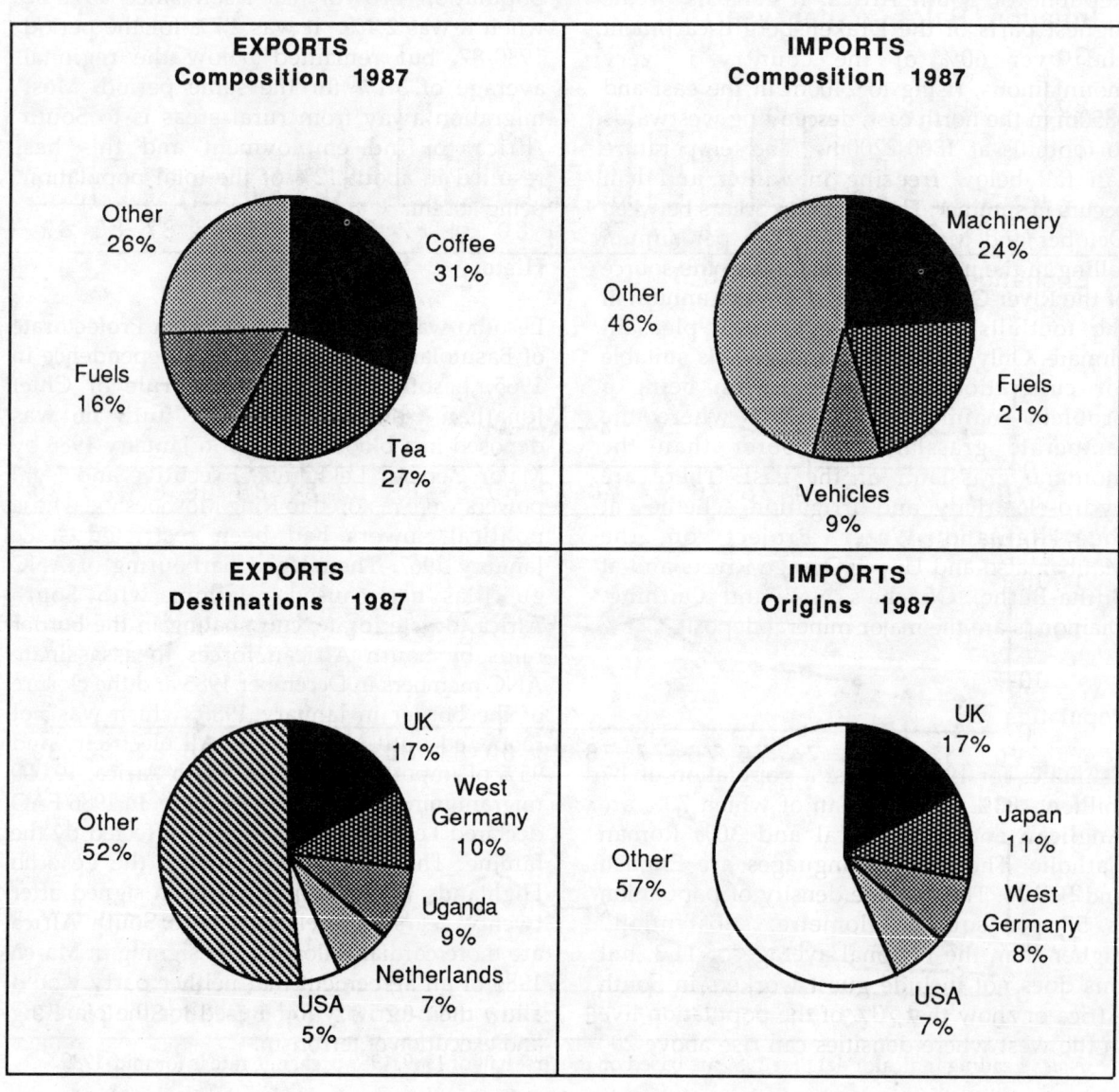

EXPORTS Composition 1987
- Coffee 31%
- Tea 27%
- Fuels 16%
- Other 26%

IMPORTS Composition 1987
- Machinery 24%
- Fuels 21%
- Vehicles 9%
- Other 46%

EXPORTS Destinations 1987
- UK 17%
- West Germany 10%
- Uganda 9%
- Netherlands 7%
- USA 5%
- Other 52%

IMPORTS Origins 1987
- UK 17%
- Japan 11%
- West Germany 8%
- USA 7%
- Other 57%

22 LESOTHO

Physical Geography and Climate

Lesotho, in the Southern African region, is a small inland state, totally surrounded by the Republic of South Africa. It consists of the highest parts of the Drakensberg escarpment and over 60% of the country is very mountainous, rising to 2400m in the east and 3350m in the north east, descending westwards to foothills at 1800-2200m. The temperature can fall below freezing in winter and hail occurs in summer. The most rain occurs between October and March, with 190cm per annum falling in the mountains which form the source of the River Orange and 65-75cm per annum in the foothills which have a more pleasant climate. Only 13% of the landspace is suitable for cultivation with soil erosion being a problem mainly in the west where the temperate grassland is poorer than the montane grassland of the east. There are hydro-electricity and irrigation schemes at the Highland Water Project on the Malibamatso and Upper Orange Rivers and at Butha-Buthe, Quacha's Nek and Quthing. Diamonds are the major mineral deposit.

Population

Estimates for 1989 suggest a population of 1.7 million. 80% are Christian of which 70% are Anglican and Evangelical and 30% Roman Catholic. The official languages are English and Sesoto. The average density of population is 57 per square kilometre, substantially higher than the regional average of 11.3, but this does not include guest workers in South Africa or show that 70% of the population live in the west where densities can rise above 200 persons per square kilometre (most of the country consists of uninhabitable mountainous areas). Urbanisation is 17% which is lower than the regional average of 23.8%. The rate of population growth has risen since 1970-82 when it was 2.4%. It was 2.7% for the period 1980-87, but remained below the regional average of 3.1% for the same period. Most migration away from rural areas is to South Africa to find employment and this has resulted in about 12% of the total population being absent.

History

Lesotho was known as the British Protectorate of Basutoland from 1868 until independence in 1965. Lesotho was under the rule of Chief Jonathan from independence until he was deposed in a bloodless coup in January 1986 by Major General Lekhanya. Executive and legal powers were restored to King Moshoeshoe whose political powers had been restricted since January 1967. The alleged harbouring of ANC guerillas has caused relations with South Africa to deteriorate, culminating in the border raids by South African forces to assassinate ANC members in December 1985 and the closure of the border in January 1986, which was not reopened until March 1986. All electricity and 95% of imports come from South Africa; 19,000 migrant miners are employed there. In 1986 FAO declared Lesotho to be seriously affected by the famine. The treaty to commence the Lesotho Highlands Water Project has been signed after twenty years and relations with South Africa are more cordial following the signing in March 1986 of an agreement that neither party would allow their territory to be used for the planning and execution of terrorism.

Lesotho

Stability

Since independence in 1965, Lesotho was ruled by Chief Jonathan until a military coup in 1986, which has subsequently suspended all political activities. There is tension with South Africa over the harbouring of ANC guerrillas, and in 1985 South African forces raided the capital, Maseru.

Economic strategy has continued to rely upon a mixed economy, with considerable dependence on South Africa, enforced by Lesotho's geographical position.

Overall, Lesotho's stability record is moderate, but long-term prospects are enhanced by the decision of South Africa in early 1990 to move toward some form of majority rule.

Economic Structure

GDP was estimated at $US 270m in 1987, and this makes it one of the smaller economies in Africa. Substantial remittances from South Africa increase income, and GNP per head was $US 370 in 1987, which puts Lesotho at the upper end of the low-income grouping.

Agriculture generates 21% of GDP, and in this respect Lesotho is less dependent on farming than is generally the case in the region. 28% of GDP is generated by the industrial sector, which is also below the regional average, but 52% is generated in the services sector. This concentration on services is a reflection of the amount of income that derives from migrant remittances and these flows from outside are vital in sustaining the economy. GDP generated $US 270m in 1989, but gross national income was estimated at $US 629m, with almost $US 360m coming from migrants' earnings.

The composition of demand is also affected by migrants' remittances, with consumption 143% of GDP, investment at 33% of GDP, exports are a mere 12% of GDP, but imports are 124% of GDP.

Lesotho's adult literacy rate was 52% in 1980, but this can be expected to improve with 115% of the primary-school age group in education. 22% of the relevant age group are in secondary education, and 2% in higher education. All these educational indicators are better than for the region generally.

Life expectancy in Lesotho is 55 years and this is better than the regional average of 50. Food supply allowed an average of 2,300 calories per person each day, and this is better than for the region which supplies 2,000 calories a person. There were 18,000 people for every doctor when the last figures, for 1977, were compiled. This is well below the average for the region, which stood at 26,000, and has probably improved in the period since 1977.

Overall, Lesotho is a small, low-income country with considerable dependence on migrant remittances. In view of its low-income status, Lesotho is able to make impressive educational provision and to provide good health services.

Economic Performance

The World Bank estimates Lesotho's GDP growth rate at 2.3% a year over the period 1980-87, and this gives a growth rate for GDP per head of -0.4% a year for the period. These performance figures were better than most in Southern Africa, where GDP fell at -3.7% a year, and GDP per head even faster at -6.6% year.

The services sector performed best, with a growth rate that, at 4.0% a year, kept pace with population expansion in the period 1980-87. Agriculture did less well expanding at 0.4% a year, while industrial sector output increased at 0.4% a year. In all these sectors performance was better than in Africa generally.

The inflation rate averaged 12.3% in the period 1980-87, and, as such, Lesotho experienced better price stability than the average for the region.

The investment ratio at 25% of GDP is high, and the relatively low growth rate for GDP implies a low capital productivity for new investment in the 1980s.

Export volumes grew at 8.3% a year in 1970-81, although there are indications that they have declined in the 1980s. Import volumes expanded at 13.2% a year in the 1970s. Unlike most African countries, ability to import has not been determined by export earnings but by the volumes of remittances from migrant workers in South Africa which cover the cost of over 60% of imports. The budget deficits stood at -2.6% of GDP in 1987, and the moderate size of the deficit contributed to the better than average price stability for the region.

Overall, Lesotho has had modest economic performance in the 1980s, with the most serious implication being falling GDP per head. Despite this, Lesotho's performance has been better than that generally experienced in the Southern Africa region. The degree of dependence on South Africa for employment and income makes Lesotho disconcertingly vulnerable to political developments in South Africa.

Recent Economic Developments

By July 1988, Lesotho had been granted a Structural Adjustment facility of $US 13m by the IMF. The conditionality of the loan included measures designed to improve Lesotho's rate of growth, reduce the deficits in the budget and on the balance of payments and lower the rate of inflation. A programme is to be introduced to reform state-owned enterprises. Performance targets were for GDP growth to average 4.0% annually by 1991, and for the budget deficit to be 4.1% of GDP.

The inflation rate appears to have deteriorated, with an annual rate of 16% being recorded for 1989.

Lesotho runs a continuous deficit on its balance of merchandise trade, with this deficit being met for the most part from remittances made by migrant workers employed in South Africa. Uncertainty over the long-term position of migrant workers in South Africa has led to concern that there should be some improvement in export earnings to narrow the trade deficit.

Lesotho's external debt in 1989 is estimated at around $US 270m, and the debt service ratio was 4.0% of exports of goods and services, which is certainly manageable.

Lesotho remains part of the Rand Monetary Union, and the Lesotho currency, the maloti, is exchanged at par with the rand, which also circulates in Lesotho. There has been substantial depreciation of the currency in the 1980s, from M 0.78 = $US 1 in 1980 to M 2.55 = $US 1 at the end of 1989.

Aid is dominated by the Highlands hydro-electric project which is costed at $US 1,800m, and is expected to involve a wide spectrum of lenders. In 1988 the UNDP committed $US 390m over three years for projects not associated with the Highlands scheme. Other loans in 1988 have included $US 37m from the IDA of the World Bank for urban development and land management.

Economic Outlook and Forecasts

The forecasts assume that there is no disturbance to Lesotho's relations with South Africa in either trade, monetary arrangements or employment of Lesotho's migrant workers.

The forecasts also assume that Lesotho manges to maintain its present level of stability, and that the liberalising reforms continue to be implemented steadily. The forecast is for expansion in GDP, but only at the rate of population expansion. The price stability record is expected to continue at around 14% annual inflation. Export volumes are expected to show positive growth, but imports are projected to grow faster as a result of earnings from the Highland Water scheme.

Lesotho: Economic Forecasts				
(average annual percentage change)				
	Actual	Forecast	
		Base	High
	1980-87	1990-95	1990-95
GDP	2.3	2.6	2.8
GDP per Head	-0.4	0.0	0.2
Inflation Rate	12.3	14.0	12.0
Export Volumes	8.3[a]	2.0	3.3
Import Volumes	13.2[a]	6.0	7.0

Note: [a] 1970-81.

LESOTHO: Comparative Data

	LESOTHO	SOUTHERN AFRICA	AFRICA	INDUSTRIAL COUNTRIES
POPULATION & LAND				
Population, mid year, millions, 1989	1.7	6.2	10.2	40.0
Urban Population, %, 1985	17	23.8	30	75
Population Growth Rate, % per year, 1980-86	2.7	3.1	3.1	0.8
Land Area, thou. sq. kilom.	30	531	486	1,628
Population Density, persons per sq kilom., 1988	56.6	11.3	20.4	24.3
ECONOMY: PRODUCTION & INCOME				
GDP, $US millions, 1986	230	2,121	3,561	550,099
GNP per head, $US, 1986	370	383	389	12,960
ECONOMY: SUPPLY STRUCTURE				
Agriculture, % of GDP, 1986	21	25	35	3
Industry, % of GDP, 1986	27	33	27	35
Services, % of GDP, 1986	52	42	38	61
ECONOMY: DEMAND STRUCTURE				
Private Consumption, % of GDP, 1986	143	67	73	62
Gross Domestic Investment, % of GDP, 1986	33	14	16	21
Government Consumption, % of GDP, 1986	35	21	14	17
Exports, % of GDP, 1986	12	34	23	17
Imports, % of GDP, 1986	124	36	26	17
ECONOMY: PERFORMANCE				
GDP growth, % per year, 1980-85	0.9	-3.7	-0.6	2.5
GDP per head growth, % per year, 1980-85	-1.8	-6.6	-3.7	1.7
Agriculture growth, % per year, 1980-85	1.6	-6.8	0.0	2.5
Industry growth, % per year, 1980-85	-3.9	-3.8	-1.0	2.5
Services growth, % per year, 1980-85	2.7	-0.9	-0.5	2.6
Exports growth, % per year, 1970-81	8.3	-5.1	-1.9	3.3
Imports growth, % per year, 1970-81	13.2	-2.5	-6.9	4.3
ECONOMY: OTHER				
Inflation Rate, % per year, 1980-86	13.1	18.1	16.7	5.3
Aid, net inflow, % of GDP, 1986	16.1	8.6	6.3	-
Debt Service, % of Exports, 1986	4.2	7.8	20.6	-
Budget Surplus (+), Deficit (-), % of GDP, 1986	-2.6	-4.1	-2.8	-5.1
EDUCATION				
Primary, % of 6-11 group, 1985	115	95	76	102
Secondary, % of 12-17 group, 1984	22	18	22	93
Higher, % of 20-24 group, 1984	2	1.3	1.9	39
Adult Literacy Rate, %, 1980	52	47	39	99
HEALTH & NUTRITION				
Life Expectancy, years, 1986	55	50	50	76
Calorie Supply, daily per head, 1985	2,299	1,998	2,096	3,357
Population per doctor, 1977	18,646	26,118	24,185	550

Notes: 'Southern Africa' and 'Africa' exclude South Africa. Dates are for the country in question, and do not always correspond with the Regional, African and Industrial averages.

LESOTHO: Leading Indicators

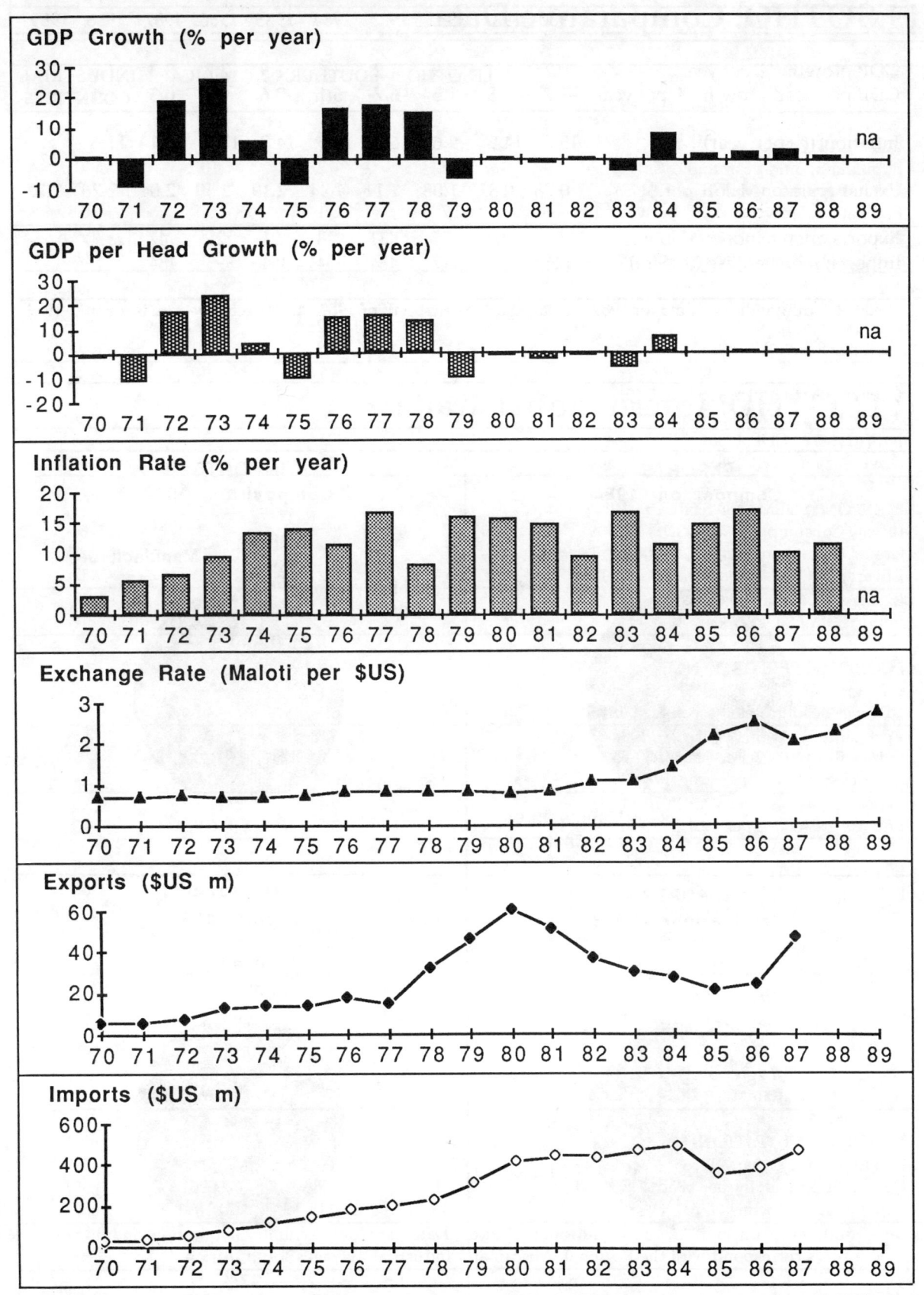

GDP Growth (% per year)

GDP per Head Growth (% per year)

Inflation Rate (% per year)

Exchange Rate (Maloti per $US)

Exports ($US m)

Imports ($US m)

Note: Leading indicator data for 1989 are based on the first half of 1989. 1989 exchange rate is for mid-1989.

LESOTHO: Leading Indicator Data

	1980	1981	1982	1983	1984	1985	1986	1987	1988	1989
GDP growth (% per year)	0.7	-1.4	0.9	-4.2	8.6	1.9	2.7	3.3	1.5	
GDP per head growth (% per year)	-1.7	-3.8	-1.5	-6.7	6.1	-0.6	0.2	0.5	-1.1	
Inflation (% per year)	15.7	14.9	9.6	16.8	11.5	14.8	17.2	10.1	11.6	
Exchange rate (Maloti per $US)	0.78	0.87	1.08	1.11	1.44	2.19	2.53	2.04	2.26	2.77
Exports, merchandise ($US m)	60	51	37	31	28	21	25	47		
Imports, merchandise ($US m)	426	454	447	484	504	364	393	484		

Note: Leading indicator data for 1989 are based on the first half of 1989. 1989 exchange rate is for mid-1989.

LESOTHO: International Trade

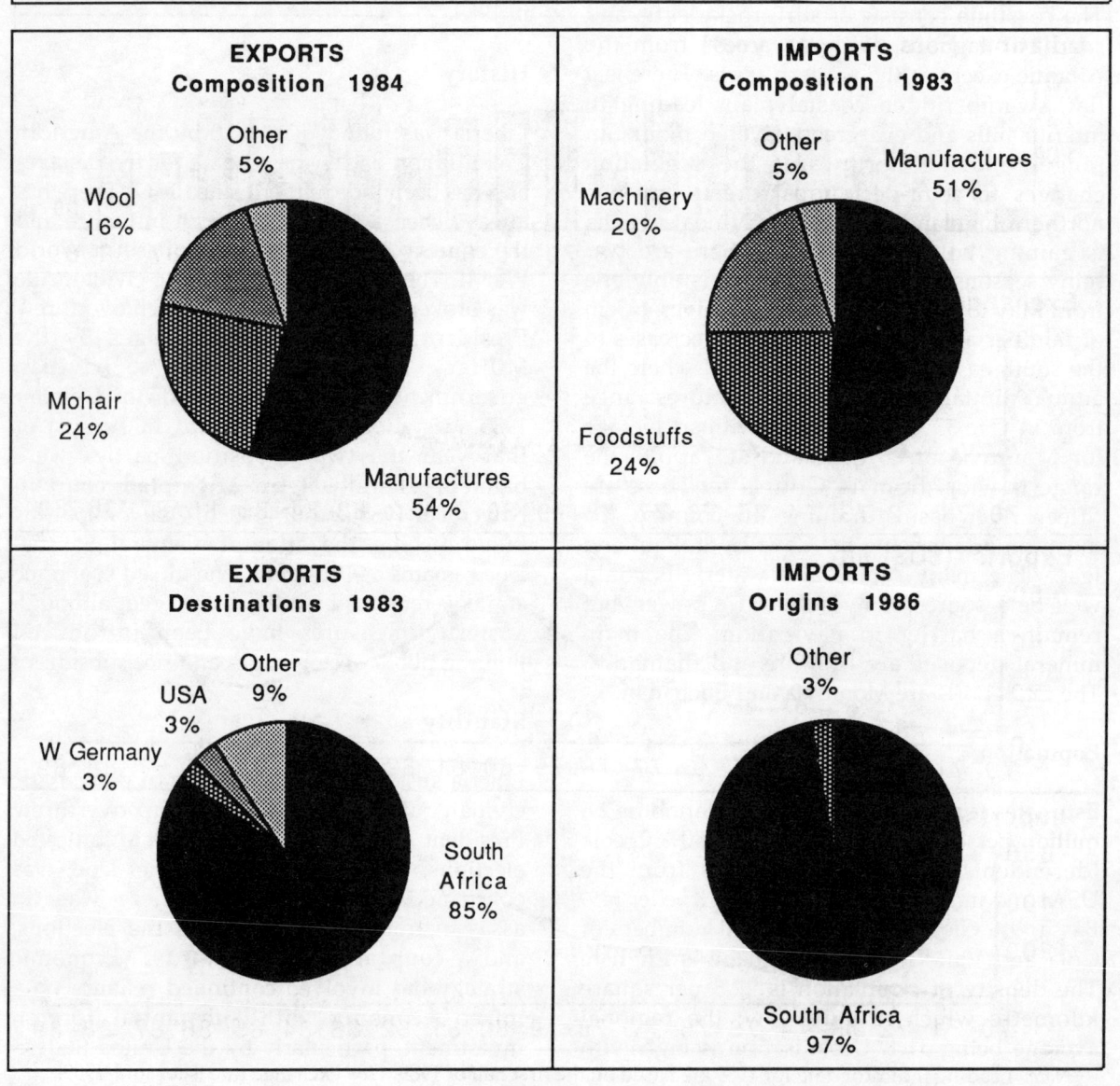

EXPORTS Composition 1984
- Other 5%
- Wool 16%
- Mohair 24%
- Manufactures 54%

IMPORTS Composition 1983
- Other 5%
- Machinery 20%
- Manufactures 51%
- Foodstuffs 24%

EXPORTS Destinations 1983
- Other 9%
- USA 3%
- W Germany 3%
- South Africa 85%

IMPORTS Origins 1986
- Other 3%
- South Africa 97%

23 LIBERIA

Physical Geography and Climate

Liberia, in West Africa, is a coastal state facing the Atlantic Ocean and is bounded by Sierra Leone, Guinea and the Côte d'Ivoire. The coastline consists of surf, rocky cliffs and sand bar lagoons, with no access from the Atlantic except at the modern ports. There is a flat, swamp ridden coastal plain leading to interior hills and evergreen covered mountain ranges; moving northwards the vegetation changes to semi-deciduous forests and the northern highlands where the Guinea savanna is gaining hold. Near Harper there are two rainy seasons, but elsewhere there is only one from May to October. Monrovia receives 465cm of rain per annum, but this figure decreases to the south east and in the hinterland where the annual rainfall is 224cm. Temperatures range from 14°C to 33°C on the coast, with an average for Monrovia of 26°C; inland at Tappita the range is wider, from 14°C to 44°C. There are fifteen river basins draining the country, the most important being the Cavalla, St. Paul and Igst. The many rapids and waterfalls could well be a source of hydro-electric power, but remain a barrier to navigation. The main mineral deposits are iron ore and diamonds. The main ports are Monrovia and Buchanan.

Population

Estimates for 1989 indicate a population of 2.6 million persons, made up of about 50% Creole (decendents of repatriated slaves from the USA) and indigenous peoples: 20% Kpelle; 14% Bassa; 9% Gio; 8% Kru and small numbers of Gola and Vai. The main language is English. The density of population is 22.5 per square kilometre which is quite low; the regional average being 31.1. Urbanisation is high with about 37% in towns, being concentrated around the capital. Population growth was 3.4% for the period 1980-87, roughly in line with the rest of the region.

Monrovia had an estimated 315,000 people in 1989.

History

Liberia was founded in 1820 by the American Colonisation Society as a home for freed slaves and has been independent since 1847. There has always been a strong American influence and the country was used as an air base in World War II. The century of democratic civilian rule was broken by a coup in 1980 which overthrew President Tolbert. Despite pledges by the Military Council to return to civilian government, the general election of October 1985 was allegedly rigged and in August of that year the two opposition parties were banned. There was an attempted coup in November 1985 and Brig. Gen. Quiwonkpa was killed fleeing the attempt. General Samuel Doe remains as President. The mixed economic strategy remains broadly unchanged, although austerity measures have been introduced limiting public sector wages and rice subsidies.

Stability

Liberia sustained over a century of democratic civilian rule until a military coup overthrew President Tolbert in 1980. In 1985, in contested elections, the coup leader, Samuel Doe, was confirmed as president. There was an assassination attempt prior to the elections, and a coup attempt afterwards. Economic strategy has involved continued reliance on a mixed economy with substantial foreign investment, particularly by the United States.

Liberia

Overall, Liberia's stability is moderate, with the poor recent record of political stability somewhat offset by continuity in economic policies. There are doubts as to whether the regime of Samuel Doe can continue to retain power in the longer term.

Economic Structure

In 1987 Liberia had a GDP of $US 990m, well below average size. The standard of living was fairly low at $US 450 per capita when the regional average was $US 510. The agricultural sector generated 37% of GDP which is lower than average in West Africa. The industrial sector (including mining) generated 28%, while the regional average was slightly lower at 26%. The service sector generated 35%.

On the demand side, 65% of GDP was private consumption, well below the regional average of 77%, while government consumed 17% which was higher than the 13% average for the region. The investment level was low, at only 10% of GDP.

Exports were equal to 43% of GDP while imports comprised 34%. Liberia's main exports are iron ore which makes up 57% of exports, while rubber makes up a further 23%. Imports include fuels 18%, food produce 16% and manufactured goods and machinery 34%. Export of merchandise brought in $US 215m while imports cost $US 224m.

Education standards in Liberia were about average for the region. In 1982 about 76% of primary-aged children were in school, with 23% of secondary-aged children in school. About 2% of 20-24 year olds are in higher education. The adult literacy rate in 1980 was 40% when the regional average was under 30%.

Life expectancy is 54 years, above the regional and African average of 50. Provision of doctors and nurses is good with more than twice as many nurses as the regional average and nearly twice as many doctors.

Overall, Liberia is a low-income country, with a heavy dependence on the trade sector where iron-ore and rubber are important sources of earnings. Education provision is average, while health provision is better than average.

Economic Performance

Over the period 1980-87 GDP contracted by -1.3%, although the regional decline was even greater at -2%. The standard of living fell over the period with GDP per head declining at -4.6% a year, a rather worse performance than elsewhere in the region and in Africa.

The agricultural sector grew from 1980-87 by 1.2%, about the regional average as Liberia tried to become more self-sufficient in food, especially rice. The industrial sector declined by -6% while the regional decline was only -3.6%, and this is partly due to falling prices and output of iron ore. While the service sector in many countries in the region has grown, in Liberia it declined slightly by -0.8%. Although export volumes declined by -2.6%, import volumes fell more rapidly by -10.2%, though these were both slower than the regional rates of decline.

The rate of inflation over the period 1980-87 was only 1.5%, the lowest in the region with the regional rate over this period at 13% and for Africa 16.7%. However, the inflation rate has increased since 1986 and had reached 11% in 1989.

Overall, the economy has deteriorated with growth being limited to the agricultural sector. Living standards have fallen and prices are starting to rise faster.

Recent Economic Developments

Although Liberia has maintained a mixed economy strategy, there has been a definite reluctance to press on with liberalising reforms in the past three years at a pace that will satisfy the IMF and World Bank. Although legislation was passed in mid-1986 to privatise eleven state-owned bodies covering airline, bus, hotel, petroleum, electricity, water, telecommunication and manufacturing enterprises, no sales have as yet occurred. A major obstacle has been the lack of accounting information on the trading and asset positions of the enterprises. In 1984 imports were liberalised with the exception of rice, poultry, beer, petroleum and arms, and income tax was cut. These reforms were followed by moves later in 1985 to introduce a 25% pay cut in the civilian public sector and some public sector fringe benefits, and mount an anti-corruption drive. The IMF and World Bank have ceased making disbursements, and it is reported that the banking system is suffering from an inability to create adequate credit and an unwillingness on the part of the public to accept cheques.

Economic performance has seriously deteriorated in the 1980s with falls in GDP occurring every year. This contrasts with modest expansion of GDP at an average rate of 1.4% a year in the 1970s, although the 1970s' performance was not enough to prevent average living standards falling at -1.6% a year. Liberia's inflation rate fell steadily from 14.7% in 1980 to close to zero in 1985, but the liberalisations of 1986 have led to a rise in prices of over 10% in 1989.

The trade balance is typically in surplus, with exports running 25% greater than imports which is more than offset by a deficit on invisibles, giving a current account deficit met by a surplus on the capital account. Exports peaked in 1980 at $US 600m, but fell to $US 323m in 1989. This has compressed imports from $US 478m in 1980 to $US 273m in 1989. Poor prospects for iron ore, Liberia's main export, are mainly responsible, but the steady deterioration of the economy in the 1980s has discouraged overseas investment in the export sector.

Liberia had a total external debt of $US 1.5b in 1988, leading to long-run debt service payments of around $US 22m annually, which even at 1989 levels of merchandise export earnings should be manageable. However, Liberia has built up arrears to its external creditors of some $US 400m, mostly of short-term credits of which $US 165m to the IMF is the main item. There are also arrears to the World Bank and various bilateral donors. These arrears have led to the suspension of further lending by these bodies, and rescheduling of debt has been shelved until the areas are cleared.

The exchange rate is officially at par with the United States dollar, and both US- and Liberian-minted currency circulates in the economy. However, exchange regulations have led to the emergence of a black market, and the US dollar commanded a 15% premium over the Liberian dollar in 1985, and this had risen to 40% in 1986 and 55% by August 1987.

There is some likelihood of Liberia ending the strict link with the US dollar, but there will be resistance to ending an arrangement that has endured for over a century.

The confusion over economic policy is reflected in the postponement of the budget in 1987, and the announcement that it would be presented in January 1988.

Historical connections have resulted in the USA being a major source of both aid and foreign investment. Arrears in repayments and dissatisfaction over the slow pace of reforms and plans for civilian elections led to the US withholding aid. Subsequently the arrears to the US have been paid and $US 10m released for roads, education and health, and a further $US 1.2m for rural communications projects. The USA is allocating $US 18m for support of the economic stabilisation programme, with the provision of 17 managers for ministries and government enterprises expected to be an important factor in implementing reforms. Other aid commitments include $US 2.7m from the European Development Fund for water projects, and $US 7m from Canada for satellite communications. There has been a run-down in foreign investment, with the Chase Manhattan Bank divesting itself of its operation in Liberia, and the Liberian American Swedish Minerals Corporation anticipating a 50% reduction in iron-ore production caused by poor world steel prices and run-down of deposits. On the positive side, Finex International of France is investing $US 12m in the palm-oil sector; Liberia will provide 15% of the output of the Mifergui-Nimba iron-project with Guinea; and Link Oil of the US is handling the privatisation of the Liberian Petroleum Refinery Company, in which 20 bidders are reported to be interested.

Economic Outlook and Forecasts

The forecasts assume that the Doe regime retains its fragile hold on power and there is no progress in implementing liberalisation reforms or coming to terms with the IMF and the World Bank.

Iron ore prices are expected to fall in real terms by 14% up to 1995. As a result, Liberia's GDP is forecast to continue declining, as is GDP per head. Inflation is expected to be above 10% a year and export volumes are projected to

Liberia

continue to decline, with import volumes declining even faster.

Liberia: Economic Forecasts
(average annual percentage change)

	Actual	Forecast Base	High
	1980-87	1990-95	1990-95
GDP	-1.3	-2.0	-1.5
GDP per Head	-4.6	-5.0	-4.5
Inflation Rate	1.5	12.0	10.0
Export Volumes	-2.6	-3.0	-2.0
Import Volumes	-10.2	-3.5	-2.5

LIBERIA: Comparative Data

	LIBERIA	WEST AFRICA	AFRICA	INDUSTRIAL COUNTRIES
POPULATION & LAND				
Population, mid year, millions, 1989	2.6	12.3	10.2	40.0
Urban Population, %, 1985	37	28.7	30	75
Population Growth Rate, % per year, 1980-86	3.3	3.2	3.1	0.8
Land Area, thou. sq. kilom.	111	384	486	1,628
Population Density, persons per sq kilom.,1988	22.5	31.1	20.4	24.3
ECONOMY: PRODUCTION & INCOME				
GDP, $US millions, 1986	990	4,876	3,561	550,099
GNP per head, $US, 1986	460	510	389	12,960
ECONOMY: SUPPLY STRUCTURE				
Agriculture, % of GDP, 1986	37	40	35	3
Industry, % of GDP, 1986	28	26	27	35
Services, % of GDP, 1986	35	34	38	61
ECONOMY: DEMAND STRUCTURE				
Private Consumption, % of GDP, 1986	65	77	73	62
Gross Domestic Investment, % of GDP, 1986	10	13	16	21
Government Consumption, % of GDP, 1986	17	13	14	17
Exports, % of GDP, 1986	43	18	23	17
Imports, % of GDP, 1986	34	20	26	17
ECONOMY: PERFORMANCE				
GDP growth, % per year, 1980-86	-1.3	-2.03	-0.6	2.5
GDP per head growth, % per year, 1980-86	-4.5	-4.7	-3.7	1.7
Agriculture growth, % per year, 1980-86	1.2	1.3	0.0	2.5
Industry growth, % per year, 1980-86	-6.0	-3.6	-1.0	2.5
Services growth, % per year, 1980-86	-0.8	-2.5	-0.5	2.6
Exports growth, % per year, 1980-86	-2.0	-4.0	-1.9	3.3
Imports growth, % per year, 1980-86	-8.8	-12.9	-6.9	4.3
ECONOMY: OTHER				
Inflation Rate, % per year, 1980-86	1.1	13.0	16.7	5.3
Aid, net inflow, % of GDP, 1986	9.6	3.7	6.3	-
Debt Service, % of Exports, 1986	6.4	21.4	20.6	-
Budget Surplus (+), Deficit (-), % of GDP, 1986	-9.0	-3.4	-2.8	-5.1
EDUCATION				
Primary, % of 6-11 group, 1982	76	75	76	102
Secondary, % of 12-17 group, 1982	23	24	22	93
Higher, % of 20-24 group, 1982	2	2.6	1.9	39
Adult Literacy Rate, %, 1980	25	30	39	99
HEALTH & NUTRITION				
Life Expectancy, years, 1986	54	50	50	76
Calorie Supply, daily per head, 1985	2,373	2,105	2,096	3,357
Population per doctor, 1981	9,340	16,199	24,185	550

Notes: 'Southern Africa' and 'Africa' exclude South Africa. Dates are for the country in question, and do not always correspond with the Regional, African and Industrial averages.

LIBERIA: Leading Indicators

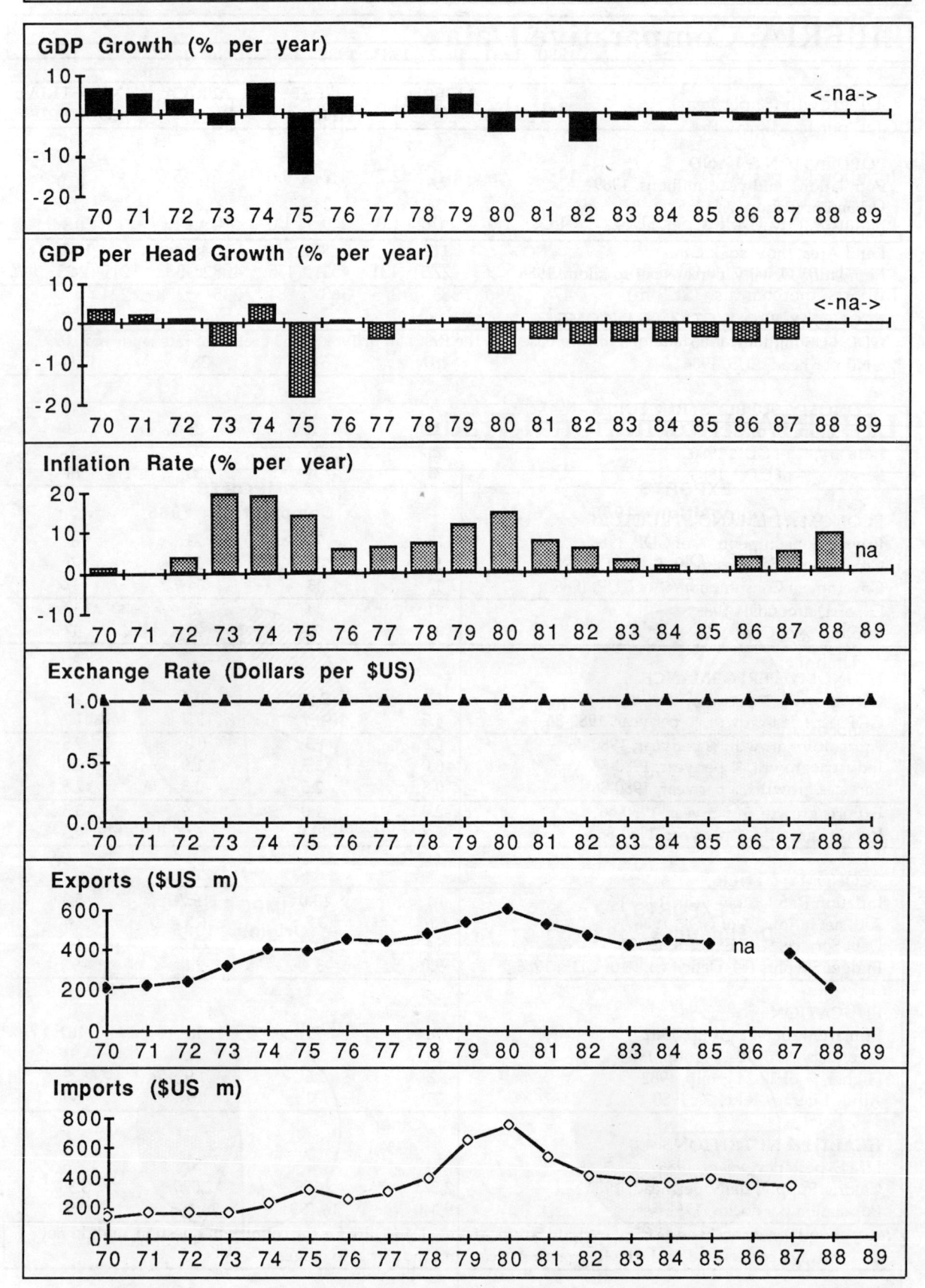

GDP Growth (% per year)

GDP per Head Growth (% per year)

Inflation Rate (% per year)

Exchange Rate (Dollars per $US)

Exports ($US m)

Imports ($US m)

Note: Leading indicator data for 1989 are based on the first half of 1989. 1989 exchange rate is for mid-1989.

LIBERIA: Leading Indicator Data

	1980	1981	1982	1983	1984	1985	1986	1987	1988	1989
GDP growth (% per year)	-4.5	-1.2	-2.0	-1.6	-1.6	-0.7	-1.7	-1.0		
GDP per head growth (% per year)	-7.7	-4.4	-5.2	-4.8	-4.8	-3.9	-4.9	-4.2		
Inflation (% per year)	14.7	7.6	6.0	2.7	1.2	-0.6	3.6	5.0	9.3	
Exchange rate (Dollar per $US)	1.0	1.0	1.0	1.0	1.0	1.0	1.0	1.0	1.0	1.0
Exports, merchandise ($US m)	600	529	477	421	452	435	408	382	215	
Imports, merchandise ($US m)	478	396	383	375	363	284	235	308	224	

Note: Leading indicator data for 1989 are based on the first half of 1989. 1989 exchange rate is for mid-1989.

LIBERIA: International Trade

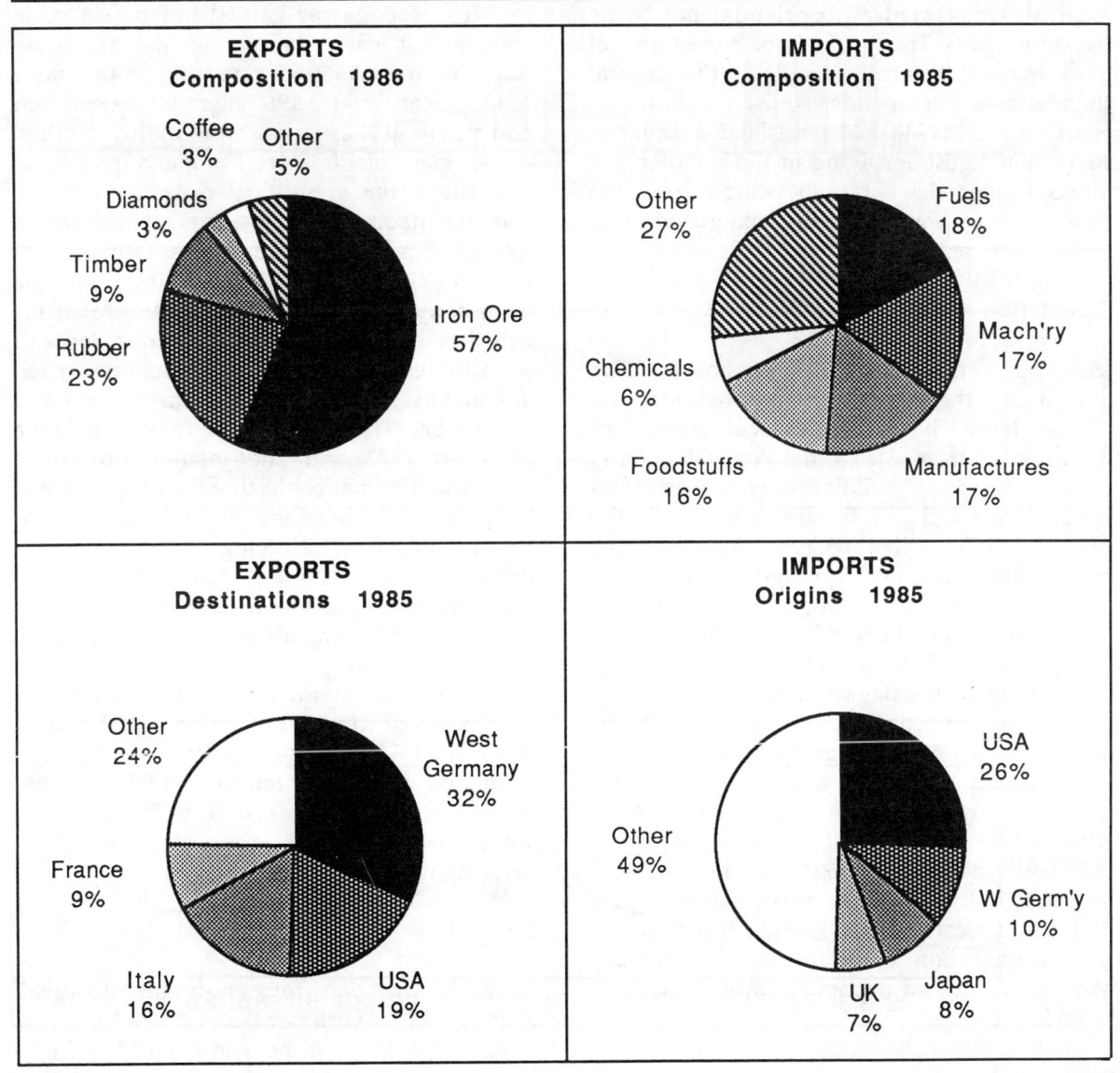

EXPORTS Composition 1986
- Coffee 3%
- Other 5%
- Diamonds 3%
- Timber 9%
- Rubber 23%
- Iron Ore 57%

IMPORTS Composition 1985
- Other 27%
- Fuels 18%
- Mach'ry 17%
- Chemicals 6%
- Foodstuffs 16%
- Manufactures 17%

EXPORTS Destinations 1985
- Other 24%
- West Germany 32%
- France 9%
- Italy 16%
- USA 19%

IMPORTS Origins 1985
- Other 49%
- USA 26%
- W Germ'y 10%
- Japan 8%
- UK 7%

24 MADAGASCAR

Physical Geography and Climate

Madagascar, in the East African region, is the fourth largest island in the world. It lies to the east of Mozambique in the Indian Ocean. The mountainous island has a narrow eastern coastal strip, central highlands and wide western plains. The east and north west are hot and humid, typically tropical; the central highlands are temperate, whilst the west and south are arid. Most of the island is savanna with rain forest surviving in the remoter wet areas, though this is rapidly being turned into savanna too, through overgrazing and poor cultivation practices.

Population

Estimates for 1989 give a population of 9.7m. Originally the island was uninhabited and there have been successive waves of immigrants from Africa, the Arab States and Indonesia. The population now compromises 26% Merina and 12% Betsileo who live in the west and other tribes 'côtiers' who live in the north, east and south. Traditional antagonism exists between the Merina and the 'côtiers', but there are also numbers of Metropolitan French, Indians and Chinese. The official languages are French and Malagasy. Population density varies from 30 people per square kilometre on the central plateau to 2 per square kilometre on the western coast. The average density has risen from 16 per square kilometre in 1982 to 19.8 in 1989. The population has increased at 3.3% over the period 1980-87. Around 37% of the people live in urban areas, rather higher than is typical in East Africa, and there is steady migration to the towns. The capital, Antananarivo had a population estimated at 1.6m in 1989.

History

Madagascar was discovered by the Portuguese in 1500 and became a French possession in 1896. It was at one time a kingdom with the last native sovereign Queen Ranavalona III dying in 1916. Madagascar gained independence in 1960 and became known as the Malagasy Republic from 1960-75 reverting to the name Madagascar. Since 1960 there have been four coups; in 1972 riots ended the civilian government of Philibert Tsiranana, preparing conditions for a military coup by General Ramanantsoa. The leadership changed following a mutiny and the assassination of Col. Ratsimandrava. The military regime eventually appointed Lt-Com. Ratsiraka; his rule has been marked by a unique form of socialism encouraging village communes called 'fokonolona', but also coup attempts, riots and retaliation. The withdrawal from the Franc Zone in 1973 and the introduction of a programme of nationalisation in 1975 reduced business confidence and discouraged foreign investment, but following IMF pressure a partial policy reversal has taken place, accompanied by devaluations, de-controlling of rice prices and termination of the government's monopoly in rice purchase. In late 1984 there were violent disturbances following the prohibition of Kung Fu resulting in 100 deaths. In March 1986 cyclones caused damage estimated at US $50 million and left 20,000 people homeless. French and US aid was accepted for the first time since the 1975 revolution.

Stability

There have been four non-constitutional changes in government since 1960. The

Madagascar

military government of Didier Ratsiraka has been in power since 1975, and has been subjected to sporadic violence, rioting and attempted coups. Economic strategy was marked by a withdrawal from the Franc Zone in 1973 and a movement towards a substantial public ownership in 1975. In 1982 there was agreement with the IMF and a return to more reliance on market forces.

Overall, Madagascar's stability is poor with internal unrest disrupting economic life and changes in economic strategy adding to the uncertainties of commercial activities. The internal unrest shows no sign of diminishing, and the economic policies introduced in 1982 have not as yet established a stable economic climate.

Economic Structure

Madagascar's GDP is smaller than average for the region at $US 2,070 in 1987 and the country's living standard of GNP of $US 210 per person puts it in the low-income group of countries. With agriculture generating 43% of GDP, industry 16% and services 42%, Madagascar's economy has a structure fairly typical of the region.

Private consumption of 79% of GDP, government consumption and investment of 14% each are also similar to the regional average.

Exports contribute 20% of GDP while imports cost 27%. Coffee made up 44% of export earnings in 1986, vanilla 16% and cloves 11%.

In 1986, the percentage of primary-aged children in school was 121%, indicating considerable adult education. A further 36% of the secondary age group were in schools, double the regional average, while 5% of 20-24 year olds were in higher education, one of the best education records in Africa.

Adult literacy was 48% in 1980, rather better than elsewhere in the region and Africa. Health facilities in the country are also good with a life expectancy of 54 years, better than the African average, and twice as many doctors than average for Africa.

Overall, Madagascar is a low-income country with substantial dependence on agriculture for generating GDP and export earnings. Health and educational provision are above average.

Economic Performance

Economic growth has been slow with GDP growing by only 0.3% a year from 1980-87. Living standards fell during the same period with GDP per head falling by -3.0% a year. Growth in the agricultural sector was 2.2%, but both the industrial and service sectors contracted, by -2.0% and -0.5% respectively.

Export volumes declined by -3.1% a year. Import volumes contracted by -2.9% a year.

Inflation over this period averaged 17.4% a year, which is less than the regional rate, but above average for Africa.

Overall, Madagascar's economic performance has been poor in the 1980s despite efforts to reform the economy. This disappointing record extends to all the producing sectors, international trade and inflation.

Recent Economic Developments

Mid-1987 saw the presentation of a World Bank package of $US 100m in support of the economic reform programme which was introduced in 1982. This has involved substantial devaluation of the currency, deregulation of domestic prices, and liberalisation of imports. Inflation performance has been variable, increasing from 18% a year in 1980 to 32% in 1982, and then declining to 10% in 1984 and 1985, and then rising again to around 14% by 1989.

The trade balance is typically in deficit, but export values peaked in 1980 at $US 437m and have declined to $US 360m in 1989, and imports have consequently been compressed from $US 764m in 1980 to $US 420m. Production of the main export, coffee, has remained stagnant. Ageing coffee tree stocks and poor roads which resulted in 20% of the crop remaining unharvested, were responsible for the poor performance.

Madagascar had $US 3.7b of outstanding external debt in 1989, giving rise to a ratio of debt service to export earnings of around 50%. Reschedulings of debts to government creditors in October 1986 over ten years with a five-year grace period and a rescheduling of commercial debt in August 1987 have reduced annual debt

Madagascar

servicing.

Madagascar withdrew from the Franc Zone in 1973, and the currency appreciated up to 1980, from which time the Malagasy franc has steadily depreciated. Two substantial devaluations in August 1986 and June 1987 have resulted in the currency depreciating from MgFr 676 = $US 1 in 1986 to MgFr 1,308 = $US 1 at the end of 1989.

In May 1989 the IMF granted an enhanced structural adjustment agreement of $US 96m over three years.

The $US 100m World Bank structural adjustment package includes Japanese and Saudi loans, and France has contributed a further $US 32m. $US 35m has been allocated to port improvements with $US 16m coming from the World Bank, US 12.6m from France, and $US 2.3m from West Germany. The African Development Bank has committed $US 12.8m to water projects, France $US 12.8m to rice and sugar schemes and $US 3.7m to exports, the World Bank $US 10m to agricultural credit, the European Investment Bank $US 17.5m to electricity, Canada $US 6.1m to oil exploration, and the United States $US 53m to agriculture.

As one of the poorest countries in Africa, Madagascar has qualified for debt cancellations and convertions agreed by a number of countries recently. France, the largest bilateral donor, agreed to write off French aid debts and the USA is expected to write off around $US 100m.

Economic Outlook and Forecasts

The forecasts assume that the poor stability record persists and that Madagascar continues with liberalising economic reforms.

The real price of coffee is forecast to fall 20% by 1995, and this is the main constraint on Madagascar's economic performance.

GDP is forecast to achieve positive growth, but GDP per head is still projected to decline. The inflation rate is expected to slow, but still to be above 10%. Export and import volumes are projected to show slow expansion.

Madagascar: Economic Forecasts
(average annual percentage change)

	Actual 1980-87	Forecast Base 1990-95	Forecast High 1990-95
GDP	0.3	1.5	2.0
GDP per Head	-3.3	-1.7	-1.2
Inflation Rate	17.4	13.0	11.0
Export Volumes	-3.4	1.0	1.5
Import Volumes	-2.9	0.9	1.2

MADAGASCAR: Comparative Data

	MADA-GASCAR	EAST AFRICA	AFRICA	INDUSTRIAL COUNTRIES
POPULATION & LAND				
Population, mid year, millions, 1989	11.5	12.2	10.2	40.0
Urban Population, %, 1985	37	30.5	30	75
Population Growth Rate, % per year, 1980-86	3.3	3.1	3.1	0.8
Land Area, thou. sq. kilom.	587	486	486	1,628
Population Density, persons per sq kilom., 1988	19.3	24.2	20.4	24.3
ECONOMY: PRODUCTION & INCOME				
GDP, $US millions, 1986	2,670	2,650	3,561	550,099
GNP per head, $US, 1986	230	250	389	12,960
ECONOMY: SUPPLY STRUCTURE				
Agriculture, % of GDP, 1986	43	43	35	3
Industry, % of GDP, 1986	16	15	27	35
Services, % of GDP, 1986	41	42	38	61
ECONOMY: DEMAND STRUCTURE				
Private Consumption, % of GDP, 1986	76	77	73	62
Gross Domestic Investment, % of GDP, 1986	14	16	16	21
Government Consumption, % of GDP, 1986	13	15	14	17
Exports, % of GDP, 1986	14	16	23	17
Imports, % of GDP, 1986	17	24	26	17
ECONOMY: PERFORMANCE				
GDP growth, % per year, 1980-86	-0.1	1.6	-0.6	2.5
GDP per head growth, % per year, 1980-86	-3.4	-1.7	-3.7	1.7
Agriculture growth, % per year, 1980-86	2.1	1.1	0.0	2.5
Industry growth, % per year, 1980-86	-3.6	1.1	-1.0	2.5
Services growth, % per year, 1980-86	-0.7	2.5	-0.5	2.6
Exports growth, % per year, 1980-86	-2.0	0.7	-1.9	3.3
Imports growth, % per year, 1980-86	-8.8	0.2	-6.9	4.3
ECONOMY: OTHER				
Inflation Rate, % per year, 1980-86	17.8	23.6	16.7	5.3
Aid, net inflow, % of GDP, 1986	9.6	11.5	6.3	-
Debt Service, % of Exports, 1986	6.4	18	20.6	-
Budget Surplus (+), Deficit (-), % of GDP, 1974	5.9	-3.0	-2.8	-5.1
EDUCATION				
Primary, % of 6-11 group, 1984	121	62	76	102
Secondary, % of 12-17 group, 1984	36	15	22	93
Higher, % of 20-24 group, 1984	5	1.2	1.9	39
Adult Literacy Rate, %, 1980	48	41	39	99
HEALTH & NUTRITION				
Life Expectancy, years, 1986	53	50	50	76
Calorie Supply, daily per head, 1985	2,452	2,111	2,096	3,357
Population per doctor, 1981	9,920	35,986	24,185	550

Notes: 'Southern Africa' and 'Africa' exclude South Africa. Dates are for the country in question, and do not always correspond with the Regional, African and Industrial averages.

MADAGASCAR: Leading Indicators

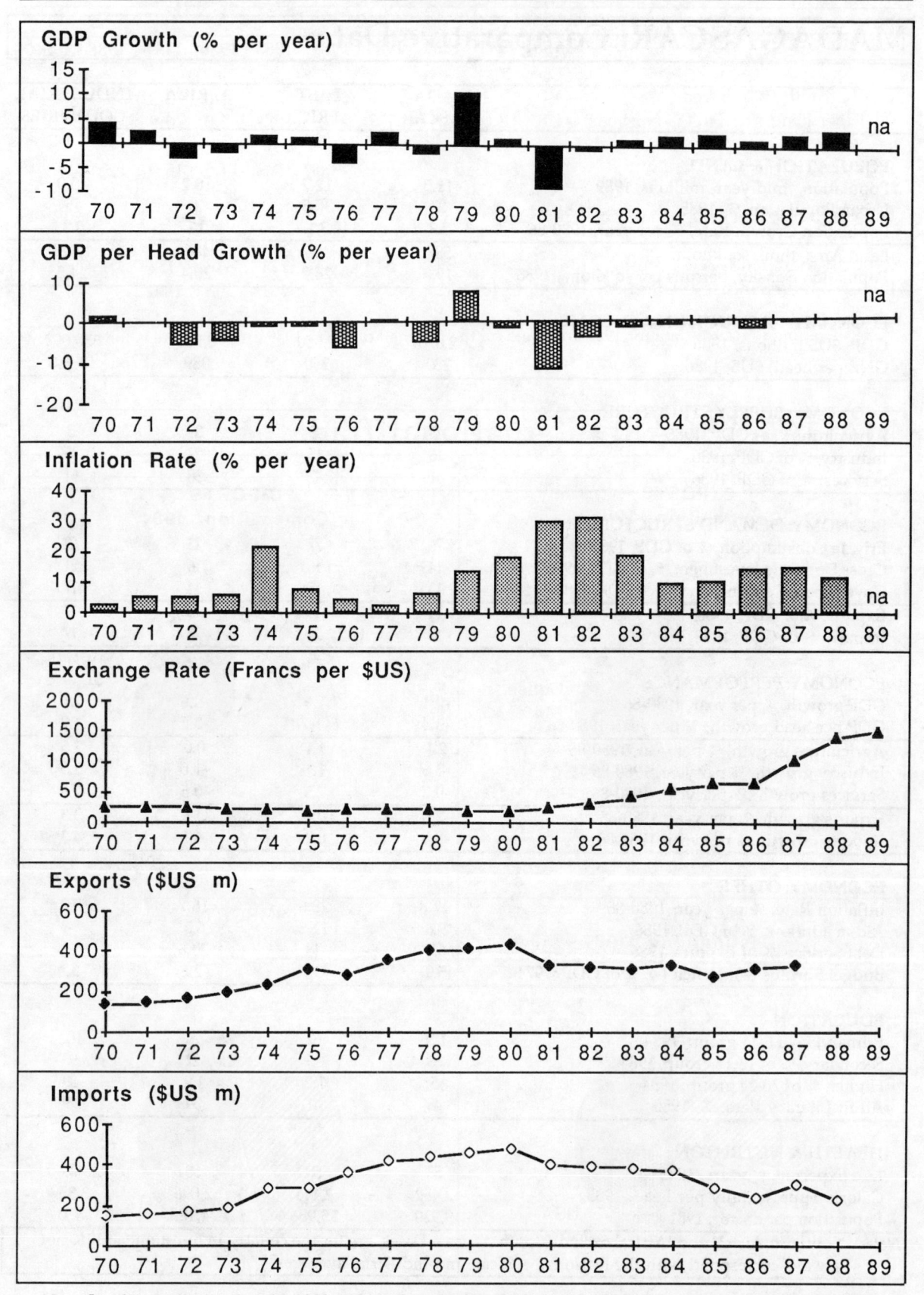

Note: Leading indicator data for 1989 are based on the first half of 1989. 1989 exchange rate is for mid-1989.

MADAGASCAR: Leading Indicator Data

	1980	1981	1982	1983	1984	1985	1986	1987	1988	1989
GDP growth (% per year)	1.0	-9.2	-1.2	1.1	1.9	2.3	1.0	2.2	3.0	
GDP per head growth (% per year)	-1.8	-12.0	-4.1	-1.8	-1.1	-0.7	-2.0	-0.9	-0.1	
Inflation (% per year)	18.2	30.5	31.8	19.3	9.8	10.6	14.5	15.0	12.0	
Exchange rate (Francs per $US)	221	272	348	431	587	663	676	1069	1407	1538
Exports, merchandise ($US m)	437	332	327	313	334	274	316	334		
Imports, merchandise ($US m)	764	546	425	387	370	401	356	340		

Note: Leading indicator data for 1989 are based on the first half of 1989. 1989 exchange rate is for mid-1989.

MADAGASCAR: International Trade

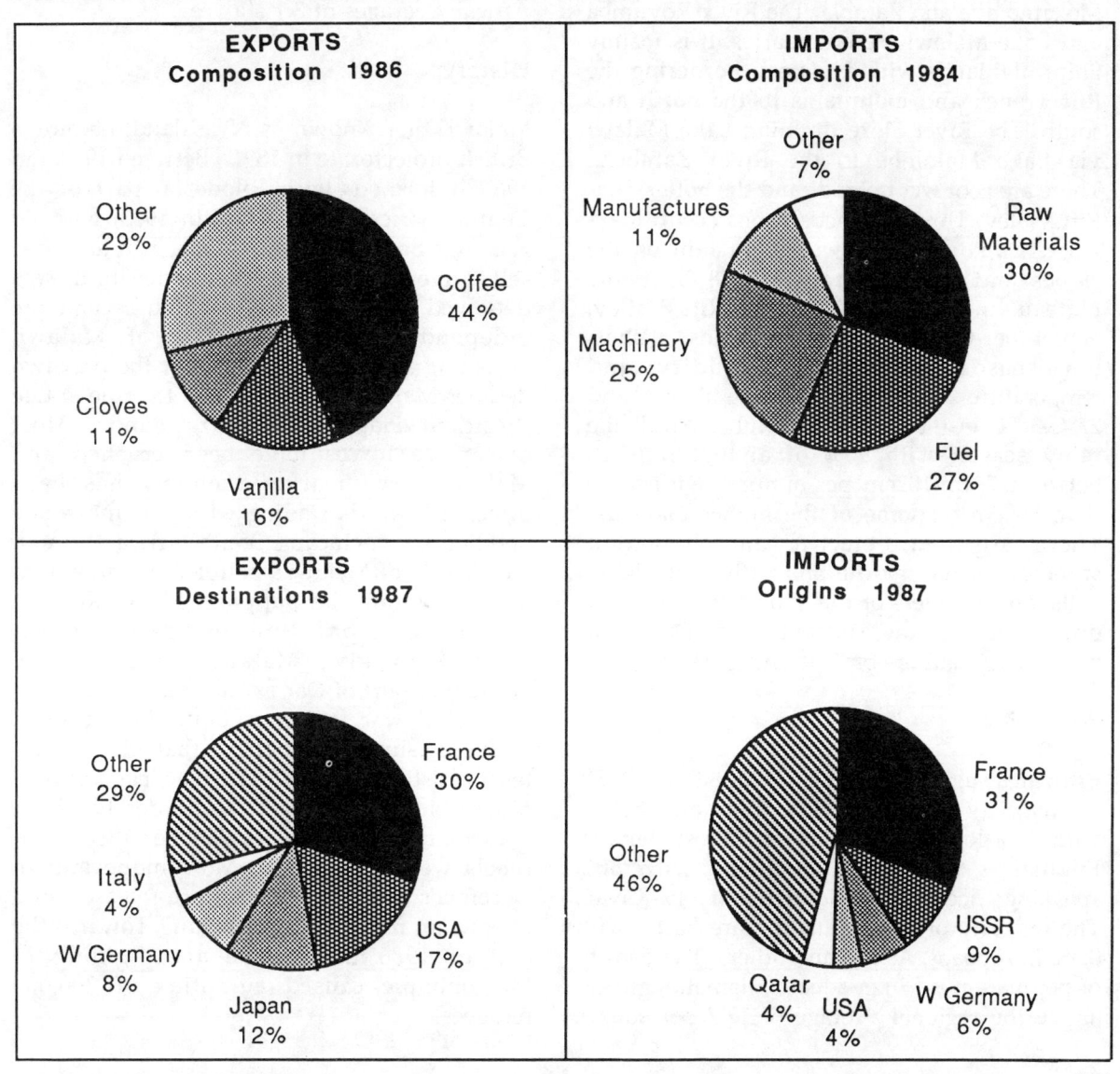

EXPORTS Composition 1986
- Coffee 44%
- Vanilla 16%
- Cloves 11%
- Other 29%

IMPORTS Composition 1984
- Raw Materials 30%
- Fuel 27%
- Machinery 25%
- Manufactures 11%
- Other 7%

EXPORTS Destinations 1987
- France 30%
- USA 17%
- Japan 12%
- W Germany 8%
- Italy 4%
- Other 29%

IMPORTS Origins 1987
- France 31%
- USSR 9%
- W Germany 6%
- USA 4%
- Qatar 4%
- Other 46%

25 MALAWI

Physical Geography and Climate

Malawi, in the Southern Africa region, is a landlocked state on the southern continuation of the Rift Valley, bounded by Tanzania, Mozambique and Zambia. The River Rovumba and Lake Malawi are boundaries. It is mainly Shire highlands with a plateau bordering the Rift Trench and mountains to the north and south. The River Shire draining Lake Malawi via Lake Malombe to the River Zambezi. There are cool wet uplands and the hotter drier Rift Valley. Three distinct seasons occur: May-August a cool season with light rain on the slopes and temperatures 15°C-18°C on the plateau and 20°C-24°C in the Rift Valley; September-October before the rains, a short hot season with high humidity and temperatures of 22°C-24°C on the plateau and 27°C-30°C in the valley; November-April the rainy season with 90% of annual rainfall, between 76cm-102cm per annum with more than 150cm on some of the higher plateaux. There are hydro-electric and irrigation schemes on the middle Shire River at Nkula Falls, on the rivers of the Viphya plateau and downstream of the Tedzani Falls. The main mineral deposits are bauxite and glass sands.

Population

Estimates suggest a population of 8.5m in 1989, the majority of whom follow traditional beliefs with 30% Roman Catholics and Presbyterians. English is an official language with 50% speaking Cicewa, 15% Ilomwe and 14% Ciyao. The majority of the population are Bantu with 0.3% Europeans, Asians and others. The density of population is 58 per square kilometre, greatly above the regional average of 12.7 per square kilometre with 12% living in the north of the country and 50% living in the south. Many are guest workers in South Africa. Urbanisation is 13% which is below the regional average of 24%. The rate of population growth was 3.8% a year, 1980-87, and this is above the regional and African averages of 3.1% a year.

History

Malawi, then known as Nyasaland, became a British protectorate in 1891. Between 1953 and 1963 it formed, with Rhodesia, part of the Central African Federation. In 1963, with Dr Hastings Banda as Prime Minister, it achieved self government and the Federation was dissolved. In July 1964 Nyasaland became the independent sovereign state of Malawi, becoming a republic in 1966. For the past two decades Malawi has been under the rule of Life President Hastings Kamuza Banda. Most disruptive forces have been crushed and Malawi's free market economy has been directed towards trading with her immediate neighbours, including South Africa through which 95% of Malawi's petroleum is imported and to whom she supplies 18,000 migrant workers. In April 1986 an agreement was reached to give Malawi access to the Tanzanian port of Dar es Salaam.

The civil war in Mozambique has strained relations, since they believe that Malawi has been assisting the MNR (Restisténcia Nacional Moçambicana) and that the border has been inadequately guarded. In October 1986 MNR rebels were expelled to Mozamique and in December MNR prisoners were released into Malawi for return to Mozambique. Towards the end of 1986 the famine and fighting in Mozambique caused an influx of 200,000 refugees.

Malawi

Stability

Dr Hastings Banda has been president since independence in 1966, and despite disruptive elements directed towards his government by dissidents based outside the country, has remained secure. Economic strategy has remained pragmatic in view of Malawi's dependence on South Africa for imports and transport links. A strong private sector has been maintained.

Overall, Malawi's stability is very good, although there is concern over whether a smooth transition to Banda's successor can be effected. Malawi is also affected by the uncertainty surrounding South Africa, and her transport problems will not improve until South Africa attains majority rule.

Economic Structure

Malawi's GDP was estimated at $US 1,110m in 1987, making it one of the smaller economies in the region and in Africa. GNP per head for 1987 was $US 160. This makes Malawi one of the half-dozen poorest countries in the world.

Agriculture generated 37% of GDP in 1987 and this is more-or-less the sector size that is to be expected in a low-income economy without significant deposits of exploitable minerals. Industry generates 18% of GDP, making Malawi one of the least industrialised countries in the Southern Africa group. Services generate 45% of GDP.

On the demand side, in 1987 private consumption took up 70% of GDP, government consumption took up 18%, and investment 14%. These proportions are broadly representative of a low-income African economy in the 1980s.

Exports represented 24% of GDP in 1987. This is rather less than the average for the Southern Africa region, but this region has extensive mineral deposits which have led to above average export dependence. The main exports are tobacco (60%), sugar (10%), and tea (10%). Imports comprise 26% of GDP, and again, although this is fairly representative of Africa, it is below the Southern Africa import ratio, which is again, just under half as much.

The last literacy survey was conducted in 1974, when the rate was estimated at 25%. This is well below the regional and African rates. Primary school enrolments were 64% in 1985, and this is a poor provision of basic education compared with the 95% enrolment ratios in the Southern Africa region. Similarly, the 4% secondary enrolment ratio is well below the 18% average for the region. Higher education enrolment is 1%, and this is half the average rate in Africa.

Life expectancy in 1987 was 46 years, below the 50 years of life expectancy in the region and Africa. Nutrition levels, indicated by daily calorie provision are good, however, being at 2,415 a day, 25% above Africa averages. Provision of medical services as indicated by doctors per head of population appears to have improved considerably in the 1980s, and at one physician per 11,560 is twice the Africa average, perhaps a reflection of the fact that the president himself was a medical practitioner. Nurses at 3,130 per head is rather worse than is general in Africa.

Overall, Malawi is a low-income country with a small economy and a heavy dependence on agriculture. Education provision is poor, and health and nutrition provision is patchy.

Economic Performance

GDP has grown at 2.6% a year in the period 1980-87, and this is rather better than the overall African economic performance where GDP has shown negligible positive growth. However, with population growing at 3.8% a year, GDP per head has fallen at -1.2% a year.

Agriculture expanded at 2.5% a year, 1980-87, this being better than Africa's agricultural performance, but below the rate of population growth. Industry expanded at 1.9% a year and services at 3.0% a year, and again these were better than was general in Africa.

Export volumes have grown at 3.4% a year in the 1980s, and Malawi is one of the few African countries to have expanded export volumes in this period. In Africa generally, export volumes have declined at -1.0% a year. Import volumes, however, constrained by export earnings that have been reduced by a terms of trade fall of 33%, 1980-87, and increasing debt service payments, have declined at -6.1% a year.

Inflation for 1980-87 was 12.4% a year, and

Malawi

this compares well with the African average of 15% for the period.

Overall, Malawi has had a good economic performance record in relation to the rest of Africa in the 1980s, with significant growth in all sectors and in export volumes, and a moderate inflation rate. Nevertheless, this has not prevented GDP per head falling, or Malawi experiencing a severe fall in import capacity.

Recent Economic Developments

Malawi has continued to rely on a mixed economy with a strong private sector. In 1981 the government launched a structural adjustment programme with IMF and World Bank support in an effort to strengthen export performance, improve efficiency in the state-controlled sector, phase out price controls and meet skill shortages in the labour force. A third structural adjustment loan of $US 99m was agreed in December 1985, and was supplemented by further lending for balance of payments support from the World Bank and bilateral donors in January 1987. A three-year IMF Extended Fund Facility ended in August 1986, and a new agreement with the Fund was made in July 1988 for an enhanced structural adjustment arrangement of $US 71.5m over three years. The IMF has been pressing for reform of agricultural marketing which is dominated by the state-owned Agricultural Marketing and Development Corporation (ADMARC), which had a monopoly of purchasing but incurred substantial losses of $US 11m in 1985/6.

Early in 1987, ADMARC lost some of its trading monopolies, with private traders allowed in all commodities except tobacco and cotton. Protection was reduced for domestic manufacturing in August, and these measures were necessary to restore IMF short-term credit to support the balance of payments.

Malawi's economic performance faltered in the early 1980s with two years of contraction in GDP in 1980 and 1981, followed by good recovery in the years 1982 to 1985. There was GDP contraction in 1986, but growth averaging 3.0% a year 1987-89. Shortages of imported inputs appear to have restricted manufacturing and export production, and this led to downturns in construction and distribution. Malawi has been considered an example of successful implementation of IMF and World Bank policies with reliance on market forces and export production.

Inflation performance has been varied in the 1980s, with roughly 20% annual increase in prices at the beginning of the decade slowing to close to 10% for 1981 to 1983. The rate doubled in 1984 and then returned to around 10% in 1985 and 1986, before exchange rate depreciation from 1987 onwards raised inflation to above 20% a year.

Malawi has $US 775m of long-term debt outstanding in 1985, and debt service payments in 1985 were estimated at $US 113m, which took up 50% of earnings from the export of goods and services.

After remaining stable for much of the 1970s, Malawi has steadily adjusted the exchange value of the kwacha from Mk 0.82 = $US 1 in 1980 to Mk 2.65 = $US 1 at the end of 1989.

Aid flows have been dominated by the structural adjustment loans to support the economic reforms, but project lending has also been considerable. The World Bank and the European Investment Bank have committed $US 20m to water supply for the new capital at Lilongwe. The World Bank is lending $US 8.8m for roads, $US 27m for education, and $US 11m for health. West Germany and the European Development Fund are allocating $US 6.1m to small enterprises, the EEC $US 3.4m to fish farming, Japan $US 14m to agriculture, and the European Development Fund $US 31.7m to secure fertiliser supplies.

Economic Outlook and Forecasts

The forecasts assume that Malawi's stability is maintained, and that there is steady progress in implementing liberalising economic reforms.

Tobacco prices in real terms are expected to fall by -12.0% up to 1995, tea prices to rise by 15% and sugar prices to fall by -5.0%.

GDP is forecast to expand, but at a rate that will just about keep pace with population growth. Export volumes are expected to expand faster than GDP, and import volumes to grow with IMF and donor support.

Inflation, as a result of exchange rate adjustments, is expected to be around 20%.

Malawi: Economic Forecasts
(average annual percentage change)

	Actual	Forecast Base	high
	1980-87	1990-95	1990-95
GDP	2.6	3.5	3.9
GDP per Head	-1.2	0.0	0.4
Inflation Rate	12.4	19.0	15.0
Export Volumes	3.4	4.0	4.5
Import Volumes	-6.1	2.0	3.0

MALAWI: Comparative Data

	MALAWI	SOUTHERN AFRICA	AFRICA	INDUSTRIAL COUNTRIES
POPULATION & LAND				
Population, mid year, millions, 1989	8.2	6.0	10.2	40.0
Urban Population, %, 1984	12	23.8	30	75
Population Growth Rate, % per year, 1980-86	3.2	3.1	3.1	0.8
Land Area, thou. sq. kilom.	119	531	486	1,628
Population Density, persons per sq kilom., 1988	66.4	11.3	20.4	24.3
ECONOMY: PRODUCTION & INCOME				
GDP, $US millions, 1986	1,100	2,121	3,561	550,099
GNP per head, $US, 1986	160	383	389	12,960
ECONOMY: SUPPLY STRUCTURE				
Agriculture, % of GDP, 1986	37	25	35	3
Industry, % of GDP, 1986	18	33	27	35
Services, % of GDP, 1986	45	42	38	61
ECONOMY: DEMAND STRUCTURE				
Private Consumption, % of GDP, 1986	75	67	73	62
Gross Domestic Investment, % of GDP, 1986	10	14	16	21
Government Consumption, % of GDP, 1986	18	21	14	17
Exports, % of GDP, 1986	22	34	23	17
Imports, % of GDP, 1986	25	36	26	17
ECONOMY: PERFORMANCE				
GDP growth, % per year, 1980-86	2.4	-3.7	-0.6	2.5
GDP per head growth, % per year, 1980-86	-0.8	-6.6	-3.7	1.7
Agriculture growth, % per year, 1980-86	2.5	-6.8	0.0	2.5
Industry growth, % per year, 1980-86	1.5	-3.8	-1.0	2.5
Services growth, % per year, 1980-86	2.8	-0.9	-0.5	2.6
Exports growth, % per year, 1980-86	1.1	-5.1	-1.9	3.3
Imports growth, % per year, 1980-86	-6.5	-2.5	-6.9	4.3
ECONOMY: OTHER				
Inflation Rate, % per year, 1980-86	12.4	18.1	16.7	5.3
Aid, net inflow, % of GDP, 1986	17.5	8.6	6.3	-
Debt Service, % of Exports, 1986	40.1	7.8	20.6	-
Budget Surplus (+), Deficit (-), % of GDP, 1985	-8.4	-4.1	-2.8	-5.1
EDUCATION				
Primary, % of 6-11 group, 1984	62	95	76	102
Secondary, % of 12-17 group, 1984	4	18	22	93
Higher, % of 20-24 group, 1984	1	1.3	1.9	39
Adult Literacy Rate, %, 1974	25	47	39	99
HEALTH & NUTRITION				
Life Expectancy, years, 1986	45	50	50	76
Calorie Supply, daily per head, 1985	2,415	1,998	2,096	3,357
Population per doctor, 1984	11,560	26,118	24,185	550

Notes: 'Southern Africa' and 'Africa' exclude South Africa. Dates are for the country in question, and do not always correspond with the Regional, African and Industrial averages.

MALAWI: Leading Indicators

GDP Growth (% per year) — 1988-89 <-na->

GDP per Head Growth (% per year) — 1988-89 <-na->

Inflation Rate (% per year) — 1989 na

Exchange Rate (CFA per $US)

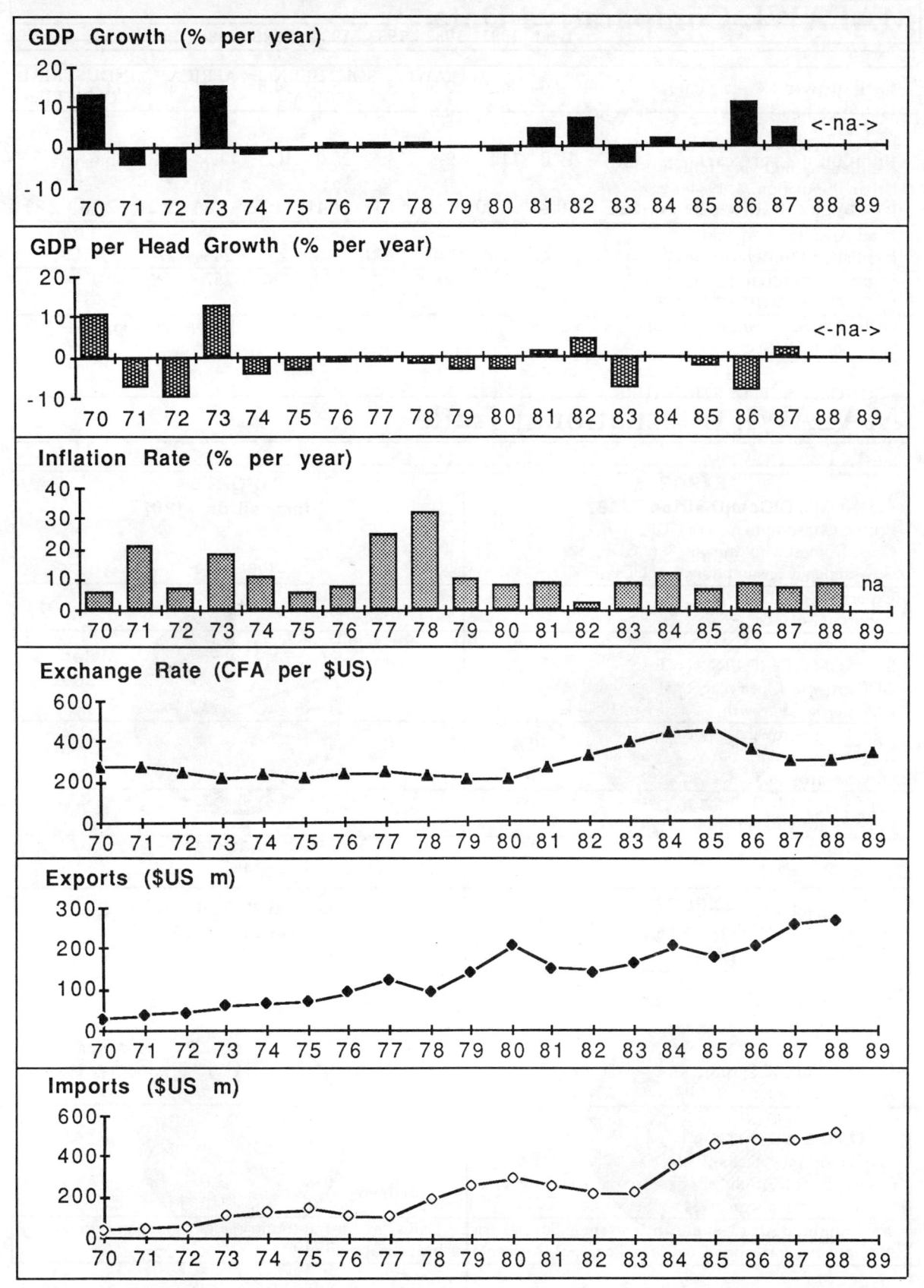

Exports ($US m)

Imports ($US m)

Note: Leading indicator data for 1989 are based on the first half of 1989. 1989 exchange rate is for mid-1989.

MALAWI: Leading Indicator Data

	1980	1981	1982	1983	1984	1985	1986	1987	1988	1989
GDP growth (% per year)	0.6	-6.2	2.5	3.8	5.5	4.5	1.2	0.4	3.6	
GDP per head growth (% per year)	-2.5	-9.3	-0.6	0.7	2.4	1.4	-1.9	-2.7	0.5	
Inflation (% per year)	19.0	11.8	9.8	13.5	20.0	10.5	11.5	25.3	35.6	
Exchange rate (Kwacha per $US)	0.82	0.90	1.06	1.17	1.41	1.72	1.86	2.21	2.56	2.83
Exports, merchandise ($US m)	284	288	240	230	316	250	248	278	297	
Imports, merchandise ($US m)	318	258	306	310	270	287	257	296	404	

Note: Leading indicator data for 1989 are based on the first half of 1989. 1989 exchange rate is for mid-1989.

MALAWI: International Trade

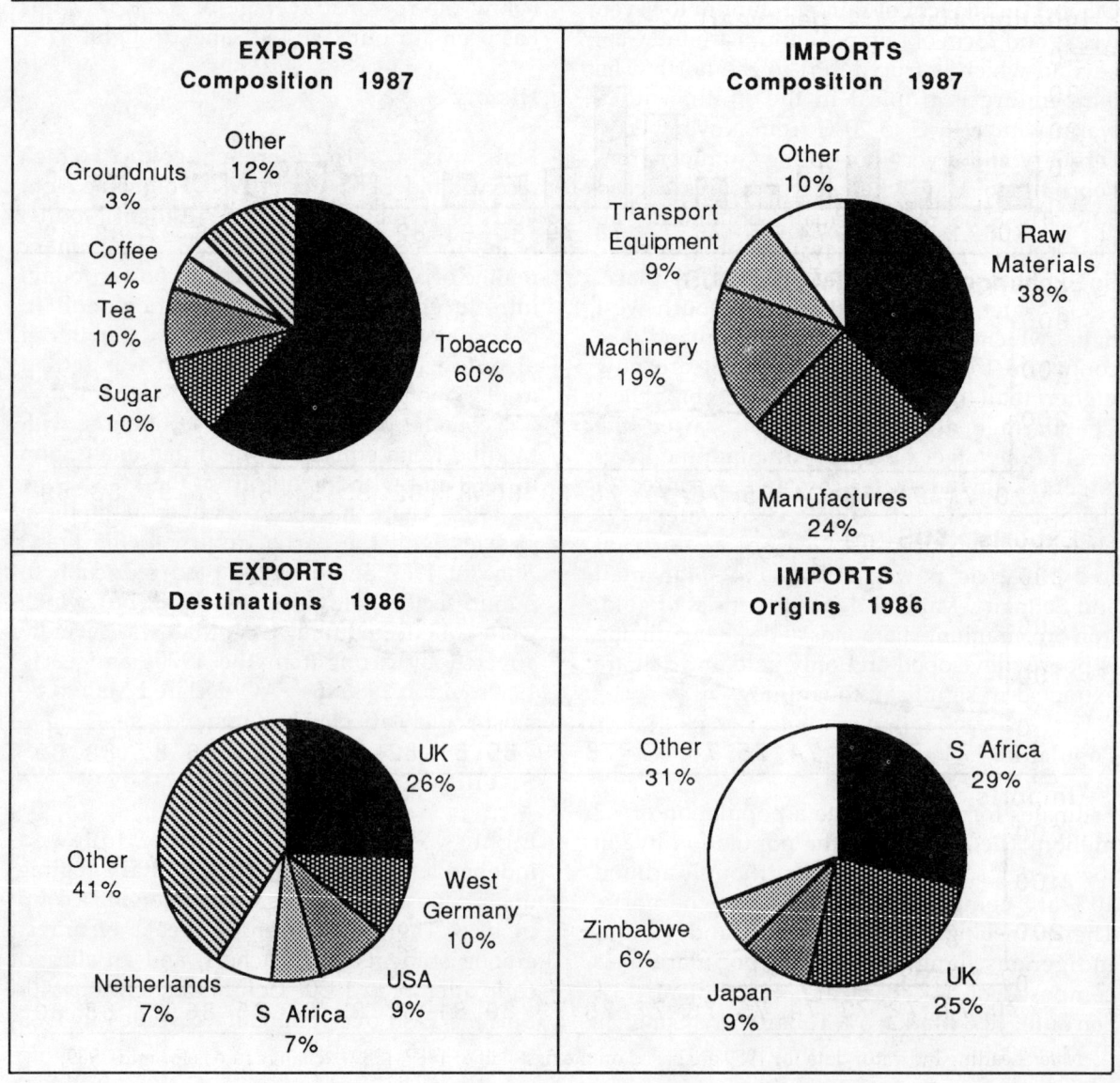

EXPORTS Composition 1987
- Tobacco 60%
- Sugar 10%
- Tea 10%
- Coffee 4%
- Groundnuts 3%
- Other 12%

IMPORTS Composition 1987
- Raw Materials 38%
- Manufactures 24%
- Machinery 19%
- Transport Equipment 9%
- Other 10%

EXPORTS Destinations 1986
- UK 26%
- West Germany 10%
- USA 9%
- S Africa 7%
- Netherlands 7%
- Other 41%

IMPORTS Origins 1986
- S Africa 29%
- UK 25%
- Japan 9%
- Zimbabwe 6%
- Other 31%

26 MALI

Physical Geography and Climate

Mali, in the West African region, is a landlocked state, bounded by Senegal, Mauritania, Algeria, Niger, Burkina Faso, Côte d'Ivoire and Guinea. The country is cut south west/northeast by the River Niger. The country is mainly dry with a rainy season varying from four to five months from June to October and 112cm of rain per annum to seven weeks and 23cm of rain per annum around Gao, beyond which is true desert to the north. The temperature is tropical in the south, with a warm winter 15°C to 20°C from November to February and a very hot spring/summer from February to May when temperatures exceed 30°C. The Sahel region remains hot and dry all year round. The Niger system provides the floods that enable irrigation to take place. Flood water is used in the 'dead' south west delta, which used to be part of an inland lake connected to the Tilemsi river, but are now higher than the 'live' flood-water branches. The barrage at Sansanding has raised the level of the Niger by 4.3m. Although the River Niger is a vital waterway and a source of fisheries, it is in effect two waterways obstructed by rapids, which are a source of hydro-electric power at Bamako, Manantali and Selingué. Mineral deposits consist of gold, iron ore, uranium, diamonds and salt but mining is poorly developed and only gold and salt are extracted in significant quantities.

Population

Estimates for 1989 indicate a population of 8.2 million. The majority of the population live in the wetter south and west, although around 10% are nomadic groups living in the north. The main languages are French and various indigenous languages. The population is composed of 34% Bamkara; 16% Fulani; 13% Sén'oufo; 10% Marka; 8% Touareg; 8% Songhai; 7% Malinké and 4% Dogan. The density of population is low at 6.5 persons per square kilometre (though it is below 2 persons per sq. km in much of the north) compared to the regional average of 20.4. The urban population at 10% is low compared to the regional average of 28.7%, but higher than one might expect considering the low density of population. The rate of population growth is 2.4% a year, well below the regional average of 3.2%, and this has been attributed to the Sahel drought.

History

Mali was a former French colony which received independence in 1960. Following eight years of civilian rule under President Modibo Keïta, there was a coup in 1968. The military regime of Moussa Traoré (still in power today) introduced limited civilian participation in government. In 1974 there was a territorial dispute with Burkina Faso which was settled in 1975 and flared up again in 1985/86.

Economic policy has been discontinuous with Modibo Keïta embarking upon nationalisation in 1960 and withdrawing from the Franc Zone in 1962. There has been a reversal of these policies with the partial return to the Franc Zone in 1967 and an attempted reduction in public sector holdings from 1968, but which were not effected until 1981. Mali was severely affected by drought in the 1970s and early 1980s and in 1986 the FAO declared Mali to be most acutely affected by famine

Stability

Eight years of civilian rule followed independence in 1960, and a military regime under Moussa Traoré has ruled following a coup in 1968. There have been outbreaks of unrest among students and teachers, and an alleged coup attempt in 1980. Continuing tension with Burkina Faso over boundary disputes in 1985

Mali

erupted into a short military conflict. These incidents have not posed serious threats to the Traoré regime. Economic policy has been uneven, with nationalisations in 1960 and withdrawal from the Franc Zone in 1962. In 1967 Mali made a partial return to the Franc Zone with convertability of its currency to the French franc, with re-entry into the West African Monetary Union (UMOA) and adoption of the CFA franc in 1984. An IMF programme involving reductions in public sector spending was introduced in 1978.

Overall, Mali's stability has been poor, the modest political stability introduced under Traoré has been offset by the reversals in economic policy since 1960. Traoré's regime seems to be reasonably secure and in favour with the main popular groups, and a consistent pursuit of liberalising economic reforms will improve the environment for business and finance.

Economic Structure

In 1987 GDP stood at $US 1,960 m, less than half the regional average size. The standard of living is very low with a GNP per capita of only $US 210. Only Burkina Faso and Guinea Bissau in the region have a lower living standard.

In 1987 agriculture generated 54% of GDP, making Mali the most dependent country in the region on this sector. The industrial sector, which has almost no manufacturing capacity is limited to processing food products and some textiles, was correspondingly smaller at 12% of GDP and only Guinea Bissau with 6% was smaller in the region. The service sector, which generated 35% of GDP is average for the region.

On the demand side, private consumption at 90% of GDP was the second highest in the region, after Cape Verde, and reflects the failure of austerity measures. Domestic investment of around 16% of GDP was above the regional average and was the same as the African average.

Exports generated 17% of GDP and although this was about average for the region, many countries were able to generate a much higher percentage of GDP from exports. Total export earnings in 1988 were around $US 270 m. In 1986, cotton was the main export, accounting for 39% of the total, with animal products accounting for a further 30%. However, cotton prices are subject to world price fluctuations, while over 75% of total export earnings depend on good climatic conditions. Gold is becoming one of Mali's main foreign exchange earners as output increases from the two main deposits.

Imports took 34% of GDP in 1987, well above the regional average. Import values reached a peak of $US 535m in 1988, an increase of 8% over the previous year, though this was due to rain failures in the 1987/88 rainy season, requiring more food imports.

Provision of education in Mali is very poor, about the worst in the region, with only 22% of the primary age group and 7% of the secondary age group in school, compared to regional averages of 75% and 24% respectively. Only 1% of the 20-24 age group are in higher education and Mali still lacks a university. In 1979, the adult literacy rate was estimated at 10%. It should be taken into account that Mali is an Islamic country where Koranic education often replaces a Western style education, and education of females is limited as they are often married at an early age. UNICEF is to spend $US 45.9m in Mali in the next few years, partly on primary education, while USAID has allocated $US 8m for primary education.

In 1984 there was only one doctor to every 25,390 of the population, compared to the regional average of one per 16,200. Many of the doctors in Mali are expatriates. Provision of nurses, however, was better than average with a ratio of 1 to every 1,350 people compared with 1 to 2,722 as the regional average.

Overall, Mali's lack of infrastructure, vulnerability to climatic fluctuations, lack of education facilities and heavy debt burden all contribute to the country's low level of development and poor living standards. Mali looks likely to continue to be heavily dependent on outside aid with only the increase in gold output providing some relief in the short term.

Economic Performance

From 1980 to 1987 GDP grew by 3.4% while on average it declined in the region and in Africa as a whole. GDP per head increased by 1.0% a year. Growth in the agricultural sector was minimal over the period 1980-87 at 0.3% though this was due to the 1983-85 drought. Record output was expected in 1989 for the second year running, though the threat of drought and locusts remain one of the country's greatest problems. Industrial growth was much more impressive at 9.8%, the fastest in the region, and in fact there was a decline of -3.2% in the region as a whole. Only Guinea Bissau in the region had a service sector which grew faster than Mali's 5.9%.

The export growth rate over 1980-87 was also one of the fastest in the region at 6.6% while import growth was 3.4% over the corresponding period.

Inflation over this period only grew by 4.2%, about one third of the regional inflation rate and one quarter of the African average.

Overall, Mali's performance up to 1987 was generally good considering the effects of drought, with low inflation and continued economic growth. Agricultural was badly affected however, and with over 80% of the population dependent on this sector, the overall standard of living of this section of the population declined.

Recent Economic Developments

In 1978 Mali introduced a programme of budgetary restraint, with revenues expected to increase faster than expenditures. In 1984 the government announced plans to privatise some 40 parastatal enterprises, and ended the monopoly of the state-owned trading body over imports. An IMF standby credit of $US 28.9m was granted in March 1985. The IMF agreement expired in March 1987, and assistance was suspended until 1988 on the grounds that Mali has been slow in introducing privatising reforms, and there is concern over the size of the budget deficit. In March 1987, however, four state-owned enterprises in construction, timber, films and publishing were closed down, and in July employment cuts were imposed in the public service.

In July 1988, however, a further credit of $US 46m was agreed over a three-year period as part of the IMF's structural adjustment facility. This was then followed by soft loans totalling over $US 330m from the World Bank to help in the structural adjustment programme, covering a total of nine projects. The programme is aimed at bringing inflation down to a level of 3.5% and achieving real GDP growth of about 4%. In order to achieve this there are to be reductions in government expenditure while increasing revenue totals; most state enterprises are to be privatised or closed down; there are to be reforms in the agricultural sector; and production levels are to be set according to market prices rather than by government control.

The growth rate of GDP has been poor in the 1980s, in part as a result of drought conditions affecting agriculture. When there are improved climatic conditions output can recover, and in 1986 GDP in fact grew at 11.5 per cent allowing a rise in living standards for the first time since 1982. Inflation performance has been good, and despite annual price increases of over 25% in 1977 and 1978, in the 1980s inflation has invariably been below 10%, and in 1989 was effectively zero.

Mali's trade gap has gradually widened since 1970 when imports were 20% greater than exports to a position where imports are currently double export values.

Mali was estimated to have $US 1,848m in external debt in 1989. Debt service was estimated at a manageable 9.9% of export earnings in 1987. However, stagnating export earnings and consequent constraints on the ability to import have caused Mali to reschedule its debts through the Paris Club group of countries for the first time.

Mali left the Franc Zone in 1962, but after a devaluation in 1967 rejoined with the Mali franc pegged at half the value of the CFA franc. In 1984, the CFA franc was reintroduced. The CFA franc has appreciated against the US dollar since 1985 when it stood at CFA 449.4 = $US 1 to CFA 289.3 = $US 1 in 1989, an increase in value of 34%. There is consequently some speculation that the CFA franc will be realigned with the French franc as the present exchange rate contributes to the balance of payments difficulties of the African Franc Zone countries.

The 1989 budget estimated total revenue of $US 685m, an increase of 66% over the previous year, while total expenditure is expected to rise by 72% to $US 866m, resulting in a 50% bigger deficit of $US 180m. Mali has a five-year plan

Mali

from 1987-91 with $US 1.8b, mainly from overseas aid, budgeted to help achieve food self-sufficiency and improve social service provision.

The Malian economy is heavily dependent on aid. In 1988/89 the largest donors were France, the IDA and USA. Total US aid for the fiscal year 1990 is budgeted for $US 22.7m, 34% more than the previous year. Other aid sources over the last year have included; $US 147m from the European Development Fund; $US 23m for rural development projects from Canada; $US 10m from Japan for agricultural projects; a $US 38m loan from Saudi Arabia for construction of a bridge across the Niger at Bamako, and $US 5.5m from West Germany for the stuctural adjustment programme.

There is considerable private interest in gold production, with Utah International of the US investing $US 1.5m in exploration, with the possibility of up to $US 100m as the deposits are developed.

Economic Outlook and Forecasts

Forecasts assume that Mali's stability shows steady improvement and that the liberalising reforms continue to be implemented.

Cotton prices are projected to remain unchanged in real terms up to 1995, while meat prices are projected to rise by 11%.

GDP is forecast to expand only moderately, and given that population growth is expected to reach 3.0% a year, GDP per head will fall by about 1% a year.

Export volumes are expected to expand a little faster than the economy generally, with import volumes rising, but more slowly than exports.

Inflation is expected to be modest, and below 10% a year.

Mali: Economic Forecasts
(average annual percentage change)

	Actual	Forecast Base	Forecast High
	1980-87	1990-95	1990-95
GDP	3.4	2.0	2.4
GDP per Head	1.0	-1.0	-0.6
Inflation Rate	4.2	6.0	4.0
Export Volumes	6.6	3.5	4.7
Import Volumes	3.4	3.0	3.5

MALI: Comparative Data

	MALI	WEST AFRICA	AFRICA	INDUSTRIAL COUNTRIES
POPULATION & LAND				
Population, mid year, millions, 1989	8.2	12.3	10.2	40.0
Urban Population, %, 1985	20	28.7	30	75
Population Growth Rate, % per year, 1980-86	2.3	3.2	3.1	0.8
Land Area, thou. sq. kilom.	1,240	384	486	1,628
Population Density, persons per sq kilom., 1988	6.5	31.1	20.4	24.3
ECONOMY: PRODUCTION & INCOME				
GDP, $US millions, 1986	1,650	4,876	3,561	550,099
GNP per head, $US, 1986	180	510	389	12,960
ECONOMY: SUPPLY STRUCTURE				
Agriculture, % of GDP, 1985	50	40	35	3
Industry, % of GDP, 1985	13	26	27	35
Services, % of GDP, 1985	37	34	38	61
ECONOMY: DEMAND STRUCTURE				
Private Consumption, % of GDP, 1986	83	77	73	62
Gross Domestic Investment, % of GDP, 1986	21	13	16	21
Government Consumption, % of GDP, 1986	13	13	14	17
Exports, % of GDP, 1986	15	18	23	17
Imports, % of GDP, 1986	32	20	26	17
ECONOMY: PERFORMANCE				
GDP growth, % per year, 1980-86	0.4	-2.03	-0.6	2.5
GDP per head growth, % per year, 1980-86	-1.9	-4.7	-3.7	1.7
Agriculture growth, % per year, 1980-86	-2.3	1.3	0.0	2.5
Industry growth, % per year, 1980-86	4.0	-3.6	-1.0	2.5
Services growth, % per year, 1980-86	3.8	-2.5	-0.5	2.6
Exports growth, % per year, 1980-86	7.2	-4.0	-1.9	3.3
Imports growth, % per year, 1980-86	3.4	-12.9	-6.9	4.3
ECONOMY: OTHER				
Inflation Rate, % per year, 1980-86	7.4	13.0	16.7	5.3
Aid, net inflow, % of GDP, 1986	22.7	3.7	6.3	-
Debt Service, % of Exports, 1986	14.2	21.4	20.6	-
Budget Surplus (+), Deficit (-), % of GDP, 1985	-9.6	-3.4	-2.8	-5.1
EDUCATION				
Primary, % of 6-11 group, 1984	23	75	76	102
Secondary, % of 12-17 group, 1984	7	24	22	93
Higher, % of 20-24 group, 1984	1	2.6	1.9	39
Adult Literacy Rate, %, 1979	10	30	39	99
HEALTH & NUTRITION				
Life Expectancy, years, 1986	47	50	50	76
Calorie Supply, daily per head, 1985	1,810	2,105	2,096	3,357
Population per doctor, 1981	26,030	16,199	24,185	550

Notes: 'Southern Africa' and 'Africa' exclude South Africa. Dates are for the country in question, and do not always correspond with the Regional, African and Industrial averages.

MALI: Leading Indicators

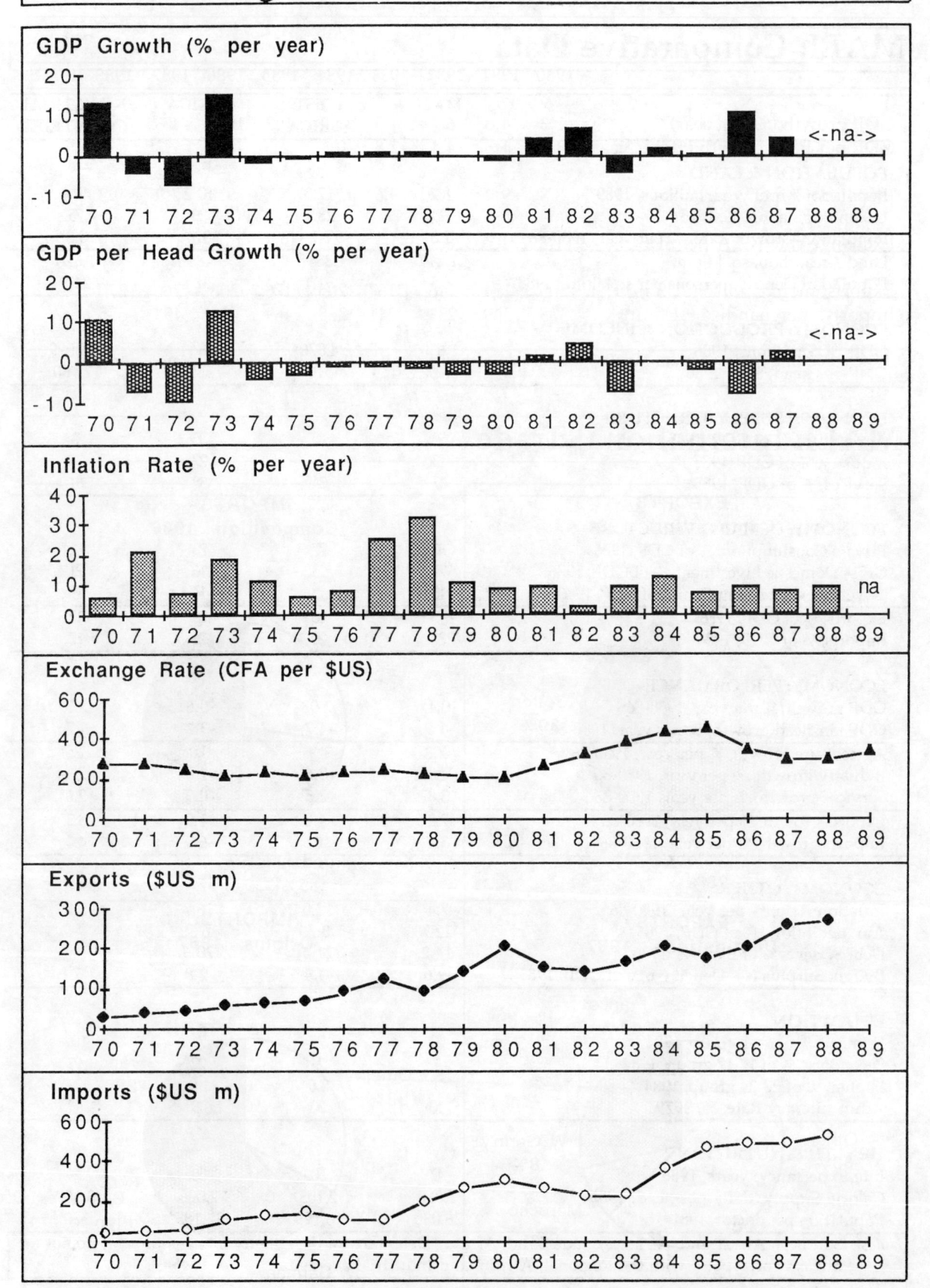

GDP Growth (% per year)

GDP per Head Growth (% per year)

Inflation Rate (% per year)

Exchange Rate (CFA per $US)

Exports ($US m)

Imports ($US m)

Note: Leading indicator data for 1989 are based on the first half of 1989. 1989 exchange rate is for mid-1989.

MALI: Leading Indicator Data

	1980	1981	1982	1983	1984	1985	1986	1987	1988	1989
GDP growth (% per year)	-1.3	4.6	6.7	-4.5	1.7	-0.2	10.6	4.6		
GDP per head growth (% per year)	-3.6	1.3	4.4	-7.8	-0.6	-2.5	-8.3	2.3		
Inflation (% per year)	8.8	9.2	2.7	9.6	12.7	7.7	9.1	7.9	9.3	
Exchange rate (CFA per $US)	211.3	271.7	328.6	381.1	437.0	449.3	346.3	300.5	297.9	327.3
Exports, merchandise ($US m)	205	154	146	167	205	176	206	260	270	
Imports, merchandise ($US m)	308	269	233	241	368	469	496	494	535	

Note: Leading indicator data for 1989 are based on the first half of 1989. 1989 exchange rate is for mid-1989.

MALI: International Trade

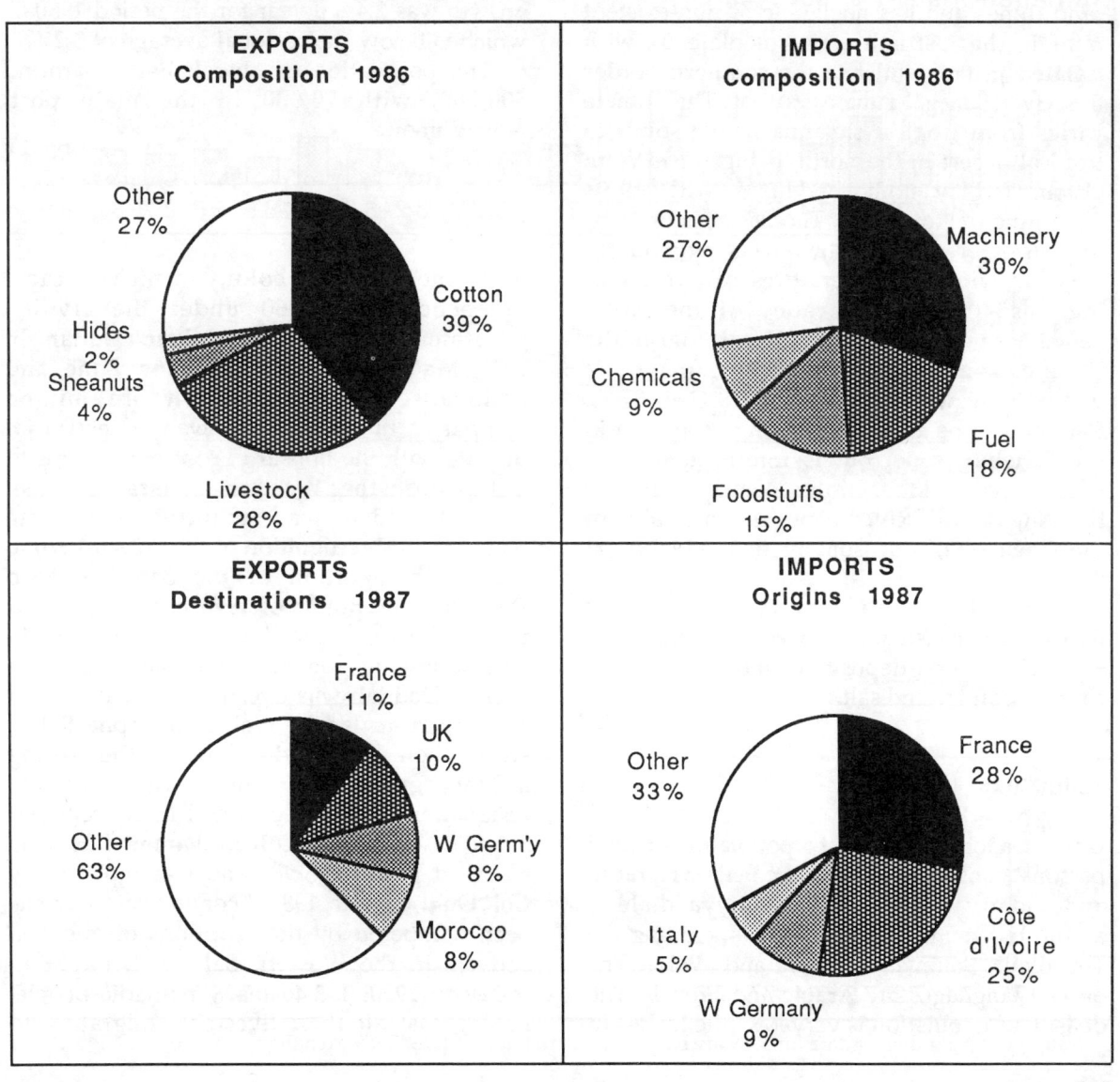

EXPORTS Composition 1986
- Cotton 39%
- Livestock 28%
- Sheanuts 4%
- Hides 2%
- Other 27%

IMPORTS Composition 1986
- Machinery 30%
- Fuel 18%
- Foodstuffs 15%
- Chemicals 9%
- Other 27%

EXPORTS Destinations 1987
- France 11%
- UK 10%
- W Germ'y 8%
- Morocco 8%
- Other 63%

IMPORTS Origins 1987
- France 28%
- Côte d'Ivoire 25%
- W Germany 9%
- Italy 5%
- Other 33%

27 MAURITANIA

Physical Geography and Climate

Mauritania, in the West African region, is an Atlantic coastal state of 1,031 sq km, bounded by Morocco, Algeria, Mali and Senegal. About 40% of the country consists of unconsolidated sand dunes and has negligible or non-existent rainfall; the rest is a series of plateaux with isolated peaks, whilst to the southern border the River Senegal runs east-west. The climate varies from tropical savanna in the south to tropical desert in the north, being warm from October to May and hot and humid with storms from June to November. The desert is dry and arid, moderated on the coast by cooling Atlantic winds. Temperatures can reach as high as 40°C in the valley of the River Senegal and can fall to 0°C at night; the climate is more moderate where the trade winds blow with ranges from 18°C to 27°C. Rainfall can be as high as 60cm per annum in the Sahelian zone, but is intermittent from year to year. Most water comes from the flooding of the River Senegal, especially in the Chemama, a seasonally flooded alluvial zone.

Systematic irrigation could be induced if dams were constructed to control the River Senegal. Mineral deposits include deposits of copper, iron ore and salt.

Population

Estimates for 1989 suggest a population of 1.9m persons, consisting of nomadic Berber Harattin and Bidan, who speak Hassaniyya dialects and in the south negroid groups including the Toucoleur, Soninké, Bambara and Wolof. The official languages are Arabic and French. The density of population is very low (the lowest in the region) at under 2 per square kilometre, compared with the regional average of 31.1 per square kilometre. There has been much migration to the towns in the recent droughts. 31% of the population is urbanised, just above the regional average of 28.7%. Population growth was 2.4% a year for the period 1980-87 which is below the regional average of 3.2%.

The population of Nouakchott is around 500,000, with 75,000 in the main port, Nouadhibou.

History

A former French colony which became independent in 1960 under the civilian government of President Moktar Daddah. In 1973 Mauritania left the Franc Zone and nationalised the major iron ore mining company. In 1975 there was a territorial dispute with the Polisaro Front over claims to a third of the Western Sahara (Spanish Sahara) which was acquired. Despite ceasefires and recognition of the Polisaro Front, the non-recognition of the Sahrawi Arab Democratic Republic of 1976 has complicated the settlement. The territory was abandoned in 1979 following guerrilla resistance. President Moktar Daddah was overthrown in 1978 by a military coup led by Col. Moustapha Salek; since then the leadership of the ruling military group has changed twice, due to resignations. Mahmoud Ould Louly replaced Salek in 1979. Lt. Col. Mohammed Hidalla assumed power in 1980 and was replaced by Col. Ould Taya in 1984. Economic activity has been hampered by the eruptions of guerrilla activity in the Western Sahara. Drought in the early 1980s led to many nomadic people, having lost all their livestock, migrating to the two main towns of Nouakchott and Nouadhibou.

Relations with Senegal have for some time been strained and in April 1989 they turned violent as widespread riots and killings took place on both sides of the border, especially in the urban centres.

Stability

After independence in 1960 there was a civilian government until a military coup in 1978. Subsequently, power has twice changed hands through resignations, culminating in a bloodless coup to bring in the regime of Col. Ould Taya in 1984. There was military involvement in the Western Sahara from 1976 until Mauritanian troops were withdrawn in 1980. Economic strategy has included leaving the Franc Zone in 1973 and the nationalisation of French, British, Italian and German interests in the iron-ore mining industry in 1974. In 1984 President Taya began to reduce public sector spending in return for IMF support, and in 1985 began a liberalising reform programme.

The racial tension between Mauritanian Moors and Senegalese black Africans turned into a serious international incident in April 1989. A few minor border clashes led to large-scale killings and looting of Mauritanian traders in Dakar and Senegalese and other black workers in Nouadhibou in particular. Mauritania has been accused of expelling large numbers of black Mauritanians (and confiscating their property), claiming them to be Senegalese. Various Arab and African leaders have attempted to mediate between the two countries since the troubles.

Many of the expelled had been accused of being members of Forces de Libération Africaine Mauritanie (FLAM), a mainly black opposition group based in Senegal. FLAM's small strength limits its threat to Mauritania, though its activity is bound to increase now.

Mauritania has recently become a member of the newly-formed Arab Maghreb Union, together with Libya, Algeria, Morocco and Tunisia.

Overall, Mauritania's stability has been very poor. The present regime is not long established and there have been two major changes in the direction of overall economic policy. The Taya regime will need to persevere with liberalising reforms and to accommodate dissident groups if the business and financial climate is to improve. Past problems with Polisario have been solved but the tension with Senegal continues.

Economic Structure

In 1987 the GDP was $US 840m, making it one of the smallest economies in Africa. GNP per head was $US 440 which is not too different from the regional average of $US 510.

Agriculture made up 37% of GDP in 1987, just below the average for the region, and employs 69% of the working population, while industry (including mining) contributed a further 26% but employed only 8%. The service sector accounted for the other 41%, higher than the regional average of 34%.

Private consumption took 73% of GDP while government consumption accounted for a further 13%, about average in West Africa. Gross domestic investment was about 20% of GDP, above average.

Exports comprised 50% of GDP while imports were 57%. The main exports were fish (49%) and iron ore (34%), while the main imports were food products (22%) due to lack of cultivable land, and fuel (9%).

In 1986 about 46% of the primary age group attended school while only 15% of secondary-aged children were at school. However, a large number of children attend Koranic schools. There is no higher education. The education system is being improved with considerable funds coming from external donors.

Life expectancy was 46 years in 1987, below the regional average of 50. There are 12,110 people per doctor when the regional ratio is 16,199 to one.

Overall, Mauritania is a low-income country, but is among the better-off in this group. The economy is heavily dependent on exports of fish and iron ore. Educational and health provision are poor.

Economic Performance

GDP grew by 1.4% a year from 1980-87 while the regional average contracted by -2.0%. GDP

Mauretania

per head contracted at -1.3% a year, but elsewhere in the region, GDP per head contracted faster.

Agricultural growth was 1.5%, about the same as the regional average of 1.3%, despite the Sahel drought. The industrial sector showed an impressive 5.1% growth when the regional average was -3.6%. However, the service sector contracted by -1.3%, though this is less than the regional decline of -2.5%.

Exports grew at 11.2% a year in Mauritania while in the region as a whole exports declined by -4.0%. Imports grew by only 1.7% while in the region they declined by -12.9%.

The inflation rate of 9.8% was below the regional rate of 13.0%.

Overall, Mauritania has had falling GDP per head in the 1980s, although industrial and export sector performance has been impressive, as has the inflation record.

Recent Economic Developments

Reforms introduced in 1984 involved devaluation of the currency, higher prices to producers, adjustment of interest rates to world levels and the phasing out of subsidies. The IMF is pressing for an increase in government revenues, and a balanced budget. Economic recovery is expected to come from better performance in parastatal enterprises, expansion of fishing output, reform of the banking sector and growth in the private sector. Mauritania was granted a $US 13.1m credit by the IMF in May of 1987, and a World Bank structural adjustment loan in June which together with programme aid from West Germany and Saudi Arabia totals $US 50m.

The IMF and World Bank have been satisfied with Mauritania's efforts to stick to reform and the country has therefore attracted considerable aid as well as having some of its debt rescheduled and cancelled. West Germany recently wrote off $US 60m in debts. Privatisation and closure of state enterprises continues, a new investment code has been introduced which will give tax benefits to both foreign and local investors, and reforms have taken place in the banking system.

The IMF agreed to a three-year enhanced structural adjustment facility of nearly $US 636m in 1989.

Mauritania had some $US 1.3b of external debt at the beginning of 1986, and debt service payments are estimated at $US 181.5m in 1987. Mauritania's debt service is around 19% of total export earnings, and with reasonably good export prospects, particularly for fish, there is no great urgency for rescheduling.

There were indications that the currency was overvalued in the early 1980s, and in 1983 when the official exchange rate was OU 54.8 = $US 1, the black market was OU 147 = $US 1. Since then the exchange rate has been devalued, and at the end of 1989 stood at OU 86.8 = $US 1. There is pressure from the IMF for further adjustment to bring the official rate more in line with market valuation.

Mauritania's most important bilateral donor is France, who committed $US 15.5m to structural adjustment lending, $US 2.8m to infrastructure and, with the World Bank, is co-financing a $US 29.2m urban electricity and water project. The US is providing $US 26.8m via food-for-water scheme, the African Development Fund $US 12.6m to agriculture, the European Investment Bank $US 5.6m to electricity, and there is a variety of donors involved in $US 60m of fishery projects.

Economic Outlook and Forecasts

Forecasts assume that stability in Mauritania continues to remain fragile, and that the programme of liberalising reforms remains in place.

Mauritania: Economic Forecasts
(average annual percentage change)

	Actual 1980-87	Forecast Base 1990-95	Forecast High 1990-95
GDP	1.4	1.8	2.0
GDP per Head	-2.0	-0.9	-0.7
Inflation Rate	9.8	7.1	6.3
Export Volumes	11.2	-2.0	-1.3
Import Volumes	1.7	1.9	2.3

Fish prices are expected to rise in real terms by 10% up to 1995, while iron-ore prices are

Mauretania

expected to fall by -14%.

GDP is forecast to rise more slowly than the rate of population increase, resulting in falling GDP per head.

Export volumes, with iron-ore production expected to remain unchanged and the fish catch to fall, are forecast to decline. Compensating aid payments are expected to allow an expansion of imports.

Inflation is expected to remain in single figures.

MAURITANIA: Comparative Data

	MAURITANIA	WEST AFRICA	AFRICA	INDUSTRIAL COUNTRIES
POPULATION & LAND				
Population, mid year, millions, 1989	1.9	12.3	10.2	40.0
Urban Population, %, 1985	31	28.7	30	75
Population Growth Rate, % per year, 1980-86	2.6	3.2	3.1	0.8
Land Area, thou. sq. kilom.	1,031	384	486	1,628
Population Density, persons per sq kilom., 1988	1.84	31.1	20.4	24.3
ECONOMY: PRODUCTION & INCOME				
GDP, $US millions, 1986	750	4,876	3,561	550,099
GNP per head, $US, 1986	420	510	389	12,960
ECONOMY: SUPPLY STRUCTURE				
Agriculture, % of GDP, 1986	34	40	35	3
Industry, % of GDP, 1986	24	26	27	35
Services, % of GDP, 1986	42	34	38	61
ECONOMY: DEMAND STRUCTURE				
Private Consumption, % of GDP, 1986	71	77	73	62
Gross Domestic Investment, % of GDP, 1986	25	13	16	21
Government Consumption, % of GDP, 1986	14	13	14	17
Exports, % of GDP, 1986	56	18	23	17
Imports, % of GDP, 1986	66	20	26	17
ECONOMY: PERFORMANCE				
GDP growth, % per year, 1980-86	1.0	-2.03	-0.6	2.5
GDP per head growth, % per year, 1980-86	-1.6	-4.7	-3.7	1.7
Agriculture growth, % per year, 1980-86	1.2	1.3	0.0	2.5
Industry growth, % per year, 1980-86	5.4	-3.6	-1.0	2.5
Services growth, % per year, 1980-86	-2.4	-2.5	-0.5	2.6
Exports growth, % per year, 1980-86	13.6	-4.0	-1.9	3.3
Imports growth, % per year, 1980-86	0.0	-12.9	-6.9	4.3
ECONOMY: OTHER				
Inflation Rate, % per year, 1980-86	9.9	13.0	16.7	5.3
Aid, net inflow, % of GDP, 1986	23.9	3.7	6.3	-
Debt Service, % of Exports, 1986	17.4	21.4	20.6	-
Budget Surplus (+), Deficit (-), % of GDP, 1979	5.3	-3.4	-2.8	-5.1
EDUCATION				
Primary, % of 6-11 group, 1982	37	75	76	102
Secondary, % of 12-17 group, 1982	12	24	22	93
Higher, % of 20-24 group, 1982	0.0	2.6	1.9	39
Adult Literacy Rate, %, 1977	17	30	39	99
HEALTH & NUTRITION				
Life Expectancy, years, 1986	47	50	50	76
Calorie Supply, daily per head, 1985	2,071	2,105	2,096	3,357
Population per doctor, 1979	14,350	16,199	24,185	550

Notes: 'Southern Africa' and 'Africa' exclude South Africa. Dates are for the country in question, and do not always correspond with the Regional, African and Industrial averages.

MAURITANIA: Leading Indicators

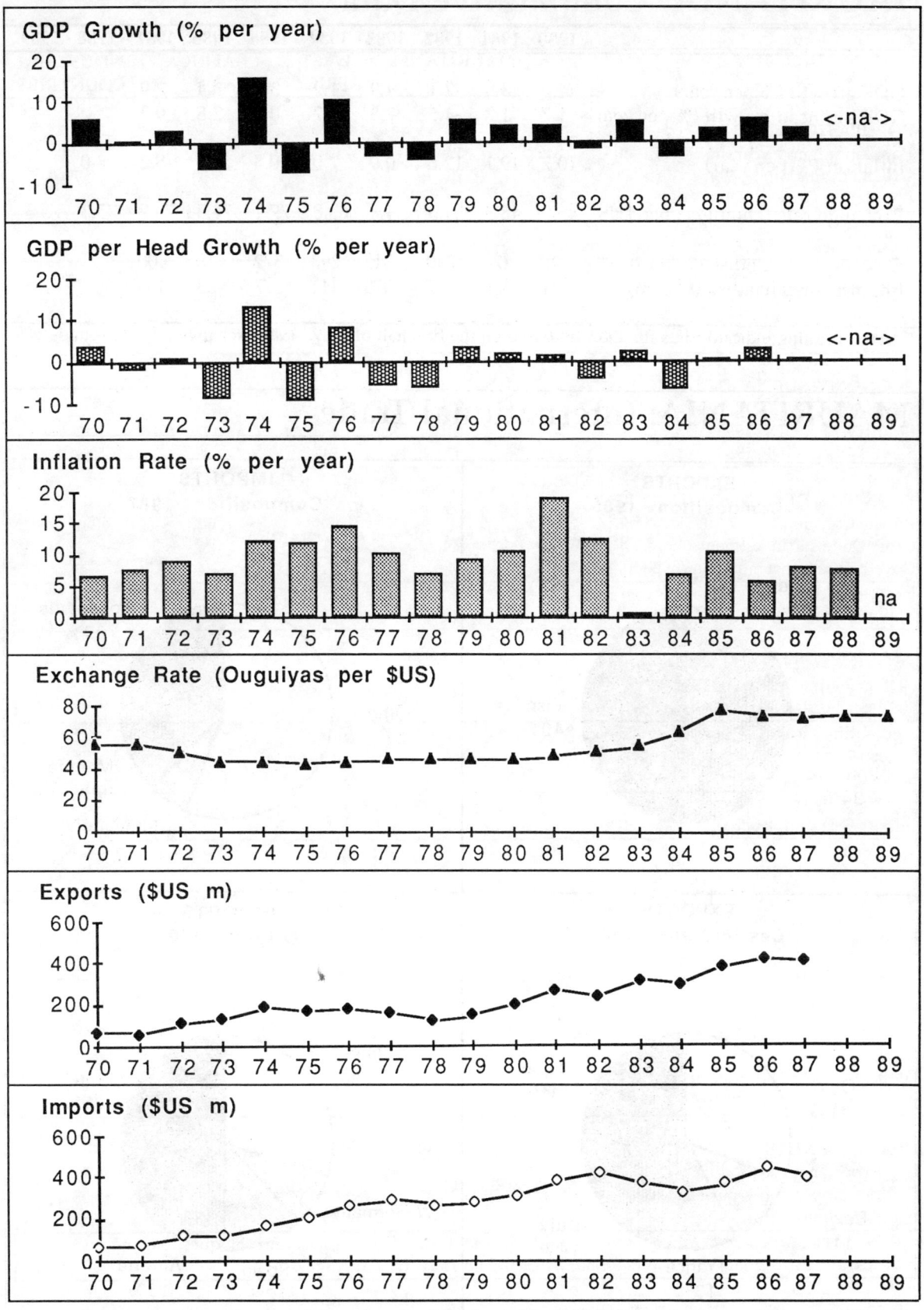

Note: Leading indicator data for 1989 are based on the first half of 1989. 1989 exchange rate is for mid-1989.

MAURITANIA: Leading Indicator Data

	1980	1981	1982	1983	1984	1985	1986	1987	1988	1989
GDP growth (% per year)	4.0	3.8	-2.1	4.9	-4.2	3.0	5.4	2.9		
GDP per head growth (% per year)	1.7	1.2	-4.5	2.5	-6.7	0.5	2.8	0.3		
Inflation (% per year)	10.7	19.1	12.6	0.9	7.0	10.5	6.1	8.2	8.0	
Exchange rate (Ouguiyas per $US)	45.9	48.3	51.8	54.8	63.8	77.1	74.38	73.9	75.3	75.3
Exports, merchandise ($US m)	196	270	240	315	293	372	418	400		
Imports, merchandise ($US m)	321	386	427	378	341	377	453	406		

Note: Leading indicator data for 1989 are based on the first half of 1989. 1989 exchange rate is for mid-1989.

MAURITANIA: International Trade

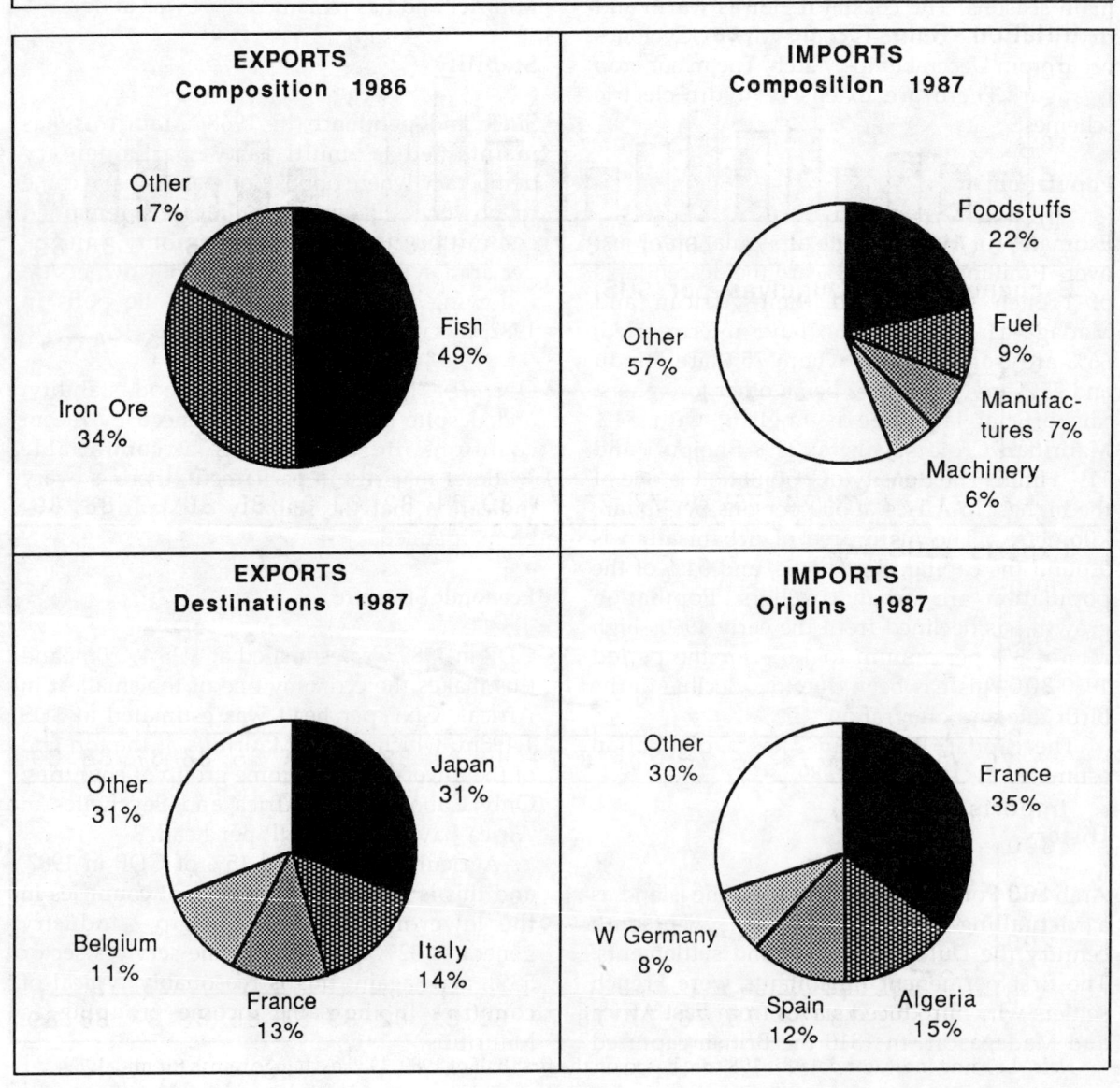

EXPORTS Composition 1986
- Fish 49%
- Iron Ore 34%
- Other 17%

IMPORTS Composition 1987
- Foodstuffs 22%
- Fuel 9%
- Manufactures 7%
- Machinery 6%
- Other 57%

EXPORTS Destinations 1987
- Japan 31%
- Italy 14%
- France 13%
- Belgium 11%
- Other 31%

IMPORTS Origins 1987
- France 35%
- Algeria 15%
- Spain 12%
- W Germany 8%
- Other 30%

28 MAURITIUS

Physical Geography and Climate

Mauritius, in the East African region, lies south of the equator in the Indian Ocean to the east of Madagascar. It consists of two main islands and two further distant groups of volcanic origin, surrounded by coral reefs. The climate is maritime sub-tropical, with two main seasons. The coastal region is warm and dry, whilst the centre is cool and wet. Cyclones occur from December to March. The major crop is sugar. There are extensive hydro-electric schemes.

Population

Estimates for 1989 indicate a population of just over 1 million. About 28% are the descendants of French settlers and their African and Madagascan slaves who have intermarried; 66% are from India of whom 75% are Hindu and 25% are Islamic; 6% being other foreigners. The official language is English, with 54% Mauritian Crole speakers; 20% Bhojpuri and 11% Hindi. The density of population is one of the highest in Africa at 500 persons per square kilometre. The main area of urbanisation is around the capital, Port Louis, and 54% of the population are urban dwellers. Population growth has declined from the early 1960s high rate of 3% per annum to 1.0% for the period 1980-87. This has been due to a decline in the birth rate and emigration.

The capital, Port Louis, has a population estimated at 140,000 in 1989.

History

Arab and Portuguese traders used the island as a victualling stop and in the seventeenth century the Dutch tried to found settlements. The first permenent inhabitants were French settlers who introduced slaves from East Africa and Madagascar. In 1810 the British captured the island, which was confirmed by the 1814 Treaty of Paris. In 1831 slavery was abolished and indentured labour was introduced from India, forming the majority of the population.

Since independence in 1968 Mauritius has had coalition civilian governments. The 1982 elections saw a more moderate coalition elected, and Mr Aneerood Jugnauth became Prime Minister and has remained in power to date.

Stability

Since independence in 1968, Mauritius has maintained a multi-party parliamentary democracy where opposition parties have come to power. Economic policy has remained committed to a mixed economy, and a declaration in favour of more public ownership following a socialist victory at the polls in 1982, was not carried out.

Overall, Mauritius has very good stability, and despite difficulties experienced by recent coalitions in maintaining a comfortable working majority in parliament, there is every indication that the stability of Mauritius will be maintained.

Economic Structure

GDP in 1987 was estimated at $US 1,480m, and this makes the economy one of the smallest in Africa. GNP per head was estimated at $US 1,490m, which places Mauritius at the top end of the lower-middle-income group of countries. Only Gabon, South Africa and Seychelles in Africa have higher GNP per head.

Agriculture generated 15% of GDP in 1987, and this is fairly representative of countries in the lower-middle-income group. Industry generates 32% of GDP, and the services sector 53%, and, again, this is reasonably typical of countries in the same income grouping as Mauritius.

Mauritius

On the demand side, consumption has comprised 60% of GDP, investment 26% and government consumption 11%. The investment ratio is slightly higher and the government consumption ratio slightly lower than the average for lower-middle-income countries.

The small size of the Mauritian economy demands that there be a heavy reliance on international markets. Exports represented 69% of GDP, and this is about triple the trade dependence of the average lower-middle-income country. The main exports are manufactures (56%) and sugar (36%).

A survey in 1976 set the literacy rate at 85%. Primary enrolments were 106%, and secondary were 51% in 1986. These aspects of educational provision are fairly representative of a lower-middle-income country. The surprise is the provision of tertiary education, where enrolments are 1% of the relevent age group. Enrolments in lower-middle-income countries are typically around the 17% level.

Life expectancy in 1987 was 67 years. There was one physician in 1983 for every 1,900 people, and one nurse per 580. These are among the best health provisions in Africa, and better than the average for the lower-middle-income group.

Overall, Mauritius is a small economy at the upper end of the lower-middle-income group. It has a small agricultural sector by African standards, and relatively large industrial and services sectors. There is a heavy trade dependence, where manufactures are the main export. Educational provision, with the exception of the tertiary sector, and health delivery are as good as, or better than, countries in the same lower-middle-income group.

Economic Performance

GDP grew at 5.5% a year in 1980-87, and a low rate of population growth at 1% a year allowed GDP per head to expand at 4.5% a year. This was the best performance in Africa apart from that of Botswana.

All sectors grew impressively, with agriculture at 5.2% a year, expanding comfortably faster than the population. The main engine of growth has been the industrial sector, growing at 8.7% a year. The services sector has grown at 4.1% a year, and this is slower than the overall rate of expansion of the economy. These performance figures are significantly better than the average for the lower-middle-income group of countries.

The performance of the trade sector has enabled the producing sectors to expand rapidly. Export volumes grew at 11.1% for the period 1980-87, and this places Mauritius among the top half dozen countries in the world for 1980s export performance. Terms of trade have moved 8% in favour of Mauritius since 1980, and this, together with the export performance, has enabled import volumes to expand by 6.7% a year.

Inflation has averaged 8.1% in the period 1980-87, and this is a better price stability performance than most of Africa, and certainly for the lower-middle-income group which contains several very high inflation South American economies.

Overall, Mauritius has had an extremely good record of economic performance ranging across all sectors of the economy, with the export and manufacturing sectors taking the palm.

Recent Economic Developments

Mauritius was in receipt of two structural adjustment loans from the World Bank in 1981 and 1983, and in 1987 received $US61m in industrial sector adjustment loans. An IMF credit of $58.2m expired in August 1986, and it is felt that the balance of payments is now strong enough not to require further IMF support. Economic policy has strengthened the reliance of Mauritius on market forces with the establishment of a successful export processing zone, considerable tariff cuts on imports and removal of sales tax. Growth of manufactured exports and tourism have been particularly strong features of good GDP performance. Inflation has steadily fallen since 1980 when prices rose by 42%, to 2% in 1986 and 1987, although it rose to 9% in 1988, and is estimated at 16% in 1989.

Merchandise imports have been greater than exports by some 20% or so in the 1980s. However, good export performance appears to have resulted in a balance of trade surplus in 1986. Imports were compressed by declining export revenues up to 1983, but thereafter have expanded as export earnings have increased.

Mauritius had $US 545m outstanding in external debts in 1987, and a ratio of debt service payments to export earnings for goods and services of 6.2%. Since then, expanding export earnings have continued to reduce the debt service ratio, and external debts appear quite manageable without rescheduling.

The rupee has steadily depreciated since 1980 when it stood at MR 7.68 = $US 1 to MR 15.4 = $US 1 in 1985. Subsequently, the exchange rate has appreciated with the weakening of the dollar and the good export performance of Mauritius to stand at MR 14.6 = $US 1 at the end of 1989. Mauritius is dismantling exchange controls, and together with the reductions in tariffs of the past year this will exert downward pressure on the value of the rupee which will tend to counteract the upward pressure of continued good performance in exports of goods and services.

In 1987 a series of tax reductions were announced. The 5% sales tax and customs and excise duties on textile and printing machinery and inputs have been abolished, and tariff reduced on imported manufactures. These measures are expected to increase the budget deficit, although the substantial World Bank industrial sector adjustment loans will offset the inflationary impact.

The World Bank is lending $US 30m for electricity and hydro-electricity schemes and a further $US 30m for sugar sector rehabilitation. France and the EEC are providing $US 12m for roads, and Japan $US 3.5m for fisheries. Saudi Arabia is lending $US 2.3m for sewerage improvements, and the US is providing $US 4.7m in wheat and rice shipments to compensate for falls in sugar earnings.

Foreign investment has been vigorous with the establishment of the export processing zone. Textiles have attracted commercial loans from the European Investment Bank and the World Bank's private sector lending arm, the International Finance Corporation of $US 16.7m. Spectrum of the US has raised $US 130m in commercial loans for two Boeing 767 aircraft for Air Mauritius; Malaysian business interests are reported to be investing $US 50m in banking, hotels, garment manufacture and palm-oil processing; Speyside Electronics from the UK is setting up a factory; there is a leathergoods enterprise, and Lonrho is establishing six industrial sites.

Economic Outlook and Forecasts

The forecasts assume that the good stability record of Mauritius is maintained. Sugar prices are expected to fall by -5% in real terms up to 1995.

It is difficult to see Mauritius maintaining the excellent growth record of the 1980s, and GDP expansion is forecast to slow down to allow GDP per head growth to be at roughly 3% a year.

Export volume growth is forecast to slow also, but continuing good investment rates should see export volumes expanding at faster than 6% a year. Import volume growth is forecast to slow as well, to around 5% a year. Inflation is expected to be higher than in the 1980s, and to be above 10%.

Mauritius: Economic Forecasts
(average annual percentage change)

	Actual	Forecast Base	High
	1980-87	1990-95	1990-95
GDP	5.5	4.0	4.5
GDP per Head	4.5	2.9	3.4
Inflation Rate	8.1	12.0	10.5
Export Volumes	11.1	6.5	7.5
Import Volumes	6.7	4.8	6.3

MAURITIUS: Comparative Data

	MAURITIUS	EAST AFRICA	AFRICA	INDUSTRIAL COUNTRIES
POPULATION & LAND				
Population, mid year, millions, 1989	1.0	12.2	10.2	40.0
Urban Population, %, 1985	54	30.5	30	75
Population Growth Rate, % per year, 1980-86	1.0	3.1	3.1	0.8
Land Area, thou. sq. kilom.	2	486	486	1,628
Population Density, persons per sq kilom., 1988	500	24.2	20.4	24.3
ECONOMY: PRODUCTION & INCOME				
GDP, $US millions, 1986	1,160	2,650	3,561	550,099
GNP per head, $US, 1986	1,200	250	389	12,960
ECONOMY: SUPPLY STRUCTURE				
Agriculture, % of GDP, 1986	15	43	35	3
Industry, % of GDP, 1986	32	15	27	35
Services, % of GDP, 1986	53	42	38	61
ECONOMY: DEMAND STRUCTURE				
Private Consumption, % of GDP, 1986	64	77	73	62
Gross Domestic Investment, % of GDP, 1986	17	16	16	21
Government Consumption, % of GDP, 1986	11	15	14	17
Exports, % of GDP, 1986	63	16	23	17
Imports, % of GDP, 1986	56	24	26	17
ECONOMY: PERFORMANCE				
GDP growth, % per year, 1980-86	4.4	1.6	-0.6	2.5
GDP per head growth, % per year, 1980-86	3.4	-1.7	-3.7	1.7
Agriculture growth, % per year, 1980-86	5.3	1.1	0.0	2.5
Industry growth, % per year, 1980-86	6.1	1.1	-1.0	2.5
Services growth, % per year, 1980-86	3.4	2.5	-0.5	2.6
Exports growth, % per year, 1980-86	10.4	0.7	-1.9	3.3
Imports growth, % per year, 1980-86	2.8	0.2	-6.9	4.3
ECONOMY: OTHER				
Inflation Rate, % per year, 1980-86	8.1	23.6	16.7	5.3
Aid, net inflow, % of GDP, 1986	4.2	11.5	6.3	-
Debt Service, % of Exports, 1986	7.3	18	20.6	-
Budget Surplus (+), Deficit (-), % of GDP, 1986	-3.5	-3.0	-2.8	-5.1
EDUCATION				
Primary, % of 6-11 group, 1985	106	62	76	102
Secondary, % of 12-17 group, 1985	51	15	22	93
Higher, % of 20-24 group, 1985	1	1.2	1.9	39
Adult Literacy Rate, %, 1976	85	41	39	99
HEALTH & NUTRITION				
Life Expectancy, years, 1986	66	50	50	76
Calorie Supply, daily per head, 1985	2,717	2,111	2,096	3,357
Population per doctor, 1981	1,820	35,986	24,185	550

Notes: 'Southern Africa' and 'Africa' exclude South Africa. Dates are for the country in question, and do not always correspond with the Regional, African and Industrial averages.

MAURITIUS: Leading Indicators

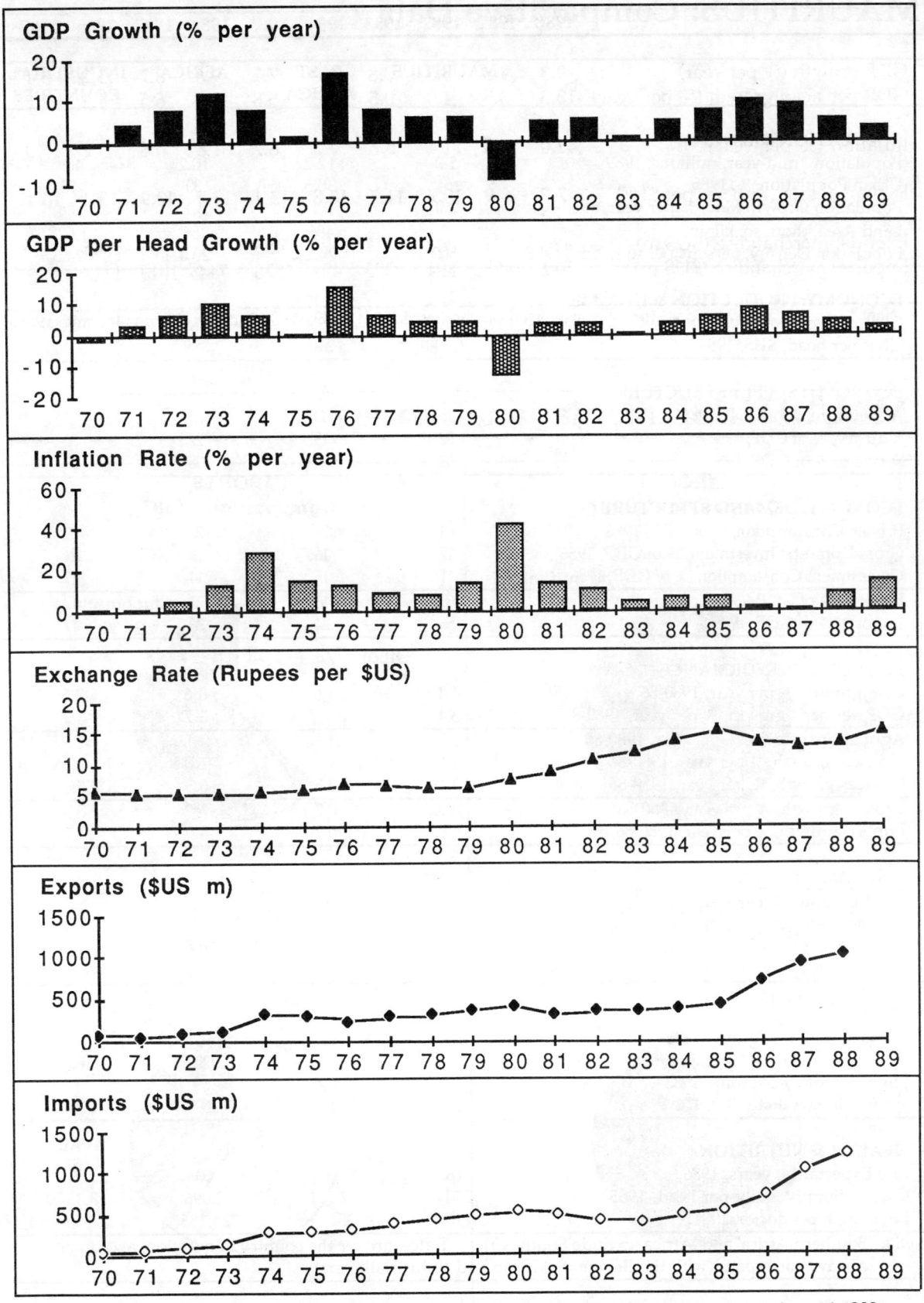

GDP Growth (% per year)

GDP per Head Growth (% per year)

Inflation Rate (% per year)

Exchange Rate (Rupees per $US)

Exports ($US m)

Imports ($US m)

Note: Leading indicator data for 1989 are based on the first half of 1989. 1989 exchange rate is for mid-1989.

MAURITIUS: Leading Indicator Data

	1980	1981	1982	1983	1984	1985	1986	1987	1988	1989
GDP growth (% per year)	-9.3	4.8	5.3	0.8	4.8	7.1	9.6	8.7	5.2	3.4
GDP per head growth (% per year)	-13.3	3.5	4.0	-0.5	3.5	5.8	8.4	6.5	4.2	2.2
Inflation (% per year)	42.0	14.5	11.4	5.6	7.3	6.7	1.8	0.6	9.2	15.0
Exchange rate (Rupees per $US)	7.7	8.9	10.7	11.7	13.8	15.4	13.5	12.9	13.4	15.4
Exports, merchandise ($US m)	430	327	364	367	398	455	731	934	1045	
Imports, merchandise ($US m)	512	475	394	385	471	526	714	1013	1217	

Note: Leading indicator data for 1989 are based on the first half of 1989. 1989 exchange rate is for mid-1989.

MAURITIUS: International Trade

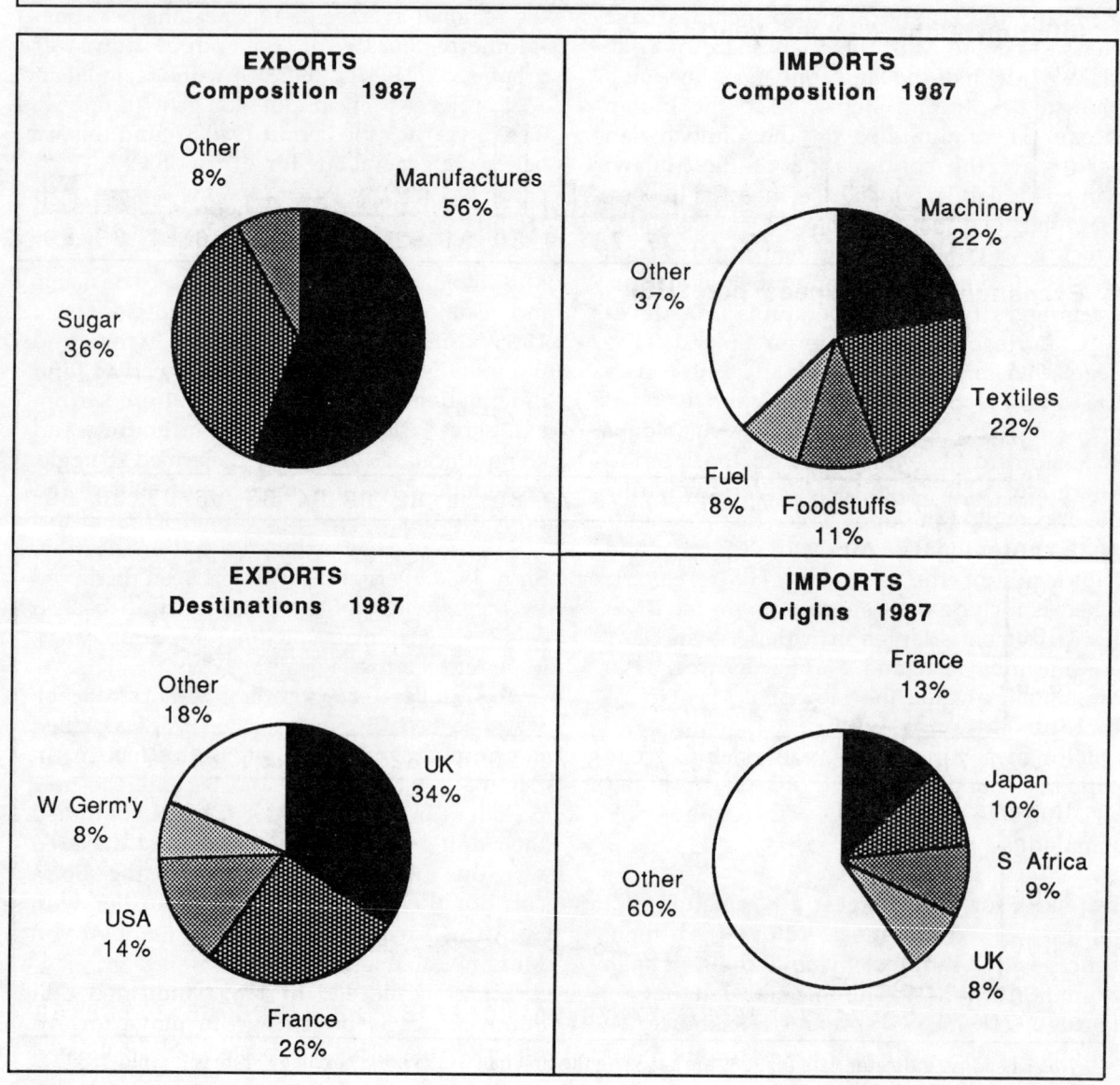

EXPORTS Composition 1987
- Other 8%
- Manufactures 56%
- Sugar 36%

IMPORTS Composition 1987
- Machinery 22%
- Textiles 22%
- Foodstuffs 11%
- Fuel 8%
- Other 37%

EXPORTS Destinations 1987
- Other 18%
- UK 34%
- France 26%
- USA 14%
- W Germ'y 8%

IMPORTS Origins 1987
- France 13%
- Japan 10%
- S Africa 9%
- UK 8%
- Other 60%

29 MOZAMBIQUE

Physical Geography and Climate

Mozambique, in the Southern Africa region, is a coastal state facing the Mozambique Channel and bounded by Tanzania, Malawi, Zambia, Zimbabwe, South Africa and Swaziland. It has a sandy shoreline bordered by lagoons, coastal islets and many river estuaries, including the Zambezi. It also includes Lake Niassa and the Mozambiquan share of Lake Malawi. Much of the land consists of low-lying plateau descending step-wise to the Indian Ocean. The mountains on the Zimbabwian border are the continuation of the Malawi Shire Highlands which rise to over 1,000m. There are two main seasons - October-March which is wet and hot with temperatures from 27°C-30°C and April-September which is warm and dry with temperatures from 18°C-20°C and cooler in the interior uplands. The vegetation ranges from lowland rainforest to broadleaved tropical woodland and wooded grassland. The River Zambezi is navigable for 460 kilometres, giving access to the interior. There are twenty five major rivers including the Incomáti, Limpopo, Save, Buzi, Pungué, Meluli, Montepuez, Messalo, Rovuma and tributaries of the Lugenda. Hydro-electric schemes include the Cabora Dam on the River Revuè, which is larger than the Aswan Dam, at Chicamba Real and Mavuzi as well as at Massinguir and on the Limpopo. Major ports include Maputo, Beira, Nacala and Quelimane. Mineral deposits include coal, copper, salt and mica.

Population

Estimates for 1989 suggest a population of 15 million, of whom most follow traditionoal beliefs with 14% Moslems and about as many Roman Catholics. The official language is Portuguese with 38% speaking Emakhuwa, 24% Sitsonga, 10% Cisena and 6% Kiswahili. North of the Zambezi the main tribes are the Makua-Lomwe; Yao, Islamised traders; Makonde; Rovuma; Nyanja and Chewa. South of the Zambezi the main groups are the Thonga, Chopi and Shona. Most of the political and economic power is concentrated in the south and east. The density of population is 18 persons per square kilometre which is considerably higher than the regional average of 11.3 persons per square kilometre. 23% of the population were urbanised in 1987, rather below the regional and African levels. The population growth rate was 2.7% a year for the period 1980-87, and this was below the general rate for Africa of 3.2%.

History

Mozambique was discovered by Vasco da Gama in 1499 and was settled by the Portuguese from 1505 onwards. The colony received independence in 1975. Between 1964 and independence, Mozambique suffered from extensive disruption of production and communications caused by the armed struggle. At independence the majority of skilled and professional Portuguese colonials left, and this was followed by extensive nationalisation. Since 1980 internal security has been disrupted by the activities of the South African backed MNR (Resisténcia Nacional Moçambicana), often referred to as Renamo.

In October 1986 President Samora Machel, who had been in power since 1975, was killed in an aircraft crash returning from a meeting in Zambia. Joachim Chissano became the new president in November 1986. MNR continued their anti-government activities and declared war on Zimbabwe, threatening the Beira corridor. Relations deteriorated further with South Africa who imposed a ban on Mozambican migrant workers.

Signs at the end of 1989 indicated that South Africa is preparing to move toward majority rule. This is expected to lead the end

Mozambique

of South African support for the MNR giving rise to hopes that peace may be possible in the not too distant future.

Stability

Frelimo, the main liberation group at independence in 1974, has continued to rule, with Joachim Chissano replacing Samora Machel as president when the latter was killed in an air crash late in 1986. Throughout the independence period there has been considerable lack of internal security as a result of the destabilising activities of the South African-backed Renamo forces, which have interrupted transport links and dislocated agricultural production. The economic strategy was directed to extensive nationalisation after independence, and the industrial and services sectors were hampered by the withdrawal of skilled Portuguese personnel. Early in 1987 the Chissano government made an agreement with the IMF which embraced a substantial programme of liberalising reforms.

Overall, Mozambique's stability record is very poor, with no internal security and two major changes in economic strategy. Internal security, agricultural production and transport links will not improve until there is majority rule in South Africa. If persevered with, the IMF liberalisations will bring about marginal improvements in the economic climate.

Economic Structure

Mozambique's economy has suffered since independence in 1974 as a result of the flight of skilled labour and the effects of civil war and drought. It had a GDP of $US 1,490 m. Living standards are low by African standards and with a GNP per head of only $US 170 per person, it is the second poorest in the region after Malawi.

Around 50% of GDP is generated by agriculture, the highest dependency on this sector in the region. The industrial sector is limited to only 12% of GDP, the lowest in the region, with manufacturing restricted by lack of skilled labour and outdated plant, as well as limited mineral exploitation. The service sector generates about 38% of GDP.

Private consumption was 90% of GDP which is second highest in the region after Lesotho. Investment is fairly high at 22%, though this may be partly accounted for by expenditure on maintaining and repairing the transport infrastructure damaged by rebels. Government consumption is 20% of GDP.

The cost of imports, most of which are food and machinery, is equal to 43% of GDP, while exports account for only 11% of GDP, one of the lowest rates in Africa. The main exports in 1987 were cashew nuts, 33% ; and shellfish, 41%; though coastal waters have been drastically overfished in recent years.

Health and education services have declined in recent years and are now reported to be among the worst in Africa. The literacy rate in 1980 was 33%. Enrolments in primary education were 82% in 1986, and in secondary education 7%. There were negligible enrolments in tertiary education.

Life expectancy in 1987 was 48 years. There is only one doctor to every 38,000 people, while many countries in the region have five times as many doctors per head.

Overall, Mozambique is a low-income country, with substantial dependence on agriculture and poor educational and health provisions.

Economic Performance

The instability in Mozambique has seriously affected the economy with the result that GDP fell by -2.6% a year from 1980-87, though the regional decline was even worse at -3.7%. Living standards have declined with GNP per head falling by -5.3% between 1980 and 1987. The agricultural sector suffered a serious decline of -11.1% as a result of instability in rural areas as well as droughts between 1983 and 1987, while the regional average was 6.8%. The industrial sector, suffering from lack of spare parts and investment for new plant, together with a loss of skilled labour, has declined by -8.4% while the regional average was -3.8%. Only the service sector saw any growth, 6.2% while the regional average was -0.9%.

The lack of stability means that there is little agricultural output for local consumption, let alone export, and the transport system has been made virtually unusable in most areas. As a result, exports fell by -10.7% a year from 1973-84 (and probably even more in recent years). The serious shortage of foreign currency and large military expenditure means imports are restricted and fell by -4.7% a year over the same period.

Inflation is a growing problem brought on by lack of foreign currency and severe shortages of even locally produced basic goods. From 1980 to 1987 it averaged 27%, the highest in the region after Zambia, and was running at 50% by 1988.

Overall, Mozambique's economic performance has been uniformly disasterous, with GDP per head, agricultural and industrial output falling, and inflation accelerating.

Recent Economic Developments

Mozambique ushered in a period of substantial economic reforms in 1987 which began to reverse the state control and intervention that had characterised economic policy since independence in 1974. The currency has been devalued regularly, price controls relaxed, producer prices increased, foreign investment encouraged, and agreements made with the IMF and World Bank for structural adjustment lending. Internal security and transport continues to remain a problem as a result of the activities of South Africa-supported Renamo forces, and a third of the population is estimated to have been hit by drought in 1987. However, aid has increased substantially as a result of international sympathy for Mozambique's difficulties and as a response to the government's willingness to introduce reforms.

Economic performance has been extremely poor in the 1980s, with positive growth of GDP recorded in 1980 and 1981, but at less than the rate of population increase, and substantial GDP falls for 1982 to 1985. There appears to have been a return to positive growth from 1986 on. Since 1980, the average living standards have fallen by an estimated 45 per cent. The bulk of the population has reverted to subsistence farming as modern sector activities have collapsed, and, even in this sector, drought and unavailability of basic agricultural imports have reduced output. There were extensive famine relief operations under way in 1987. The official figures for inflation are quite unreliable as they reflect regulated prices at which goods are simply unavailable, and most transactions are carried out at black market rates. The World Bank has estimated the rate of price increases at 25.8% per year over the period 1980-85. Although price rises in 1987 of 300-500% for certain commodities have been announced, much of this will be recognising the black market rates at which the goods were previously available. A 170 per cent inflation rate for 1987 has been reported by the government.

Mozambique's balance of trade has deteriorated as production has declined. Export values peaked at $US 298m at independence in 1974, and have subsequently shown a declining trend such that they were below one-third of this level in 1989. In the early 1970s merchandise values were twice the value of exports of goods, but the gap began to widen from 1977 as exports fell by 30 per cent but imports doubled, such that imports are seven times the value of exports, and the trade gap was some $US 674m in 1988.

Mozambique is estimated to have $US 4,300m outstanding in external debt in 1988, and it is quite beyond the present capacity of the economy to service this. It is expected that debt service payments will need to be rescheduled for the foreseeable future.

The official exchange rate remained virtually unchanged up to 1986, while the black market rate rapidly depreciated with Mozambique's deteriorating balance of payments situation. From MT 40.4 = $US 1 in 1986, there have been successive adjustments that have brought the currency to MT 812.3 = $US 1 at the end of 1989.

The 1989 budget raised expenditure by 29 per cent, but with revenue set to rise 33 per cent, the deficit has sstopped growing for the present. Up to 40 per cent of current expenditure is allocated for defence. A major area of uncertainty surrounds the rate of domestic price increase, and it is difficult to ascertain the degree of real expenditure increase. While prices of some commodities have increased fivefold, salaries are expected to increase between 50-100%. The devaluations have increased the local currency value of imports by a factor of ten since the budget. Higher tax levels have been introduced on private sector

Mozambique

incomes, and there have been increases in taxes on beer and cigarettes, while fuel and utility prices have been raised. In the face of such radical changes, although it seems clear that the government will increase the real level of spending, it is difficult to ascertain the amount.

Mozambique has received substantial aid commitments since 1986. There are loans for balance of payments support from the IMF of $US 54.7m and from the World Bank of $US 90m. A wide variety of bilateral donors have contributed to famine relief, energy projects, agriculture, health and infrastructure rehabilitation. A major focus of aid has been the efforts to restore the three sets of rail and port facilities. The Maputo and Beira links have been operating at 50% of capacity, and the rail line to Malawi has been closed.

There has been revived interest in foreign investment. A US firm is involved in titanium mining, BP and Amoco are exploring off-shore oil deposits, China is in a joint venture to mine granite, and UK's Lonrho is investing in cotton ginning and agriculture.

Economic Outlook and Forecasts

The forecasts assume no significant improvement in Mozambique's stability, but that the government persists with liberalising economic reforms.

GDP is forecast to expand, but only at the rate of population expansion giving no significant improvements in GDP per head.

Export volumes are projected to continue to decline, although aid should enable import volumes to expand.

Inflation is expected to accelerate to around 30% a year.

Mozambique: Economic Forecasts
(average annual percentage change)

	Actual	Forecast Base	Forecast High
	1980-87	1990-95	1990-95
GDP	-2.6	3.2	3.5
GDP per Head	-5.3	0.0	0.3
Inflation Rate	27.0	33.0	28.0
Export Volumes	-10.7[a]	-4.5	-3.5
Import Volumes	-4.7[a]	3.7	4.3

Note: [a] 1973-84.

MOZAMBIQUE: Comparative Data

	MOZAMBIQUE	SOUTHERN AFRICA	AFRICA	INDUSTRIAL COUNTRIES
POPULATION & LAND				
Population, mid year, millions, 1989	15.0	6.0	10.2	40.0
Urban Population, %, 1985	19	23.8	30	75
Population Growth Rate, % per year, 1980-86	2.7	3.1	3.1	0.8
Land Area, thou. sq. kilom.	802	531	486	1,628
Population Density, persons per sq kilom., 1988	18.3	11.3	20.4	24.3
ECONOMY: PRODUCTION & INCOME				
GDP, $US millions, 1986	4,300	2,121	3,561	550,099
GNP per head, $US, 1986	210	383	389	12,960
ECONOMY: SUPPLY STRUCTURE				
Agriculture, % of GDP, 1986	35	25	35	3
Industry, % of GDP, 1986	12	33	27	35
Services, % of GDP, 1986	53	42	38	61
ECONOMY: DEMAND STRUCTURE				
Private Consumption, % of GDP, 1986	86	67	73	62
Gross Domestic Investment, % of GDP, 1986	9	14	16	21
Exports, % of GDP, 1986	3	34	23	17
Imports, % of GDP, 1986	13	36	26	17
ECONOMY: PERFORMANCE				
GDP growth, % per year, 1980-86	-9.0	-3.7	-0.6	2.5
GDP per head growth, % per year, 1980-86	-11.7	-6.6	-3.7	1.7
Agriculture growth, % per year, 1980-86	-15.9	-6.8	0.0	2.5
Industry growth, % per year, 1980-86	-13.3	-3.8	-1.0	2.5
Services growth, % per year, 1980-86	0.2	-0.9	-0.5	2.6
Exports growth, % per year, 1973-84	-10.7	-5.1	-1.9	3.3
Imports growth, % per year, 1973-84	-4.7	-2.5	-6.9	4.3
ECONOMY: OTHER				
Inflation Rate, % per year, 1980-86	28.1	18.1	16.7	5.3
Aid, net inflow, % of GDP, 1986	9.8	8.6	6.3	-
Debt Service, % of Exports, 1984	4.4	7.8	20.6	-
Budget Surplus (+), Deficit (-), % of GDP, 1986	-	-4.1	-2.8	-5.1
EDUCATION				
Primary, % of 6-11 group, 1985	84	95	76	102
Secondary, % of 12-17 group, 1985	7	18	22	93
Higher, % of 20-24 group, 1985	0.0	1.3	1.9	39
Adult Literacy Rate, %, 1980	33	47	39	99
HEALTH & NUTRITION				
Life Expectancy, years, 1986	48	50	50	76
Calorie Supply, daily per head, 1985	1,617	1,998	2,096	3,357
Population per doctor, 1981	36,970S	26,118	24,185	550

Notes: 'Southern Africa' and 'Africa' exclude South Africa. Dates are for the country in question, and do not always correspond with the Regional, African and Industrial averages.

MOZAMBIQUE: Leading Indicators

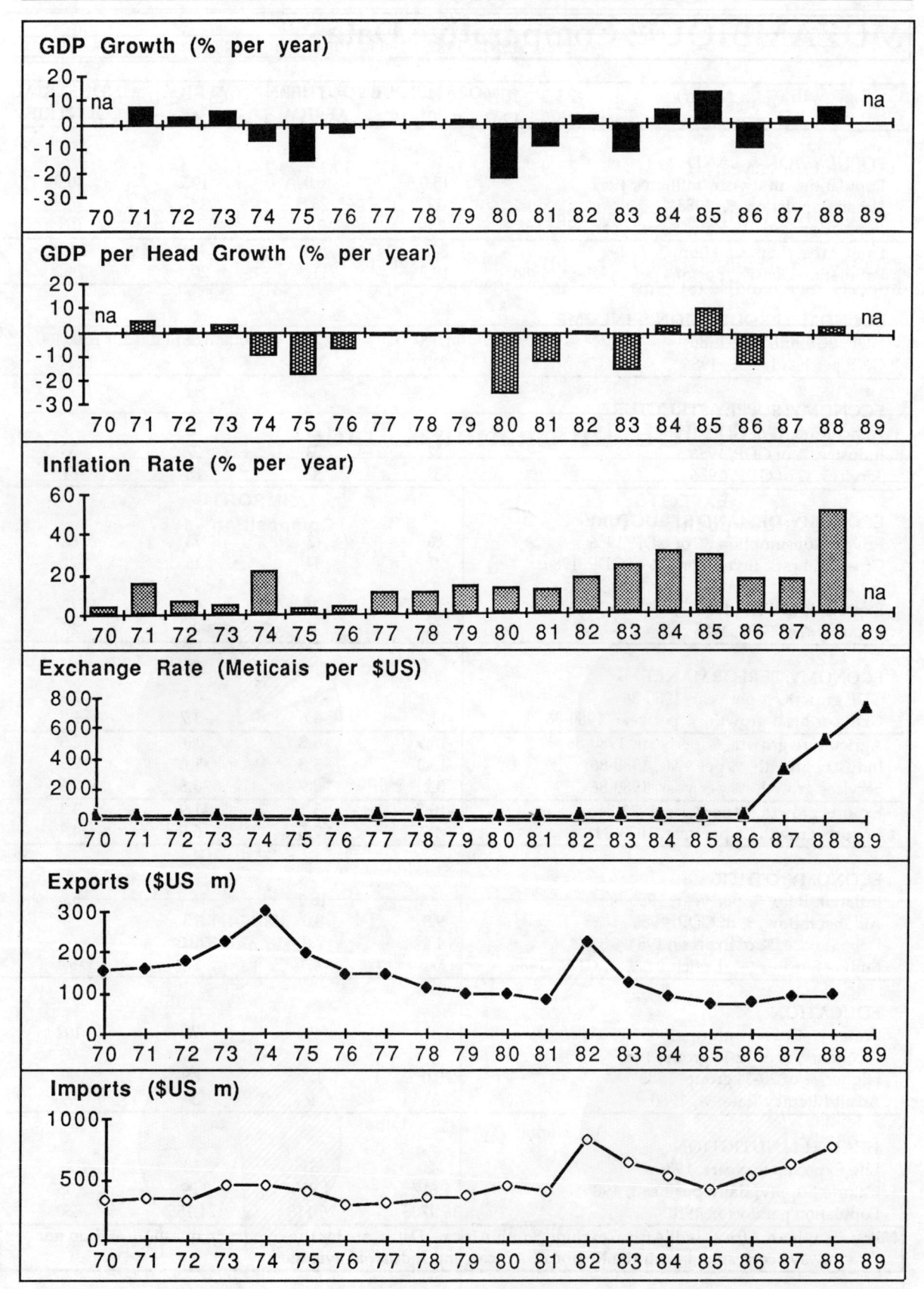

Note: Leading indicator data for 1989 are based on the first half of 1989. 1989 exchange rate is for mid-1989.

MOZAMBIQUE: Leading Indicator Data

	1980	1981	1982	1983	1984	1985	1986	1987	1988	1989
GDP growth (% per year)	-23.3	-10.1	2.6	-12.8	5.4	13.1	-10.8	2.1	6.0	
GDP per head growth (% per year)	-25.9	-12.7	-0.1	-15.5	2.7	10.3	-13.6	-0.3	3.1	
Inflation (% per year)		12.9	11.9	18.0	24.0	30.3	29.2	16.9	16.6	50
Exchange rate (Meticais per $US)	28.0	28.5	38.8	40.2	42.4	43.2	40.4	327.2	525.0	738.6
Exports, merchandise ($US m)	101	85	229	131	96	77	79	97	99	
Imports, merchandise ($US m)	465	408	836	636	540	424	543	625	773	

Note: Leading indicator data for 1989 are based on the first half of 1989. 1989 exchange rate is for mid-1989.

MOZAMBIQUE: International Trade

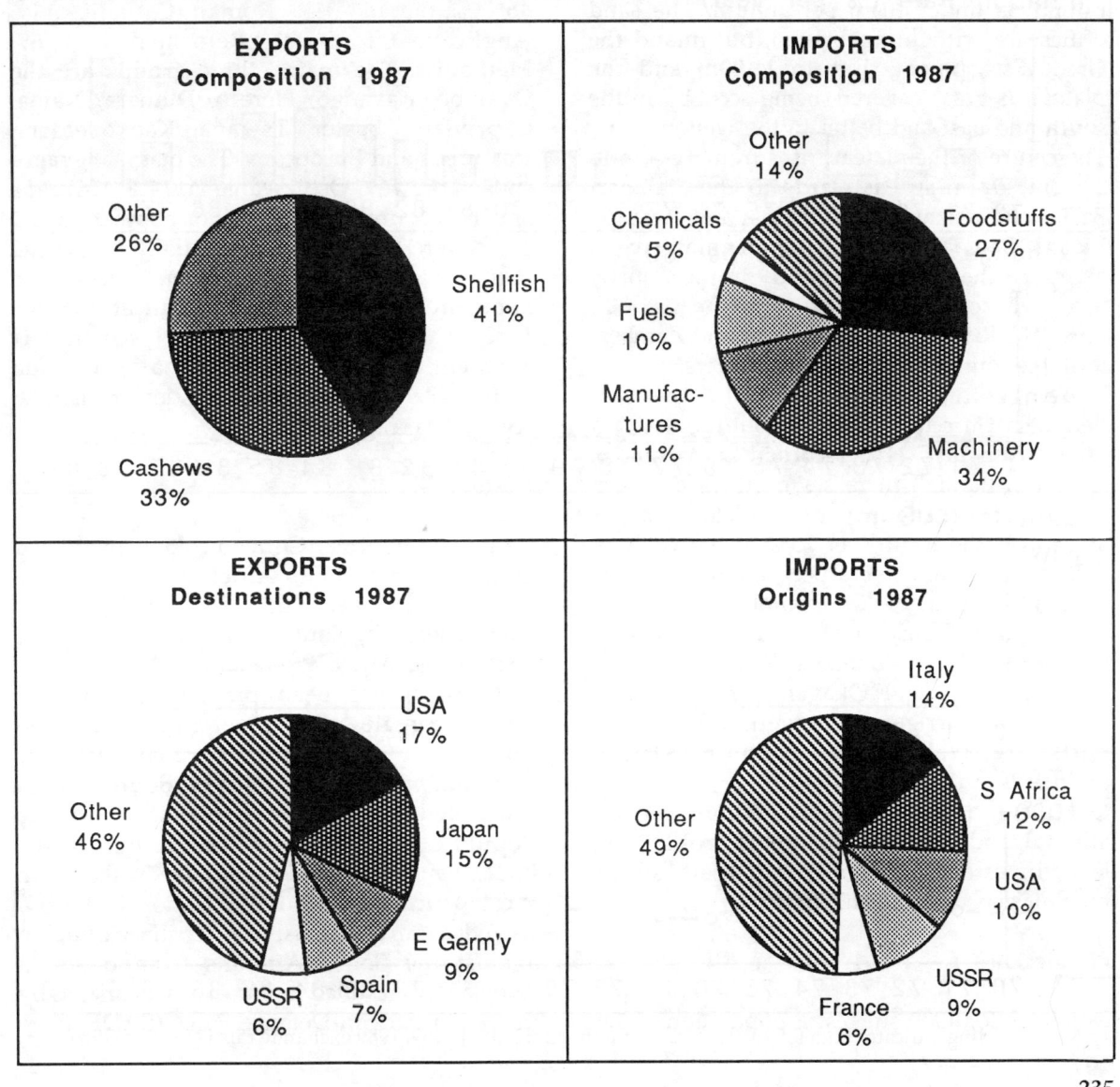

EXPORTS Composition 1987
- Shellfish 41%
- Cashews 33%
- Other 26%

IMPORTS Composition 1987
- Foodstuffs 27%
- Machinery 34%
- Manufactures 11%
- Fuels 10%
- Chemicals 5%
- Other 14%

EXPORTS Destinations 1987
- USA 17%
- Japan 15%
- E Germ'y 9%
- Spain 7%
- USSR 6%
- Other 46%

IMPORTS Origins 1987
- Italy 14%
- S Africa 12%
- USA 10%
- USSR 9%
- France 6%
- Other 49%

30 NAMIBIA

Physical Geography and Climate

Namibia, in the Southern Africa region, is a coastal territory bounded by Angola, Zambia, Zimbabwe, Botswana and South Africa. The Namib Desert extends to the Atlantic coast and rainfall is under 10cm per annum. The sand dunes are without vegetation, but inland the Great Escarpment rises to 1,100m and the plateau is grass covered, being scrubby in the south and east and better in the wetter north. The centre of the plateau rises from 1,500m to 2,500m, draining eastwards to the Kalahari Basin and northwards to the flat and swampy Etosha Pan. There are no perennial rivers, except in the north east mostly in the Caprivi Strip where the Okavango and the Cuando flow. The Rivers Orange, Kunene and Zambezi form the southern and northern borders.

Temperatures are modified by the cool Benguela Current and also by altitude with a range from 14°C-24°C recorded at Windhoek (1,700m). Rainfall increases northwards, from under 10cm per annum in the south to 50 cm per annum in the north, yet 75% of the country receives less than 40cm per annum. There is some water available from boreholes.

A hydro-electric power station has been built on the River Kunene at Ruacana and the Rivers Orange and Okavango have hydro-electric and irrigation potential. The main ports are Walvis Bay, a South African enclave, and Lüderitz.

Mineral deposits include cadmium, copper, alluvial and dredged diamonds, gold, iron, lead, limestone, manganese, rocksalt, silver, tin, tungsten, uranium and zinc.

Population

Estimates for 1989 suggest a population of 1.4 million. Over half of the population, including two thirds of the Africans, live in the wetter northern third of the plateau. The majority of Europeans live in the south, especially around the capital Windhoek. About 90% of the population are Christian, 1980 figures being 46% Lutheran, 31% Roman Catholic, 11% Anglican, 6% Dutch Reformed and 6% Methodist. The main ethnic groups are the Ovambo, Kavango, Herero, Dumara, Nama, Caprivians, Baster, Tswana, Kaokovelders, Coloureds and Europeans. The main languages spoken are 46% Oshikwanama, 14% Afrikaans and 13% Nama. The density of population is 2 per square kilometre, substantially below the regional average of 11.3 persons per square kilometre. Urbanisation was put at 48% for 1982. The rate of population growth has apparently dropped from 3.4% a year in the 1970s to 2.8% for 1980-84 and is lower than the regional average of 3.3%.

History

Namibia is a former German colony, mandated by the Union of South Africa at the end of the First World War. Until April 1989, it was administered by South Africa under the name South West Africa.

Namibia has been plagued by internal armed struggles since 1966 which have been directed toward achieving majority rule government. During 1968/9 South Africa proceeded to incorporate Namibia into the Republic of South Africa. Prior to the implementation of UN resolution 435 the South West Africa People's Organisation (SWAPO) waged a fairly intensive guerrilla campaign against the South Africans from bases in Angola. This resulted in numerous incursions by the South African Defence Force (SADF) deep

Namibia

into Angola, supported by UNITA Angolan rebels, often for several weeks, thereby extending the conflict more widely in the region.

On 1 April 1989 the United Nations began the supervision of a ceasefire and withdrawal of SADF forces from Namibia as part of the UN Resolution 435 to start the process leading to full independence.

The UN-supervised elections in November resulted in a majority of votes for SWAPO but not the two thirds majority needed to bring about constitutional changes.

Stability

There has been armed struggle in Namibia in pursuit of independence from South Africa since 1966.

With independence scheduled early in 1990, there is considerable speculation about future international relations with Namibia. SWAPO has said that it will not have any relationship with South Africa, but this seems impossible bearing in mind that most imports at present come from there and in particular the lack of alternative trade routes while South Africa retains control of Walvis Bay. This small enclave is of considerable economic and strategic value as the only deep water port on the coast. South Africa maintains that it will not hand the territory over and has recently expanded its military strength there. The new Namibian administration may be forced to develop a new port, probably by dredging at Swakopmund.

SWAPO has promised to develop a mixed economy and will have to seek considerable foreign investment and aid.

Overall, Namibia's stability has been very poor, but should improve considerably with independence. Stability will only begin to improve further with majority rule in South Africa.

Economic Structure

In 1982 GDP was estimated at $US 1,379m. With a small population and considerable natural resources, Namibia could be expected to have a high standard of living and indeed in that year income was estimated at $US 1,253 per capita, one of the highest in Africa. However, there are large disparities in income levels, with the majority of blacks living in rural areas at subsistance level.

In 1982 the agricultural sector contributed 11% of GDP making the country one of the least dependent on agriculture in Africa. Conversely, the industrial sector is one of the largest in Africa at 51%, and this includes the large mining sector (presently about 25% of GDP). Manufacturing is limited to processing raw materials, especially fish.

Services contributed 37%, though in 1989 this sector should be larger as the demands of international organisations supervising the lead up to independence are met.

On the demand side, private consumption was 58% of GDP in 1982, though this has been growing in recent years. Investment was 18% but this figure has since declined. Government consumption was 16% of GDP. Exports represented 87% of GDP, and in 1986 were made up of uranium 38%, diamonds 31%, with other minerals making up 18%. Spending on imports was equal to 78% of GDP.

Imports are predominantly foodstuffs at 36% of the total, with fuels comprising 30%, and machinery and transport equipment 28%. Namibia appears to have negligible external debt.

The adult literacy rate was last estimated in 1970 at 34%. Since then it can be expected to have improved to above 50%, with primary enrolments at 103% in 1981. Secondary enrolments in 1981 were well above the regional and African averages at 31%. However, there was negligible enrolment in higher education.

Life expectancy was 43 years in 1984, and this is significantly below the regional and African average of 50. Population per doctor was 6,600 in 1979, and this is about four times better than the average in Africa. Average daily calorie intake is roughly comparable with levels elsewhere in Africa. The apparent paradox of good medical provision and adequate calorie supply co-existing with low life expectancy is explained by the dual nature of the economy. Medical care and superior nutrition are concentrated on the high-income urban sector, whereas provision is much poorer in the rural areas.

Namibia

Overall, Namibia is a lower middle-income country, with a structure of production which relies heavily on mining, and very little on agriculture. There is heavy dependence on trade, with exports predominantly minerals. Educational enrolments are good at the primary and secondary levels, and health provision is well above average, but very unequally distributed.

Economic Performance

GDP is estimated to have grown by 0.9% a year in the period 1980-87, and this is better than the average elsewhere in the region and Africa generally, where GDP has declined. Although GDP per head has declined at -2.1% a year, this is a better performance than on average in Africa, where GDP per head has declined even faster.

The latest figures for sector growth rates are for 1973-82, and in this period agriculture contracted at -1.7% a year, and industry contracted even faster at -5.0% a year. The services sector expanded at 4.3% a year.

Export volumes are estimated to have fallen at -10% a year, and import volumes at -6.3% a year in the period 1980-87, and this is a rather faster rate of decline than elsewhere in Africa.

Inflation has averaged 12.8% a year for 1980-87, and this is about two thirds the annual rate of price increase in the region.

Overall, Namibia has experienced some growth although GDP per head has fallen. The service sector appears to have been the main source of GDP growth, however, expansion in export and import volumes has fallen rapidly. Inflation has been moderate, although in double figures.

Recent Economic Developments

There has been some pressure since 1987 for non-South African mining firms to quit Namibia, but as yet there is no indication that this will occur. There has been a general unwillingness to expand economic operations in Namibia as a result of the insecure internal situation, uncertainty over Namibian independence and aversion to doing business while the economy is under South African control. After steady growth in the 1970s, the expansion of GDP slowed after 1980, being negative for 1983 to 1985, showing a recovery to positive growth in 1986. However, population growth at 3% a year has caused average living standards to fall steadily since 1980. Inflation has fluctuated between 10% and 15% a year, with an acceleration to 18% in 1986. Exports, dominated by the mining sector, peaked in 1981 at $US 2,082m, but have rapidly declined to $US 719m in 1985 with strengthening diamond prices and increased diamond production yielding a recovery in 1986 to $US 850m. Namibia has typically run a trade surplus, but deteriorating export performance has compressed imports from $US 1,381m in 1980 to $US 562m in 1985.

As Namibia is administered by South Africa, there is no clear indication of her external debt position. The 1985/6 budget, however, indicated that Namibia's external debt was $US 377m, with debt servicing running at $US 32m, a manageable 4% of visible export earnings.

Namibia circulates the South African rand, and the exchange rate has steadily depreciated from R 0.78 = $US 1 in 1980 to R 2.53 = $US 1 in 1986. At the end of 1989, the exchange rate stood at R 2.55 = $US 1.

Namibia's relationship with South Africa virtually excludes conventional aid flows. The mineral wealth of the region has continued to attract some potential investors, however, despite the political uncertainty. North Sea oil companies are involved in offshore gas exploration, and Anglo American are considering a gold mining operation.

Economic Outlook and Forecasts

Namibia's prospects have improved immensely with the orderly progression to independence, and the announcement early in 1990 that South Africa is preparing to move toward majority rule.

The forecasts assume a stable political scene. It is expected that a coalition will emerge between SWAPO and the next largest party, the more conservative Democratic Turnhalle Alliance (DTA) so that the two-thirds majority required to adopt the constitution can be obtained. It is expected that the coalition with the DTA will be instrumental in preventing expansion of state ownership.

GDP is forecast to expand at about the rate of population growth in the period up to 1995. This will arrest the fall in GDP per head which has been experienced in the 1980s.

The inflation rate is projected to accelerate under the pressures of withdrawal of South African subsidies and the anticipated expansion of the government sector. The annual rate of price increase could be double that experienced in the 1980s.

Export volumes are expected to increase, but that major expansion will not be experienced before the latter part of the 1990s. Import volumes are forecast to expand more rapidly than exports as aid flows are expanded.

Namibia: Economic Forecasts
(average annual percentage change)

	Actual	Forecast Base	High
	1980-87	1990-95	1990-95
GDP	0.9	4.2	4.5
GDP per Head	-3.2	0.1	0.4
Inflation Rate	12.7	25.0	20.0
Export Volumes	-10.0	2.5	2.8
Import Volumes	-6.3	5.2	6.5

NAMIBIA: Comparative Data

	NAMIBIA	SOUTHERN AFRICA	AFRICA	INDUSTRIAL COUNTRIES
POPULATION & LAND				
Population, mid year, millions, 1989	1.4	6.0	10.2	40.0
Urban Population, %, 1982	48	23.8	30	75
Population Growth Rate, % per year, 1970-81	3.4	3.1	3.1	0.8
Land Area, thou. sq. kilom.	824	531	486	1,628
Population Density, persons per sq kilom., 1988	1.9	11.3	20.4	24.3
ECONOMY: PRODUCTION & INCOME				
GDP, $US millions, 1982	1,379	2,121	3,561	550,099
GNP per head, $US, 1982	1,253	383	389	12,960
ECONOMY: SUPPLY STRUCTURE				
Agriculture, % of GDP, 1982	11	25	35	3
Industry, % of GDP, 1982	51	33	27	35
Services, % of GDP, 1982	37	42	38	61
ECONOMY: DEMAND STRUCTURE				
Private Consumption, % of GDP, 1982	58	67	73	62
Gross Domestic Investment, % of GDP, 1982	18	14	16	21
Government Consumption, % of GDP, 1982	16	21	14	17
Exports, % of GDP, 1982	87	34	23	17
Imports, % of GDP, 1982	78	36	26	17
ECONOMY: PERFORMANCE				
GDP growth, % per year, 1973-82	3.9	-3.7	-0.6	2.5
GDP per head growth, % per year, 1973-82	0.5	-6.6	-3.7	1.7
Agriculture growth, % per year, 1973-82	-1.7	-6.8	0.0	2.5
Industry growth, % per year, 1973-82	-5.0	-3.8	-1.0	2.5
Services growth, % per year, 1973-82	4.3	-0.9	-0.5	2.6
Exports growth, % per year, 1973-82	3.4	-5.1	-1.9	3.3
Imports growth, % per year, 1973-82	5.7	-2.5	-6.9	4.3
ECONOMY: OTHER				
Inflation Rate, % per year, 1973-82	8.9	18.1	16.7	5.3
Aid, net inflow, % of GDP, 1981	..	8.6	6.3	-
Debt Service, & of Exports, 1981	..	7.8	20.6	-
Budget Surplus (+), Deficit (-), % of GDP, 1984	-3.3	-4.1	-2.8	-5.1
EDUCATION				
Primary, % of 6-11 group, 1981	103	95	76	102
Secondary, % of 12-17 group, 1981	31	18	22	93
Higher, % of 20-24 group, 1981	..	1.3	1.9	39
Adult Literacy Rate, %, 1973	34	47	39	99
HEALTH & NUTRITION				
Life Expectancy, years, 1984	43	50	50	76
Calorie Supply, daily per head, 1983	2,025	1,998	2,096	3,357
Population per doctor, 1979	6,600	26,118	24,185	550

Notes: 'Southern Africa' and 'Africa' exclude South Africa. Dates are for the country in question, and do not always correspond with the Regional, African and Industrial averages.

NAMIBIA: Leading Indicators

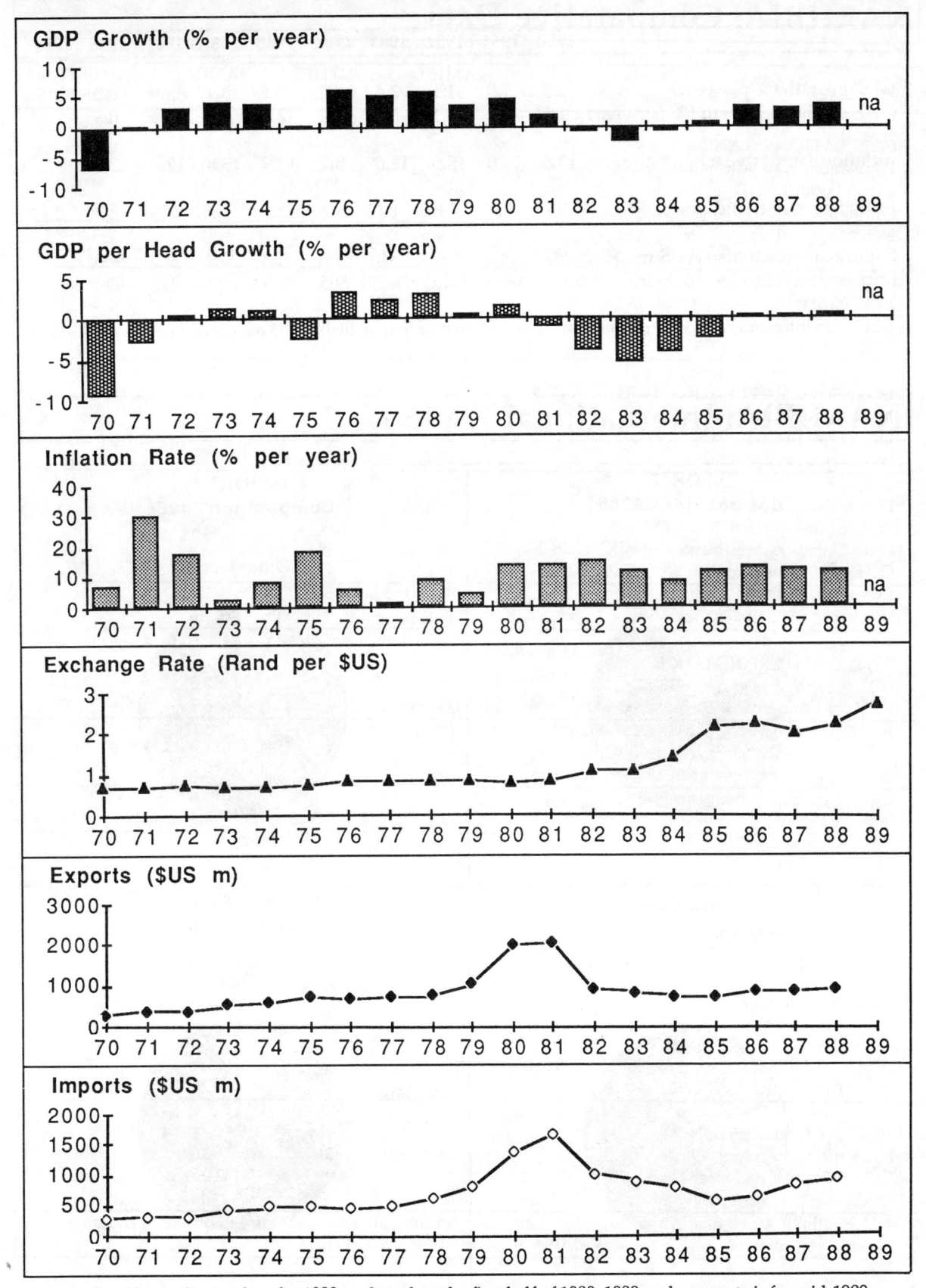

GDP Growth (% per year)

GDP per Head Growth (% per year)

Inflation Rate (% per year)

Exchange Rate (Rand per $US)

Exports ($US m)

Imports ($US m)

Note: Leading indicator data for 1989 are based on the first half of 1989. 1989 exchange rate is for mid-1989.

NAMIBIA: Leading Indicator Data

	1980	1981	1982	1983	1984	1985	1986	1987	1988	1989
GDP growth (% per year)	4.5	1.9	-1.1	-2.6	-1.2	0.4	3.1	2.9	3.5	
GDP per head growth (% per year)	1.5	-1.1	-4.1	-5.6	-4.2	-2.6	0.1	0.1	0.5	
Inflation (% per year)	13.6	14.0	15.5	12.0	8.7	11.9	13.4	12.6	12.2	
Exchange rate (Rand per $US)	0.78	0.87	1.08	1.11	1.44	2.19	2.28	2.04	2.26	2.77
Exports, merchandise ($US m)	2008	2082	922	846	766	728	878	889	960	
Imports, merchandise ($US m)	1381	1675	1012	906	805	565	638	842	920	

Note: Leading indicator data for 1989 are based on the first half of 1989. 1989 exchange rate is for mid-1989.

NAMIBIA: International Trade

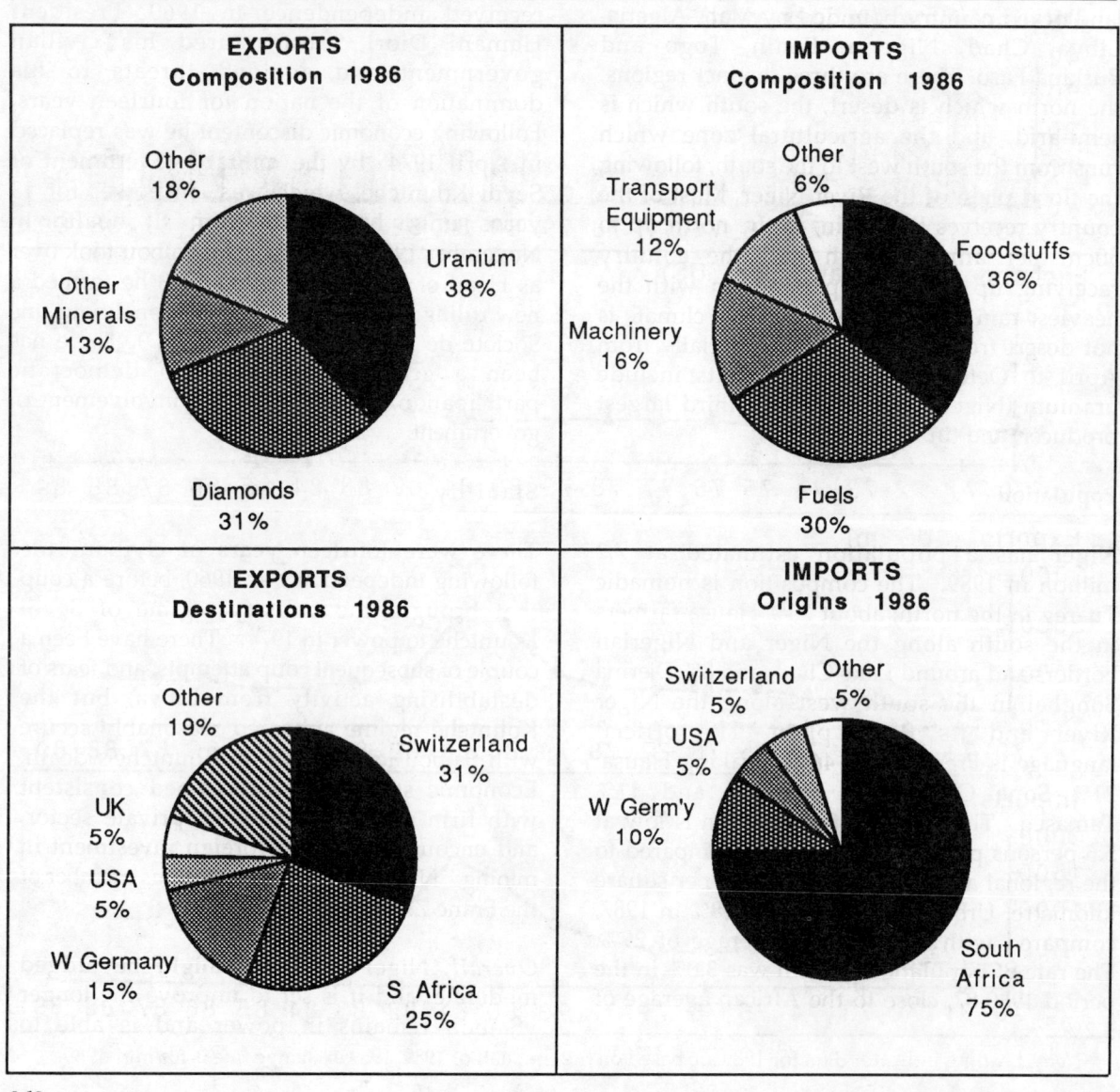

EXPORTS Composition 1986
- Uranium 38%
- Diamonds 31%
- Other Minerals 13%
- Other 18%

IMPORTS Composition 1986
- Foodstuffs 36%
- Fuels 30%
- Machinery 16%
- Transport Equipment 12%
- Other 6%

EXPORTS Destinations 1986
- Switzerland 31%
- S Africa 25%
- W Germany 15%
- USA 5%
- UK 5%
- Other 19%

IMPORTS Origins 1986
- South Africa 75%
- W Germ'y 10%
- USA 5%
- Switzerland 5%
- Other 5%

31 NIGER

Physical Geography and Climate

Niger, in the West African region, is a landlocked country bounded by Mali, Algeria, Libya, Chad, Nigeria, Benin, Togo and Burkina Faso. There are three distinct regions, the north which is desert, the south which is semi-arid, and the agricultural zone which runs from the south west to the south, following the flood plain of the River Niger. Most of the country receives little rain, in the north up to 50cm per annum, with half the country receiving up to 190cm per annum with the heaviest rain falling in August. The climate is hot desert from 24°C to 34°C, especially from April to October. Mineral deposits include uranium (Niger is now the joint third largest producer) and tin.

Population

Niger has a population estimated at 7.2 million in 1989. The composition is nomadic Tuareg in the north; about 25% Hausa farmers in the south along the Niger and Nigerian borders and around Lake Chad, and 8% Djerma Songhai in the south west along the Niger River and its flood plain. The official language is French with 46% speaking Hausa; 21% Sona Ciini; 14% Fulfulde and 11% Tamaseq. The density of population is low at 5.6 persons per square kilometre, compared to the regional average of 31.1 persons per square kilometre. Urbanisation is low at 18% in 1987, compared with the regional average of 29%. The rate of population growth was 3.0% in the period 1980-87, close to the African average of 3.2%.

History

Niger was a former French colony which received independence in 1960. President Hamani Diori restructured his civilian government and avoided threats to his domination of the nation for fourteen years. Following economic discontent he was replaced in April 1974 by the military government of Seyni Kountché, which was in power for 13 years until his death from ill health in November 1987 when Brig. Ali Saïbou took over as Head of State. In August 1988 he formed a new ruling party, Mouvement National pour une Société de Développement (MNSD). There has been a gradual movement to democratic participation and more civilian involvement in government.

Stability

There were fourteen years of civilian rule following independence in 1960, before a coup that brought the military regime of Seyni Kountché to power in 1974. There have been a couple of subsequent coup attempts, and fears of destabilising activity from Libya, but the Kountché regime appeared reasonably secure with Saïbou taking over on Kountché's death. Economic strategy has remained consistent with firm commitment to the private sector, and encouragement for foreign investment in mining. Niger has been a continuous member of the Franc Zone.

Overall, Niger's stability might be judged moderate, and it is set to improve the longer Saïbou remains in power and is able to

Niger

incorporate civilian elements into his government.

Economic Structure

GDP was estimated at $US 2,160m in 1987, less than half the regional average. GNP per head was estimated at $US 260, again about half the regional average.

Agriculture generated 34% of GDP in 1987, below the regional average, while the industrial sector provided 24%, also just below the regional average. The service sector is larger than average for the region, generating 42% of GDP.

On the demand side, private consumption was 84% of GDP in 1987 when the regional average was about 77% and the African average 73%. Gross domestic investment was only 9% in Niger, while regional and African averages were 13% and 16% respectively. Government consumption was 12%, similar to the regional and African averages of 13% and 14%. Exports represented 19%, about average for the region, while imports were also 19%.

Niger's main export is uranium which is sold to France at above world market prices and represents 77% of export earnings. Most of the other exports are agricultural products which make up over 15% of exports. Main imports in 1987 were machinery, 31%; food, 18%; raw materials, 11%; and fuels 6%. There is substantial unofficial cross-border trade in petroleum products from Nigeria. It should also be noted that imports and exports of food and livestock products fluctuate greatly, depending on climatic conditions in the Sahel.

Debt service payments took up 33.5% of export earnings in 1987, more than twice the African average.

Education is poor with only 29% of primary-aged children in school in 1986 when the regional and African averages were around 75%. Only 6% of the secondary school age range were receiving an education, compared to 24% in the region as a whole and 22% in Africa. It should be borne in mind that Niger is predominantly Islamic and therefore females of this age would be considered more eligible for marriage than further education. Furthermore, a considerable number of children attend Koranic schools and therefore receive some form of education. Only 1% of the 20-24-year-old age group were in higher education, while the regional average was 2.6% and the African average 1.9%.

Life expectancy at birth in Niger is low, only 45 years, while on average a person in the region or Africa can expect to live five years longer than that. With only one doctor to 39,000 people, Niger has one of the highest ratios of people to doctors in the region. On the positive side, however, there is one nurse for every 450 people, better than any other country in the region and well above the regional ratio of 1 to 2,722.

Overall, Niger is a low-income country. The economy is heavily reliant on the agreement with France for the supply of uranium and climatic conditions for agriculture. Educational and health provision are poor.

Economic Performance

GDP in Niger fell by -1.9% from 1980-87, about the same as the -2.0% fall in the region as a whole. Despite periods of serious drought, the agricultural sector, which generates about one third of GDP, showed a growth of 2.8%, better than the African average of 1.2%. The industrial sector, however, showed a decline of -4.3%, while the service sector fell even further, by -8.0%. This compares with regional falls of -3.6% and -2.5% in these two sectors respectively.

A fall in export volumes of -4.8% a year was similar to the regional average of -4.0%. Niger reduced import volumes by -6.2% a year, 1980-87, compared with Africa generally where they fell by -5.8% a year.

The inflation rate for this period was only 4.1% a year, compared with the African average of 15.2%.

Overall, Niger has experienced declining GDP and GDP per head in the 1980s, with the industry and services sectors showing the fastest contraction. Export and import volumes declined at rates faster than the average. Only in inflation was performance above average.

Recent Economic Developments

Niger attempted to respond to falling uranium

prices in the 1980s and the effects of drought in 1984 and 1985 with a structural adjustment programme monitored by the IMF and the World Bank. Lending for this purpose included $US 31m from the IMF in late 1986, and a further structural adjustment loan of $US 80m from the World Bank in July 1987, following the $US 62m World Bank loan of February 1986. The programme envisaged reductions in public sector spending, the phasing out of price controls, liberalisation of agricultural marketing, privatisation of 22 state-owned enterprises and the reform of 32 others. Private investment is to be encouraged, and rural development and agriculture are to have priority in government spending, with increasing emphasis on health, education, housing, and water provision. In June of 1987, there was a 30% cut in indirect taxes, and there were price falls affecting some 3,000 commodities. In December 1988 the IMF approved SDR 50.55m ($US 68.4m) through the Enhanced Structural Adjustment Facility (ESAF), replacing the previous SAF arrangement. In order to secure the facility, the government agreed to promote private sector growth, make improvements in the public sector and reforms in agriculture. This also falls in line with the government's 1987-95 five-year plan which aims to reduce debt servicing, increase privatisation and reduce public expenditure.

GDP expanded continuously from 1974 to 1980, but falling export revenues and drought caused falls in GDP between 1981 and 1984. Positive growth was recorded again in 1985 and 1986 though drought caused a decline again in 1987. However, it increased in 1988. There was a steady fall in GDP per head throughout the 1980s, and this was particularly severe as a result of the drought in 1984, amounting to a drop of -19%. From 1985 to 1988 it increased except in 1987 when drought again caused a fall of 8.8% Inflation performance has been variable, with the highest annual rate of price increase being recorded in the 1980s in 1981 with a 23% rise. Otherwise inflation has been in single figures, with price falls recorded in 1983, 1985 and 1986. Niger's policy of setting balanced budgets and then sometimes failing to make all the projected investment expenditure has led to low budget deficits and helped price stability.

Export revenue reached a peak in 1980 at $US 576m, but had fallen by 1988 to $US371m. Niger has typically run a deficit on the balance of trade, with merchandise imports some 11% greater than exports in 1987. Falling export revenues have compressed the value of imports from $US 677m in 1980 to $US 401m in 1987.

Niger had approximately $US 1,700m of outstanding external debt in 1989. Niger's debts with government creditors were rescheduled in 1983, but, despite this, debt service payments were projected to rise from $US 57m in 1984 to $US 128m in 1988, before beginning to fall again. This is something of a problem, as at present levels of exports this represents a debt service to export earnings ratio of close to 50%.

Niger is a member of the Franc Zone, and the CFA franc depreciated from CFA 211 = $US 1 in 1980 to CFA 449 = $US 1 in 1985. Thereafter, the CFA franc has appreciated with the weakness of the dollar, and at the end of 1989 stood at CFA 289.3 = $US 1. This appreciation, which has occurred at a time when most CFA countries are experiencing falling export revenues and widening current deficits has led to speculation that the CFA franc might be devalued against the French franc.

Niger has typically presented a balanced budget, and expenditure has gradually declined in nominal terms. Recurrent expenditure has fallen at the expense of investment expenditure, with the latter rising 15% in 1986/7. There is a divergence, however, between the budget projections and the actual outcome, and up to a quarter of developments for which finance is in place do not get implemented.

Niger's willingness to introduce a reform programme and adverse external factors have led to a supportive response among donors. The 1987-9 two-year investment plan presented in June 1987 appears to have attracted the required funding of $US 350m a year. Major projects include $US 31.3m for an Islamic university, $US 11m from the European Development Bank, and $US 7m from France for urban development. France is allocating $US 9.8m to electricity projects, Japan $US 2m to nutrition, and the World Bank $US 5.5m to reform of the parastatal sector.

Niger

Economic Outlook and Forecasts

The forecasts assume that stability continues to be consolidated by the current regime, and that liberalising reforms continue to be implemented. It is also assumed that there will be no major drought in the forecast period.

Uranium prices are expected to fall in real terms up to 1995, but Niger can be expected to maintain volumes of uranium exports as a result of the agreements with France.

With the return of more favourable climatic conditions and the benefits of the reform programme, GDP is expected to expand at a rate that will allow some improvement in GDP per head. Inflation is expected to remain modest at around 5.0% a year. Export volumes are expected to stagnate, but not decline, while import volumes expand as a result of programme aid for the economic reforms.

Niger: Economic Forecasts
(average annual percentage change)

	Actual	Forecast Base	Forecast High
	1980-87	1990-95	1990-95
GDP	-1.9	4.5	5.6
GDP per Head	-4.9	1.3	2.4
Inflation Rate	4.1	5.5	4.0
Export Volumes	-4.8	0.0	0.5
Import Volumes	-6.2	1.5	2.0

NIGER: Comparative Data

	NIGER	WEST AFRICA	AFRICA	INDUSTRIAL COUNTRIES
POPULATION & LAND				
Population, mid year, millions, 1989	7.2	12.3	10.2	40.0
Urban Population, %, 1985	15	28.7	30	75
Population Growth Rate, % per year, 1980-86	3.0	3.2	3.1	0.8
Land Area, thou. sq. kilom.	1,267	384	486	1,628
Population Density, persons per sq kilom., 1988	5.5	31.1	20.4	24.3
ECONOMY: PRODUCTION & INCOME				
GDP, $US millions, 1986	2,080	4,876	3,561	550,099
GNP per head, $US, 1986	260	510	389	12,960
ECONOMY: SUPPLY STRUCTURE				
Agriculture, % of GDP, 1986	46	40	35	3
Industry, % of GDP, 1986	16	26	27	35
Services, % of GDP, 1986	39	34	38	61
ECONOMY: DEMAND STRUCTURE				
Private Consumption, % of GDP, 1986	82	77	73	62
Gross Domestic Investment, % of GDP, 1986	11	13	16	21
Government Consumption, % of GDP, 1986	11	13	14	17
Exports, % of GDP, 1986	18	18	23	17
Imports, % of GDP, 1986	22	20	26	17
ECONOMY: PERFORMANCE				
GDP growth, % per year, 1980-86	-2.6	-2.03	-0.6	2.5
GDP per head growth, % per year, 1980-86	-5.6	-4.7	-3.7	1.7
Agriculture growth, % per year, 1980-86	2.8	1.3	0.0	2.5
Industry growth, % per year, 1980-86	-4.3	-3.6	-1.0	2.5
Services growth, % per year, 1980-86	-8.0	-2.5	-0.5	2.6
Exports growth, % per year, 1980-86	-13.4	-4.0	-1.9	3.3
Imports growth, % per year, 1980-86	-4.4	-12.9	-6.9	4.3
ECONOMY: OTHER				
Inflation Rate, % per year, 1980-86	6.6	13.0	16.7	5.3
Aid, net inflow, % of GDP, 1986	15.2	3.7	6.3	-
Debt Service, % of Exports, 1986	40.3	21.4	20.6	-
Budget Surplus (+), Deficit (-), % of GDP, 1984	-4.8	-3.4	-2.8	-5.1
EDUCATION				
Primary, % of 6-11 group, 1985	28	75	76	102
Secondary, % of 12-17 group, 1985	6	24	22	93
Higher, % of 20-24 group, 1984	1	2.6	1.9	39
Adult Literacy Rate, %, 1980	10	30	39	99
HEALTH & NUTRITION				
Life Expectancy, years, 1986	44	50	50	76
Calorie Supply, daily per head, 1985	2,276	2,105	2,096	3,357
Population per doctor, 1979	38,796	16,199	24,185	550

Notes: 'Southern Africa' and 'Africa' exclude South Africa. Dates are for the country in question, and do not always correspond with the Regional, African and Industrial averages.

NIGER: Leading Indicators

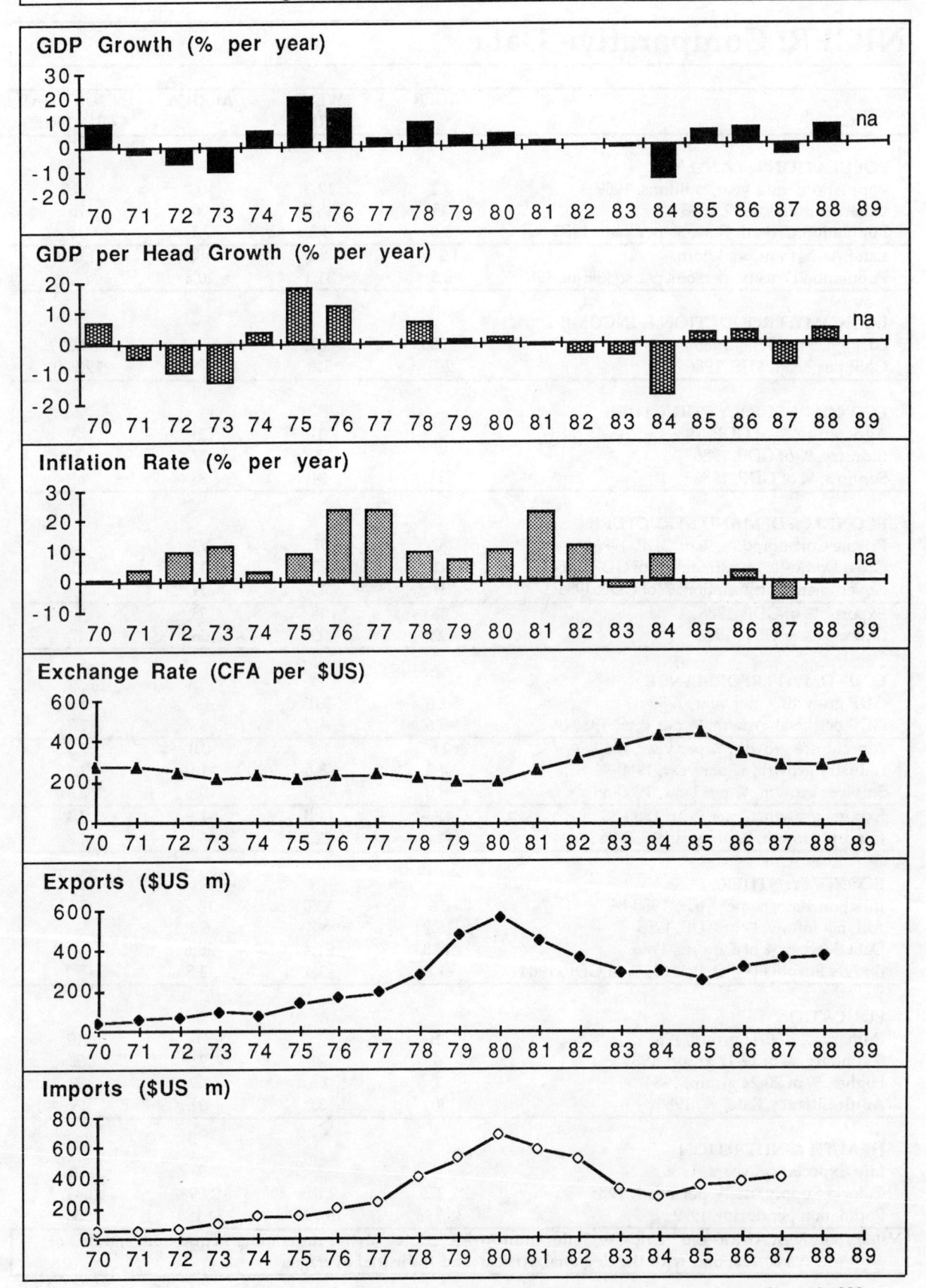

Note: Leading indicator data for 1989 are based on the first half of 1989. 1989 exchange rate is for mid-1989.

NIGER: Leading Indicator Data

	1980	1981	1982	1983	1984	1985	1986	1987	1988	1989
GDP growth (% per year)	4.8	1.2	-1.2	-1.8	-14.7	5.9	6.5	-4.9	7.5	
GDP per head growth (% per year)	1.8	-1.8	-4.2	-4.9	-17.8	2.8	3.4	-8.1	4.3	
Inflation (% per year)	10.3	22.9	11.6	-2.5	8.4	-0.9	3.2	-6.7	-1.4	
Exchange rate (CFA per $US)	211.3	271.7	328.6	381.1	437.0	449.3	346.3	300.5	297.9	327.3
Exports, merchandise ($US m)	576	455	367	299	304	259	330	361	371	
Imports, merchandise ($US m)	677	582	517	324	270	346	364	401		

Note: Leading indicator data for 1989 are based on the first half of 1989. 1989 exchange rate is for mid-1989.

NIGER: International Trade

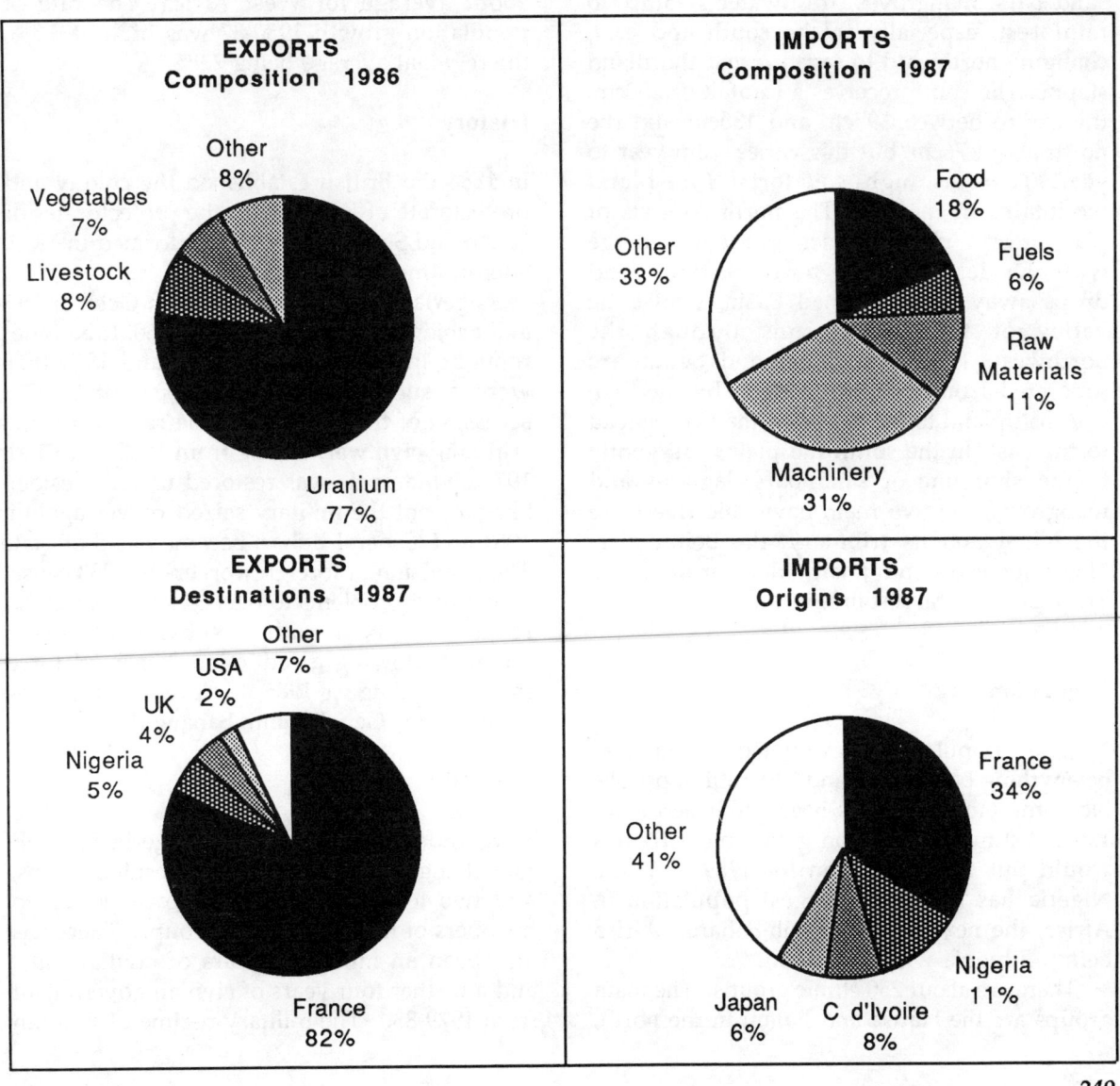

EXPORTS Composition 1986
- Uranium 77%
- Livestock 8%
- Vegetables 7%
- Other 8%

IMPORTS Composition 1987
- Food 18%
- Fuels 6%
- Raw Materials 11%
- Machinery 31%
- Other 33%

EXPORTS Destinations 1987
- France 82%
- Nigeria 5%
- UK 4%
- USA 2%
- Other 7%

IMPORTS Origins 1987
- France 34%
- Nigeria 11%
- C d'Ivoire 8%
- Japan 6%
- Other 41%

32 NIGERIA

Physical Geography and Climate

Nigeria is a large coastal state, in the West African region, bordered by Benin, Niger, Chad and Cameroon. It consists of a wide range of climatic and environmental zones, from coastal sand bars, mangroves, freshwater swamp, to rainforest, especially in the south and east, changing northward to savanna and thornland steppe. The south receives a rainfall of 250cm, the centre between 95cm and 135cm and the north about 76cm, but this varies from year to year. There is high equatorial forest and mountains in the east. The north consists of plateau and plains, which gradually merge with the desert. In the north east the land drops away into the Chad Basin, whilst the valley of the Sokoto runs through the northwest. The rivers Niger and Benue are separated from the coastal delta by the Oyo Yoruba upland to the west and the Udi plateau to the east. In the south the plains fall gently to the shoreline of sand bars, lagoons and mangrove. The two main navigable rivers are the Niger and its tributary, the Benue. The Niger opens out into a large delta at the coast. The main mineral resource is oil.

Population

Nigeria's population is estimated variously to be anything between 85 and 120 million people. No complete census has been published since 1963. Estimates based on government figures would put the population for 1989 at 114m. Nigeria has by far the largest population in Africa, the next largest in sub-Saharan Africa being Ethiopia with 47 million.

There are about 250 ethnic groups. The main groups are the Hausa and Fulani in the north, and below the Niger and Benue, to the east Ibo, and to the west Yoruba. The density of population is extremely high at 119 persons per square kilometre, compared with the regional average of 31.1 per square kilometre. About 30% of the population live in urban areas, about average for West Africa. The rate of population growth 1980-87 was high at 3.3%, the regional average being 2.9%.

History

In 1886 the British established the colony and protectorate of Lagos, with the protectorates of North and South Nigeria being formed in 1900, later uniting with Lagos in 1906.

Nigeria became a federation in October 1954 and achieved Independence in 1960. It became a republic in 1963. Between 1966 and 1979 there were a succession of military regimes. The secession of the province of Biafra resulted in a crippling civil war, lasting from 1967 to 1970. In 1979 civilian rule was restored under President Shagari, but the military seized power again in 1983 and General Buhari became Head of State. The expulsion of foreign workers in 1983 caused adjustment problems to the economy which has remained mixed with a substantial private sector. Following a reduction in armed forces salaries in August 1985, there was a bloodless coup by Maj. Gen. Ibrahim Babangida.

Stability

Since Independence in 1960, Nigeria has had four changes of government as a result of coups, and two leaders killed with power going to members of the same ruling group. There has also been an initial six years of civilian rule, and a further four years of civilian government from 1979-83. The military regime of Ibrahim

Babangida has been in power since 1985, and is committed to a return to civilian rule by 1992. Between 1967 and 1970 there was a debilitating civil war with Biafra. Economic policy has remained firmly committed to a mixed economy, and in 1986 this was reinforced by a liberalising economic reform programme.

Overall, Nigeria's stability has been very poor, and it has been the inability to provide continuity in government, together with the depressing pervasiveness of corruption in public life, that has discouraged foreign investment and business participation, despite the immense potential of Nigeria. The Babangida regime thwarted a coup attempt late in 1985, and political stability appears as elusive as ever. Over the longer term, it will be hoped that the liberalising economic reforms will reduce corruption and encourage investment.

Economic Structure

Nigeria's GDP in 1987 was $US 24,390m (the largest in sub-Saharan Africa, excluding South Africa) according to the World Bank, while the regional average was only $US 4,876m. As the price of oil has fallen since the oil boom of the 1970s, the standard of living in Nigeria has dropped rapidly and by 1987 GNP per head was only $US 370, below the regional average of $US 510.

Agriculture contributed only 30% of GDP in 1987, 10% less than the regional average. Nigeria has had little success in trying to achieve its aim of food self-sufficiency for its massive population. Although food output may be stimulated by declining food imports due to foreign currency restrictions, the benefits may be offset by continued falling world commodity prices for crops such as cocoa. Industry in Nigeria is relatively well developed and this sector (which includes oil production) provides 43% of GDP, well above the regional figure of 26%. The service sector is proportionately smaller than average, contributing only 27% of GDP.

About 69% of GDP was taken by private consumption, less than the West African average of 77%. Gross domestic investment equalled 13% of GDP while government consumption accounted for 11%.

Exports represented 31% of GDP in 1987 when the regional average was only 18%. Imports were equal to 27%, 7% higher than average. The main export is oil which accounts for 95% of all exports and makes Nigeria extremely vulnerable to world oil price fluctuations. A further 3% of export earnings comes from cocoa. Imports consist of machinery and other manufactured goods, 65%, chemicals 16%, and food, 9%.

Debt servicing took up 11.7% of export earnings in 1987 which is well below the regional average of 21.4%. Only Sierra Leone, Gambia, Liberia and Mali have lower debt service ratios in the region.

Nigeria's education system is probably the best in the region with 92% of primary-aged and 29% of secondary-aged children in school. A further 3% of 20-24 year olds are in higher education, equalled only by Côte d'Ivoire in West Africa. These figures are all well above the regional averages. The adult literacy rate for Nigeria was 34% in 1980, surprisingly low for a country which has had a well developed education system since before independence. There are great regional variations as the northern states were largely left undeveloped before 1960, much of the earlier education being carried out by the churches in the Christian-dominated south of the country.

Life expectancy of 50 years is about average for the region. Average daily calorie supply in 1986 was 2,185, slightly better than in the rest of Africa. Nigeria has twice as many doctors as the regional average. Only Cape Verde and Guinea-Bissau in the region have more favourable ratios of doctors to population.

Overall, Nigeria is a low-income country despite the presence of a substantial oil sector, which has led to a larger than average industry share in production, and smaller agriculture and services sectors. Exports, almost all oil, are a large proportion of GDP. Educational provision is good, although literacy is poor. Health provision is good.

Economic Performance

GDP in Nigeria fell by -1.7% from 1980-87, slightly less than the regional average fall of -2.0%. GDP per head declined at -5.1% a year. However, since 1980 only 1985 showed any noticible improvement while the rest of the decade has seen GNP per head fall or remain stagnant as oil revenues fell and the Naira has depreciated. The agricultural sector, which contributes about a quarter of GDP, and which the government has been trying for some time to develop, grew only marginally by 0.6%, as against 1.3% for the region. The industrial sector declined by 4.4% despite increased oil output, mainly because of falling world oil prices. This compares with a similar fall of -3.6% for the region. Nigeria's four oil refineries were all damaged by fires or explosions in 1989, causing domestic shortages of petrolium products. Manufacturing growth has been constrained by lack of foreign exchange limiting imports of raw materials, components and capital equipment. The service sector also declined slightly, by -0.3%, though the overall regional decline was faster at -2.5% a year.

Export volumes, of which petroleum products make up 97%, fell by -5.1% due to the depressed world oil market. However, the government's austerity measures meant that Nigeria reduced import volumes by -14.0% a year, more than the regional average of -12.9% and more than twice the African average of -6.9%.

Inflation averaged 10.1% for 1980-87, but by 1989 it was estimated to have risen to around 45%.

Overall, Nigeria's economic performance in the 1980s has been poor with falling GDP and GDP per head, resulting primarily from steadily contracting industrial output. Export and import volumes have also declined. Inflation, although only just in double figures in the 1980s has accelerated alarmingly toward the end of the decade.

Recent Economic Developments

In June 1986, Nigeria announced plans for a structural adjustment programme, the main measure being the introduction, in September 1986, of an auction system for foreign exchange. Other important changes have included the abolition of commodity boards which has led to a fourfold rise in producer prices for cocoa, plans for privatisation of public enterprises but not key utilities, the lifting of controls on interest rates, and moves to end subsidies on petrol, gas and fertilisers. The government is examining the scope for reductions in tariff protection for manufacturing.

Economic performance in the 1980s has been dominated by the fortunes of the oil sector, and it is noticeable in the Nigerian economy that expansion in GDP follows increases in export earnings. Since 1980 oil earnings have declined, and from 1982 to 1984 GDP fell with a return to positive growth in 1986 of 2.5%, but then negative growth again in 1986. The year 1980 was the last one that GDP per head rose, and in the subsequent six years GDP per head has declined by 34%. The inflation record has been variable, with annual rates of price increase above 20% in 1981, 1983 and 1984, single-figure inflation in 1982 and 1985, but the 1985 figure accelerating to 17% in 1986 and 1987, before reaching 45% in 1989.

Export revenues peaked in 1980 with earnings of $US 25b, of which 94% comprised crude petroleum sales. Subsequently, export revenues have fallen, firstly as a result of lower oil production resulting from tighter OPEC quotas, and then halving in 1986 as a result of a disastrous decline in oil prices. Export revenues hovered around $US 12b 1982 to 1985, and were $US 6.1b in 1986. Imports tend to lag a year behind export earnings, and import expenditure peaked in 1981 at $US 18.8b, but had been compressed by export earnings to $US 5.5b by 1986. Oil production, after running at 1.05m barrels a day, was back up close to 1.5m barrels a day in 1989. In addition, it was hoped that non-oil exports might expand with the initiatives of the reform programme toward $US 1b a year. These two features could result in a substantial improvement of up to 50% in export revenues compared with 1986.

Nigeria's external debt continues to be the subject of concern. There was estimated to be $US 30b outstanding in 1989, of which $US 5b is short-term trade arrears.

An encouraging development is the interest being shown in exchange of debt for equity in Nigerian enterprises. Such swaps will depend on continued progress in the reform programme, and assurance that profits can be repatriated. It is clear that international firms are viewing Nigeria more favourably, and with the improved conditions provided by the foreign exchange auctions, investment and production in Nigeria is becoming more attractive. It is significant that most major international firms operating in Nigeria have experienced increased profits since the reform programme was instituted.

In September 1986 Nigeria instituted a weekly auction system for foreign exchange. A priority rate was allowed for debt servicing and the government's affiliation goes to certain international institutions, at N 1.25 = $US 1. The naira immediately began to depreciate, and after six months stood at N 3.8 = $US 1. The auction was changed to a fortnightly system in March 1987, with deals being settled at the higher actual bid rates rather than the marginal bid which exhausted the foreign currency available in a particular auction. In July of 1987 the priority rate was abolished. By the end of 1989, the rate was N 7.45 = $US 1.

On the aid front, the African Development Bank is allocating $US 31m to water supply and $US 87m to agricultural credit, meanwhile the European Investment Bank is considering a $US 469m loan to rehabilitate water supplies in Lagos. East Germany is to allocate $US 200m to education and electrification, and the EEC is donating $US 227m for agriculture, water and the environment. The World Bank is lending $US 150m for roads, and is funding the feasibility studies for a $US 2,000m petrochemicals complex.

The private foreign investment scene is mixed. Leyland have decided to sell their assembly plant. Shell, Agip and Elf are considering substantial investments in a $US 4,000m liquified natural gas scheme. American corporations are reported to be interested in agricultural investments.

Economic Outlook and Forecasts

The forecasts assume that Nigeria continues to consolidate its stability up to 1995, and that the advent of civilian government thereafter does not herald a return to the instability of the Shagari period. It is also assumed that Nigeria perseveres with the current liberalising economic reform programme.

Oil prices are expected to increase only marginally up to 1995 in real terms.

The forecasts show growth of GDP returning to positive rates, but at around the projected rate of population growth, leaving GDP per head unchanged. Export volumes are expected to expand, and this factor, together with increasing flows of aid and foreign investment, will allow slightly faster growth in import volumes.

Nigeria: Economic Forecasts
(average annual percentage change)

	Actual 1980-87	Forecast Base 1990-95	Forecast High 1990-95
GDP	-1.7	3.0	3.2
GDP per Head	-5.1	0.0	0.2
Inflation Rate	10.1	30.0	25.0
Export Volumes	-5.1	2.0	2.5
Import Volumes	-14.0	3.0	3.5

NIGERIA: Comparative Data

	NIGERIA	WEST AFRICA	AFRICA	INDUSTRIAL COUNTRIES
POPULATION & LAND				
Population, mid year, millions, 1989	113.9	12.3	10.2	40.0
Urban Population, %, 1985	30	28.7	30	75
Population Growth Rate, % per year, 1980-86	3.3	3.2	3.1	0.8
Land Area, thou. sq. kilom.	924	384	486	1,628
Population Density, persons per sq kilom., 1988	119.0	31.1	20.4	24.3
ECONOMY: PRODUCTION & INCOME				
GDP, $US millions, 1986	49,110	4,876	3,561	550,099
GNP per head, $US, 1986	640	510	389	12,960
ECONOMY: SUPPLY STRUCTURE				
Agriculture, % of GDP, 1986	41	40	35	3
Industry, % of GDP, 1986	29	26	27	35
Services, % of GDP, 1986	36	34	38	61
ECONOMY: DEMAND STRUCTURE				
Private Consumption, % of GDP, 1986	78	77	73	62
Gross Domestic Investment, % of GDP, 1986	12	13	16	21
Government Consumption, % of GDP, 1986	12	13	14	17
Exports, % of GDP, 1986	14	18	23	17
Imports, % of GDP, 1986	16	20	26	17
ECONOMY: PERFORMANCE				
GDP growth, % per year, 1980-86	-3.2	-2.03	-0.6	2.5
GDP per head growth, % per year, 1980-86	-6.5	-4.7	-3.7	1.7
Agriculture growth, % per year, 1980-86	1.4	1.3	0.0	2.5
Industry growth, % per year, 1980-86	-5.1	-3.6	-1.0	2.5
Services growth, % per year, 1980-86	-4.0	-2.5	-0.5	2.6
Exports growth, % per year, 1980-86	-6.0	-4.0	-1.9	3.3
Imports growth, % per year, 1980-86	-17.2	-12.9	-6.9	4.3
ECONOMY: OTHER				
Inflation Rate, % per year, 1980-86	10.5	13.0	16.7	5.3
Aid, net inflow, % of GDP, 1986	0.1	3.7	6.3	-
Debt Service, % of Exports, 1986	23.4	21.4	20.6	-
Budget Surplus (+), Deficit (-), % of GDP, 1978	-1.8	-3.4	-2.8	-5.1
EDUCATION				
Primary, % of 6-11 group, 1984	92	75	76	102
Secondary, % of 12-17 group, 1984	29	24	22	93
Higher, % of 20-24 group, 1984	3	2.6	1.9	39
Adult Literacy Rate, %, 1980	34	30	39	99
HEALTH & NUTRITION				
Life Expectancy, years, 1986	51	50	50	76
Calorie Supply, daily per head, 1985	2,139	2,105	2,096	3,357
Population per doctor, 1980	9,400	16,199	24,185	550

Notes: 'Southern Africa' and 'Africa' exclude South Africa. Dates are for the country in question, and do not always correspond with the Regional, African and Industrial averages.

NIGERIA: Leading Indicators

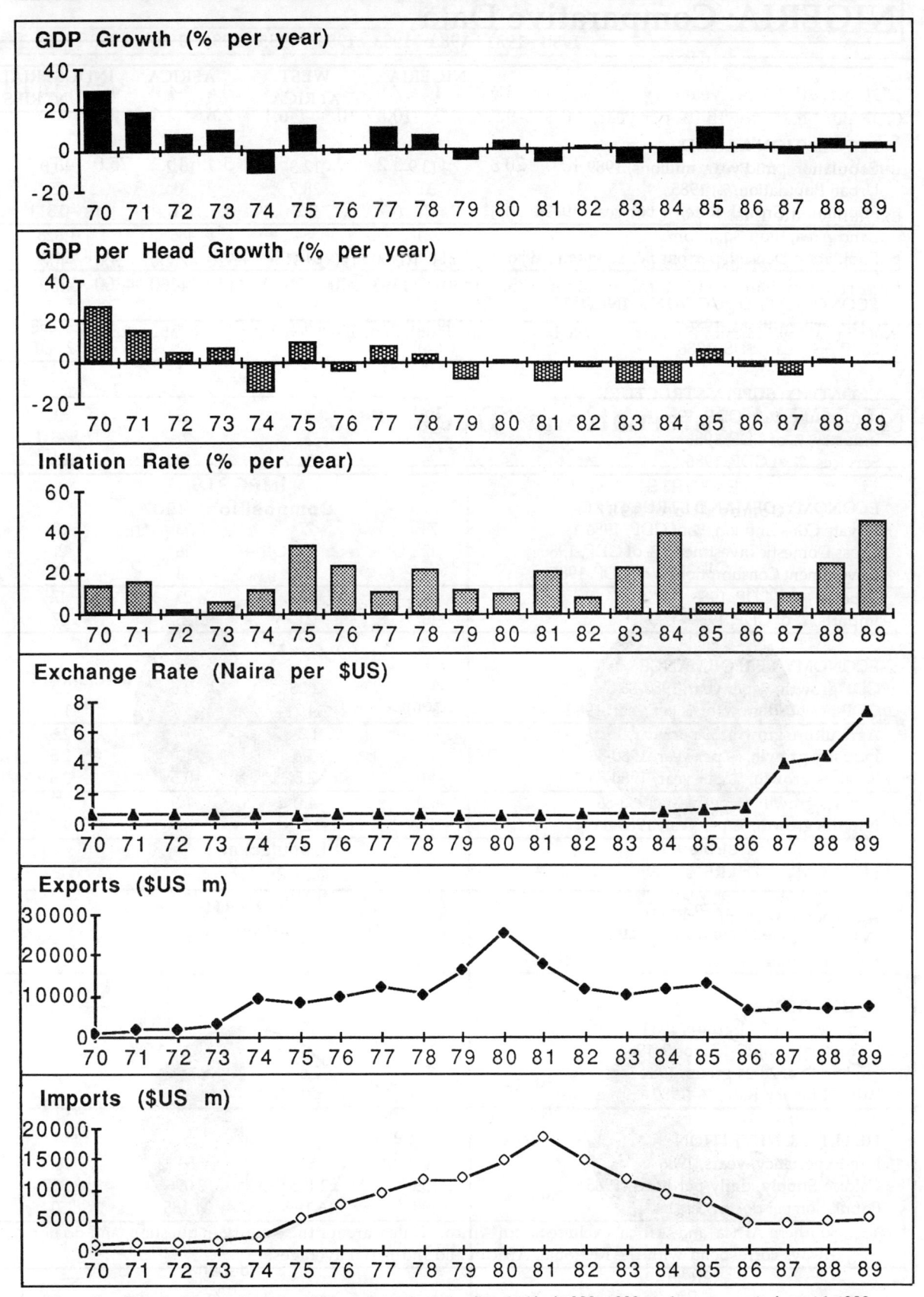

Note: Leading indicator data for 1989 are based on the first half of 1989. 1989 exchange rate is for mid-1989.

NIGERIA: Leading Indicator Data

	1980	1981	1982	1983	1984	1985	1986	1987	1988	1989
GDP growth (% per year)	3.4	-6.6	-0.4	-7.8	-7.6	9.3	0.4	-4.0	4.1	2.0
GDP per head growth (% per year)	0.6	-9.5	-3.3	-10.8	-10.6	6.1	-2.8	-7.3	0.7	-1.4
Inflation (% per year)	10.0	20.8	7.7	23.2	39.6	5.5	5.4	10.2	25.0	45.0
Exchange rate (Naira per $US)	0.55	0.61	0.67	0.72	0.76	0.89	1.13	4.01	4.48	7.33
Exports, merchandise ($US m)	25741	17961	12088	10309	11900	13140	6620	7550	6900	7400
Imports, merchandise ($US m)	14636	18872	14801	11393	9400	7920	4440	4460	4600	5200

Note: Leading indicator data for 1989 are based on the first half of 1989. 1989 exchange rate is for mid-1989.

NIGERIA: International Trade

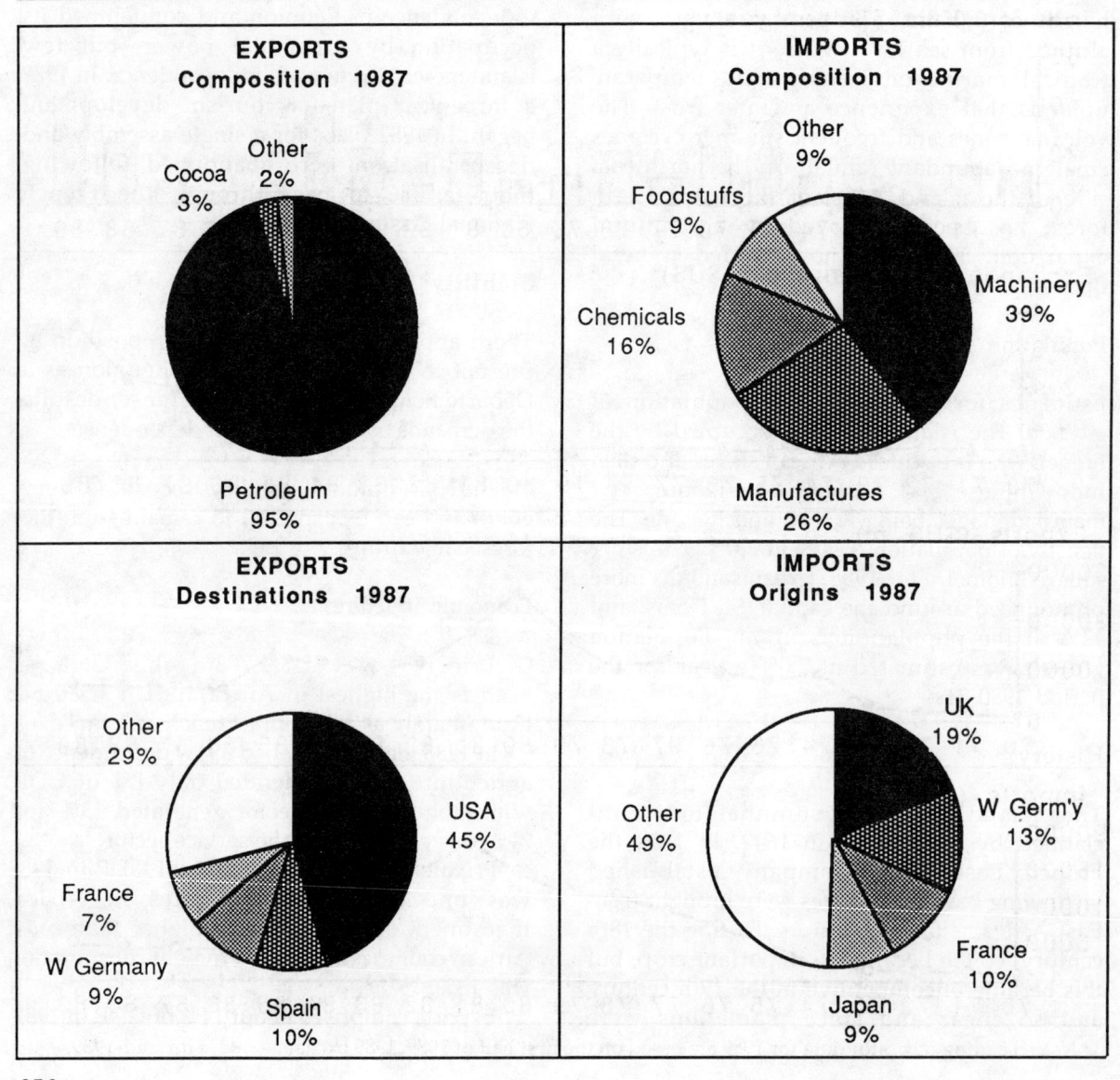

EXPORTS Composition 1987
- Petroleum 95%
- Cocoa 3%
- Other 2%

IMPORTS Composition 1987
- Machinery 39%
- Manufactures 26%
- Chemicals 16%
- Foodstuffs 9%
- Other 9%

EXPORTS Destinations 1987
- USA 45%
- Other 29%
- Spain 10%
- W Germany 9%
- France 7%

IMPORTS Origins 1987
- Other 49%
- UK 19%
- W Germ'y 13%
- France 10%
- Japan 9%

33 REUNION

Physical Geography and Climate

Réunion is a small island, in the East African region, located in the Indian Ocean to the east of Madagascar and to the south west of Mauritius. It is regarded by France as a Département and for statistical purposes part of Metropolitan France. Volcanoes stretch diagonally across the island and rise to a height of 3,000m. Temperatures vary with altitude from sea level where it is typically a tropical marine environment, to temperate uplands that experience a winter frost. The volcanic cones and frequent summer cyclones create an abundant rainfall on the north east side of the island. Most of the natural rain forest has been destroyed by agricultural expansion. There are no recorded mineral deposits.

Population

Estimates for 1989 suggest a population of 604,000. The island was first occupied by the French who introduced African slaves and then Indo-Chinese and East African labour. The main languages being French and Reyone. The density of population is high at 237 persons per square kilometre for 1988. Urbanisation is more pronounced around the capital, St. Denis, and 42% of the population is urban. Population growth was estimated at 2.0% a year for the period 1980-84.

History

The island remained uninhabited until claimed by the French in 1642. In 1654 the French East India Company established victualling bases and slaves were brought from East Africa as farm labourers. During the 18th century, coffee became an important crop, but this became uncompetitive in the 19th century and so sugar and spice plantations were developed. In 1848 France abolished slavery and indentured labour was brought from Indo-China and East Africa. In 1946 France ceased to rule and the island was given Département status. In 1972 a series of strikes were called resulting in the general strike of 1973-74 against unemployment and the rise of rice prices. In 1974-76 a sugar modernisation plan was introduced. In 1978 the OAU called for the independence of Réunion and condemned its occupation by a colonial power, but few islanders seemed to want independence. In 1979 a three-year plan for tourism development began. In 1982 plans for a single assembly and decentralisation were abandoned following their failure to pass through the French National Assembly.

Stability

There appears to be no serious opposition at present to the current status of Réunion as a Département of Metropolitan France, despite the demands by the OAU for independence

Overall, Réunion's stability record is very good, and can be expected to remain so in the foreseeable future.

Economic Structure

GDP in 1984 was $US 1,708m while GNP per head is the highest in Africa at $US 3,580 per head, mainly as a result of French assistance.

In 1983 the economy had a low dependence on agriculture which generated only 8% of GDP, while the industrial sector generated 13% and 79% was generated by the service sector.

Private consumption at 99% of GDP in 1983 was one of the highest rates in Africa. Investment of 22% was also higher than most African countries while government consumption was 18%.

Exports comprised about 14% of GDP in 1983,

Reunion

and this is about the average for the region, though below the African average. Imports, however, were 53% of GDP in 1983, and this is possible by virtue of financial transfers from France and the considerable invisible earnings on services, mainly tourism. Merchandise exports are mainly sugar, which made 75% in 1984, and rum which made up 4%. Imports in 1984 comprised foodstuffs 25%, machinery 25%, manufactures 23%, intermediate goods 16% and fuel 11%. France takes 61% of Réunion's exports and provides 65% of her imports.

Educational provision, as might be expected in an upper middle-income country, is good, with primary enrolment rates in 1981 of 116%, secondary at 51%, and 6% enrolled in higher education. These rates are markedly better than elsewhere in Africa.

Health provision would also appear to be good, with one doctor for every 1,360 people in 1981. This is seventeen times better provision than the African average.

Overall, Réunion is a small economy, classified in the upper middle-income group, with the highest average living standards in Africa. Most of GDP is generated in the services sector where tourism is important, and there is a considerable deficit on the balance of trade. Educational and health provision are good.

Economic Performance

Figures are only available for the four years 1980-83, and in this period GDP grew at 4.6% a year. This implies a rate of expansion of GDP per head at 2.6% a year, and this is considerably better than elsewhere in the region.

All sectors appear to have expanded steadily, with agriculture growing at 4.7% a year in the period 1974-83, industry at 4.4% a year and services at 5.4%. These growth rates are all very much better than elsewhere in Africa.

Export volumes expanded at 5.0% a year in the 1974-83 period, again, well above the African average. Import volumes as a result of expanding tourism revenues and financial transfers from France have expanded at 11.5% a year in the same period.

Inflation has been modest, at 8.6% a year for 1980-87.

Overall, Réunion has had good economic performance, with high rates of growth for GDP, GDP per head, all producing sectors and exports, being experienced in the context of single digit inflation.

Recent Economic Developments

GDP and GDP per head growth rate have been lower in the first part of the 1980s compared with the latter half of the 1970s. However, inflation performance has been better in the 1980s, and the most recent figures for 1986 and 1987 show the impact of recent monetary and fiscal discipline imposed by France with annual rate of price increase below 3%.

The French franc is the currency used, and this depreciated against the US dollar in the early 1980s, being f 4.23 = $US 1 in 1980 and f 8.99 = $US 1 in 1985. Subsequently, the weakness of the dollar has brought about an appreciation, with the exchange rate standing at f 5.79 = $US 1 at the end of 1989.

Merchandise exports in dollar terms reached their peak in 1979, but declined by 34% up to 1984 despite the depreciation of the franc which would have raised domestic prices to producers. Merchandise imports peaked in dollar terms in 1980, and they fell by 12% up to 1984.

Economic Outlook and Forecasts

The forecasts assume that Réunion remains a Département of France, and that present stability is maintained without any disruptive agitation for independence.

The price of sugar, Réunion's main export, is expected to decline in real terms by -5.0% in the six years up to 1995.

The base forecasts show GDP and GDP per head growth slowing slightly compared with the first four years of the 1980s.

The expansion of export volumes is expected to slow down slightly as well, and import volumes to grow at a much slower rate, closer to the overall growth rate of the economy.

Inflation is expected to be maintained at around the rate of the middle 1980s, close to 5.0% a year.

Réunion : Economic Forecasts
(average annual percentage change)

	Actual 1980-83	Forecast Base 1990-95	Forecast High 1990-95
GDP	4.6	4.0	4.7
GDP per Head	2.6	2.0	2.7
Inflation Rate	8.6[a]	5.0	3.5
Export Volumes	5.0[b]	4.5	4.8
Import Volumes	11.5[b]	5.2	5.7

Note: [a] 1980-87 [b] 1974-83.

REUNION: Comparative Data

	REUNION	EAST AFRICA	AFRICA	INDUSTRIAL COUNTRIES
POPULATION & LAND				
Population, mid year, millions, 1989	0.604	12.2	10.2	40.0
Urban Population, %, 1982	42	30.5	30	75
Population Growth Rate, % per year, 1974-83	1.6	3.1	3.1	0.8
Land Area, thou. sq. kilom.	2.5	486	486	1,628
Population Density, persons per sq kilom., 1988	237.6	24.2	20.4	24.3
ECONOMY: PRODUCTION & INCOME				
GDP, $US millions, 1984	1,708	2,650	3,561	550,099
GNP per head, $US, 1984	3,580	250	389	12,960
ECONOMY: SUPPLY STRUCTURE				
Agriculture, % of GDP, 1983	7.9	43	35	3
Industry, % of GDP, 1983	13.4	15	27	35
Services, % of GDP, 1983	78.9	42	38	61
ECONOMY: DEMAND STRUCTURE				
Private Consumption, % of GDP, 1983	98.8	77	73	62
Gross Domestic Investment, % of GDP, 1983	21.6	16	16	21
Government Consumption, % of GDP, 1983	18.4	15	14	17
Exports, % of GDP, 1983	14.4	16	23	17
Imports, % of GDP, 1983	53.3	24	26	17
ECONOMY: PERFORMANCE				
GDP growth, % per year, 1974-83	5.2	1.6	-0.6	2.5
GDP per head growth, % per year, 1974-83	3.6	-1.7	-3.7	1.7
Agriculture growth, % per year, 1974-83	4.7	1.1	0.0	2.5
Industry growth, % per year, 1974-83	4.4	1.1	-1.0	2.5
Services growth, % per year, 1974-83	5.4	2.5	-0.5	2.6
Exports growth, % per year, 1974-83	5.0	0.7	-1.9	3.3
Imports growth, % per year, 1974-83	11.5	0.2	-6.9	4.3
ECONOMY: OTHER				
Inflation Rate, % per year, 1974-83	9.4	23.6	16.7	5.3
Aid, net inflow, % of GDP, 1986	-	11.5	6.3	-
Debt Service, % of Exports, 1986	-	18	20.6	-
Budget Surplus (+), Deficit (-), % of GDP, 1986	-	-3.0	-2.8	-5.1
EDUCATION				
Primary, % of 6-11 group, 1981	116	62	76	102
Secondary, % of 12-17 group, 1981	51	15	22	93
Higher, % of 20-24 group, 1981	6	1.2	1.9	39
Adult Literacy Rate, %, 1981	-	41	39	99
HEALTH & NUTRITION				
Life Expectancy, years, 1986	-	50	50	76
Calorie Supply, daily per head, 1985	-	2,111	2,096	3,357
Population per doctor, 1977	1,360	35,986	24,185	550

Notes: 'Southern Africa' and 'Africa' exclude South Africa. Dates are for the country in question, and do not always correspond with the Regional, African and Industrial averages.

REUNION: Leading Indicators

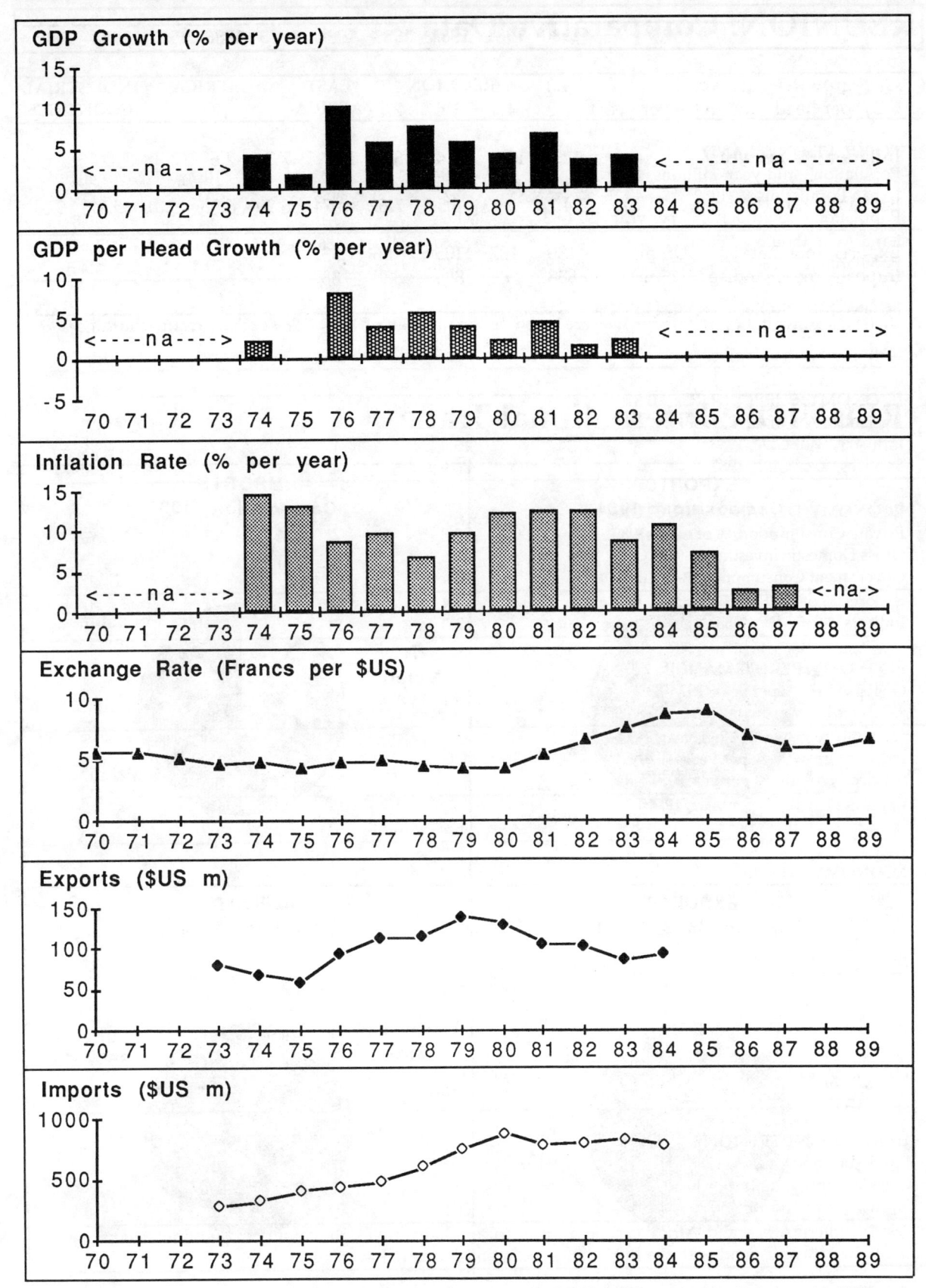

GDP Growth (% per year)

GDP per Head Growth (% per year)

Inflation Rate (% per year)

Exchange Rate (Francs per $US)

Exports ($US m)

Imports ($US m)

Note: Leading indicator data for 1989 are based on the first half of 1989. 1989 exchange rate is for mid-1989.

REUNION: Leading Indicator Data

	1980	1981	1982	1983	1984	1985	1986	1987	1988	1989
GDP growth (% per year)	4.2	6.6	3.4	4.1						
GDP per head growth (% per year)	2.2	4.4	1.4	2.1						
Inflation (% per year)	12.0	12.2	12.4	8.6	10.7	7.2	2.6	2.9		
Exchange rate (Francs per $US)	4.23	5.43	6.57	7.62	8.74	8.99	6.93	6.01	5.92	6.64
Exports, merchandise ($US m)	130	107	105	86	93					
Imports, merchandise ($US m)	888	786	804	838	788					

Note: Leading indicator data for 1989 are based on the first half of 1989. 1989 exchange rate is for mid-1989.

REUNION: International Trade

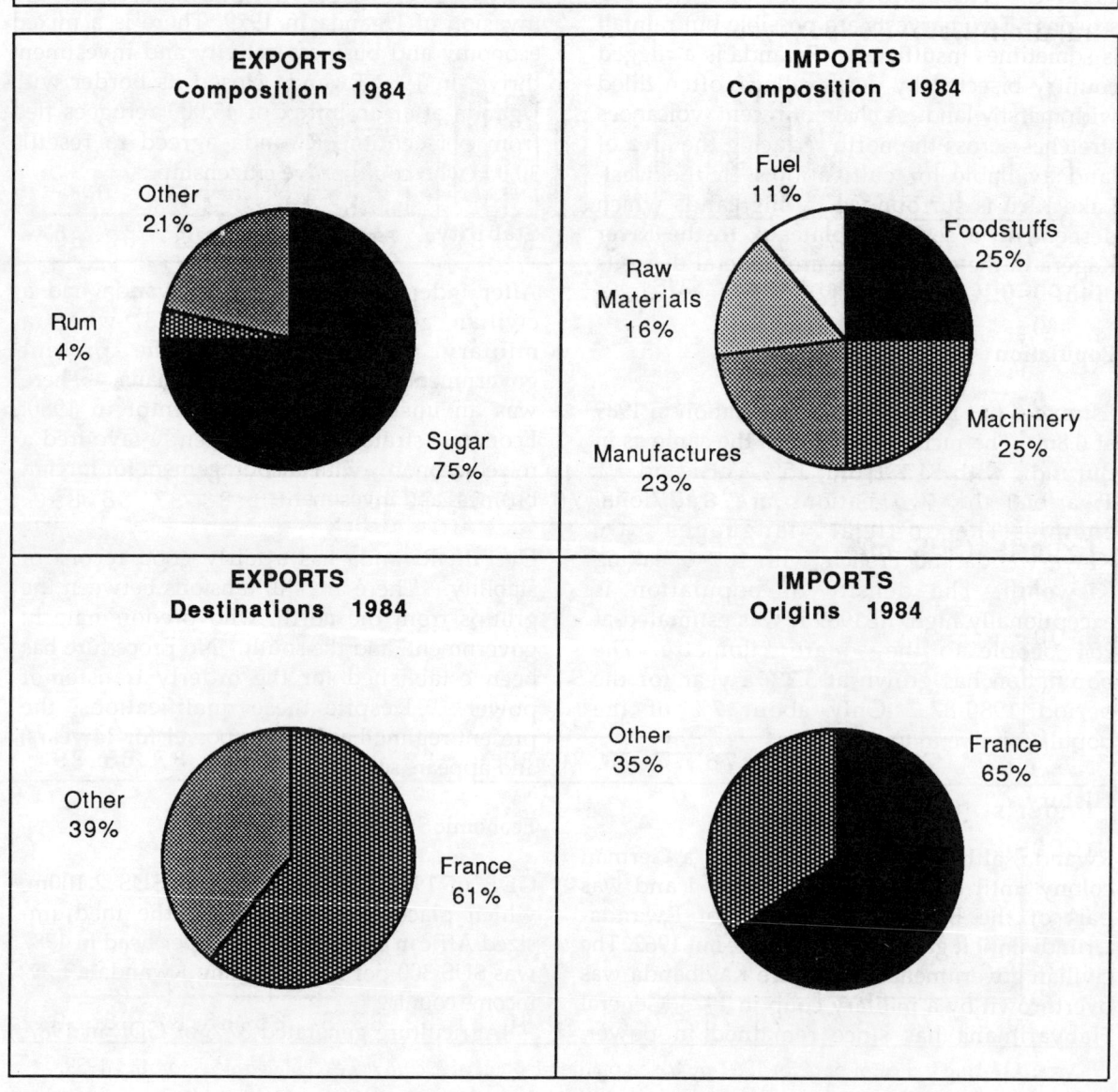

EXPORTS Composition 1984
- Sugar 75%
- Other 21%
- Rum 4%

IMPORTS Composition 1984
- Foodstuffs 25%
- Machinery 25%
- Manufactures 23%
- Raw Materials 16%
- Fuel 11%

EXPORTS Destinations 1984
- France 61%
- Other 39%

IMPORTS Origins 1984
- France 65%
- Other 35%

34 RWANDA

Physical Geography and Climate

Rwanda is a small, densely populated, poor, landlocked country in East Africa bounded by Burundi, Zaïre, Uganda and Tanzania. The climate has an average temperature of 19°C and 100cm of rain annually, but this varies with altitude. Temperatures can vary as much as 14°C. There are two wet and two dry seasons. Two harvests are possible but rainfall is sometimes insufficient. Rwanda is a rugged country bisected by deep valleys, often filled with marshy land. A chain of recent volcanoes stretches across the north reducing the area of land available for cultivation. In the west, Lake Kiru is surrounded by highlands which descend in a series of plateaux to the River Kagera in the east. There are mineral deposits of tin.

Population

Estimates put the size of the population in 1989 at 6.8m. The racial mix is much the same as in Burundi, with 84% Hutu, 15% Tutsi and 1% Twa but the two nations are traditional enemies. The official languages are Kinyarwanda and French, with 10% speaking Kiswahili. The density of population is exceptionally high, in 1989 it was estimated at 254 people to the square kilometre. The population has grown at 3.3% a year for the period 1980-87. Only about 7% of the population live in towns.

History

Rwanda, although a kingdom, was a German colony until the end of World War I and was part of the Belgian trusteeship of Rwanda-Urundi until it gained independence in 1962. The civilian government of Grégoire Kayibanda was overthrown by a military coup in 1973. General Habyarimana has since remained in power, despite coup attempts, reinforcing his leadership of the one party Mouvement Révolutionnaire National pour le Développement (MRND) by one-candidate presidential elections and limited legislative elections. Despite the establishment of the Economic Community of the Great Lakes in 1976, trade and communications were disrupted by the activities of the Amin regime and Tanzanian invasion of Uganda in 1979. There is a mixed economy and business activity and investment thrive. In 1982 Rwanda closed its border with Uganda after an influx of 45,000 refugees fled from persecution. Rwanda agreed to resettle 30,000 who could prove citizenship.

Stability

After independence in 1962, Rwanda had a civilian government until 1973, when a military coup brought in the present government of General Habyarimana. There was an unsuccessful coup attempt in 1980. Economic strategy has consistently favoured a mixed economy with encouragement for foreign business and investment.

Overall, Rwanda has a fairly good record of stability. There are still tensions between the groups from the north, who predominate in government, and the south. No procedure has been established for the orderly transfer of power. Despite these qualifications, the present regime has been in power for 14 years, and appears secure.

Economic Structure

GDP in 1987 was estimated at $US 2,100m, which places Rwanda among the medium-sized African economies. GNP per head in 1987 was $US 300 per head, making Rwanda a low-income country.

Agriculture generated 37% of GDP in 1987,

Rwanda

industry 23% of which 16% was manufacturing, and services generated 40%. All this represents a slightly larger industry sector and a slightly smaller agricultural sector than we might expect.

On the expenditure shares of income, private consumption comprised 83% of GDP in 1987, government consumption was 12%, and domestic savings 5%. Gross domestic investment was 17% of GDP, and the gap between this and domestic savings, equivalent to 12% of GDP, was covered by net inflow of resources from overseas. These figures imply a larger committment to consumption, and lower savings than the average in Africa.

Exports of goods and services were 8% of GDP in 1987, and imports 20% of GDP. This is a rather lower dependence on international trade than usual in Africa. 82% of merchandise exports were coffee in 1987, and 11% were tea. Merchandise imports were 26% machinery in 1983, 19% fuel, 15% raw materials, 11% manufactures, and 11% foodstuffs.

The adult literacy rate was 50% in 1981, and this is good by African standards, where the average is closer to 40%. The primary school enrolment rate was 67% in 1986, with a 3% enrolment rate for secondary education, but negligible tertiary enrolments. Although primary enrolments are comparable with Africa generally, the secondary and tertiary enrolments are significantly lower.

Life expectancy is about the Africa average at 49 years. There were 34,680 persons per doctor and 3,650 persons per nurse in 1983, and these provision rates are below the African average. Average daily calorie supply at 1,830 in 1986 was 14% below the Africa average.

Overall, Rwanda is of medium economic size and has low-income levels. It has a larger industrial sector than might be expected. This reflects a greater degree of self-sufficiency than is common elsewhere in Africa, and export and import dependence are consequently low as well. Secondary and tertiary educational provision are poor as are nutrition and health services.

Economic Performance

GDP grew by 2.4% from 1980-87, above the regional and African averages of 1.6% and -0.6% respectively. Nevertheless, this implied falling GDP per head at -0.7% a year. Despite a policy of aiming for self-sufficiency in food production, the agricultural sector grew by only 1.1%, the same as the regional average. Industrial growth reached 4.8% while the regional average was only 1.1%. The service sector grew by 3.9% compared with a regional growth rate of 0.7%. Export volumes expanded by 2.5% a year, 1980-87, better than elsewhere in Africa, where export volumes declined. Import volumes increased by 5.4% while in the region on average they only increased by 0.2%.

The inflation rate was 4.5% which compares favourably with the regional and African rates of 23.6% and 16.7% respectively.

Overall, bearing in mind the fast-growing population, little unused land and a poorly developed infrastructure, Rwanda has not performed too badly, although GDP per head has declined in the 1980s. Industrial sector growth and price stability have been above average.

Recent Economic Developments

Unlike most of Africa, economic policy in Rwanda has not undergone a change in favour of state withdrawal from marketing and manufacturing, or a liberalisation of tariffs, price controls and exchange rates. The main emphasis has been on food self-sufficiency in the face of the rapid rate of population growth at 3.5% a year, and government spending has emphasised infrastructure to aid agriculture. Economic progress has been negligible in the 1980s, with GDP exhibiting substantial year-to-year fluctuations as a result of the impact of weather on agricultural output and of international prices on export earnings.

Rwanda's exports peaked at $US 203m in 1979, since which time they have declined to a reported $US 90m in 1989. Increased aid has enabled Rwanda to increase its levels of imports, and these have risen from $US 196m in 1980 to $US 380m in 1989.

Rwanda was estimated to have $US 544m of outstanding external debt in 1987, none of this being with commercial lenders. As a result of the long repayment periods on most of

Rwanda's debt and the low interest charges, Rwanda's debt servicing ran at a manageable 11.3% of merchandise export revenues in 1987.

Rwanda has an exchange rate which has fluctuated between RF 90 = $US 1 and RF 100 = $US 1 up to 1985. In 1986, the weakness of the American dollar allowed the Rwanda franc to appreciate, and by the end of 1989, it stood at RF 77.6 = $US 1. Although Rwanda has had a modest rate of domestic inflation since 1970, there is little doubt that the Rwanda franc is overvalued by the maintenance of the fixed exchange rate policy, and any continuing balance of payments problems which lead to requests for IMF credit will involve pressure for depreciation and a more flexible system.

Rwanda has typically run a budget deficit of around -2.0% of GDP, and this has contributed substantially to the good record of price stability.

Infrastructure has featured heavily in aid commitments to Rwanda, with France committing $US 4.4m to telecommunications, and Japan $US 2.5m to roads. A major water development is to receive $US 15m from the World Bank, $US 16m from France, $US 11.1m from the African Development Bank, $US 7m from Austria, $US 7.5m from the Arab Bank for Economic Development in Africa, and $US 8.3m from Switzerland.

Economic Outlook and Forecasts

The forecasts assume that the regime of General Habyarimana continues in power, and that there are no moves to implement a programme of liberalising economic reforms.

The price of coffee is expected to fall in real terms up to 1995 by -8%, while the price of tea is forecast to rise by 15% in real terms.

The forecasts show GDP expanding at around 2.0% a year, but this will imply falls in GDP per head of around -1.0% a year.

Export volumes are expected to grow more slowly at around 2.0% a year. Import expansion, as a result of aid sqeezes to pressure Rwanda into an IMF agreement, will slow to around 2.5% a year.

Price stability is expected to worsen as the government faces budgetary problems, but should remain under 10%.

Rwanda: Economic Forecasts
(average annual percentage change)

	Actual 1980-87	Forecast Base 1990-95	Forecast High 1990-95
GDP	2.4	2.0	2.2
GDP per Head	-0.7	-1.3	-1.1
Inflation Rate	4.5	7.5	6.0
Export Volumes	2.5	2.0	2.3
Import Volumes	5.4	2.5	2.7

RWANDA: Comparative Data

	RWANDA	EAST AFRICA	AFRICA	INDUSTRIAL COUNTRIES
POPULATION & LAND				
Population, mid year, millions, 1989	6.8	12.2	10.2	40.0
Urban Population, %, 1985	5	30.5	30	75
Population Growth Rate, % per year, 1980-86	3.3	3.1	3.1	0.8
Land Area, thou. sq. kilom.	26	486	486	1,628
Population Density, persons per sq kilom., 1988	253.8	24.2	20.4	24.3
ECONOMY: PRODUCTION & INCOME				
GDP, $US millions, 1986	1,850	2,650	3,561	550,099
GNP per head, $US, 1986	290	250	389	12,960
ECONOMY: SUPPLY STRUCTURE				
Agriculture, % of GDP, 1986	40	43	35	3
Industry, % of GDP, 1986	23	15	27	35
Services, % of GDP, 1986	37	42	38	61
ECONOMY: DEMAND STRUCTURE				
Private Consumption, % of GDP, 1986	71	77	73	62
Gross Domestic Investment, % of GDP, 1986	19	16	16	21
Government Consumption, % of GDP, 1986	20	15	14	17
Exports, % of GDP, 1986	12	16	23	17
Imports, % of GDP, 1986	22	24	26	17
ECONOMY: PERFORMANCE				
GDP growth, % per year, 1980-86	1.8	1.6	-0.6	2.5
GDP per head growth, % per year, 1980-86	-1.5	-1.7	-3.7	1.7
Agriculture growth, % per year, 1980-86	0.9	1.1	0.0	2.5
Industry growth, % per year, 1980-86	4.8	1.1	-1.0	2.5
Services growth, % per year, 1980-86	1.1	2.5	-0.5	2.6
Exports growth, % per year, 1980-86	1.3	0.7	-1.9	3.3
Imports growth, % per year, 1980-86	6.5	0.2	-6.9	4.3
ECONOMY: OTHER				
Inflation Rate, % per year, 1980-86	5.6	23.6	16.7	5.3
Aid, net inflow, % of GDP, 1986	11.5	11.5	6.3	-
Debt Service, % of Exports, 1986	7.6	18	20.6	-
Budget Surplus (+), Deficit (-), % of GDP, 1980	-1.7	-3.0	-2.8	-5.1
EDUCATION				
Primary, % of 6-11 group, 1985	64	62	76	102
Secondary, % of 12-17 group, 1985	2	15	22	93
Higher, % of 20-24 group, 1985	0	1.2	1.9	39
Adult Literacy Rate, %, 1981	50	41	39	99
HEALTH & NUTRITION				
Life Expectancy, years, 1986	48	50	50	76
Calorie Supply, daily per head, 1985	1,935	2,111	2,096	3,357
Population per doctor, 1981	32,150	35,986	24,185	550

Notes: 'Southern Africa' and 'Africa' exclude South Africa. Dates are for the country in question, and do not always correspond with the Regional, African and Industrial averages.

RWANDA: Leading Indicators

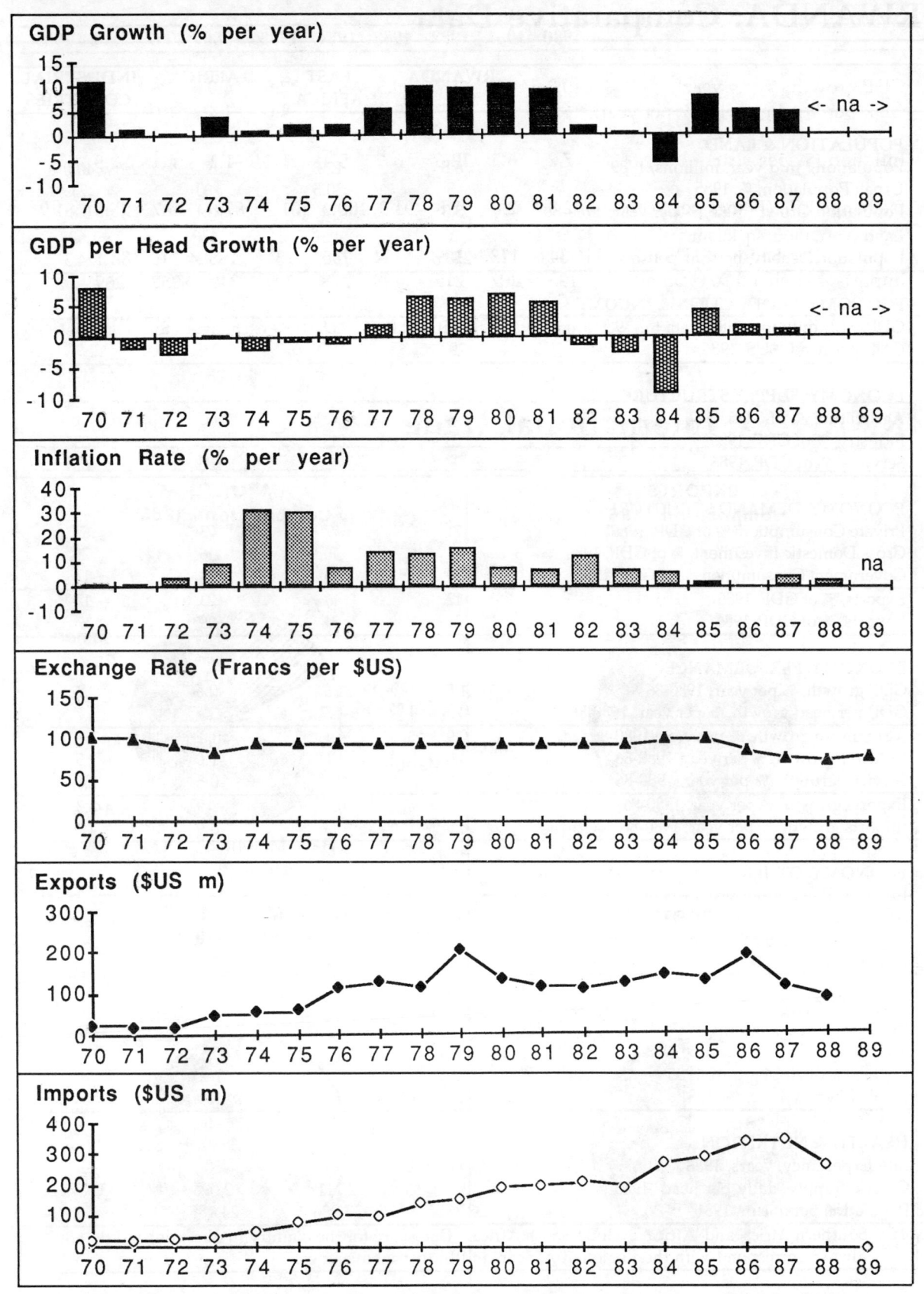

Note: Leading indicator data for 1989 are based on the first half of 1989. 1989 exchange rate is for mid-1989.

RWANDA: Leading Indicator Data

	1980	1981	1982	1983	1984	1985	1986	1987	1988	1989
GDP growth (% per year)	10.2	8.8	1.7	0.3	-6.0	7.5	4.9	4.6		
GDP per head growth (% per year)	6.8	5.4	-1.7	-3.2	-9.5	4.0	1.4	1.0		
Inflation (% per year)	7.2	6.5	12.6	6.6	5.4	1.7	-1.1	4.1	2.9	
Exchange rate (Francs per $US)	92.8	92.8	92.8	94.3	100.2	101.3	87.6	79.7	76.5	81.9
Exports, merchandise ($US m)	134	113	109	124	145	131	188	112	83.1	
Imports, merchandise ($US m)	196	207	215	198	278	299	349	352	269	

Note: Leading indicator data for 1989 are based on the first half of 1989. 1989 exchange rate is for mid-1989.

RWANDA: International Trade

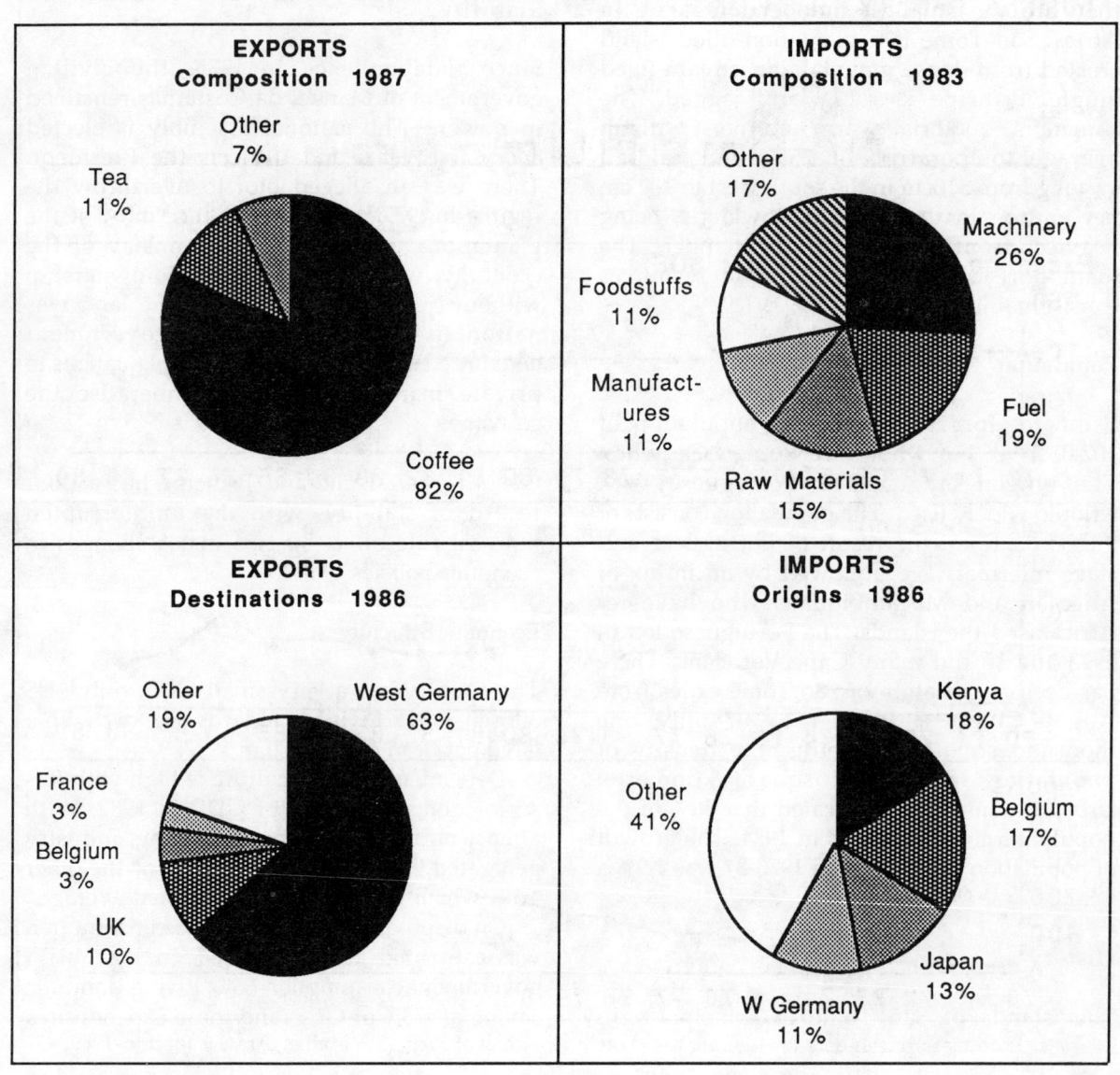

EXPORTS Composition 1987
- Coffee 82%
- Tea 11%
- Other 7%

IMPORTS Composition 1983
- Machinery 26%
- Fuel 19%
- Other 17%
- Raw Materials 15%
- Foodstuffs 11%
- Manufactures 11%

EXPORTS Destinations 1986
- West Germany 63%
- Other 19%
- UK 10%
- France 3%
- Belgium 3%

IMPORTS Origins 1986
- Other 41%
- Kenya 18%
- Belgium 17%
- Japan 13%
- W Germany 11%

35 SAO TOME AND PRINCIPE

Physical Geography and Climate

São Tomé and Príncipe, in the Central African region, is an archipelago in the Gulf of Guinea, following a range of extinct volcanoes and lying just north of the Equator. After the Seychelles they are the smallest independent state in Africa. São Tomé is a plantation-filled island created from dense mountainous stream-filled jungle. Príncipe is rocky and jagged. The climate is equatorial, warm and moist with an average temperature of 25°C and rainfall varying from 510cm in the south west to 102 cm per annum in the northern lowlands, being heaviest from June to mid-September. The main port is São Tomé. Mineral resources consist mainly of lime

Population

Estimates for 1989 suggest a population of 107,000, 90% of whom are Roman Catholics. The official language is Portuguese with Krioulo widely used. The population consists of 'filhos da terra' the result of Portuguese and slave intermarriage, followed by an influx of Angolan and Moçambiquans who have re-Africanised the islands. The Portuguese left in 1975 and so did many Cape Verdeans. There has been repatriation of São Tomé exiles from Angola. With a small area of only one thousand square kilometres, the density of population is 107 per square kilometre. Urbanisation figures indicated that 36% of the population lived in towns in 1984. The growth of population for the period 1970-81 was 1.9% a year.

History

The islands of São Tomé et Príncipe were discovered by the Portuguese in 1471. From 1522 they constituted a province of Portugal. They received full independence in 1975 and remain under the leadership of President da Costa, despite threats to his regime.

Stability

Since independence in 1975, the civilian government of Manuel da Costa has remained in power. The national assembly is elected every five years and it elects the President. There was an alleged plot to overthrow the regime in 1977. At independence most of the plantations which were the mainstay of the economy, were taken into public ownership without compensation. 90% of the land was nationalised. In 1985 the government announced a programme to return plantations to private management and to liberalise the economy.

Overall, Saō Tomé and Príncipe has a poor record of stability, with the uninterrupted political rule offset by two major changes in economic policies.

Economic Structure

The country has a very small GDP, only $US 30m in 1984. Living standards are low with a GNP per head of $US 280 in 1987.

Dependence on agriculture is high with this sector generating 51% of GDP in 1982. With almost no manufacturing or mining, industry generated 9%. Services accounted for the other 40%, which is similar to the regional average.

On disposal of income, private consumption was estimated at 62% of GDP in 1982, with government consumption 66%, giving domestic saving at -28% of GDP, and these expenditures

Sao Tome and Principe

are enabled by substantial financial transfers from overseas. The government's share of expenditure is the highest in Africa.

Exports of goods and services were 47% of GDP in 1981, and imports comprised 97%. Merchandise exports were 98% cocoa in 1987. Merchandise imports were 36% machinary, 22% intermediates, 20% foodstuffs, 14% manufactures and 7% fuels in 1986.

Adult literacy was 57% in 1987, and this compares well with the African average of 40% at roughly that time. Primary enrolments were 116% in 1977, secondary 27%, but there was negligible enrolment in higher education.

Life expectancy was 65 years in 1986, well above the African average of 50. There was one doctor for every 1,952 people in 1987, twelve times the African average provision. Average daily calorie supply was 2,535, a fifth better than the African average.

Overall, Saõ Tomé and Príncipe is a small economy in the low-income group. The economy is heavily dependent on agriculture, and there is a large government sector. There is heavy dependence on international trade. Basic education provision is good. Health and nutrition provision is well above average.

Economic Performance

GDP is estimated to have fallen by -0.6% a year, 1980-87, but there are large variations from year to year as a result of the heavy dependence on agriculture which is vulnerable to climatic changes. In particular, there was a -27% fall in GDP in 1981 followed by a 26% recovery in 1982. GDP per head has fallen by -2.5% a year, 1980-87.

Data for growth in the producing sectors is only available for 1970-81, and in this period industry grew fastest at 2.3% a year, with services growing at 1.9% a year and agriculture at 0.9%.

Export volumes declined at -3.5% a year, 1970-81, while import volumes, supported by financial flows from overseas, expanded at 4.3% a year.

Inflation has been modest, estimated at 5.3% a year, 1980-86, but again there are wide variations from year to year as the weather affects harvests and prices of foodstuffs.

Overall, Saõ Tomé and Príncipe's economic performance has been poor, with declining GDP, GDP per head and export volumes. Only in price stability has performance been good.

Recent Economic Developments

Early in 1985 Saõ Tomé and Príncipe began to introduce reforms to encourage economic growth and to end the comparative isolation of the island since independence in 1975. The programme anticipated a liberalisation of imports, privatisation of shops and housing, and moves to introduce private management into the plantation sector. An IMF loan in support of the programme of $US 2.6m was granted in March 1985. In June of 1987, $US 17.9m of structural adjustment lending from the World Bank, IMF, and the Africa Development Fund was scheduled in anticipation of public sector spending cuts, price de-control, incentives for the private sector, phasing out of subsidies and rises in indirect taxes.

An IMF structural adjustment facility of worth $US 3.5m was agreed in 1989 after the government showed that efforts were being made to tackle the economic crisis, for example by reducing civil service spending. A loan of $US 4.8m was made available by the World Bank towards implementing the structural adjustment programme. The IMF and World Bank had been seeking a 60% devaluation of the currency but seem to be satisfied with a compromise. Rescheduling of debts by the Paris Club in 1990 is now likely as a result.

In the 1980s, the trade balance began to move toward a substantial deficit. Export revenue peaked at $US 27m in 1979, but with the neglect of the cocoa and palm oil plantations, which provide the bulk of exports, and falls in cocoa prices, exports fell to $US 5m recovering to $US 11m in 1989. Imports have been maintained by aid inflows, and have grown from $US 19m in 1980 to $US 29m in 1984 before falling to $US 20m in 1989.

In the 1989 budget the government set expenditure at DB 2,677m ($US 22m), with 20% of this allocated for investment, and revenue was estimated at DB 1,915m ($US 16m).

External debt in 1989 was estimated at $US 95m, which was equivalent to 316% of

This debt is long-term and at low interest rates, and gives rise to debt-servicing at 52% of merchandise export earnings.

The value of the currency has depreciated internationally at a steady rate from independence in 1975 when the dobra stood at DB 25.6 = $US 1. It was DB 139 = $US 1 at the end of 1989. In 1986 there was speculation that São Tomé and Príncipe would join the Franc Zone, but as yet there have been no further developments in this direction.

Aid has concentrated on rehabilitation of the cocoa and palm oil export sectors, and efforts to establish a tourism industry. The European Development Fund is contributing $US2.5m and the European Investment Bank $US 2.4m towards a $US 7.8m palm oil rehabilitation project. The World Bank, the Arab Bank for Economic Development in Africa, France, the European Investment Bank, USAID and the International Fund for Argicultural Development are combining to contribute $US 23.7m to rehabilitation of cocoa plantations. The World Bank is donating $US 0.7m to electricity supply, the African Development Bank $US 2m to telecommunications, and France $US 6.2m to airport improvements.

Economic Outlook and Forecasts

The forecasts assume that the government of President da Costa remains in power, and that the current programme of liberalising economic reforms remains in place.

Under the impact of the reforms and structural adjustment support, GDP is forecast to expand up to 1995, but at a slower rate than population growth so that GDP per head continues to fall, but at a slower rate.

Export volumes are projected to expand slowly, and import volumes to expand faster than exports, at around 4.0% a year as a result of aid inflows.

Although the projected budget deficit is -20% of GDP, it is expected that much of this will be covered by financial transfers from overseas. Although inflation is forecast to accelerate, it will remain in single figures.

São Tomé and Príncipe: Economic Forecasts
(average annual percentage change)

	Actual	Forecast Base	Forecast High
	1980-87	1990-95	1990-95
GDP	-0.6	1.5	1.7
GDP per Head	-2.5	-0.4	-0.2
Inflation Rate	5.3[a]	6.0	7.0
Export Volumes	-3.5[b]	1.0	1.3
Import Volumes	4.3[b]	4.0	4.1

Note: [a] 1980-86 [b] 1970-81.

SAO TOME & PRINCIPE: Comparative Data

	SAO TOME & PRINCIPE	CENTRAL AFRICA	AFRICA	INDUSTRIAL COUNTRIES
POPULATION & LAND				
Population, mid year, millions, 1989	0.107	7.3	10.2	40.0
Urban Population, %, 1984	36	38.6	30	75
Population Growth Rate, % per year, 1970-81	1.9	3.0	3.1	0.8
Land Area, thou. sq. kilom.	1	638	486	1,628
Population Density, persons per sq kilom., 1988	105	11.1	20.4	24.3
ECONOMY: PRODUCTION & INCOME				
GDP, $US millions, 1984	30	4,146	3,561	550,099
GNP per head, $US, 1986	340	395	389	12,960
ECONOMY: SUPPLY STRUCTURE				
Agriculture, % of GDP, 1982	51	18	35	3
Industry, % of GDP, 1982	9	41	27	35
Services, % of GDP, 1982	40	41	38	61
ECONOMY: DEMAND STRUCTURE				
Private Consumption, % of GDP, 1981	62	62	73	62
Gross Domestic Investment, % of GDP, 1981	21	25	16	21
Government Consumption, % of GDP, 1981	66	14	14	17
Exports, % of GDP, 1981	47	35	23	17
Imports, % of GDP, 1981	97	35	26	17
ECONOMY: PERFORMANCE				
GDP growth, % per year, 1970-81	1.2	4.2	-0.6	2.5
GDP per head growth, % per year, 1970-81	-0.7	-0.8	-3.7	1.7
Agriculture growth, % per year, 1970-81	0.9	0.5	0.0	2.5
Industry growth, % per year, 1970-81	2.3	7.9	-1.0	2.5
Services growth, % per year, 1970-81	1.9	3.2	-0.5	2.6
Exports growth, % per year, 1970-81	-3.5	4.5	-1.9	3.3
Imports growth, % per year, 1970-81	4.3	0.4	-6.9	4.3
ECONOMY: OTHER				
Inflation Rate, % per year, 1980-86	5.3	17	16.7	5.3
Aid, net inflow, % of GDP, 1984	65	5.4	6.3	-
Debt Service, % of Exports, 1984	21.9	28	20.6	-
Budget Surplus (+), Deficit (-), % of GDP, 1977	-2.5	-1.4	-2.8	-5.1
EDUCATION				
Primary, % of 6-11 group, 1977	116	93	76	102
Secondary, % of 12-17 group, 1977	27	42	22	93
Higher, % of 20-24 group, 1977	..	1.9	1.9	39
Adult Literacy Rate, %, 1977	57	52	39	99
HEALTH & NUTRITION				
Life Expectancy, years, 1986	65	52	50	76
Calorie Supply, daily per head, 1983	2,538	2,115	2,096	3,357
Population per doctor, 1977	1,952	13,835	24,185	550

Notes: 'Southern Africa' and 'Africa' exclude South Africa. Dates are for the country in question, and do not always correspond with the Regional, African and Industrial averages.

SAO TOME & PRINCIPE: Leading Indicators

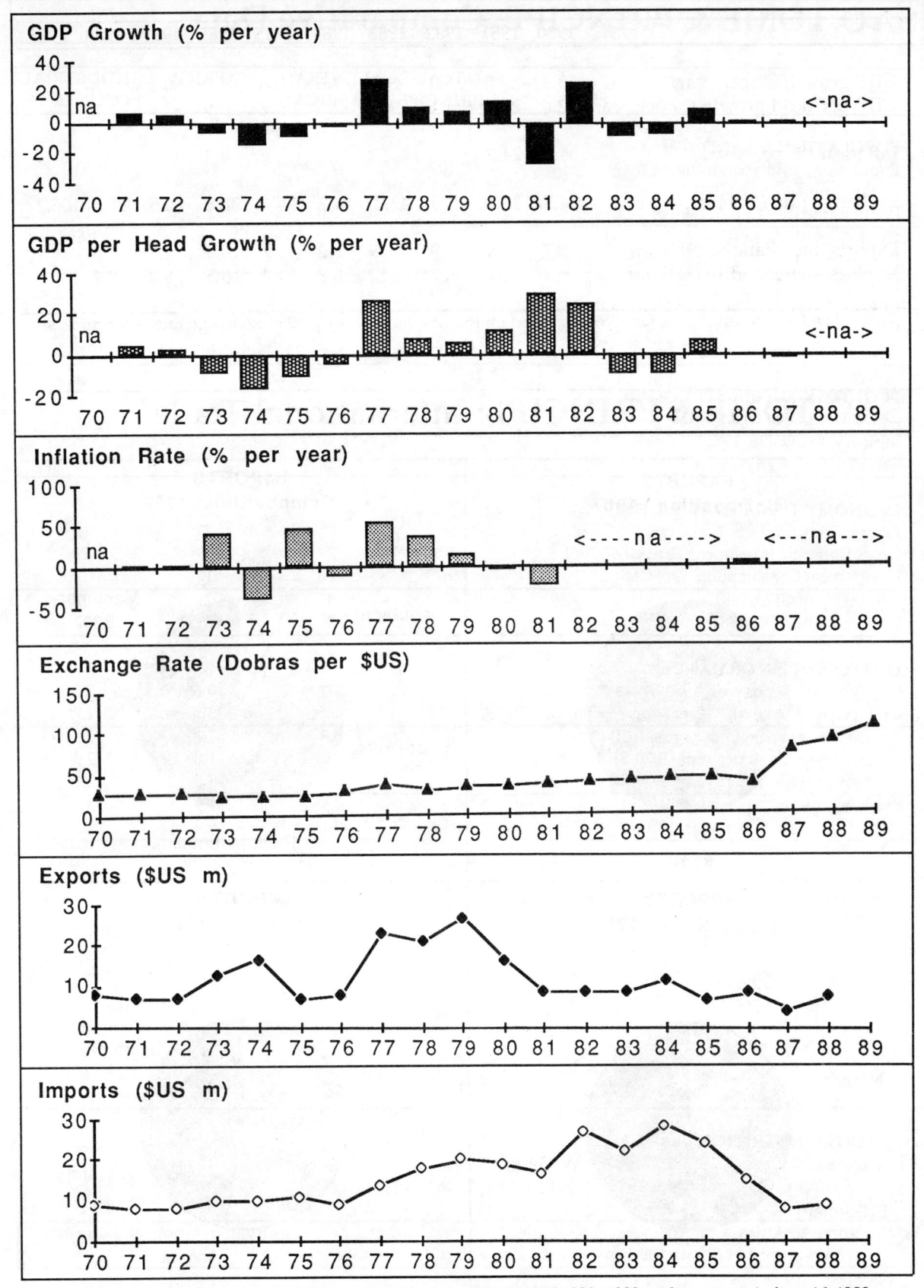

Note: Leading indicator data for 1989 are based on the first half of 1989. 1989 exchange rate is for mid-1989.

SAO TOME & PRINCIPE: Leading Indicator Data

	1980	1981	1982	1983	1984	1985	1986	1987	1988	1989
GDP growth (% per year)	13.9	-27.5	26.0	-8.9	-8.6	8.6	1.0	-0.5		
GDP per head growth (% per year)	12.0	29.4	24.1	-9.8	-9.5	6.7	-0.9	-2.4		
Inflation (% per year)	-4.2	-24.7					4.2			
Exchange rate (Dobras per $US)	34.77	38.40	41.0	43.3	44.2	44.6	38.6	80.8	88.5	106.2
Exports, merchandise ($US m)	17	9	9	9	12	7	8.9	4.8	7.8	
Imports, merchandise ($US m)	19	17	27	22	28.8	24.2	15.5	8.2	9.2	

Note: Leading indicator data for 1989 are based on the first half of 1989. 1989 exchange rate is for mid-1989.

SAO TOME & PRINCIPE: International Trade

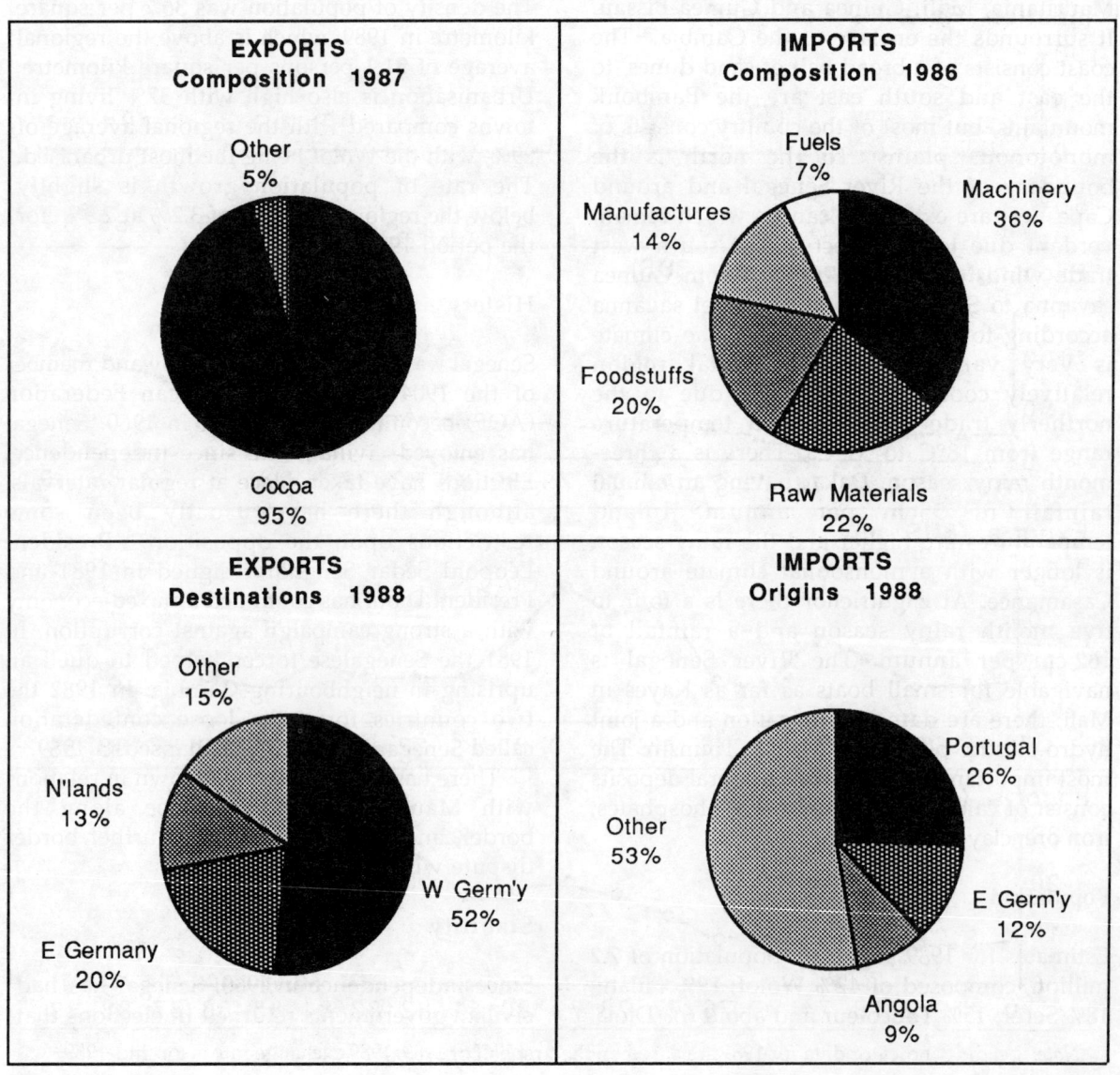

EXPORTS Composition 1987
- Other 5%
- Cocoa 95%

IMPORTS Composition 1986
- Fuels 7%
- Machinery 36%
- Manufactures 14%
- Foodstuffs 20%
- Raw Materials 22%

EXPORTS Destinations 1988
- Other 15%
- N'lands 13%
- E Germany 20%
- W Germ'y 52%

IMPORTS Origins 1988
- Portugal 26%
- E Germ'y 12%
- Angola 9%
- Other 53%

36 SENEGAL

Physical Geography and Climate

Senegal is a medium-sized Atlantic coastal state in the West African region, bounded by Mauritania, Mali, Guinea and Guinea-Bissau. It surrounds the enclave of the Gambia. The coast consists of a broad belt of sand dunes, to the east and south east are the Bambouk mountains, but most of the country consists of monotonous plains. To the north is the boundary of the River Senegal and around Cape Vert are extinct volcanoes which remain verdant due to the effect of the south west trade winds. Vegetation ranges from Guinea savanna to Sudan savanna to Sahel savanna according to the rainfall pattern. The climate is very varied with the coastal region relatively cool for its latitude, due to the northerly trade winds with a temperature range from 18°C to 31°C. There is a three-month rainy season, Dakar having an annual rainfall of 55cm per annum. Inland temperatures are higher and the rainy season is longer with a monsoonal climate around Casamance. At Ziguirichor there is a four to five month rainy season and a rainfall of 162 cm per annum. The River Senegal is navigable for small boats as far as Kayes in Mali; there are dams for irrigation and a joint hydro-electric plant with Mali at Tounsin. The most important port is Dakar. Mineral deposits consist of calcium and aluminium phosphates, iron ore, clay and sea salt.

Population

Estimates for 1989 suggest a population of 7.2 million, composed of 42% Wolof; 19% Fulani; 18% Serer; 15% Toucoleur and about 6% Diola. The official language is French with 44% speaking Wolof, 21% Pulaar, 16% Seereer and 6% Mandinka. Most of the north is Moslem, whilst the south retains its traditional beliefs. The density of population was 36.7 per square kilometre in 1989, which is above the regional average of 31.1 persons per square kilometre. Urbanisation is also high with 37% living in towns compared with the regional average of 29%, with the Wolof being the most urbanised. The rate of population growth is slightly below the regional average of 3.2% at 2.9% for the period 1980-87.

History

Senegal was a former French colony and member of the 1904 French West African Federation (AOF) becoming independent in 1960. Senegal has enjoyed civilian rule since independence. Elections have taken place at regular intervals, although there have usually been some restrictions upon the opposition. President Léopold Sédar Senghor resigned in 1981 and President Diouf has continued a mixed economy with a strong campaign against corruption. In 1981 the Senegalese forces helped to quell an uprising in neighbouring Gambia. In 1982 the two countries formed a loose confederation called Senegambia, but this collapsed in 1989.

There has also been a breakdown in relations with Mauritania with violence along the border, and the beginnings of a further border dispute with Guinea-Bissau.

Stability

Since independence in 1960, Senegal has had civilian governments returned in elections that

Senegal

have been contested by opposition parties, although some opposition groups have been subject to restrictions. The Parti Socialiste has remained in office throughout the period, and when the first President resigned in 1980, power passed smoothly to his successor, Abdou Diouf. Strong links are maintained with France and French troops are stationed near the capital. Economic strategy has been consistent, with reliance on a mixed economy, and encouragement for foreign, especially French, investment and business interests. Senegal has remained a member of the Franc Zone.

Overall, Senegal has a very good stability record, and there is no reason to expect this to change in the foreseeable future, despite the recent strained relations with Mauritania and Guinea-Bissau.

Economic Structure

GDP in 1987 stood at $US 4,720m, about the same as the regional average, as was the GNP per head of $US 520.

Agriculture contributed only 22% of GDP in 1987. This is well below the regional average of 40% and reflects the greater importance of the service sector which generates 52%, the third largest in the region after the Gambia (highly dependent on tourism) and Togo. The industrial sector accounts for 27%, about average.

Private consumption represents 77% of GDP, which is average for the region, as is the figure of 13% for investment. Government consumption is slightly higher than average at 17%.

Exports represented 28% of GDP (10% above average for the region). Imports comprised 35% of spending, well above the regional average of 20%, partly reflecting the large demand for food imports and other consumer goods in urban areas. Exports to Mauritania have been seriously affected by the trouble between the two countries.

Senegal's main exports are fish (26%) and fuels (18%), with the latter involving imported grade which is refined and exported to the neighbouring countries. Other exports include phosphates, cotton and groundnuts. Imports included food products (19%) mainly for the urban consumers, and fuels (18%). Being on the margins of the Sahel, Senegal's exports are partly dependent on climatic stability.

The debt service ratio was about average for West Africa with debt payments equal to 22% of exports in 1987.

The most recent adult literacy figures are for 1974, when the estimate was 10%. In 1986 55% of the primary age group attended school, about one third less than the regional average. Of the secondary age group, 13% attend school, about half the regional average. These figures do not include children at Koranic schools. About 2% of the 20-24-year age group are in higher education.

Life expectancy in 1987 was 48%, similar to the regional average. With one doctor for every 13,450 of the population and one nurse to every 2,090 Senegal has slightly better health worker provision than the region in general. Average daily calorie supply at 2,350 a person in 1986 was 12% above the Africa average.

Overall, Senegal is a medium-sized economy at the bottom end of the lower-middle-income group, with a relatively large service sector. The structure of expenditure is fairly typical of a lower-middle-income African country. Educational provision is much worse than might be expected. Health and nutritional provision are about average.

Economic Performance

GDP in Senegal grew by 3.3% from 1980-87 while GDP per head grew at 0.4% a year. Agriculture, which accounts for about 25% of GDP, grew by 4.2% compared with a regional growth average of 1.3%. The industrial sector grew by 4.3% which compares favourably with a regional average of -3.6%. The service sector also showed an increase of 2.4% while the regional average was -2.5%.

Exports during this period grew by 6.7% a year, while the region averaged a decline of -4.0% a year. Imports also increased by 2.7% a year while on average in the region they fell by 12.9%.

Inflation over the same period averaged 9.1%, below both the regional and African averages of 13.0% and 16.7% respectively.

Overall, Senegal has had satisfactory economic performance with GDP and GDP per head expanding. Most growth has been generated by the agriculture and industry sectors. Export performance has been excellent and price stability good.

Recent Economic Developments

In the 1980s, Senegal has accepted structural adjustment lending from the IMF and World Bank to support a programme of economic reforms. The main measures being introduced are reductions in public expenditure and cuts in government employment levels, together with efforts to increase the level of government revenues. By the beginning of 1987, some 2,000 public sector jobs had been cut, and plans were in hand to cut a further 5,000. Of the 63 enterprises in the public sector, 15 are to be completely privatised, and the main operations for sale are in textiles, sack manufacture and vehicle assembly. Eighteen parastatals are to be partially privatised, and they include 5 banks, ship repair, salt manufacture and housing enterprises. 30 enterprises are to remain in the public sector and the main operations involved are in fertiliser manufacture and in wholesale and retail distribution. In 1988 the IMF granted an enhanced structural adjustment facility of $US 194m over three years.

Senegal has typically had a deficit on its visible trade, but this has been wider in the 1980s, with merchandise imports running at almost twice the value of exports in 1987. The trade deficit has narrowed slightly in the last two years however, with merchandise exports valued at 58% of merchandise imports in 1989. There is no perceptible trend to merchandise imports in dollar values, but in real terms they have fallen since 1980.

Total external debt stood at $US 3,068m in 1987, and debt service payments were 21% of total exports of goods and services.

Senegal is a member of the Franc Zone, and the value of the CFA franc steadily depreciated against the dollar, from CFA 211 = $US 1 in 1980, to CFA 449 = $US 1 in 1985. Since then the CFA franc has appreciated, standing at CFA 289 = $US 1 at the end of 1989. This recent appreciation, in circumstances of poor export performance by many Franc Zone countries has led to speculation that the CFA franc might be devalued against the French franc, with which it is freely convertible.

Major recent aid commitments include $US 4m from the Arab Bank for Economic Development in Africa for irrigation, and $US 14m from the US for the phasing out of fertiliser subsidies, and $US 11.6m to support structural adjustment. Canada is allocating $US 11.5m to railways, and a further $US 21.2m for fisheries, forestry, conservation, oil exploration and education. The EEC is allocating $US 113.5m to a major agricultural scheme in the Senegal River Basin, and the World Bank, the Africa Development Bank and France are lending a total of $US 112m for electrification projects.

Economic Outlook and Forecasts

The forecasts assume that Senegal maintains its good stability record and that the dispute with Mauritania does not escalate. It is also assumed that Senegal continues with its programme of liberalising economic reforms, and that weather conditions are normal.

Groundnut prices are expected to fall by -11% up to 1995 in real terms, cotton prices to remain unchanged and phosphate prices to rise by 10%.

GDP growth is forecast to accelerate, with GDP per head growth improving to 1% a year.

Export volumes are projected to expand more slowly as the appreciating CFA franc makes it difficult to maintain producer prices. Import volumes are expected to expand slightly faster as a result of structural adjustment balance of payments support.

Inflation is forecast to improve marginally to a rate of around 8.0% a year.

Senegal: Economic Forecasts
(average annual percentage change)

	Actual	Forecast Base	High
	1980-87	1990-95	1990-95
GDP	3.3	4.2	4.4
GDP per Head	0.4	1.1	1.3
Inflation Rate	9.1	8.5	8.0
Export Volumes	6.7	3.9	4.1
Import Volumes	2.7	3.0	3.2

SENEGAL: Comparative Data

	SENEGAL	WEST AFRICA	AFRICA	INDUSTRIAL COUNTRIES
POPULATION & LAND				
Population, mid year, millions, 1989	7.2	12.3	10.2	40.0
Urban Population, %, 1985	36	28.7	30	75
Population Growth Rate, % per year, 1980-86	2.9	3.2	3.1	0.8
Land Area, thou. sq. kilom.	196	384	486	1,628
Population Density, persons per sq kilom., 1988	35.7	31.1	20.4	24.3
ECONOMY: PRODUCTION & INCOME				
GDP, $US millions, 1986	3,740	4,876	3,561	550,099
GNP per head, $US, 1986	420	510	389	12,960
ECONOMY: SUPPLY STRUCTURE				
Agriculture, % of GDP, 1986	22	40	35	3
Industry, % of GDP, 1986	27	26	27	35
Services, % of GDP, 1986	51	34	38	61
ECONOMY: DEMAND STRUCTURE				
Private Consumption, % of GDP, 1986	77	77	73	62
Gross Domestic Investment, % of GDP, 1986	14	13	16	21
Government Consumption, % of GDP, 1986	17	13	14	17
Exports, % of GDP, 1986	28	18	23	17
Imports, % of GDP, 1986	36	20	26	17
ECONOMY: PERFORMANCE				
GDP growth, % per year, 1980-86	3.2	-2.03	-0.6	2.5
GDP per head growth, % per year, 1980-86	0.3	-4.7	-3.7	1.7
Agriculture growth, % per year, 1980-86	2.3	1.3	0.0	2.5
Industry growth, % per year, 1980-86	4.0	-3.6	-1.0	2.5
Services growth, % per year, 1980-86	3.2	-2.5	-0.5	2.6
Exports growth, % per year, 1980-86	8.7	-4.0	-1.9	3.3
Imports growth, % per year, 1980-86	1.8	-12.9	-6.9	4.3
ECONOMY: OTHER				
Inflation Rate, % per year, 1980-86	9.5	13.0	16.7	5.3
Aid, net inflow, % of GDP, 1986	16.0	3.7	6.3	-
Debt Service, % of Exports, 1986	20.2	21.4	20.6	-
Budget Surplus (+), Deficit (-), % of GDP, 1983	-6.0	-3.4	-2.8	-5.1
EDUCATION				
Primary, % of 6-11 group, 1985	55	75	76	102
Secondary, % of 12-17 group, 1985	13	24	22	93
Higher, % of 20-24 group, 1985	2	2.6	1.9	39
Adult Literacy Rate, %, 1974	10	30	39	99
HEALTH & NUTRITION				
Life Expectancy, years, 1986	47	50	50	76
Calorie Supply, daily per head, 1985	2,418	2,105	2,096	3,357
Population per doctor, 1981	13,070	16,199	24,185	550

Notes: 'Southern Africa' and 'Africa' exclude South Africa. Dates are for the country in question, and do not always correspond with the Regional, African and Industrial averages.

SENEGAL: Leading Indicators

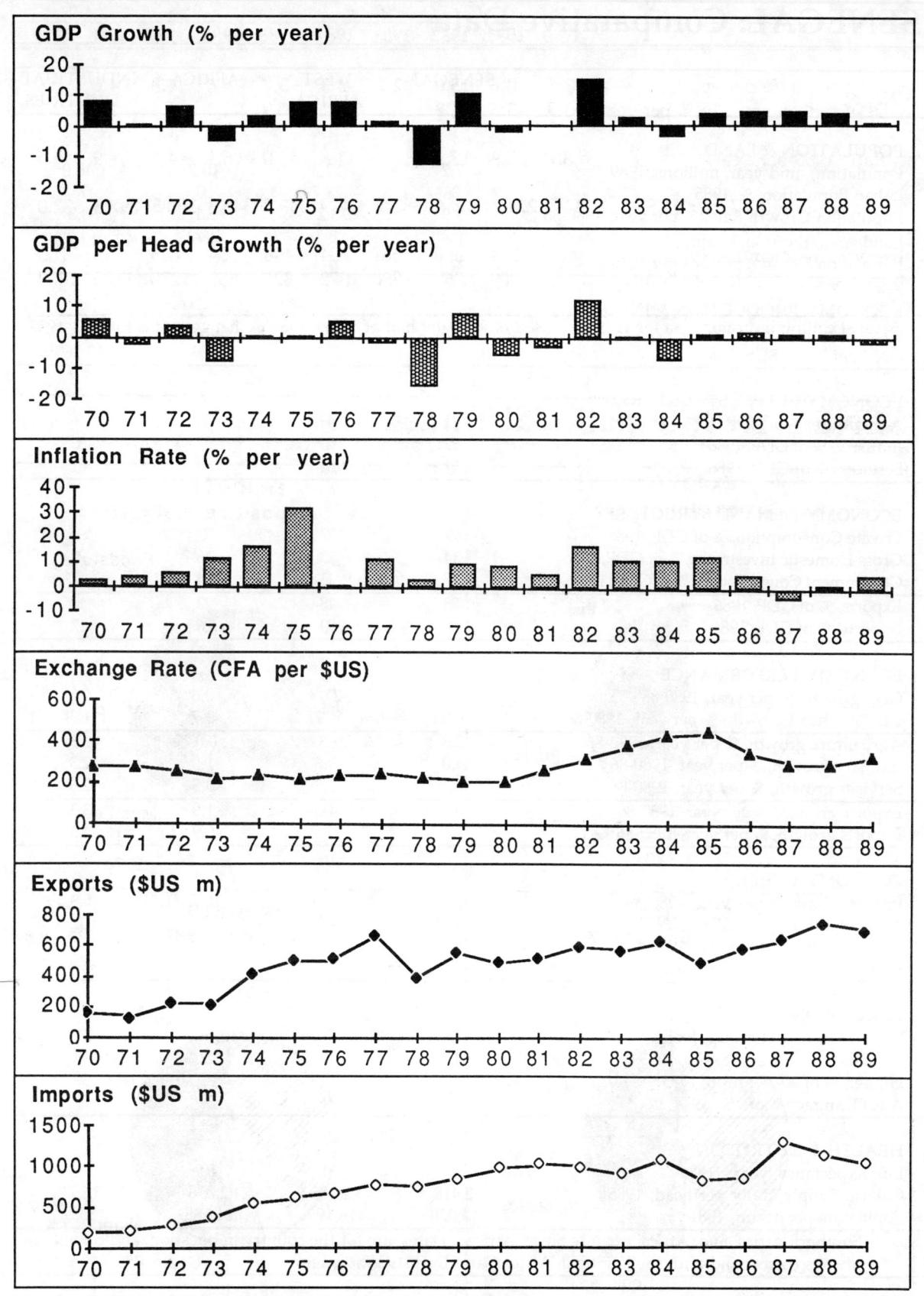

Note: Leading indicator data for 1989 are based on the first half of 1989. 1989 exchange rate is for mid-1989.

SENEGAL: Leading Indicator Data

	1980	1981	1982	1983	1984	1985	1986	1987	1988	1989
GDP growth (% per year)	-3.1	-0.6	15.0	2.7	-4.3	3.7	4.6	4.3	4.0	0.6
GDP per head growth (% per year)	-5.9	-3.4	12.2	-0.1	-7.1	0.9	1.7	1.4	1.1	-2.3
Inflation (% per year)	8.7	5.9	17.3	11.7	11.8	13.0	6.1	-4.1	-1.9	5.5
Exchange rate (CFA per $US)	211.3	271.7	328.6	381.1	437.0	449.3	346.3	300.5	297.8	332.0
Exports, merchandise ($US m)	481	511	590	569	634	498	584	645	750	700
Imports, merchandise ($US m)	973	1009	968	880	1072	826	852	1270	1100	1000

Note: Leading indicator data for 1989 are based on the first half of 1989. 1989 exchange rate is for mid-1989.

SENEGAL: International Trade

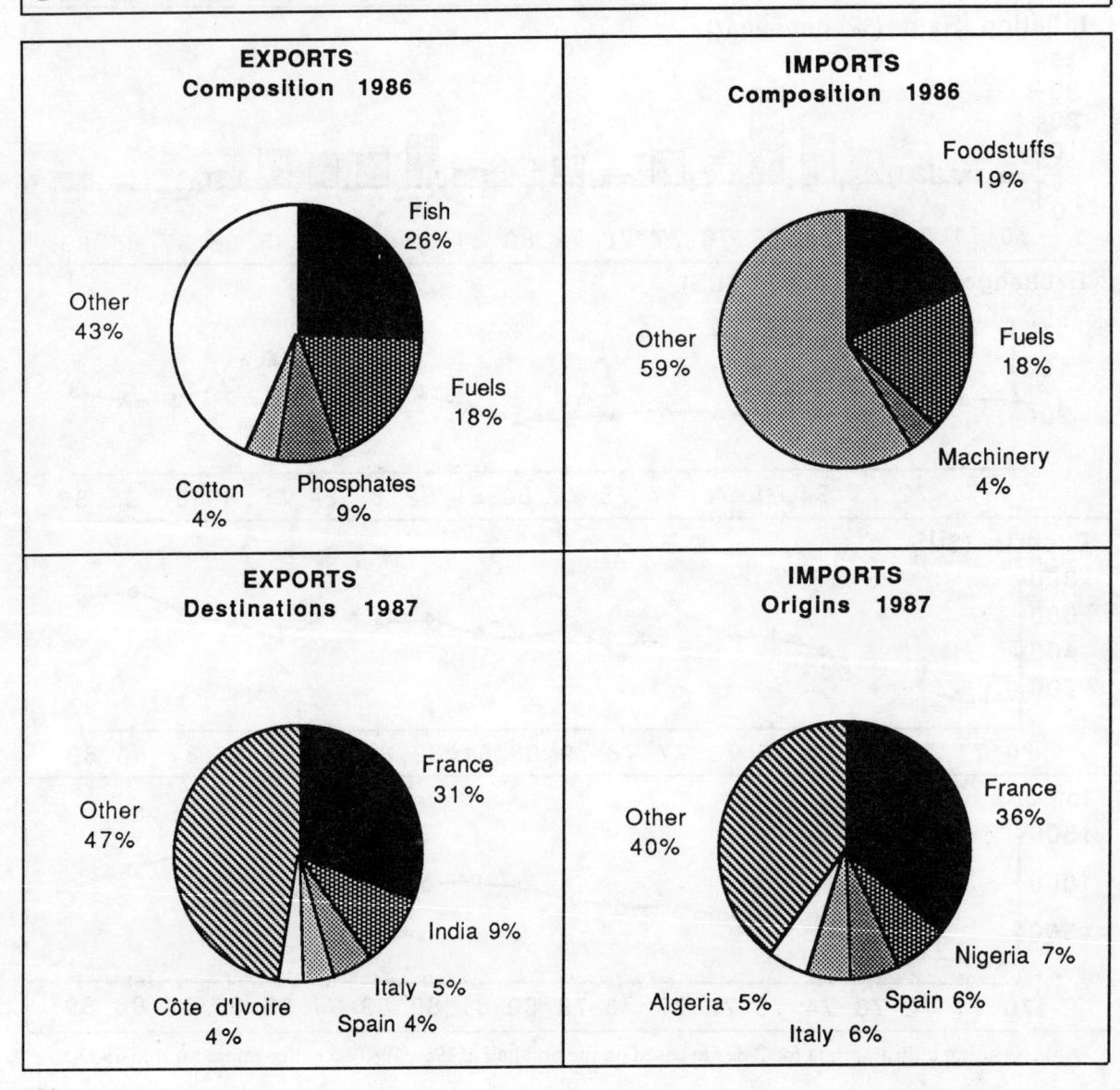

EXPORTS Composition 1986
- Fish 26%
- Fuels 18%
- Phosphates 9%
- Cotton 4%
- Other 43%

IMPORTS Composition 1986
- Foodstuffs 19%
- Fuels 18%
- Machinery 4%
- Other 59%

EXPORTS Destinations 1987
- France 31%
- India 9%
- Italy 5%
- Spain 4%
- Côte d'Ivoire 4%
- Other 47%

IMPORTS Origins 1987
- France 36%
- Nigeria 7%
- Spain 6%
- Italy 6%
- Algeria 5%
- Other 40%

37 SEYCHELLES

Physical Geography and Climate

The Seychelles, in the East African region, consist of a scattered group of 115 islands and atolls in the western Indian Ocean. Despite being close to the equator the climate is pleasant, varying at sea level from 24°C to 29°C. There is a hot season from December to May and a cool season from June to November, due to the south east trade winds. The islands lie beyond the cyclone belt, with rainfall varying within the group; most falling in the hot season. The average annual rainfall in the capital, Victoria, being 236cm. The islands are scenically beautiful and rise steeply from the surrounding coral reefs.

Population

Estimates for 1989 indicate a population of 69,000. The population is a mixture of the descendants of French and British colonists, African slaves and Indian immigrant workers. The majority of the inhabitants speak Seychellois Creole, with French and English. The density of population averages 151 persons per square kilometre which is high. Urbanisation is high around the capital, Victoria, and averages 26% for the islands as a whole. Population growth is low by African standards at 1.5% a year for the period 1973-82, mainly due to emigration.

History

The islands were first sighted by Vasco da Gama in 1502. The French explored the islands and claimed possesssion in 1756, but they remained uninhabited until the end of the century. In 1794 the French garrison surrendered to a British naval force, but token French administration continued until the 1814 Treaty of Paris confirmed British possession of the Seychelles. In 1903 the islands became a Crown Colony. Independence was achieved in 1976 under President Machan, but there was a successful coup in 1977 by members of the Seychelles People's United Party (SPUP) and Prime Minister Albert René became President. There have been further coup attempts since then, including the 1979 attempt by South African mercenaries and a 1982 mutiny. In 1985 the leader of the Mouvement pour Résistance (MRP), Gérard Hoareau was assassinated in London. Attempts by President René, who was elected for a third term of office in 1989, to introduce socialist reforms and the expectations that these would be reversed in a coup has damaged business confidence and deterred foreign investment.

Stability

After independence in 1976, Seychelles had a coalition government that was ousted by a coup in 1977. Subsequently President Albert René has established single-party rule, and succeeded in remaining in power despite several coup attempts. In 1989 he was elected President for a third term.

Economic strategy has effectively remained unchanged, with consistent reliance on a mixed economy, despite a declaration of President René in favour of more socialist policies.

Overall, Seychelles' stability record must be judged poor. There is continued concern that one of the regular coup attempts mounted from outside the country will be successful, although there is now less anxiety that the René government will introduce radical changes in the economy.

Seychelles

Economic Structure

With a GDP of $US 150m in 1983, the islands have one of the smallest economies in Africa. However, GNP per head was around $US 3,120 in 1987. Only Réunion in Africa has a higher living standard.

Agriculture, which includes fishing, generated only 7% of GDP in 1984, compared to 16% for industry, with limited manufacturing, fish processing being the most important development in this sector. The service sector, dominated by the tourist industry, generates 77% of GDP.

On the demand side, private consumption was 60% of GDP in 1984. Investment was 21%, while government consumption at 31% was one of the highest in the region.

Exports were 64% of GDP in 1984 and this is well above both the regional and African averages of 16% and 23% respectively. Imports were equivalent to 76% of GDP while the regional and African averages are about one third of that figure. Such a high dependence on trade is inescapable when a small domestic market does not allow manufacturing and heavy industry at a sufficiently large scale to be efficient. In 1987, the main commodity exports were fish, 80%, and copra, 9%.

Education has been a priority with an adult literacy rate of 58% in 1971. Primary enrolment was 113% in 1982, one of the highest in Africa. The percentage of secondary-aged population in school was 31%, double the regional average, while 16% of the 20-24 age group were in higher education, by far the highest ratio in Africa.

Provision of health services is also very good with a life expectancy of 70 years, one of the best in Africa. Provision of doctors and nurses in 1975 was also extremely good with 2,700 people per doctor and one nurse for every 433 people. Daily average calorie supply was 2,351 in 1983, about 12% better than the African average.

Overall, Seychelles is a small economy in the upper-middle-income group, with most GDP generated by services which has a large tourist component. Dependence on international trade is high. Educational, health and nutrition provision are good.

Economic Performance

GDP growth for 1980-87 was 0.2% a year, implying GDP per head growth of -1.3%. The industrial sector grew fastest during 1973-82, at 8%, a high growth rate for an African economy, as was the 6.9% a year growth in the service sector. Agricultural growth was slightly negative, -0.6, though in recent years the government has made efforts to stimulate growth in this sector and the fishing industry has grown considerably.

Export volume growth over the 1973-82 period was also good, with a 10.1% annual increase, while imports grew by 10.5%, though government austerity measures aim to reduce this figure.

Inflation over the 1980-87 period ran at an average of 4.6% a year, one of the lowest rates in Africa.

Overall, Seychelles has had disappointing economic performance in the 1980s, with GDP increasing marginally, and GDP per head falling. Price stability has been good.

Recent Economic Developments

The government has increased its participation in the economy since 1977 by taking majority shareholdings in certain enterprises. There are a number of state farms, some hotels have been acquired, and parastatals have been set up to run utilities, market commodities and control imports. State acquisition and intervention, however, appears to be extended in an *ad hoc* manner, and there is considerable encouragement to private foreign investment. Two minor parastatals were abolished in 1987, and some printing parastatals rationalised. At the same time, investment in other parastatals has increased, notably Air Seychelles and the Seychelles Marketing Board. The development strategy is mainly concerned to maintain balance on external payments, and to ensure that debt service can be maintained. To this end the tourist industry is of vital importance as the main foreign exchange earner, and private foreign investment is an important element in sustaining Seychelles' position as an attractive tourist destination.

Seychelles has found it difficult in the

1980s to sustain the rapid growth of GDP achieved in the 1970s. Much depends on the fortunes of the tourist sector which is very sensitive to rumours of political instability and decisions of major airlines concerning both scheduled and charter flights. GDP growth was consequently negative in the period 1980-83 following the reports of coup attempts by South African mercenaries, but has shown some recovery, with record tourist arrivals for longer-stay vacations over 1984-89 leading to a 5.5% annual GDP growth rate.

Inflation performance has improved. Annual price rises were consistently above 10% up to 1982, but inflation has sharply declined subsequently, being estimated at 1% in 1989.

Seychelles runs a substantial trade deficit, with earnings for merchandise exports estimated at $US 15m in 1989, while imports were $US 160m.

Seychelles' external debt was estimated at $US 124m in 1987. As Seychelles is an upper-middle-income country, this is mostly commercial debt or loans not made at the highly concessionary rates extended to the very poorest countries. Debt servicing is low at 6.5% of exports of goods and services.

The exchange rate depreciated gently from 1980 when it stood at R 6.39 = $US 1 to R 7.13 = $US 1 in 1985. Thereafter, the weakening of the US dollar has seen an appreciation to R 5.3 = $US 1 at the end of 1989.

Seychelles' upper-middle income status restricts the amount of concessionary lending it receives. Main lending commitment have been $US 12.7m from the African Development Bank for water supply, and $US 3.3m from France for sewerage. France gave over half of the total $US 20.1m bilateral aid in 1987 with only another $US 4.7m coming from multilateral sources.

Private investment has included the establishment of an international airline, which has raised $US 2.9m, most of it locally. The International Finance Corporation of the World Bank is to lend $US 7.6m for hotel development, with another $US 1.7m from the Bank of Baroda. Intercontinental Hotels are to manage the project, and are investing $US 0.5m. Enterprise Oil of the UK have taken up an option to explore for oil off-shore to the south of the Seychelles.

Economic Outlook and Forecasts

The forecasts assume that stability continues to be consolidated by the René regime, and that the cautious programme of liberalising economic reforms remains in place.

Copra prices are expected to rise by 11% in real terms up to 1995.

The forecasts show GDP growth at around 4%, allowing a steady increase in GDP per head.

As commodity exports are growing from such a low base, there is scope for continuing expansion of volumes at reasonably rapid rates, although the growth of import volumes is expected to slow down.

Inflation is expected to be under 5%.

Seychelles: Economic Forecasts
(average annual percentage change)

	Actual 1980-87	Forecasts Base 1990-95	Forecasts High 1990-95
GDP	0.2	4.0	4.2
GDP per Head	-1.3	2.5	2.7
Inflation Rate	4.6	4.5	4.0
Export Volumes	10.1	6.1	7.2
Import Volumes	10.5	5.0	5.6

SEYCHELLES: Comparative Data

	SEYCHELLES	EAST AFRICA	AFRICA	INDUSTRIAL COUNTRIES
POPULATION & LAND				
Population, mid year, millions, 1989	0.069	12.2	10.2	40.0
Urban Population, %, 1983	26	30.5	30	75
Population Growth Rate, % per year, 1973-82	1.5	3.1	3.1	0.8
Land Area, thou. sq. kilom.	0.45	486	486	1,628
Population Density, persons per sq kilom., 1988	151.1	24.2	20.4	24.3
ECONOMY: PRODUCTION & INCOME				
GDP, $US millions, 1983	150	2,650	3,561	550,099
GNP per head, $US, 1983	2,300	250	389	12,960
ECONOMY: SUPPLY STRUCTURE				
Agriculture, % of GDP, 1984	7	43	35	3
Industry, % of GDP, 1984	16	15	27	35
Services, % of GDP, 1984	77	42	38	61
ECONOMY: DEMAND STRUCTURE				
Private Consumption, % of GDP, 1984	60	77	73	62
Gross Domestic Investment, % of GDP, 1984	21	16	16	21
Government Consumption, % of GDP, 1984	31	15	14	17
Exports, % of GDP, 1984	64	16	23	17
Imports, % of GDP, 1984	76	24	26	17
ECONOMY: PERFORMANCE				
GDP growth, % per year, 1973-82	6.3	1.6	-0.6	2.5
GDP per head growth, % per year, 1973-82	4.8	-1.7	-3.7	1.7
Agriculture growth, % per year, 1973-82	-0.6	1.1	0.0	2.5
Industry growth, % per year, 1973-82	8.0	1.1	-1.0	2.5
Services growth, % per year, 1973-82	6.9	2.5	-0.5	2.6
Exports growth, % per year, 1973-82	10.1	0.7	-1.9	3.3
Imports growth, % per year, 1973-82	10.5	0.2	-6.9	4.3
ECONOMY: OTHER				
Inflation Rate, % per year, 1980-86	3.8	23.6	16.7	5.3
Aid, net inflow, % of GDP, 1983	5.1	11.5	6.3	-
Debt Service, % of Exports, 1984	4.3	18	20.6	-
Budget Surplus (+), Deficit (-), % of GDP, 1980	7.0	-3.0	-2.8	-5.1
EDUCATION				
Primary, % of 6-11 group, 1982	116	62	76	102
Secondary, % of 12-17 group, 1982	31	15	22	93
Higher, % of 20-24 group, 1982	16	1.2	1.9	39
Adult Literacy Rate, %, 1971	58	41	39	99
HEALTH & NUTRITION				
Life Expectancy, years, 1986	70	50	50	76
Calorie Supply, daily per head, 1983	2,351	2,111	2,096	3,357
Population per doctor, 1975	2,762	35,986	24,185	550

Notes: 'Southern Africa' and 'Africa' exclude South Africa. Dates are for the country in question, and do not always correspond with the Regional, African and Industrial averages.

SEYCHELLES: Leading Indicators

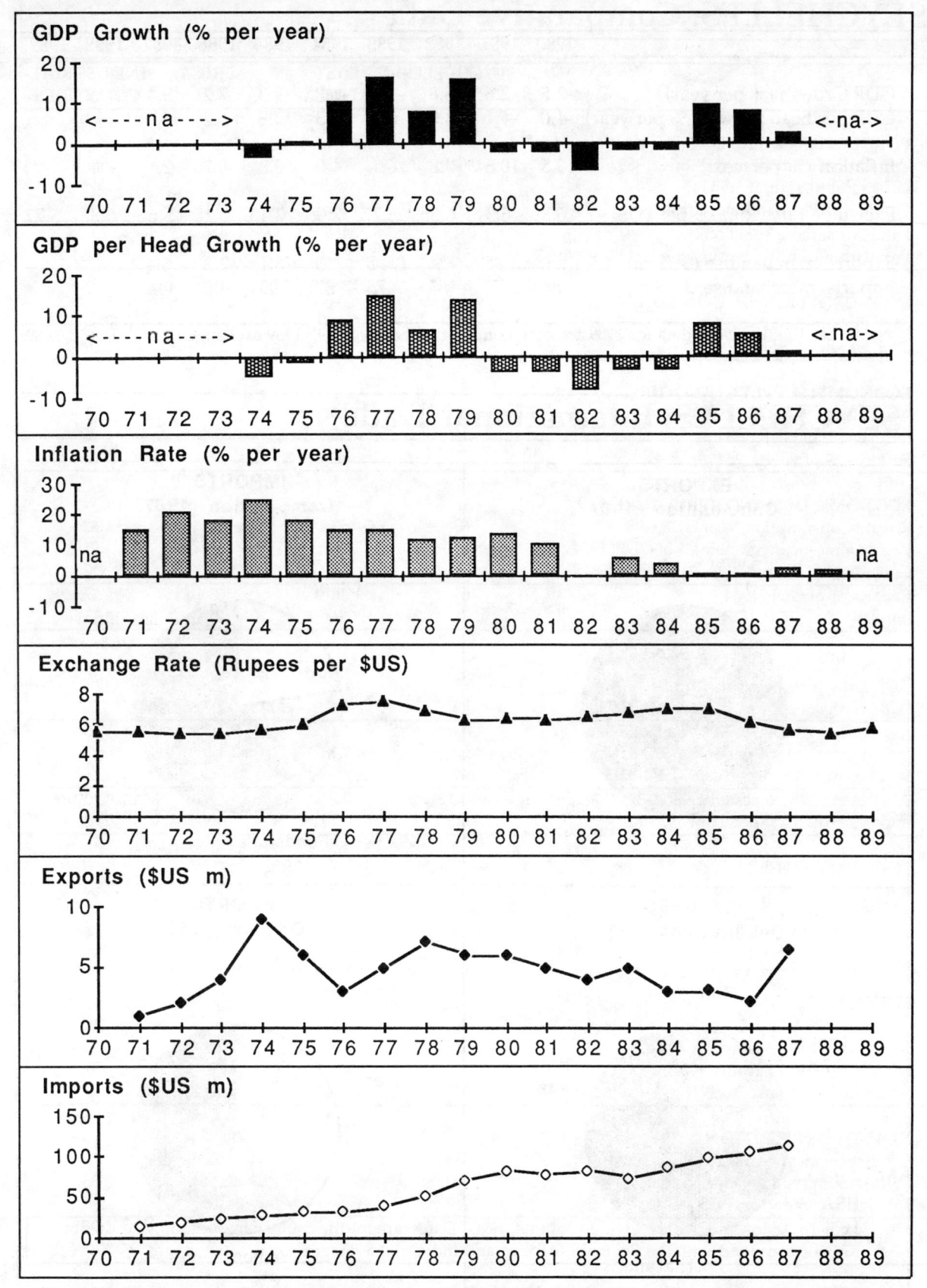

GDP Growth (% per year)

GDP per Head Growth (% per year)

Inflation Rate (% per year)

Exchange Rate (Rupees per $US)

Exports ($US m)

Imports ($US m)

Note: Leading indicator data for 1989 are based on the first half of 1989. 1989 exchange rate is for mid-1989.

SEYCHELLES: Leading Indicator Data

	1980	1981	1982	1983	1984	1985	1986	1987	1988	1989
GDP Growth (% per year)	-2.5	-2.6	-6.8	-1.9	-1.8	9.3	7.7	2.4		
GDP per head growth (% per year)	-4.0	-4.1	-8.3	-3.4	-3.3	7.8	5.2	0.9		
Inflation (% per year)	13.5	10.6	-0.8	6.1	4.0	0.8	0.7	2.5	2.1	
Exchange rate (Rupees per $US)	6.39	6.31	6.55	6.77	7.06	7.13	6.18	5.6	5.36	5.77
Exports, merchandise ($US m)	6	5	4	5	3	3.1	2.2	6.4		
Imports, merchandise ($US m)	84	79	83	75	87	99	106	114		

Note: Leading indicator data for 1989 are based on the first half of 1989. 1989 exchange rate is for mid-1989.

SEYCHELLES: International Trade

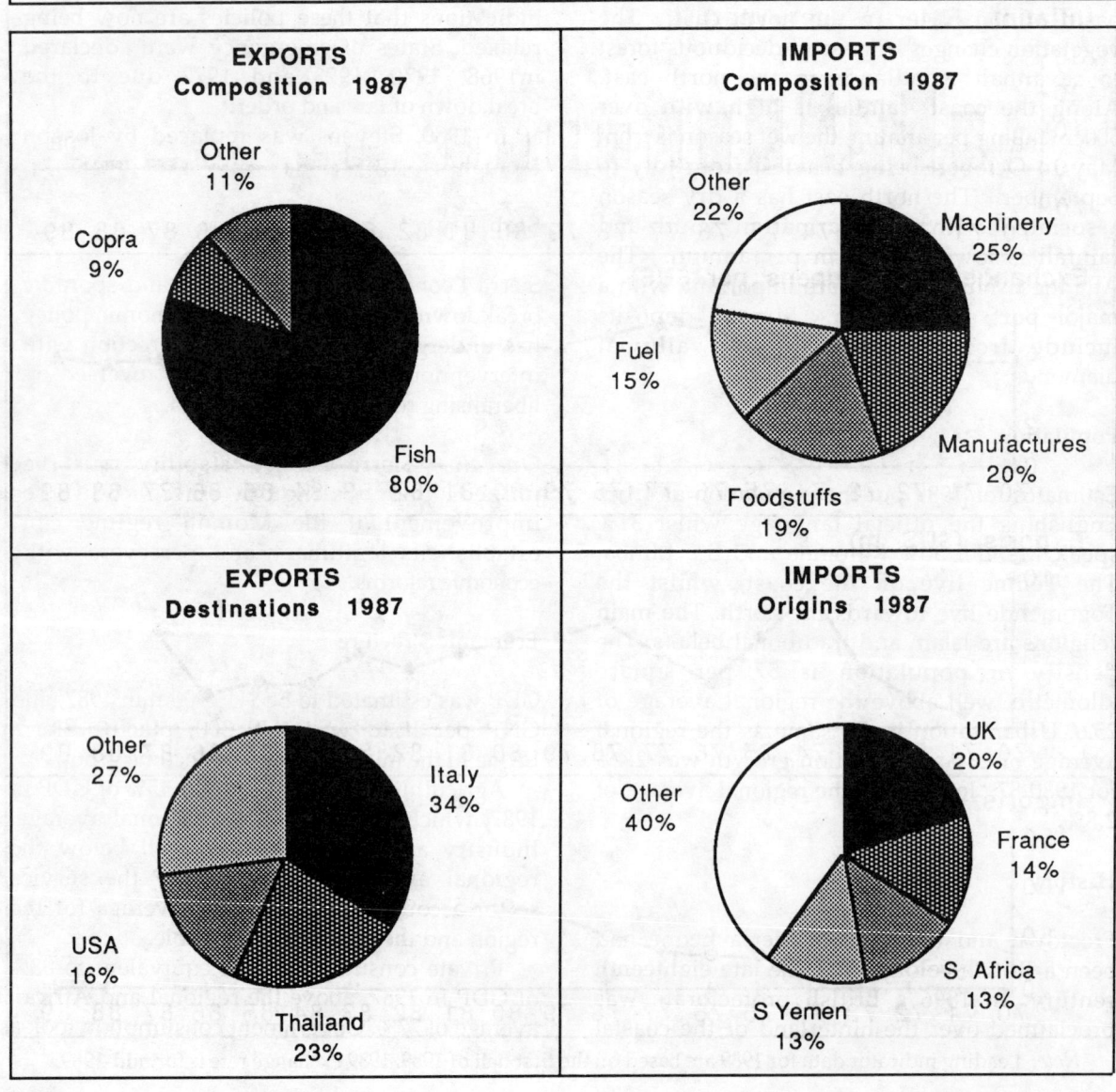

EXPORTS Composition 1987
- Fish 80%
- Other 11%
- Copra 9%

IMPORTS Composition 1987
- Machinery 25%
- Other 22%
- Manufactures 20%
- Foodstuffs 19%
- Fuel 15%

EXPORTS Destinations 1987
- Italy 34%
- Other 27%
- Thailand 23%
- USA 16%

IMPORTS Origins 1987
- Other 40%
- UK 20%
- France 14%
- S Africa 13%
- S Yemen 13%

38 SIERRA LEONE

Physical Geography and Climate

Sierra Leone is a small Atlantic coastal state bounded by Guinea and Liberia, in the West African region. The south west consists of surf beaches, whilst the north west has mangrove swamps and dense rain forest. The land rises in a series of irregular plateaux to form the base of the Futa Djallon mountains and the basin of the Niger in the north east. The vegetation changes from semi-deciduous forest to savannah woodlands in the north east. Along the coast, rainfall is high with over 500cm falling per annum, the wet season is from May to October, being heaviest from July to September. The north east has a dry season associated with the Harmattan wind and rainfall is lower at 190cm per annum. The Atlantic fishing grounds are important, with a major port at Freetown. Mineral deposits include iron ore, bauxite and alluvial diamonds.

Population

Estimates for 1989 put the population at 4.1m. English is the official language whist 31% speak Mende; 30% Kafemne and 8% Limba. The Temne live on the coast, whilst the Togomende live towards the north. The main religions are Islam and traditional beliefs. The density of population is 57 per square kilometre, well above the regional average of 25.6. Urbanisation is the same as the regional average of 23%. Population growth was 2.4% for 1980-87, lower than the regional average of 2.8%.

History

Freetown and the coast of Sierra Leone had been a British colony since the late eighteenth century. In 1896 a British protectorate was proclaimed over the hinterland of the coastal colony. Both regions united and became independent as Sierra Leone in 1961. In 1967 there was a military coup which lasted a year, following six years of civilian rule. A second coup restored the previously elected government of Siaka Stevens. In 1968 he nationalised the mining companies and intervened in the economy by means of direct controls and food subsidies. There are indications that these policies are now being relaxed. States of emergency were declared in1968, 1970, 1973 and 1977 due to the breakdown of law and order.

In 1980, Stevens was replaced by Joseph Momoh.

Stability

Sierra Leone has had two coups and sporadic breakdown in law and order. Economic policy has undergone two changes in direction with intervention under Stevens being reversed by liberalising reforms under Momoh.

Overall, Sierra Leone's stability must be judged poor, although there will be improvement if the Momoh regime can establish its legitimacy and persevere with economic reforms.

Economic Structure

GDP was estimated to be $US 900m in 1987, and GNP per head was $US 300, placing Sierra Leone in the middle of the low-income group.

Agriculture generated around 45% of GDP in 1987, which is 5% above the regional average. Industry accounted for 19%, well below the regional average of 26%, while the service sector accounts for 36%, about average for the region and the continent as a whole.

Private consumption was equivalent to 83% of GDP in 1987, above the regional and African average of 77%. Government consumption took a

Sierra Leone

further 7%, about half the regional average, while 9% went on investment, well below the regional figure of 13%. Domestic saving was equivalent to 10% of GDP.

Exports were 9% of GDP in 1987, and imports were 8%. This dependence on international trade is well below the average for Africa, although it is thought there is much smuggling of diamonds which lowers official trade figures.

Main exports are well diversified, with rutile, a mineral used in producing non-stick surfaces at 31%, diamonds 23%, bauxite at 15%, cocoa at 14% and coffee at 12%. Imports comprise the normal mix of manufactures, machinary and fuel, with the foodstuffs a significant category at 20 % of total value.

The education system in Sierra Leone is poorly developed. Adult literacy was 15% in 1975. Only 45% of primary-aged children attended school in 1982, with a further 14% of the secondary age group in full-time education and 1% of 20-24 year olds in higher education.

Life expectancy in 1987 was only 41 years while the average for Africa and the region is 50 years. However, there are better than average ratios of doctors and nurses per thousand people than regionally. Average daily calorie supply in 1986 was 1,855, which is 13% below the average in Africa.

Overall, Sierra Leone is a low-income country with a high proportion of GDP generated by agriculture. Exports are well diversified, and smuggling obscures the actual reliance on international trade. Educational provision is poor as are nutritional standards, although there is a good provision of health care.

Economic Performance

GDP grew only marginally from 1980-87, by 0.7%, though this is better than the regional average of -2.0%. This implied a fall in GDP per head of -1.7% a year. The agricultural sector, which generates 45% of GDP, grew by 1.6%, similar to the regional average of 1.3%. The service sector grew by 1.3% while in the region as a whole this sector declined by -2.5%. However, the industrial sector declined by -2.3% (though the regional decline was -3.6%).

Exports declined by -2.1%, though regional export decline was even greater at -4.0%. This was partly due to widespread smuggling of diamonds which has now led to greater government control over cross-border movements. Government measures to reduce the country's reliance on imports has led to an overall decline of -15.1% a year, which is better than the regional and African average reductions of -12.9% and -6.9% respectively.

Inflation continued to rise between 1980 and 1987 at an average of 50%, well above the regional and African averages of 13.0% and 16.7% respectively.

Overall, Sierra Leone's economic performance has been poor with marginal rises in GDP, but falls in GDP per head. Agriculture and services have shown positive growth, but industrial output has declined. Export performance has been poor, and inflation has been disconcertingly high.

Recent Economic Developments

President Momoh introduced a programme of liberalising economic reforms in 1986, together with an IMF agreement. However, IMF and World Bank support lapsed because of repayment arrears in 1988.

Inflation has accelerated significantly in the latter part of the last decade, from around 11% a year in 1980 to 65% a year in 1989.

The exchange rate has undergone relentless depreciation from L 1.05 = $US 1 in 1980 to L 63.30 = $US 1 at the end of 1989.

Merchandise export revenues peaked in 1986 at $US 232m, but by 1989 they had fallen to $US 110m. Merchandise imports reached their peak in 1980 at $US 386m but had fallen to under half that value by 1989.

Economic Outlook and Forecasts

The forecasts assume the Momoh government remains in power and gradually improves stability while cautiously implementing liberalising economic reforms.

The price of bauxite is expected to rise in real terms by 11% up to 1995, real cocoa prices are expected to fall by -19%, and real coffee prices to fall by -8%.

Aid flows are expected to be stagnant up to 1992, but to improve thereafter with a new agreement with the IMF.

GDP growth is forecast to be slightly higher,

Sierra Leone

but this, as are the rest of the projections, is critically dependent on an IMF agreement and renewed aid flows. Nevertheless, GDP per head will continue to decline.

Export volumes are expected to continue to decline, but at a slower rate. Import volumes are expected to decline over the first half of the forecast period, but to recover in the second half to give an overall expansion of around 1% to 1.5%.

Sierra Leone: Economic Forecasts
(average annual percentage change)

	Actual 1980-87	Forecast Base 1990-95	Forecast High 1990-95
GDP	0.7	1.0	1.2
GDP per Head	-1.7	-1.6	-1.4
Inflation Rate	50.0	50.0	40.0
Export Volumes	-2.1	-1.0	-0.8
Import Volumes	-15.1	1.0	1.5

SIERRA LEONE: Comparative Data

	SIERRA LEONE	WEST AFRICA	AFRICA	INDUSTRIAL COUNTRIES
POPULATION & LAND				
Population, mid year, millions, 1989	4.1	12.3	10.2	40.0
Urban Population, %, 1985	25	28.7	30	75
Population Growth Rate, % per year, 1980-86	2.4	3.2	3.1	0.8
Land Area, thou. sq. kilom.	72	384	486	1,628
Population Density, persons per sq kilom., 1988	55.6	31.1	20.4	24.3
ECONOMY: PRODUCTION & INCOME				
GDP, $US millions, 1986	1,180	4,876	3,561	550,099
GNP per head, $US, 1986	310	510	389	12,960
ECONOMY: SUPPLY STRUCTURE				
Agriculture, % of GDP, 1986	45	40	35	3
Industry, % of GDP, 1986	22	26	27	35
Services, % of GDP, 1986	33	34	38	61
ECONOMY: DEMAND STRUCTURE				
Private Consumption, % of GDP, 1984	86	77	73	62
Gross Domestic Investment, % of GDP, 1984	9	13	16	21
Government Consumption, % of GDP, 1984	7	13	14	17
Exports, % of GDP, 1984	17	18	23	17
Imports, % of GDP, 1984	19	20	26	17
ECONOMY: PERFORMANCE				
GDP growth, % per year, 1980-86	0.4	-2.03	-0.6	2.5
GDP per head growth, % per year, 1980-86	-2.0	-4.7	-3.7	1.7
Agriculture growth, % per year, 1980-86	0.5	1.3	0.0	2.5
Industry growth, % per year, 1980-86	-2.4	-3.6	-1.0	2.5
Services growth, % per year, 1980-86	1.5	-2.5	-0.5	2.6
Exports growth, % per year, 1980-86	-3.1	-4.0	-1.9	3.3
Imports growth, % per year, 1980-86	-16.5	-12.9	-6.9	4.3
ECONOMY: OTHER				
Inflation Rate, % per year, 1980-86	33.5	13.0	16.7	5.3
Aid, net inflow, % of GDP, 1986	7.0	3.7	6.3	-
Debt Service, % of Exports, 1986	8.2	21.4	20.6	-
Budget Surplus (+), Deficit (-), % of GDP, 1986	-8.9	-3.4	-2.8	-5.1
EDUCATION				
Primary, % of 6-11 group, 1982	45	75	76	102
Secondary, % of 12-17 group, 1982	14	24	22	93
Higher, % of 20-24 group, 1982	1	2.6	1.9	39
Adult Literacy Rate, %, 1975	15	30	39	99
HEALTH & NUTRITION				
Life Expectancy, years, 1986	41	50	50	76
Calorie Supply, daily per head, 1985	1,784	2,105	2,096	3,357
Population per doctor, 1981	19,130	16,199	24,185	550

Notes: 'Southern Africa' and 'Africa' exclude South Africa. Dates are for the country in question, and do not always correspond with the Regional, African and Industrial averages.

SIERRA LEONE: Leading Indicators

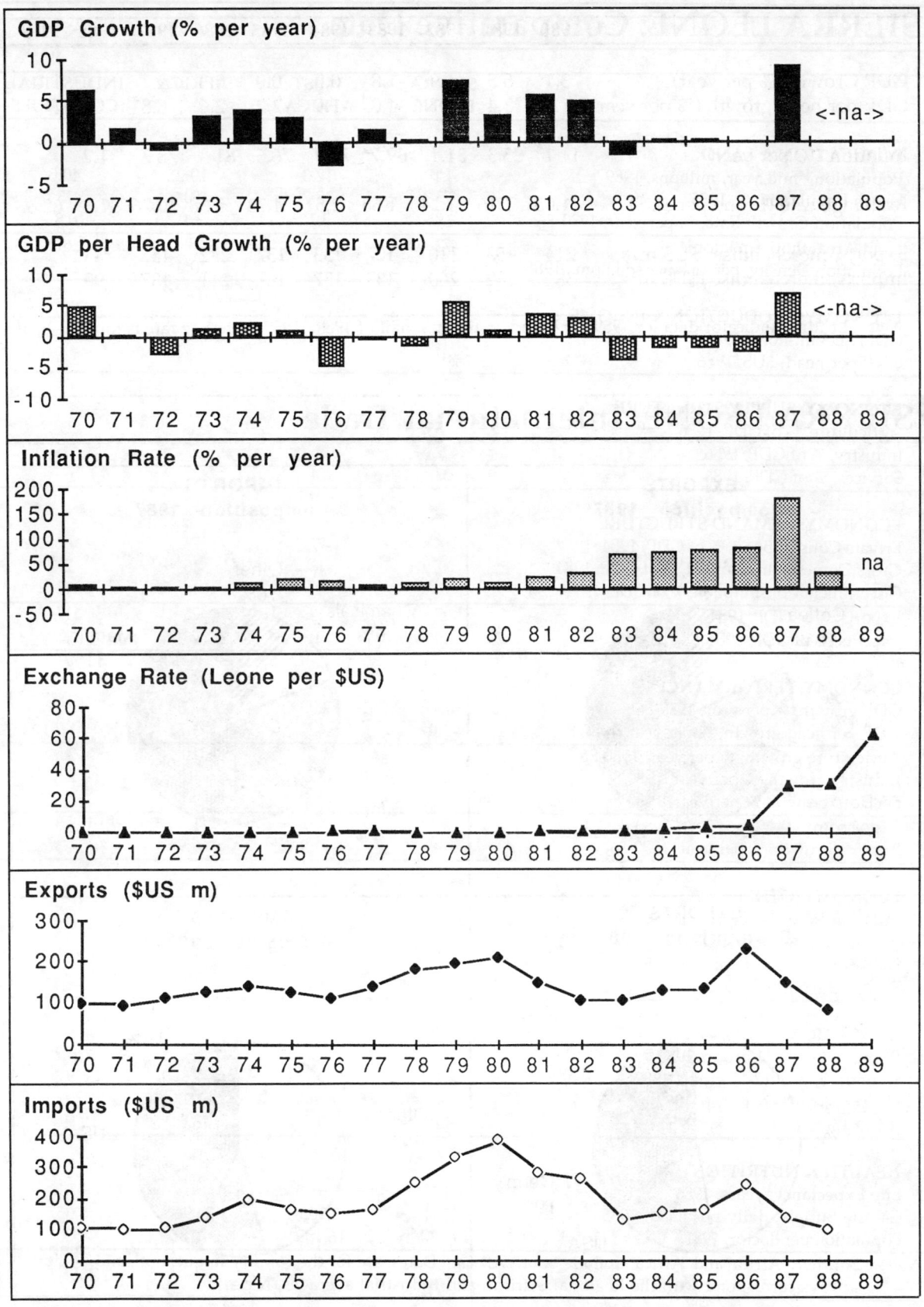

GDP Growth (% per year)

GDP per Head Growth (% per year)

Inflation Rate (% per year)

Exchange Rate (Leone per $US)

Exports ($US m)

Imports ($US m)

Note: Leading indicator data for 1989 are based on the first half of 1989. 1989 exchange rate is for mid-1989.

SIERRA LEONE: Leading Indicator Data

	1980	1981	1982	1983	1984	1985	1986	1987	1988	1989
GDP Growth (% per year)	3.1	5.5	4.9	-1.8	0.3	0.0	-0.3	9.1		
GDP per head growth (% per year)	1.0	3.4	2.7	-4.0	-1.9	-2.2	-2.6	6.8		
Inflation (% per year)	11.1	23.3	31.1	69.7*	72.9	76.6	81	178.7	31.2	
Exchange rate (Leone per $US)	1.05	1.16	1.24	1.68	2.51	4.73	5.08	30.76	31.3	62.8
Exports, merchandise ($US m)	214	153	110	107	133	137	232	152	83	
Imports, merchandise ($US m)	386	282	260	133	157	167	241	137	99	

Note: Leading indicator data for 1989 are based on the first half of 1989. 1989 exchange rate is for mid-1989.

SIERRA LEONE: International Trade

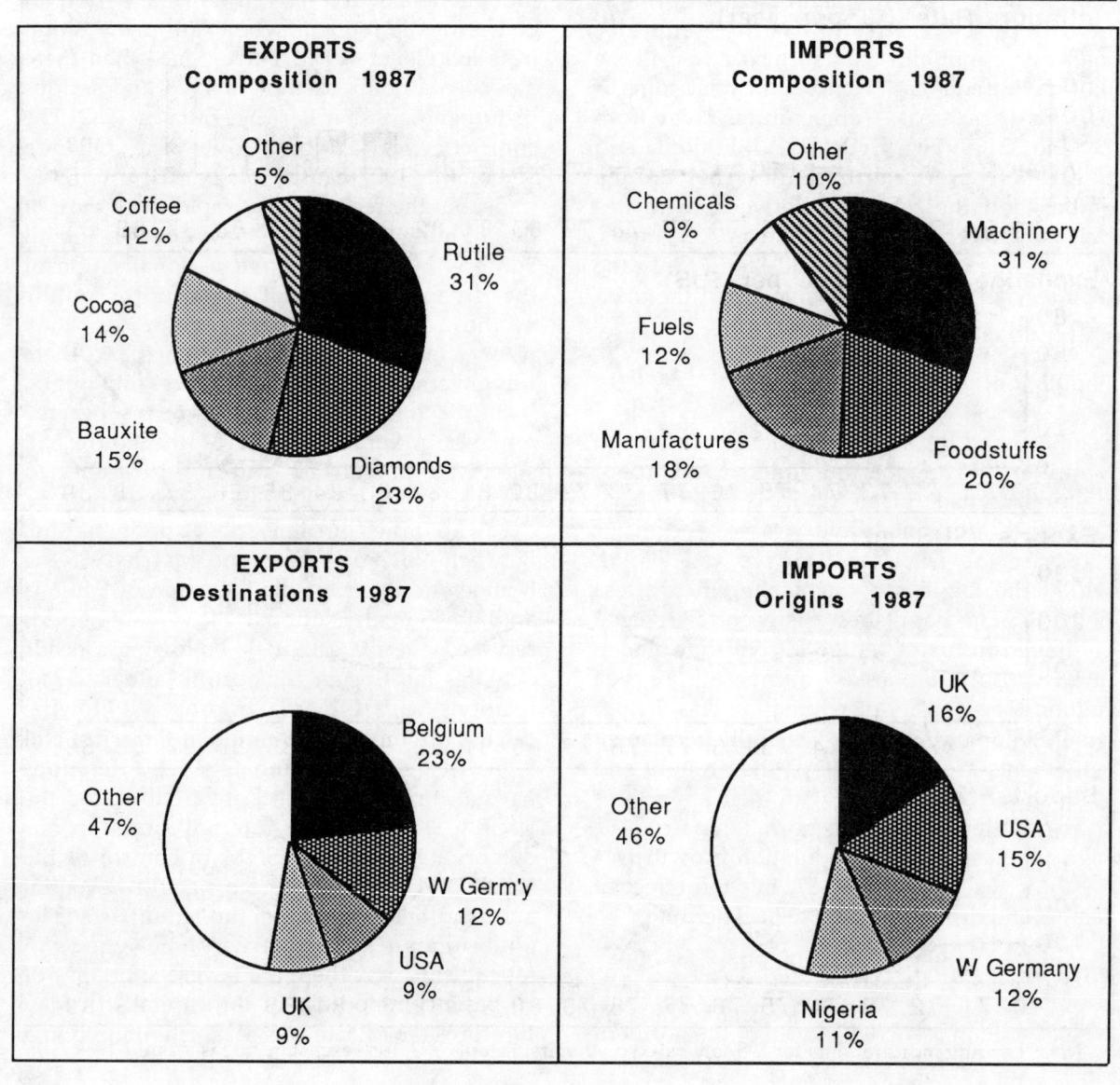

EXPORTS Composition 1987
- Rutile 31%
- Diamonds 23%
- Bauxite 15%
- Cocoa 14%
- Coffee 12%
- Other 5%

IMPORTS Composition 1987
- Machinery 31%
- Foodstuffs 20%
- Manufactures 18%
- Fuels 12%
- Chemicals 9%
- Other 10%

EXPORTS Destinations 1987
- Belgium 23%
- W Germ'y 12%
- USA 9%
- UK 9%
- Other 47%

IMPORTS Origins 1987
- UK 16%
- USA 15%
- W Germany 12%
- Nigeria 11%
- Other 46%

39 SOMALIA

Physical Geography and Climate

Somalia is a large coastal state, in the East African region, situated on the north-east 'Horn of Africa' facing the Arabian peninsular and Indian Ocean and bounded by Kenya, Ethiopia and Djibouti. It consists of dry savanna plains with mountains towards the north coast. The temperature is hot, but modified by altitude with generally inadequate rainfall - only 50cm per year in the most favoured areas. Hence it includes some of Africa's driest areas. Temperatures range from 27°C to 32°C. Two rivers rise in Ethiopia and bisect the country north/south. They are the Shibeli and the Juba and between them lies rich irrigated agricultural land and pastures. Beyond this is more marginal land. Only the Juba reaches the sea at Kisimayu, but the main port is Mogadishu.

Population

Estimates put the size of population for 1989 at 6 million. The population is made up of pastoral nomads of Bantu origin. The official languages are Somali and Arabic. The major clans are the Dir, Isag, Hawiye and Darod, whilst the Digil and the Rahanwin are less nomadic. The population has been raised by a continued influx of refugees from the Ogaden and surrounding areas, numbering over a million. The density of population was 9.4 per square kilometre in 1989. Generally permanent settlements are scattered, but the drought and ensuing refugees staying with their kin groups have swollen the urban population to 34%, high for East Africa. Population growth was 2.9% for the period 1980-87, but the effect of famine and drought remain to be determined.

History

In 1886 Britain assumed control of the northern area of Somalia in order to ensure a supply of mutton to her colony of Aden and encouraged the Italians to occupy the southern regions to deny access to the French. Between 1900 and 1920 the Somalis waged wars for independence against the British, French and Italian colonists. In 1950 Italy returned Somalia, as trusteeship authority, to the UN prior to independence in 1960. In 1969 a period of civilian rule ended when an army coup installed General Said Barre. Since then there has been a coup attempt in 1978 and serious disturbances in northern Somalia in 1982. The conflict with Ethiopia over the Ogaden Somalis in 1977 caused major disruptions to trade, but the recovery was rapid since most of the fighting took place beyond the Somali borders. In 1977 extensive nationalisation of the financial, medical, transport, public utilities, educational and trade sectors took place. Land became publicly owned and there was government regulation of prices and wages.

In 1979-80 a million Ethiopeans became refugees in Somalia. By 1981 the figure had stabilised at 700,000 and by 1984-85 had reduced to 107,000 but represented an overwhelming burden for a poor nation. Between November 1984 and March 1985 the Democratic Front for the Salvation of Somalia and the Somali National Movement (SNM) resumed guerilla activities. Following a heads of states meeting in Djibouti in January 1986, relations with Ethiopia improved following discussions on the prevention of terrorism and Somalia's claims to the Ogaden. Relations have also resumed with Kenya, Libya and the USSR. However, the internal situation has deteriorated, with the SNM in control of the North, and other rebel groups active in central and southern parts of the country. The government is effectively besieged in Mogadishu, and there are severe shortages of all basic commodities in the capital. In May 1986 President Barre was seriously injured in a

Somalia

road accident near Mogadishu and Lt. General Samater became acting President. Despite rumours of poor health, Barre has remained in control.

Stability

Since independence in 1960, Somalia had nine years of civilian government before a military coup brought Said Barre to power in 1969. There was a coup attempt in 1978, and a mutiny and riots in 1982. A series of disputes with Kenya and Ethiopia over territory occupied by Somalis erupted in 1977 into an armed conflict with Somalia over the Ogaden. Currently there is civil war with rebel groups controlling most of the country.

Economic strategy underwent a radical change in 1975 when public ownership was widely extended, land was nationalised and wages and prices set by the government. A reform programme introduced in 1980 with IMF support reversed some of these measures.

Overall, Somalia has a very poor stability record. Although the reform programme had improved the atmosphere for business and foreign investment, there can be no economic progress while the current civil conflicts continue. Said Barre is now over 70 years old, and was injured in a 1986 road accident. There is considerable uncertainty concerning the succession.

Economic Structure

GDP in 1987 was estimated at $US 1,890m, and GNP per head was $US 290, placing Somalia in the low-income group.

The agriculture sector generated 65% of GDP in 1987, with industry generating 9% and services 26%. The contribution of the agriculture sector is over twice the average for low-income countries, and the industry and service sector shares are correspondingly low.

On the demand side, private consumption was equivalent to 89% of GDP in 1987, with government consumption 11%, and domestic saving about 1% of GDP. However, gross domestic investment was 35% of GDP, made possible by substantial financial transfers from overseas equivalent to 57% of GNP in 1987. This investment rate is over twice that observed on average elsewhere in Africa.

Exports of goods and services comprised 11% of GDP in 1987, and imports were 45% of GDP. The main exports in 1986 were livestock 60%, bananas 12% and hides 3%. In the import category, machinery comprised 28%, foodstuffs 20% and fuel 4%.

The adult literacy rate in 1979 was 39%. Primary school enrolment rates were 20% in 1985, while secondary enrolments were 15%, with 4% enrolment in higher education. These figures indicate good educational provision for one fifth of the population and very little for the rest.

Life expectancy was 47 years in 1987, below the African average of 51.

There was one doctor for every 16,090 persons and one nurse for every 1,530, and these provisions are rather better than general Africa provision, although it is to be expected that distribution of medical care is very unequal. Average daily calorie supply was 2,138, about the African average.

Overall, Somalia is a low-income country with heavy dependence on agriculture to generate GDP and export revenues. There is heavy reliance on overseas aid to finance imports. Educational provision is poor and unequally distributed as are health and nutrition.

Economic Performance

GDP grew by 2.2% from 1980-87, which is better than the regional average of 1.6%. Nevertheless, GDP per head fell by -0.7% a year. Despite serious drought and some exporting difficulties, the agricultural sector grew by 2.8% while the regional average was only 1.1%. Industrial growth of 1.0% was on a par with the regional average, though 0.9% growth in the service sector was below the regional average of 2.5%.

Export volumes declined rapidly by -7.7% a year as drought and import restrictions by Saudi Arabia, a major livestock buyer, reduced the number and quality of animals sold. In Africa exports declined by only -1.9%, while regionally they increased marginally by 0.7%. Import volumes fell by -1.3% a year, compared with the regional average which increased by 0.2%.

Somalia

Inflation averaged 37.8% for 1980-87. Although it reached over 90% in 1984, it was brought back down to a more manageable 28% in 1987, though it rose again to 63% in 1988. By conparison, the average rate of inflation in the region for 1980-87 was 23.6%, and the whole of Africa 16.7%.

Overall, Somalia has had poor economic performance in the 1980s, when, despite increases in GDP, GDP per head fell. All sectors grew more slowly than population growth, exports and imports declined and inflation has been high.

Recent Economic Developments

In 1980 Somalia began to introduce reforms which reversed some of the measures dating from 1975 which extended government ownership and control in the economy. The currency was devalued, though not by as much as was urged by the IMF. In 1985 the government agreed to liberalise imports and dismantle controls in agriculture and marketing. In September 1986, an auction system for foreign exchange was introduced, but by early 1987 there were reservations that the implementation of reforms had been slow. In June it was announced that the government would allow private traders in hides and skins, but would retain its monopoly over the export of these goods. In August it was announced that the state monopoly of banking would be ended. In July the IMF agreed $US 70m of programme lending to support the reform programme. Agriculture was hit by drought when the early rains of 1987 failed. High domestic prices and shortages of petrol are thought to have provoked a coup attempt in July.

Somalia is estimated to have $US 2,700m outstanding in external debt in 1989, most of it to Arab creditors, which has created problems over rescheduling. The debt-service payments are clearly beyond Somalia's capacity and even after a Paris Club rescheduling were estimated to be $US 138m in 1987, rising to $US 146m in 1988. After the July 1987 Paris Club rescheduling, it was hoped that Arab creditors would reschedule on similar terms. At present levels, debt servicing will take up all of Somalia's earnings from exports of goods and services, and further enforced reschedulings are inevitable.

Somalia's exchange rate remained virtually unchanged up to 1981, but steadily depreciated from the SS 6.30 = $US 1 of 1981 to SS 408.1 = $US 1 at the end of 1989.

Aid projects have concentrated on infrastructure and agriculture. The World Bank has loaned $US 24.4m for ports and $US 26.1m for nomad settlement and agricultural projects. The African Development Bank is allocating $US 9m to agricultural extension, and the EEC $US 2.1m to hospitals. Several oil companies are engaged in oil exploration, both on- and off-shore.

Economic Outlook and Forecasts

The forecasts assume that Somalia's seemingly intractable internal instability continues, and that the donors continue to be reluctant to extend aid to the Barre regime.

No increase in GDP is expected, and a steady slide in GDP per head is in prospect. Export volumes are projected to continue falling, and poor export earnings and reduced aid flows are expected to lead to faster falling import volumes.

Lower foreign funding is expected to increase inflationary pressure, with the annual rate of price increase accelerating to around 50%.

Somalia: Economic Forecasts
(average annual percentage change)

	Actual	Forecast Base	High
	1980-87	1990-95	1990-95
GDP	2.2	-0.5	0.0
GDP per Head	-0.7	-3.5	-3.0
Inflation Rate	37.8	50.0	45.0
Export Volumes	-7.7	-5.0	-4.0
Import Volumes	-1.3	-4.0	-3.0

SOMALIA: Comparative Data

	SOMALIA	EAST AFRICA	AFRICA	INDUSTRIAL COUNTRIES
POPULATION & LAND				
Population, mid year, millions, 1989	6.0	12.2	10.2	40.0
Urban Population, %, 1985	34	30.5	30	75
Population Growth Rate, % per year, 1980-86	2.9	3.1	3.1	0.8
Land Area, thou. sq. kilom.	638	486	486	1,628
Population Density, persons per sq kilom., 1988	9.1	24.2	20.4	24.3
ECONOMY: PRODUCTION & INCOME				
GDP, $US millions, 1985	2,320	2,650	3,561	550,099
GNP per head, $US, 1986	280	250	389	12,960
ECONOMY: SUPPLY STRUCTURE				
Agriculture, % of GDP, 1985	58	43	35	3
Industry, % of GDP, 1985	9	15	27	35
Services, % of GDP, 1985	34	42	38	61
ECONOMY: DEMAND STRUCTURE				
Private Consumption, % of GDP, 1985	93	77	73	62
Gross Domestic Investment, % of GDP, 1985	15	16	16	21
Government Consumption, % of GDP, 1985	12	15	14	17
Exports, % of GDP, 1985	7	16	23	17
Imports, % of GDP, 1985	28	24	26	17
ECONOMY: PERFORMANCE				
GDP growth, % per year, 1980-85	4.9	1.6	-0.6	2.5
GDP per head growth, % per year, 1980-85	1.9	-1.7	-3.7	1.7
Agriculture growth, % per year, 1980-85	7.9	1.1	0.0	2.5
Industry growth, % per year, 1980-85	-5.1	1.1	-1.0	2.5
Services growth, % per year, 1980-85	3.6	2.5	-0.5	2.6
Exports growth, % per year, 1980-85	-7.9	0.7	-1.9	3.3
Imports growth, % per year, 1980-85	-1.7	0.2	-6.9	4.3
ECONOMY: OTHER				
Inflation Rate, % per year, 1980-85	45.4	23.6	16.7	5.3
Aid, net inflow, % of GDP, 1986	27.8	11.5	6.3	-
Debt Service, % of Exports, 1986	62.1	18	20.6	-
Budget Surplus (+), Deficit (-), % of GDP, 1981	1.1	-3.0	-2.8	-5.1
EDUCATION				
Primary, % of 6-11 group, 1984	25	62	76	102
Secondary, % of 12-17 group, 1984	17	15	22	93
Higher, % of 20-24 group, 1984	4	1.2	1.9	39
Adult Literacy Rate, %, 1979	39	41	39	99
HEALTH & NUTRITION				
Life Expectancy, years, 1986	47	50	50	76
Calorie Supply, daily per head, 1985	2,074	2,111	2,096	3,357
Population per doctor, 1981	17,460	35,986	24,185	550

Notes: 'Southern Africa' and 'Africa' exclude South Africa. Dates are for the country in question, and do not always correspond with the Regional, African and Industrial averages.

SOMALIA: Leading Indicators

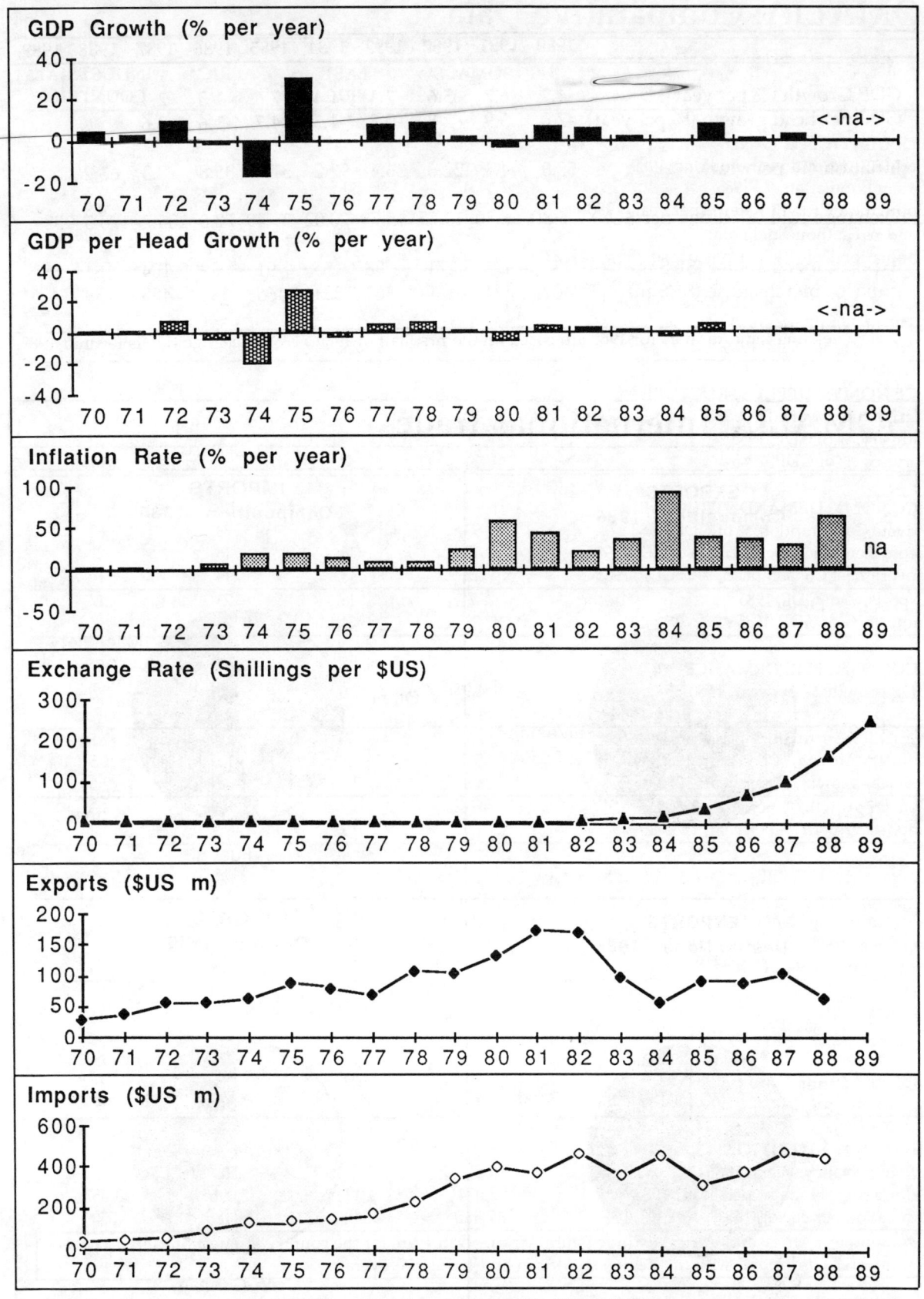

Note: Leading indicator data for 1989 are based on the first half of 1989. 1989 exchange rate is for mid-1989.

SOMALIA: Leading Indicator Data

	1980	1981	1982	1983	1984	1985	1986	1987	1988	1989
GDP Growth (% per year)	-4.2	6.7	5.6	2.1	-1.4	7.6	1.3	2.4		
GDP per head growth (% per year)	-7.0	3.9	2.8	-0.8	-4.3	4.7	-1.6	-0.6		
Inflation (% per year)	58.8	44.4	22.6	36.4	92.2	37.8	35.8	28.3	63.2	
Exchange rate (Shillings per $US)	6.30	6.30	10.75	15.79	20.02	39.49	72.0	105.2	170.5	252
Exports, merchandise ($US m)	133	175	171	98	55	91	89	104	64	
Imports, merchandise ($US m)	402	371	471	362	331	466	383	485	458	

Note: Leading indicator data for 1989 are based on the first half of 1989. 1989 exchange rate is for mid-1989.

SOMALIA: International Trade

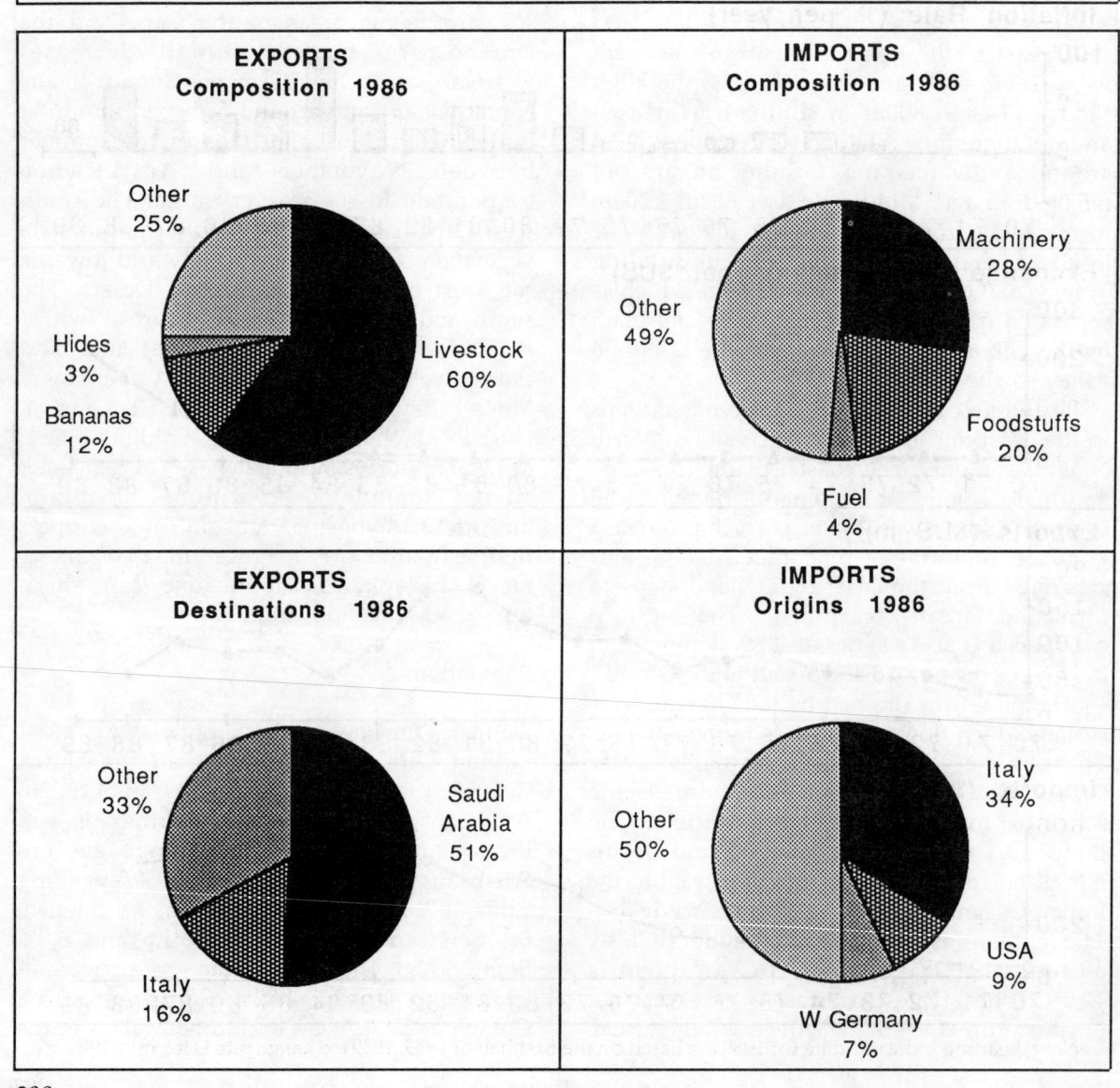

EXPORTS Composition 1986
- Livestock 60%
- Other 25%
- Bananas 12%
- Hides 3%

IMPORTS Composition 1986
- Other 49%
- Machinery 28%
- Foodstuffs 20%
- Fuel 4%

EXPORTS Destinations 1986
- Saudi Arabia 51%
- Other 33%
- Italy 16%

IMPORTS Origins 1986
- Other 50%
- Italy 34%
- USA 9%
- W Germany 7%

40 SOUTH AFRICA

Physical Geography and Climate

South Africa is a country with a long coastline, lying mostly south of the Tropic of Capricorn with access to both the Atlantic and Indian Oceans. It borders Namibia, Botswana, Zimbabwe, Mozambique and Swaziland. It consists of an escarpment bounding a vast plateau which is mostly at 900m but rises to 2,000m. The plateau dips from the east and south east to the Kalahari Basin in the north west. There are three main regions: the High Veld at 1,200-1,800m in southern Transvaal and Orange Free State; an area at 1,500m around Witwatersrand forming an area of major drainage; Middle Veld at 600m-1,200m forming the remainder of the plateau. The lowland margin includes the undulating Transvaal Low Veld at 150m-600m which is separated from the Mozambique coastal plain by the Lebombo Mountains and the Limpopo valley to the east.

The south east coastal belt descends in steps to the coast and is cut by river valleys. North Natal is the only true coastal plain. The Cape mountain ranges are remnants flanking the south and south-west plateau which produce a knot of mountains reaching 2,000m and separated from the coast by lowland steps to form the Great Escarpment. The western coastal belt is also a series of steps sloping from the foot of the Great Escarpment to the coast.

The plateau is drained by the Orange River system which flows westwards to the Atlantic. The Orange River Project regulates water supply providing irrigation and hydro-electric schemes to aid water conservation. The Limpopo rises in Witwatersrand and drains north and centrally across the Transvaal to the Indian Ocean. None of the rivers is navigable. The climate is sub-tropical, moderated by altitude, i.e. Cape Town 16.7°C, Pretoria 17.2°C. The greatest contrast is east-west not north-south, due to the effects of the warm Mozambique current in the west and the cold Benguela current in the east. In Durban, on the east coast, the average temperature for January is 24.4°C compared to Port Nolloth which averages 15.6°C.

The range of temperatures increase as one moves away from the coast. Rainfall is above 65cm per annum in the east and southern Cape Province; rising to 150cm per annum in the seaward facing areas of the Cape and the Drakensberg ranges; rainfall decreases westwards to below 25cm per annum in the Namib Desert and around 5cm per annum at Port Nolloth. Most rain falls in the summer between November and April when evaporation losses are very high. The south west Cape has a shrubby Mediterranean style vegetation becoming increasingly arid towards the west ending in the Namib Desert. The south and east coasts include forests whilst most of the rest of the land is grassland and thorny veld. Harbours include Cape Town, Mossel Bay, Port Elizabeth, East London, Durban, Richard's Bay and Saldanha Bay. Mineral resources include deposits of coal, gold, silver, diamonds, platinum, uranium, antimony, asbestos, chromium, copper, fluorspar, iron ore, magnesium, manganese, nickel, phosphates, kaolin, lime, salt, silica, tin, vanadium, and natural gas.

Population

Estimates for 1989 suggest a population of 34.4 million with 2.4m belonging to the Dutch Reformed Church; 2.4m Roman Catholics; 2m Anglicans; 0.8m Evangelical Lutherans and local churches. The official languages are Afrikaans and English with 19% speaking Zulu; 18% Afrikaans; 18% Xhosa; 9% English; 8% Setswana; 7.5% North Sotho and 6.8% South Sotho. The main ethnic groups were in 1980 68% African; 18.2% European; 10.5% Coloured; 3.3% Asian (including Malays in the

South Africa

Cape). The main groups in the Cape when the first settlers arrived were the Khoisan: Bushmen, Hottentot and Borgdamara; with Nguni: Zulu, Swazi, Ndebele, Pondo, Tembu, Xhosa further inland. European settlers included : Dutch, French Huguenots, Germans (the nationalities from which the Afrikaners descended) and British. Other racial groups today include: Cape coloureds, Indians and coloureds in the Cape, and around Natal and Witwatersrand. There are also the so-called 'homelands' of Transkei, Ciskei, Venda and Bophuthatswana whose populations are mainly black and have grown rapidly as the South African government has forcibly 'repatriated' people to their 'place of orgin', putting considerable pressure on the limited resources of these 'homelands'. The density of population is 28 persons per square kilometre which is considerably above the regional average of 12.7 persons per square kilometre. Urbanisation is 57% which is more than double the regional average of 24.1%. The rate of population growth was 2.3%, 1980-87, below the regional average of 3.3%.

History

The Dutch established a supply station at Table Bay in 1652 to serve the East India trade and their descendants formed the Afrikaners. In 1814 the Cape came under British control, provoking opposition from the Afrikaners. In 1838 the Boers (Afrikaners) made the 'Great Trek' north from the Cape over the Great Escarpment to the veld. In 1824 a new British settlement was founded at Natal. Some of the Trek Boers settled in North Natal, whilst the remainder crossed the River Vaal, defeated the Zulus and in 1852 formed the South African Republic (Transvaal). In 1854 the Orange Free State was established.

The discovery of diamonds at Kimberley led to non-Afrikaner immigration and the friction resulted in the Boer War of 1899-1902. The Republics were annexed and in 1910 the four colonies became the Union of South Africa. Namibia (South West Africa) was mandated to South Africa in 1919. After a referendum among the white electorate in 1960, South Africa became a republic and left the Commonwealth. During the 1970s and 1980s tribal homelands were established, but these were not recognised by the UN.

Up to 1990 South Africa had a secure internal regime based on minority white government. Severe restrictions (the apartheid system) were imposed on the non-white population (80% of the total). The African National Congress (ANC) is the main black political group, but was banned in South Africa. Nelson Mandela and other senior ANC leaders were detained. Disruption by armed dissidents pressing for majority rule was contained. The economy has maintained a vigorous private sector with limited intervention and withstood the effects of sanctions by other nations and alterations to trading patterns resulting from changes of regime in neighbouring states. Both efficient agricultural production and extensive mineral wealth contribute to a diversified economy. In September 1986 the EEC failed to agree to a ban on South African coal imports, whilst President Reagan vetoed a bill of economic sanctions passed by the US Senate and House of Representatives. Even so many US multinationals technically withdrew from the region by means of local management buy-outs.

A nationwide state of emergency was imposed in June 1986 and was renewed for the fourth year running in 1989. During this period, minor reforms were made to the apartheid restrictions.

F. W. de Klerk became president in late 1989, and surprised the world early in 1990 by announcing that it was his government's intention to move toward majority rule. The ban on the ANC was lifted, and Nelson Mandela was released.

Stability

Since 1910 South Africa has existed as an independent state with political control in the hands of the white minority. Increasingly there has been pressure for majority rule from armed dissidents operating mostly from neighbouring countries. South Africa has found itself engaged in supporting military activities to destabilise the regimes of southern African countries supporting her opponents. Internal rioting intensified in the late 1980s, and emergency measures were introduced.

Economic strategy has relied upon a strong

private sector with considerable overseas business participation and investment. In the last few years, however, there has been pressure for firms to divest themselves of holdings, and several key American and British companies have done so.

Overall, South Africa's stability has been very good until comparatively recently, and the regime and the economy have been robust enough to survive internal opposition and external disapproval. However, opposition to the government's internal policies escalated to a degree that seriously impaired South Africa's stability, and provided severe discouragement to business expansion and investment in the republic. Majority rule holds out the prospect of solving some of these stability difficulties. There is much uncertainty over the negotiations in prospect, and it is possible that other stability problems will arise as the result of conflicts between factions in the black majority.

Economic Structure

South Africa has by far the largest economy in sub-Saharan Africa with a GDP of $74,260m in 1987. The standard of living on average was higher than any country in Africa other than Gabon at $US 1,890 per head, though this figure fails to take into account the large disparity in incomes (including differentiated wage structures) between races. However, black South African workers have better living standards than Africans in many other countries and as a result migrant workers are attracted from most neighbouring countries and considerable numbers of Ugandans, Zaïreans, Ghanaians and other Africans (often professionals) are now working in South Africa and its 'independent homelands', despite their governments' opposition.

In 1987 only 6% of GDP was generated by the agricultural sector which has declined in relative importance in the last few years, especially during the drought of 1983-86. Compared to the rest of Africa, where dependence on this sector is often above 50%, South Africa's economy is not largely dependent on agriculture but it should be noted that except during the recent drought South Africa has been a net exporter of food. The industrial sector accounts for 44% of GDP and only Botswana (with its large mining industry) has a larger contribution from this sector in Africa. However, it should be noted for comparison that although some other countries, such as Gabon and Nigeria also have large industrial contributions to the economy, this is largely from oil. South Africa is heavily dependent on gold, and has the largest and most advanced manufacturing sector in Africa. The service sector is also large, generating 50% of GDP.

Private consumption was equivalent to 53% of GDP in 1987, and only Gabon in Africa has a lower commitment to consumption. Government consumption is equivalent to 19% of GDP, and this is high compared with Africa's average of 16%. The level of investment was high in 1987 at 20% of GDP, with gross domestic saving even higher at 28% of GDP, this latter being over twice the African average. The excess of domestic saving over investment implies a substantial surplus on the external account.

Exports in 1987 were 29% of GDP and imports were 21%. This is a higher dependence than the rest of Africa on international trade for markets, but a lower dependence on trade for supplies of goods. Gold made up the largest part of exports with 41% of merchandise export earnings, with other metals and minerals contributing a further 26%. Machinery comprised 41% of imports, with fuel 13% and chemicals 11%.

There are no recent figures for adult literacy or educational provision. The latest literacy figures available are for 1960, when the level was estimated at 57% of the population. Educational provision figures for 1970 indicate enrolment levels of 99% in primary education and 18% in secondary education with 4% enrolled in higher education in 1965. It is to be expected that all these educational indicators have improved in the intervening period to place South Africa among the countries with better educational attainments and provision in Africa, although there is considerable disparity between the white community and the rest.

Life expectancy was 60 in 1987, and the average daily calorie supply was 3,132 per head. In 1973 there was one doctor for every 1,967 citizens. All these health indicators are

South Africa

well above the average for the rest of Africa, although again there is a marked disparity between the white and non-white comunities.

Overall, South Africa is Africa's largest economy, with lower-middle-income status. The economy is well industrialised with little reliance on agriculture, but heavy dependence on minerals, most of which are exported. Educational and health provisions are above the African average, but provision is markedly better for the white community than for the rest.

Economic Performance

GDP grew by only 1.0% a year from 1980-87, a reflection of the country's political instability and the effects of sanction and disinvestment measures. Although the standard of living is high, incomes in South Africa have fallen at -1.3% a year in this period.

The agricultural sector grew at 0.3% a year, 1980-87, slower than the 1.2% a year expansion recorded for Africa. This relative stagnation is due to a fall in demand for South African produce for political reasons, compounded by the 1983-86 drought. The industrial sector, hit by disinvestment and reduced export markets declined by -0.3%, although in Africa contraction was faster at -1.2% a year. Only the service sector showed any significant growth at 2.3% a year.

Comparative growth figures for South African imports and exports are not available. The South African Customs Union (SACU), which includes Namibia, Botswana, Lesotho and Swaziland, of which South Africa is the largest member, saw declines in both imports and exports of -8.8% and -0.1% respectively over the 1980-87 period, though this negative growth excludes intra-SACU trade. The decline is largely due to increased sanctions on trade with South Africa by western nations and the 1983-86 drought throughout the region.

The inflation rate 1980-87 reached 13.8%, though this was still below the 15.2% average for Africa.

Overall, economic performance has been poor in the 1980s, with falling GDP per head, contractions in agricultural and industrial output, and declining export performance. The inflation record has shown deterioration.

Recent Economic Developments

Economic policy has taken a distinctly short-term perspective with the government reacting to external pressures and internal events with the major preoccupation being the need to maintain internal security and stability. External influences have been dominated by the steady divestment by US and European companies of their South African assets in response to political pressure from the opponents of apartheid. More than 150 firms have sold their South African assets since 1984. Of the 250 US firms operating in South Africa, 38 have withdrawn. However, it is the largest firms that have come under the most pressure, and the withdrawals are dominated by multinationals such as IBM, Ford, Barclays and Exxon, and the impact is somewhat greater than the actual numbers of corporations involved.

Over the short term, the withdrawals have a considerable psychological impact, but little effect on output and employment. The operations are sold, usually at a discount, to South African buyers, and trading continues as before. In the longer term the economy will suffer from the absence of investment flows from overseas, and the technological improvements that are introduced by large corporations with substantial research and development programmes. Increasingly, government expenditure has been directed toward recurrent expenditure on law and order and on purchases of military equipment to bolster internal security, and to pursue destabilising policies against South Africa's hostile neighbours.

South Africa's economic performance shows considerable deterioration after 1981. Growth of GDP has been negative or negligible for the past five years with the exception of 1984. Recent reports indicate a return to positive growth in 1987 and 1988 at roughly the rate of current population expansion of 2.5% a year, arresting the fall in average living standards.

Inflation performance has gradually deteriorated, continuing the steady acceleration perceived in the 1970s where it averaged around 10% to be around 15% in the 1980s.

In the balance of payments, merchandise exports reached a peak in 1980 at $US 25.7b, and have subsequently declined to $US 16.b in 1986, before recovering to $US 22.5b in 1988. There is typically a surplus on the trade balance, but declining export performance has forced down imports which reached a peak in 1981 at $ 20.6b, but had fallen to $US 10.3b in 1985, before recovering to $US 17.5b in 1988.

South Africa's external debt has been the subject of considerable attention since the government imposed a freeze on an estimated $US14b of short-term debt in 1985. Overall external debt was estimated at $US 21b in 1988, all of it to commercial creditors at commercial interest rates. Of this total, $US 8b is currently still frozen, and South Africa has negotiated to repay $US 1.5b by 1993. However, of the $US 12b that is un-frozen there is $US 6b of repayments falling due by 1993 in the absence of any re-scheduling. Debt servicing will thus place strains on the balance of payments, compressing imports significantly.

The exchange rate has steadily depreciated from R 0.78 = $US 1 in 1980 to R 2.55 = $US 1 at the end of 1989. Depreciation of the currency can be expected to continue until trade sanctions are lifted and investment flows are restored.

Economic Outlook and Forecasts

The forecasts assume that, despite the optimism at the end of 1989 for a political settlement leading to majority rule, there is no significant improvement in the atmosphere of uncertainty that limits trade flows and foreign investment.

Gold prices are forecast to fall in real terms by 22% up to 1995, and this can be expected to provide a significant constraint on South Africa's expansion.

GDP is forecast to expand at around 1.5% a year, which will imply falls in living standards of around -0.8% a year. Export volumes are expected to stagnate. Import volumes are projected to continue falling under the impact of further terms of trade deteriorations and debt servicing demands, though not as fast as in the early 1980s.

Inflation is expected to accelerate to average 15% a year up to 1995.

South Africa: Economic Forecasts
(average annual percentage change)

	Actual 1980-86	Forecast Base 1990-95	Forecast High 1990-95
GDP	1.0	1.5	1.7
GDP per Head	-1.3	-0.8	-0.6
Inflation Rate	13.8	15.0	16.0
Export Volumes	-0.1[a]	0.0	0.3
Import Volumes	-8.8[a]	-1.5	-1.3

Note: [a] South African Customs Union.

SOUTH AFRICA: Comparative Data

	SOUTH AFRICA	SOUTHERN AFRICA	AFRICA	INDUSTRIAL COUNTRIES
POPULATION & LAND				
Population, mid year, millions, 1989	34.4	6.0	10.2	40.0
Urban Population, %, 1985	56	23.8	30	75
Population Growth Rate, % per year, 1980-86	2.2	3.1	3.1	0.8
Land Area, thou. sq. kilom.	1,221	531	486	1,628
Population Density, persons per sq kilom., 1988	27.6	11.3	20.4	24.3
ECONOMY: PRODUCTION & INCOME				
GDP, $US millions, 1986	56,370	2,121	3,561	550,099
GNP per head, $US, 1986	1,850	383	389	12,960
ECONOMY: SUPPLY STRUCTURE				
Agriculture, % of GDP, 1986	6	25	35	3
Industry, % of GDP, 1986	46	33	27	35
Services, % of GDP, 1986	49	42	38	61
ECONOMY: DEMAND STRUCTURE				
Private Consumption, % of GDP, 1986	51	67	73	62
Gross Domestic Investment, % of GDP, 1986	19	14	16	21
Government Consumption, % of GDP, 1986	19	21	14	17
Exports, % of GDP, 1986	33	34	23	17
Imports, % of GDP, 1986	23	36	26	17
ECONOMY: PERFORMANCE				
GDP growth, % per year, 1980-86	0.8	-3.7	-0.6	2.5
GDP per head growth, % per year, 1980-86	-1.4	-6.6	-3.7	1.7
Agriculture growth, % per year, 1980-86	-1.3	-6.8	0.0	2.5
Industry growth, % per year, 1980-86	-0.5	-3.8	-1.0	2.5
Services growth, % per year, 1980-86	-1.7	-0.9	-0.5	2.6
Exports growth, % per year, 1980-86	-0.4	-5.1	-1.9	3.3
Imports growth, % per year, 1980-86	-9.5	-2.5	-6.9	4.3
ECONOMY: OTHER				
Inflation Rate, % per year, 1980-86	13.6	18.1	16.7	5.3
Aid, net inflow, % of GDP, 1986	..	8.6	6.3	-
Debt Service, % of Exports, 1986	18.0	7.8	20.6	-
Budget Surplus (+), Deficit (-), % of GDP, 1986	-4.5	-4.1	-2.8	-5.1
EDUCATION				
Primary, % of 6-11 group, 1970	99	95	76	102
Secondary, % of 12-17 group, 1970	18	18	22	93
Higher, % of 20-24 group, 1970	..	1.3	1.9	39
Adult Literacy Rate, %, 1960	57	47	39	99
HEALTH & NUTRITION				
Life Expectancy, years, 1986	61	50	50	76
Calorie Supply, daily per head, 1985	2,967	1,998	2,096	3,357
Population per doctor, 1973	1,967	26,118	24,185	550

Notes: 'Southern Africa' and 'Africa' exclude South Africa. Dates are for the country in question, and do not always correspond with the Regional, African and Industrial averages.

SOUTH AFRICA: Leading Indicators

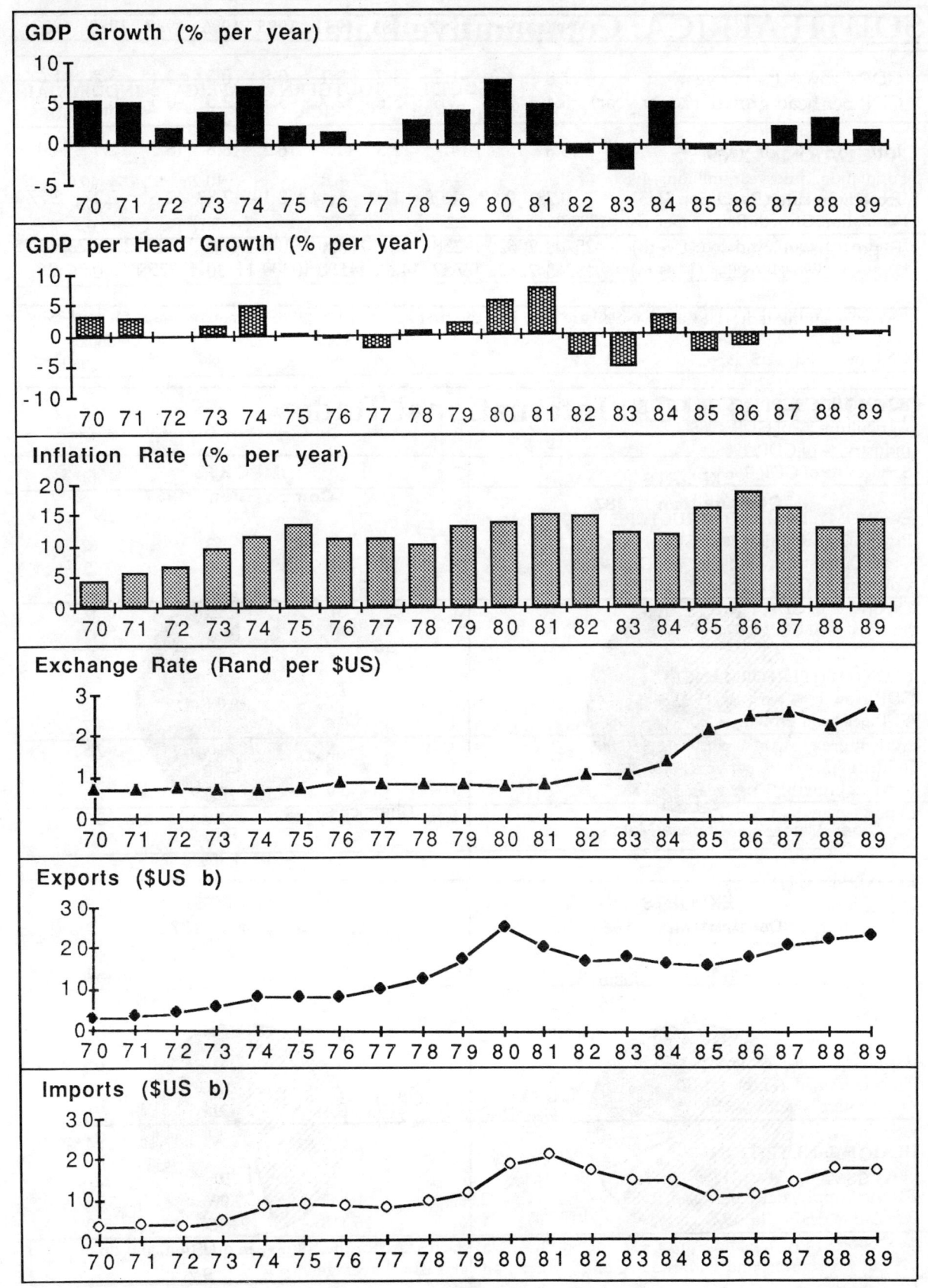

GDP Growth (% per year)

GDP per Head Growth (% per year)

Inflation Rate (% per year)

Exchange Rate (Rand per $US)

Exports ($US b)

Imports ($US b)

Note: Leading indicator data for 1989 are based on the first half of 1989. 1989 exchange rate is for mid-1989.

SOUTH AFRICA: Leading Indicator Data

	1980	1981	1982	1983	1984	1985	1986	1987	1988	1989
GDP Growth (% per year)	7.8	4.9	-1.2	-3.2	5.1	-0.8	0.3	2.1	3.2	1.6
GDP per head growth (% per year)	5.4	7.5	-3.6	-5.6	2.7	-3.3	-2.5	-0.5	0.7	-0.9
Inflation (% per year)	13.8	15.2	14.7	12.3	11.7	16.2	18.6	16.5	12.9	14.0
Exchange rate (Rand per $US)	0.78	0.87	1.08	1.11	1.44	2.19	2.53	2.61	2.27	2.77
Exports, merchandise ($US m)	25708	20620	17358	18245	16700	16100	18330	21100	22500	23563
Imports, merchandise ($US m)	18268	20622	16682	14202	14500	10300	11130	13925	17200	16826

Note: Leading indicator data for 1989 are based on the first half of 1989. 1989 exchange rate is for mid-1989.

SOUTH AFRICA: International Trade

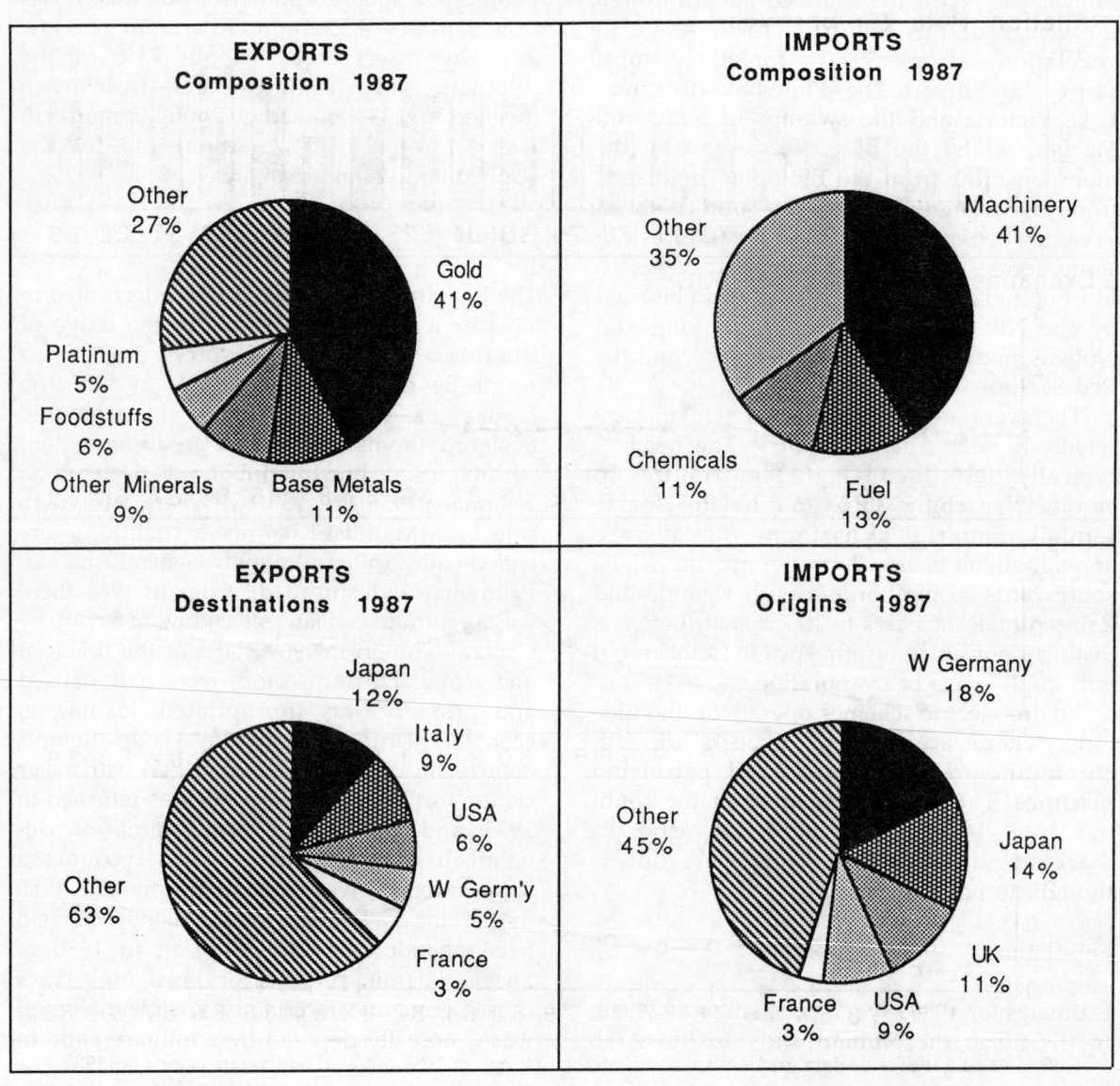

EXPORTS Composition 1987
- Gold 41%
- Other 27%
- Base Metals 11%
- Other Minerals 9%
- Foodstuffs 6%
- Platinum 5%

IMPORTS Composition 1987
- Machinery 41%
- Other 35%
- Fuel 13%
- Chemicals 11%

EXPORTS Destinations 1987
- Other 63%
- Japan 12%
- Italy 9%
- USA 6%
- W Germ'y 5%
- France 3%

IMPORTS Origins 1987
- Other 45%
- W Germany 18%
- Japan 14%
- UK 11%
- USA 9%
- France 3%

41 SUDAN

Physical Geography and Climate

Sudan, in the East African region, is the largest state in Africa in terms of area, situated south of Egypt with coastal access to the Red Sea. It lies to the east of Ethiopia and west of Libya, Chad and the Central African Republic and bordered to the south by Zaïre, Uganda and Kenya. The Nile runs south to north through the centre of the country with a range of vegetation - tropical rain forest, swamps, savanna and desert. The White Nile rises from Lake Victoria and the swamps of Sudd and Machar, whilst the Blue Nile is fed by the monsoon rains from the Ethiopian highlands from July to mid-October. The south is moist savanna vegetation, with tropical forest in some areas, whilst the north is arid desert. Most of Sudan consists of a level plain bisected by the Nile system, with the Imatong and Nubian mountain ranges to the west and the Red Sea hills to the east.

The average temperature and rainfall vary widely in such a large country. The north is typically desert with high temperatures in summer, reaching 35°C and having sparse rainfall. South of Khartoum the average annual rainfall is only 20cm per annum, whilst southwards to the borders with Uganda and Zaïre rainfall increases to 100cm and there is a distinct rainy season from April to October, but with high levels of evaporation.

Hydro-electric schemes operate on the Blue Nile. There are mineral exports of salt, chromium ore, manganese and petroleum products. The mineral potential of the south has not been fully explored due to inaccessibility and the civil war, but is thought to be fairly high.

Population

Estimates for 1989 give a population of 24.6m. In the north the Nubian and Muslim Arab groups predominate, whilst in the south Nilotic groups such as the Nuer, Dinka and Shilluk predominate. The official language is Arabic with 10% Dinka speakers in the south. Arab culture permeates the north where Islam predominates whilst the south is mainly Christian, which has led to internal conflict. Population density is low and averages 9.5 people per square kilometre, but this varies from densities of 55 around Khartoum to as low as 3 in desert areas. About 71% of the population are rural dwellers, 18% urban dwellers and 11% nomadic. Population growth was 3.1% for 1980-87. About 21% of the population lives in urban areas.

History

The Egyptian rule of Sudan was interrupted in 1898 by a Mahdist revolt. After the Battle of Khartoum in 1898 the country came under Anglo-Egyptian administration. In 1955 the House of Representatives unanimously declared themselves to be an independent sovereign state and a republic was declared on 1 January 1956. In 1958 two years of civilian rule by a Mahdist-Khatmiyya coalition were ended with a military coup by General Abboud. Following widespread disorder in 1964 there was a return to civilian rule ending in a coup by General Numeri in 1969. Parts of the financial and manufacturing sectors were nationalised and property was appropriated, leading to economic disruption. Communist coup attempts occurred in 1971, 1975, 1976 and 1981 with major rioting in 1979. Some property was returned in 1973 and by 1977 a programme of de-nationalisation had begun which encouraged some foreign investment to resume. By 1980 most of the government trading monopolies had been abandoned. The adoption in 1980 of Sharia Islamic reforms of laws and taxes caused economic uncertainty. President Numeri was eventually deposed by a military coup in

Sudan

April 1985, resulting in a Military Council assuming power led by Lieutenant-Colonel Boterse. In September 1985 there was an attempted coup at Omdurman. In December 1985 General Swar el Dahab of the Military Council relinquished power to a civilian five-man Supreme Council. Elections were held in April 1986; the majority party was Umma and Dr Sadiq el Mahdi became Prime Minister. Government policy became more strongly non-aligned, the Islamic tax system was abandoned and a western system was adopted. El Mahdi was removed in a military coup in 1989 and replaced by General Omar el Bashir.

The civil war in the south which began in 1964 continues, despite the granting of regional autonomy in 1972 to the three southern provinces. The Sudan People's Liberation Army (SPLA) led by Colonel John Garang continues to fight for greater control of the south. The famines beginning in 1983 brought an influx of refugees from Ethiopia, Chad and Uganda.

Stability

Since independence in 1956 Sudan has experienced regular attempted and successful coups. There have been seven years of civilian rule in 34 years with military governments at other times

Problems with the south of the country have persisted since Independence, and there is continual armed conflict in the region. Politically successive governments have found it difficult to deal with divisions in the population over the adoption of Islamic laws, and proposals to adopt such laws have infuriated the southern Sudanese.

Economic strategy has oscillated considerably, with widespread nationalisations beginning in 1969, and the state taking over substantial amounts of private property. In 1973 some of these seized properties were returned, in 1977 the state privatised some of its enterprises, and in 1980 ended some state trading monopolies. In 1983 Islamic laws on taxation and banking practices were introduced, but abandoned in 1986.

Overall, Sudan's stability is very poor, with the problems in the south seemingly insoluble and the present regime not well established. Economic policies show no continuity, and any perseverance with market-oriented policies is complicated by the issue of Islamic economic laws. It is not possible to be confident of any improvement in Sudan's stability in the foreseeable future.

Economic Structure

The size of the economy is large with a GDP of $US 8,210m in 1987 - only Cameroon, Nigeria and South Africa have larger economies in sub-Sahara Africa. GNP per head in 1987 was $US 330, and this is about the African average.

Agriculture generated 37% of GDP in 1987, industry 15% and services 48%. This is a rather larger dependence on agriculture and a less industrialised production structure than is general in Africa.

Private consumption was the equivalent of 79% of GDP in 1987, with investment 11%, government spending 15%, and domestic saving only 6% of GDP. Government spending is about the same proportion as elsewhere in Africa, but the structure of the rest of expenditure has a higher commitment to consumption and a lower allocation to saving and investment. The savings ratio is half that elsewhere in Africa. The gross investment ratio of 11% of GDP will scarcely maintain the existing capital stock.

Exports of goods and services were equivalent to 8% of GDP and imports were 13%. This is well under half the dependence on international trade observed on average in Africa. Cotton is the main source of export revenue, making up 30% of earnings in 1987, with gum arabic contributing 18% and sorghum 17%. Imports were made up of machinery 33%, manufactures and chemicals 26% and fuels 19%.

Educational provision is very uneven, with higher enrolment rates in the urban areas and the north. Overall enrolment rates in primary education were 50% in 1985, significantly below the African average of 66%. Secondary enrolments at 20% were better than average. Higher education enrolment, at 2%, is about the Africa average. Adult literacy in 1981 was 32%, and this is poor by African standards.

There was one doctor for every 10,110 citizens in 1984 and one nurse for every 1,250 citizens, and this provision is around twice as good as that in the rest of Africa. However, health provision is very uneven, being very poor in the south as a result of the civil war. Life expectancy, at 50 in 1987, is about typical for Africa.

Overall, Sudan is a low-income country with a small industrial sector and a large services sector. The commitment to private consumption is high, and mostly at the expense of savings. There is a low dependence on international trade. Educational and health provision appear reasonable, but there is a wide disparity between the better served north, and the south.

Economic Performance

Sudan's GDP declined by -0.1% a year over the period 1980-87, which implies that GDP per head declined at -3.2%. This is a slightly worse performance than elsewhere in Africa.

The agriculture sector expanded at 0.8% over 1980-87, which was not enough to keep pace with population increasing at 3.1% a year. Industry performed best of all the sectors with a 2.1% annual rate of expansion, and manufacturing, a sub-sector of industry, expanded at 1.6% a year. The sevices sector showed a decline at -1.3% a year. The industrial sector showed a decline at -1.35 a year. The industrial sector performance was better than the average elsewhere in Africa, although the performances in agriculture (affected by drought) and services, were worse.

Export volumes expanded by 4.2% a year over 1980-87, encouraged by substantial depreciation of the currency which raised the domestic price of exported goods. This was a much better performance than the African average. However, Sudan's terms of trade declined by 16% over these seven years, and this, combined with a slowing of development assistance as a result of political instability and the civil war, has caused import volumes to decline by -8.7% a year.

Inflation has steadily accelerated in the 1980s, being around 25% a year at the start of the decade, and over 80% at the close, averaging 31.7% for the period 1980-87, roughly twice the African average.

Overall, Sudan's economic performance, hampered by instability and drought, has been poor, with falling GDP per head. the industrial sector has shown the best performance, and expansion of export volumes has been impressive in the general context of Africa's trade performance. Inflation performance has been poor, and shows a deteriorating trend.

Recent Economic Developments

Reform of the exchange regime, food pricing policy and of state-owned enterprises met resistance from the coalition government established in 1986, leading to continued political instability. The IMF suspended lending in February 1987 when $US 252m in arrears had accumulated. By the end of 1987 Sudan owed the IMF $US 937m, one of the highest debts in Africa. In August of 1987, the IMF appeared prepared to consider renewed finances provided there were moves to unify Sudan's multi-tiered exchange system, reduce the budget deficit, cut overall government spending and reform state-owned enterprises. The proposals fell short of embracing the full package of market-orientated reforms that would include ending the flour and sugar subsidies, adopting some form of flexible exchange regime and privatising state-owned enterprises. At the end of 1987 Sudan tried to introduce its own structural adjustment programme aimed at stimulating economic growth, reducing inflation, trying to attain agricultural self-sufficiency and reducing the trade deficit. The fragile nature of the civilian coalition then in power made it difficult for the government to introduce the tough measures urged by the IMF. For example, a 40% devaluation of the Sudanese pound in 1987 led to riots in Khartoum. As of the end of 1989, there had been no progress in formulating an agreement with the IMF.

Sudan had a surplus on the balance of trade in the early 1970s and this has deteriorated to a position where imports of goods were twice the value of exports in 1981. The value of exports peaked in 1981 at $US 793m, but had fallen to $US 350m in 1989. A major factor in Sudan's balance of payments has been remittances from workers employed elsewhere in the Middle East. These remittances have helped to finance the growing trade imbalance in the need to service mounting external

Sudan

indebtedness. However, most of the remitted sums are exchanged at the more favourable black market rates, and it is thought that 80% of remittances do not pass through the official exchange system. Remittances are thought to run at $US 360m a year, more than was earned from merchandise exports in 1989.

Sudan has an external debt estimated at close to $US 11,000m, giving rise to $US 937m in debt servicing in 1987. This is clearly beyond the ability of Sudan to meet, given declining export revenues and remittances exchanged on the black market. There are some $US 2,600m in payment arrears. Sudan's debts were rescheduled in 1981 and 1985, but a further formal rescheduling has been hampered by arrears to the IMF.

The exchange rate underwent sporadic devaluation in the 1980s after remaining virtually unchanged in the 1970s. In February 1985 Sudan introduced a three-tier system with an official rate set at £S 2.5 = $US 1, a rate for agricultural products of £S 2.03 = $US 1 and £S 4 = $US 1 for financial transactions, that is, migrants' remittances and tourists. The financial rate is an attempt to discourage black-market transactions. By the end of 1989, the official rate had depreciated to £S 11.4 = $US 1. The black market maintains a premium on foreign exchange, and in mid-1988, this was reported at over 700%.

Most of Sudan's aid comes from Middle East sources. Saudi Arabia and the World Bank are collaborating on a $US 80m sugar rehabilitation scheme, while the Saudi Fund for Development is providing $US 27m for airport work. Kuwait is allocating $US 10m for health, irrigation and education projects. The Islamic Development Bank is involved in social projects for the homeless, and for vocational training. The World Bank and the African Development Fund are involved in a $US 52m electrification project, the US is allocating $US 9m to forestry and renewable energy schemes, and Italy is providing $US 13m for railway rehabilitation. Other recent aid includes $US 22m in grant form from Japan for water and communications projects and a $US 10m loan from OPEC for petroleum imports. The USA gives a considerable amount of assistance (especially military). USAID's Sudan budget for 1988-89 was around $US 120m. Substantial emergency aid was given following the devastating floods of 1988, including $US 300m from various western countries, multilaterals and development funds. In addition, $US 26.5m was offered by Japan and $US 75m from the World Bank.

Private foreign investment in the oil sector is at a standstill because of the armed conflict in the south, and there is speculation that the White Nile Petroleum Company which involves Chevron, Royal Dutch Shell and Saudi Arabia's Apicorp in a $US 1,000m pipeline scheme is to go into liquidation.

Economic Outlook and Forecasts

The forecasts assume that Sudan's seemingly intractable problems with the south remain unchanged and that this situation continues to cause instability in national affairs. These factors are expected to prevent the introduction of the mix of austerity and liberalisation that will need to presage an agreement with the IMF.

Sudan: Economic Forecasts
(average annual percentage change)

	Actual	Forecast	
		Base	High
	1980-87	1990-95	1990-95
GDP	-0.1	-0.4	-0.2
GDP per Head	-3.2	-3.1	-3.0
Inflation Rate	31.7	90.0	80.0
Export Volumes	4.2	2.2	2.9
Import Volumes	-8.7	-2.3	-2.0

Cotton prices are not expected to increase in real terms up to 1995, and sorghum prices are projected to fall by 15% in real terms up to the mid-1990s. GDP is projected to continue to decline, and this decline is expected to be faster in the 1990s than it was in the 1980s. However, because the rate of population growth is expected to slow, the rate of decline in GDP per head is expected to be about the same as experienced in the 1980s.

Export volumes are projected to continue growing under the stimulus of the 1989 exchange rate adjustment, although at a slower rate than

Sudan

in the 1980s. This expansion of export volumes will be dependent on normal weather conditions being experienced. Import volumes, as a result of restricted aid flows and adverse terms of trade movements, are expected to continue to fall, although at a slower rate than in the 1980s.

Inflation is projected to continue at around the rate it achieved at the end of the 1980s.

SUDAN: Comparative Data

	SUDAN	EAST AFRICA	AFRICA	INDUSTRIAL COUNTRIES
POPULATION & LAND				
Population, mid year, millions, 1989	24.6	12.2	10.2	40.0
Urban Population, %, 1985	21	30.5	30	75
Population Growth Rate, % per year, 1980-86	2.8	3.1	3.1	0.8
Land Area, thou. sq. kilom.	2,506	486	486	1,628
Population Density, persons per sq kilom., 1988	9.5	24.2	20.4	24.3
ECONOMY: PRODUCTION & INCOME				
GDP, $US millions, 1986	7,470	2,650	3,561	550,099
GNP per head, $US, 1986	320	250	389	12,960
ECONOMY: SUPPLY STRUCTURE				
Agriculture, % of GDP, 1986	35	43	35	3
Industry, % of GDP, 1986	15	15	27	35
Services, % of GDP, 1986	50	42	38	61
ECONOMY: DEMAND STRUCTURE				
Private Consumption, % of GDP, 1986	83	77	73	62
Gross Domestic Investment, % of GDP, 1986	12	16	16	21
Government Consumption, % of GDP, 1986	14	15	14	17
Exports, % of GDP, 1986	9	16	23	17
Imports, % of GDP, 1986	17	24	26	17
ECONOMY: PERFORMANCE				
GDP growth, % per year, 1980-86	0.3	1.6	-0.6	2.5
GDP per head growth, % per year, 1980-86	-2.4	-1.7	-3.7	1.7
Agriculture growth, % per year, 1980-86	0.4	1.1	0.0	2.5
Industry growth, % per year, 1980-86	2.1	1.1	-1.0	2.5
Services growth, % per year, 1980-86	-0.3	2.5	-0.5	2.6
Exports growth, % per year, 1980-86	6.9	0.7	-1.9	3.3
Imports growth, % per year, 1980-86	-4.0	0.2	-6.9	4.3
ECONOMY: OTHER				
Inflation Rate, % per year, 1980-86	32.6	23.6	16.7	5.3
Aid, net inflow, % of GDP, 1986	12.8	11.5	6.3	-
Debt Service, % of Exports, 1986	7.7	18	20.6	-
Budget Surplus (+), Deficit (-), % of GDP, 1982	-4.6	-3.0	-2.8	-5.1
EDUCATION				
Primary, % of 6-11 group, 1984	49	62	76	102
Secondary, % of 12-17 group, 1984	19	15	22	93
Higher, % of 20-24 group, 1984	2	1.2	1.9	39
Adult Literacy Rate, %, 1981	32	41	39	99
HEALTH & NUTRITION				
Life Expectancy, years, 1986	49	50	50	76
Calorie Supply, daily per head, 1985	2,168	2,111	2,096	3,357
Population per doctor, 1981	9,810	35,986	24,185	550

Notes: 'Southern Africa' and 'Africa' exclude South Africa. Dates are for the country in question, and do not always correspond with the Regional, African and Industrial averages.

SUDAN: Leading Indicators

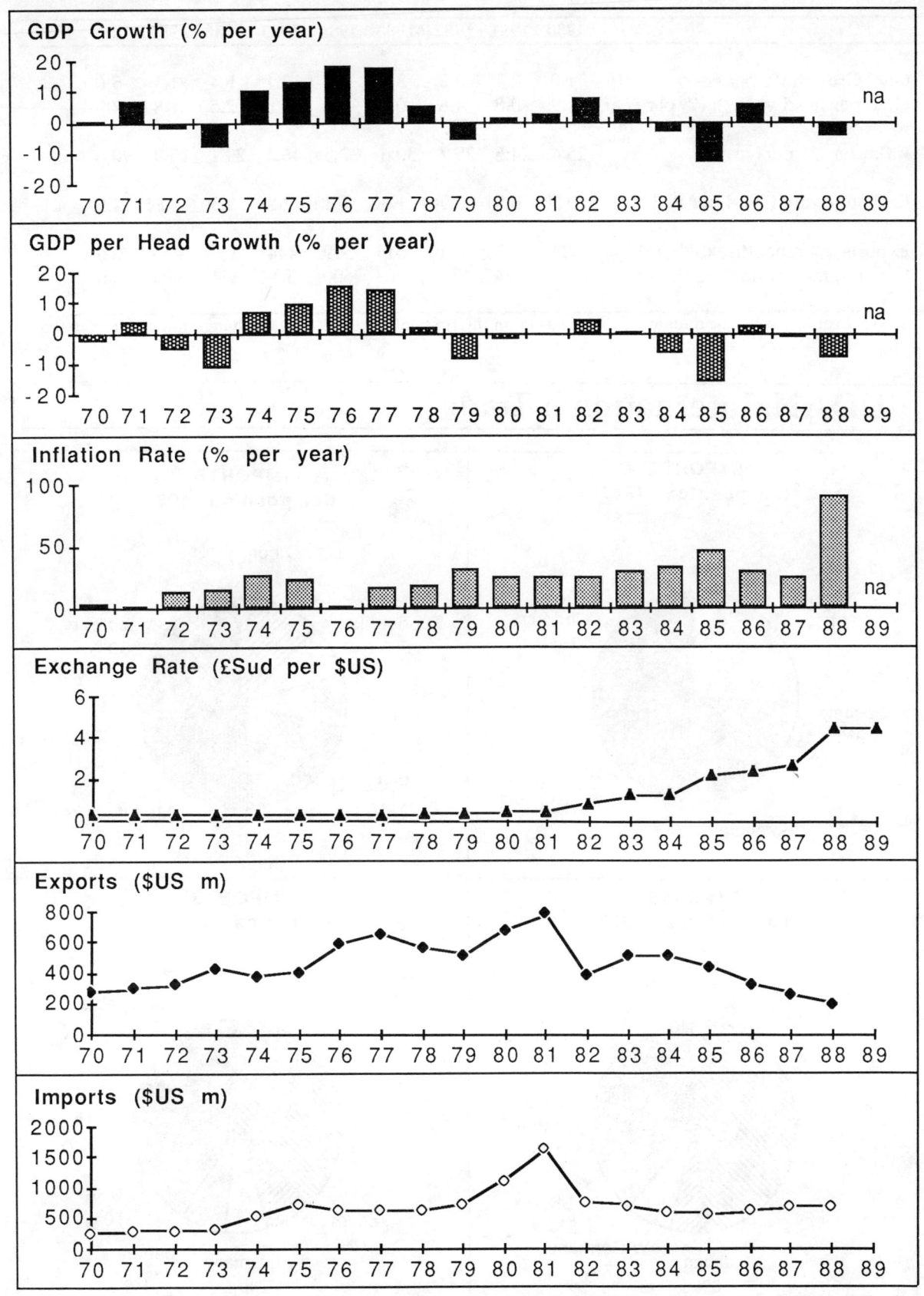

GDP Growth (% per year)

GDP per Head Growth (% per year)

Inflation Rate (% per year)

Exchange Rate (£Sud per $US)

Exports ($US m)

Imports ($US m)

Note: Leading indicator data for 1989 are based on the first half of 1989. 1989 exchange rate is for mid-1989.

SUDAN: Leading Indicator Data

	1980	1981	1982	1983	1984	1985	1986	1987	1988	1989
GDP Growth (% per year)	1.0	2.1	7.5	3.4	-3.7	-13.1	5.4	1.1	-5.0	
GDP per head growth (% per year)	-1.9	-0.8	4.6	0.5	-6.6	-16.0	2.5	-1.8	-7.9	
Inflation (% per year)	25.4	24.6	25.7	30.6	37.5	46.3	29.1	25.0	90.0	
Exchange rate (£Sud per $US)	0.50	0.54	0.94	1.30	1.30	2.29	2.50	2.81	4.5	4.5
Exports, merchandise ($US m)	689	793	401	514	519	444	327	265	200	
Imports, merchandise ($US m)	1127	1634	750	703	600	579	634	695	700	

Note: Leading indicator data for 1989 are based on the first half of 1989. 1989 exchange rate is for mid-1989.

SUDAN: International Trade

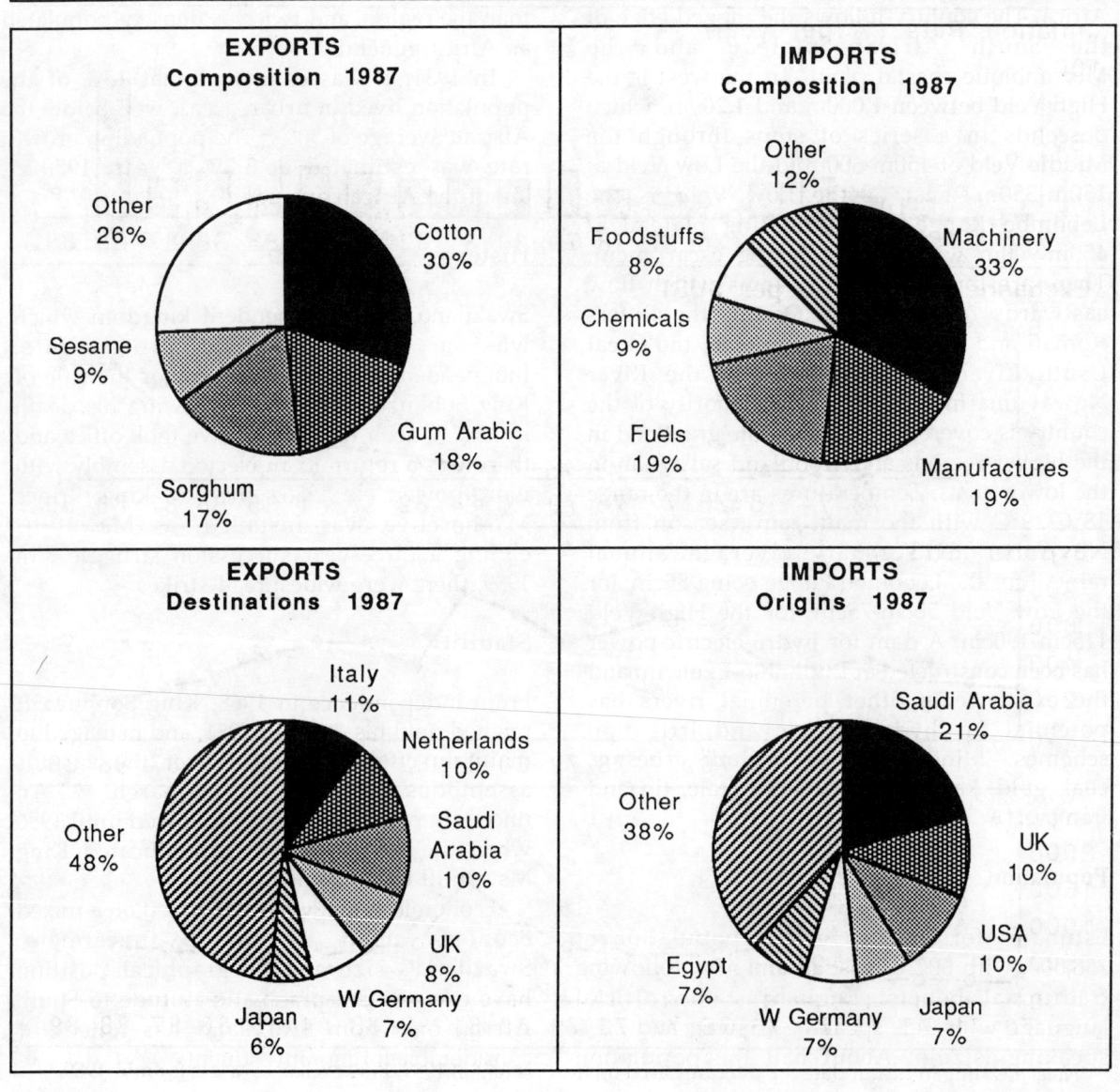

EXPORTS Composition 1987
- Cotton 30%
- Gum Arabic 18%
- Sorghum 17%
- Sesame 9%
- Other 26%

IMPORTS Composition 1987
- Machinery 33%
- Manufactures 19%
- Fuels 19%
- Chemicals 9%
- Foodstuffs 8%
- Other 12%

EXPORTS Destinations 1987
- Italy 11%
- Netherlands 10%
- Saudi Arabia 10%
- UK 8%
- W Germany 7%
- Japan 6%
- Other 48%

IMPORTS Origins 1987
- Saudi Arabia 21%
- UK 10%
- USA 10%
- Japan 7%
- W Germany 7%
- Egypt 7%
- Other 38%

42 SWAZILAND

Physical Geography and Climate

Swaziland, in the Southern Africa region, is a small landlocked kingdom in the south east of Africa, bounded by Mozambique and South Africa. The country follows the ragged edge of the South African plateau and the Mozambique coastal plain. To the west is the High Veld between 1,000m and 1,200m, which descends in a series of steps through the Middle Veld at 450m-600m to the Low Veld at 150m-350m. East of the Low Veld is the Lebombo Range, an undulating plateau at 450m-900m with a west facing escarpment. There are four drainage systems which flow eastwards to the Indian Ocean, the Rivers Komati and Umbeluzi in the north, the Great Usuth River in the centre and the River Ngwavuma in the south. The majority of the country is covered with temperate grassland in the higher regions and woodland savannah in the lower areas. Temperatures are in the range 18°C-29°C with the main rainy season from November to February: average annual rainfall for the Lebombo Range being 85cm; for the Low Veld 50cm-75cm; for the High Veld 115cm-190cm. A dam for hydro-electric power has been constructed at Luphohlo-Ezulwini and the existence of other perennial rivers has potential for hydro-electric and irrigation schemes. Mineral deposits include asbestos, coal, gold, kaolin, quarry stone, talc, tin and iron ore.

Population

Estimates for 1989 suggest a population of 756,000, with 60% Christian and 40% following traditional beliefs. English is the official language with 91% speaking Isiswati and 2.2% speaking Isizulu. About half the population live in the Middle Veld. Around 20,000 migrant workers are employed in South Africa.

The density of the population in 1989 was 44.5 persons per square kilometre, making Swaziland four times more densely populated than the region, and twice as densely populated as Africa generally.

In 1981, it was estimated that 15% of the population lived in urban areas, well below the African average of 30%. The population growth rate was estimated at 3.2% a year, 1970-81, about the African average.

History

Swaziland is an independent kingdom which was a former British protectorate. Independence was achieved under the rule of King Sobhuza II in 1968. Following his death in 1982, Queen Regent Dzeliwe took office and there was a return to an elected assembly with more power. In 1986, the new king, Prince Makhosetive was installed, as Mswati III ending a three-year succession struggle. In 1989, there were widespread strikes.

Stability

From independence in 1968, King Sobhuza II reigned until his death in 1981, and managed to maintain effective control over the various assemblies advising the monarch. An uncertain period of regency followed until 1986 when the 18-year-old British-educated King Mswati III was crowned.

Economic strategy has depended on a mixed economy open to foreign investment. Swaziland's size and geographical position have compelled a pragmatic attitude to South Africa, on whom Swaziland has relied for considerable foreign investment.

Swaziland

Overall, Swaziland's stability record has been good. There must be some concern over Swaziland's ability to maintain this record with the young King faced with competing factions among his advisers coupled with industrial unrest.

Economic Structure

Swaziland's economy reflects the small land area and population, with a GDP of only $US 540m in 1987. However, living standards are relatively high with a GNP per capita of $US 700 in 1987, nearly double the regional average income, placing Swaziland in the lower-middle-income group.

In 1981 agriculture generated 20% of GDP, well below the African average of 33%. It should be noted, however, that much of the industrial output is processing of agricultural and timber produce. The industrial sector (which includes mining) generated 24% of GDP while services generated a further 57%. The service sector has grown in importance in recent years as Swaziland has developed its tourist industry.

Private consumption was equivalent to 78% of GDP in 1983, and this is slightly above the African average. Government consumption was 29% of GDP, and this is almost double the average for Africa. Gross domestic investment was running at 26% of GDP, and this is half as great again as the average elsewhere in Africa. Domestic saving was thus equivalent to -33% of GDP, and this is largely due to the considerable inflow of resources comprising remittances from workers employed in South Africa.

Exports in 1983 were equivalent to 64% of GDP, and imports were 97% of GDP. This heavy dependence on trade is a result of Swaziland's small economy and the consequent inability to produce goods for the domestic market in large enough volumes to be efficient. The trade imbalance is largely financed by workers' remittances. The main merchandise exports were sugar 34% and wood pulp 16%. Merchandise imports were made up of manufactures 32%, machinery 21%, fuels 21% and food 16%.

The literacy rate in 1978 was estimated at 65%, well above the African average of around 40%. The primary enrolment rate was 111% in 1983, the secondary enrolment rate was 44%, and 3.3% of the tertiary age group were in higher education. These rates are considerably above the African average, and are among the best in the continent.

Life expectancy was 55 years, above the African average of 50 years. In 1980 there was one doctor for every 17,360 persons and calorie supply per person per day in 1983 was 2,258, and both these indicators are above the African average.

Overall, Swaziland has a small economy in the lower-middle-income group with a small agricultural sector and a large services sector. There is substantial inflow of resources from the remittances of migrant workers. Trade dependence is considerable. Educational provision is particularly good, and health provision is above average for Africa.

Economic Performance

GDP grew at a rate of 2.0% a year in the period 1980-87, and this is well above the African average of 0.4% a year. However with population expanding at 3.2% a year, GDP per head fell at -1.2% a year.

Sectoral and merchandise trade growth rates are only available for 1970-81. Agriculture expanded at 3.1% a year, industry at 3.8% and services at 5.7%. These rates will all have fallen in the 1980s. However, the performance in the 1970s was significantly better than the average for Africa in each of these sectors.

Export volumes grew at 2.8% a year 1970-81, and import volumes at 5.7%, and, again, both these expansion rates were faster than the African average for this period.

Inflation averaged 10.2% a year for 1980-87, and this was lower than the African average for this period.

Overall, Swaziland has shown above average economic performance in the 1980s, although GDP per head has fallen.

Recent Economic Developments

Economic strategy has not undergone any major changes, although the increasing concern with public sector performance in Africa is reflected in emphasis on restraining government spending and reducing the size of the budget deficit. A public sector industrial corporation has been privatised.

Economic performance shows considerable year-to-year variations, but apart from an 8.2% increase in 1981, GDP expansion has not managed to keep pace with population growth. Drought and depressed export markets as well as the effects of a cyclone in 1984 have contributed to hamper economic progress. Inflation in the 1980s has fluctuated in the 10% to 20% range, with a 20% price increase in 1985 followed by an improvement to 12% inflation in 1989.

Swaziland's export performance has been affected by drought and the 1984 cyclone. Merchandise exports peaked at $US 388m in 1981, and have subsequently fallen to $US 166m in 1985 but recovered to around $US 410m in 1989. In the 1980s Swaziland has begun to run a deficit on the balance of trade, and imports have been sustained at a level between 20% and 50% greater than export revenue. Falling export receipts reduced imports, and they have fallen from $US 517m in 1980 to $US 256m in 1985 before recovering to $US 480m in 1989.

Swaziland's external debt was $US 273m in 1987, giving rise to debt servicing comprising less than 5.8% of export earnings.

The Swazi currency, the lilangeni was circulated with the South African rand, with which it was exchanged at par under the Rand Monetary Agreement, up to June 1986. This arrangement has now been replaced by the Tripartite Monetary Agreement, under which Swaziland will be able to determine its own exchange rate. Up to the present, however, Swaziland has chosen to keep the lilangeni at par with the rand. Thus the lilangeni has steadily depreciated from SL 0.78 = $US1 in 1980 to SL 2.55 = $US1 at the end of 1989.

Aid is dominated by road projects, with the World Bank supporting six schemes to restore roads damaged by the 1984 cyclone comprising a total of $US 7.2m, and the African Development Bank and the African Development Fund are providing $US 9.6m.

On the private foreign investment front, the Swaziland Industrial Development Company has taken over the assets of the public sector National Industrial Development Corporation of Swaziland. The shareholders include a variety of overseas banks. In all $US 25.7m has been raised, and $US 70m of projects are being negotiated.

South Africa's SA Maganese Amcar is planning to invest $US 7m in a ferrochrome plant, providing agreement can be reached on competitive electricity rates.

Economic Outlook and Forecasts

The forecasts assume that Swaziland shows moderately impaired political instability up to 1995, and that the economy will not be affected by the move toward majority rule in South Africa.

Swaziland: Economic Forecasts
(average annual percentage change)

	Actual 1980-86	Forecast Base 1990-95	Forecast High 1990-95
GDP	2.0	2.5	2.7
GDP per Head	-1.2	-0.7	-0.5
Inflation Rate	10.2	11.0	12.0
Export Volumes	2.8[a]	3.2	3.5
Import Volumes	5.7[a]	4.1	4.2

Note: [a]1970-81.

It is expected that the price of sugar will fall by -5.0% up to 1995, and that timber prices will rise by 7.0%.

The forecasts indicate a modest rise in the GDP growth rate of 2.5% a year, with the prospect of political instability and the slowdown in the flow of tourists mainly responsible for preventing a faster rate of expansion.

This will imply GDP per head falling at around -0.7% a year. Export volumes are forecast to expand at a slightly faster rate of

Swaziland

3.2% a year, but import volumes to grow more slowly at around 4.0% a year, as a result of slower growth in invisible export earnings.

The inflation rate is forecast to remain at around 12% a year, slightly higher than in the 1980s.

SWAZILAND: Comparative Data

	SWAZILAND	SOUTHERN AFRICA	AFRICA	INDUSTRIAL COUNTRIES
POPULATION & LAND				
Population, mid year, millions, 1989	0.756	6.0	10.2	40.0
Urban Population, %, 1981	14.6	23.8	30	75
Population Growth Rate, % per year, 1970-81	3.2	3.1	3.1	0.8
Land Area, thou. sq. kilom.	17	531	486	1,628
Population Density, persons per sq kilom., 1988	43.1	11.3	20.4	24.3
ECONOMY: PRODUCTION & INCOME				
GDP, $US millions, 1984	627	2,121	3,561	550,099
GNP per head, $US, 1986	690	383	389	12,960
ECONOMY: SUPPLY STRUCTURE				
Agriculture, % of GDP, 1981	20	25	35	3
Industry, % of GDP, 1981	24	33	27	35
Services, % of GDP, 1981	57	42	38	61
ECONOMY: DEMAND STRUCTURE				
Private Consumption, % of GDP, 1983	78	67	73	62
Gross Domestic Investment, % of GDP, 1983	26	14	16	21
Government Consumption, % of GDP, 1983	29	21	14	17
Exports, % of GDP, 1983	64	34	23	17
Imports, % of GDP, 1983	97	36	26	17
ECONOMY: PERFORMANCE				
GDP growth, % per year, 1970-81	4.5	-3.7	-0.6	2.5
GDP per head growth, % per year, 1970-81	1.3	-6.6	-3.7	1.7
Agriculture growth, % per year, 1970-81	3.1	-6.8	0.0	2.5
Industry growth, % per year, 1970-81	3.8	-3.8	-1.0	2.5
Services growth, % per year, 1970-81	5.7	-0.9	-0.5	2.6
Exports growth, % per year, 1970-81	2.8	-5.1	-1.9	3.3
Imports growth, % per year, 1970-81	5.7	-2.5	-6.9	4.3
ECONOMY: OTHER				
Inflation Rate, % per year, 1980-85	9.6	18.1	16.7	5.3
Aid, net inflow, % of GDP, 1984	1.4	8.6	6.3	-
Debt Service, % of Exports, 1984	4.4	7.8	20.6	-
Budget Surplus (+), Deficit (-), % of GDP, 1983	-2.4	-4.1	-2.8	-5.1
EDUCATION				
Primary, % of 6-11 group, 1983	111	95	76	102
Secondary, % of 12-17 group, 1983	44	18	22	93
Higher, % of 20-24 group, 1983	3.3	1.3	1.9	39
Adult Literacy Rate, %, 1978	65	47	39	99
HEALTH & NUTRITION				
Life Expectancy, years, 1986	55	50	50	76
Calorie Supply, daily per head, 1983	2,258	1,998	2,096	3,357
Population per doctor, 1980	17,360	26,118	24,185	550

Notes: 'Southern Africa' and 'Africa' exclude South Africa. Dates are for the country in question, and do not always correspond with the Regional, African and Industrial averages.

SWAZILAND: Leading Indicators

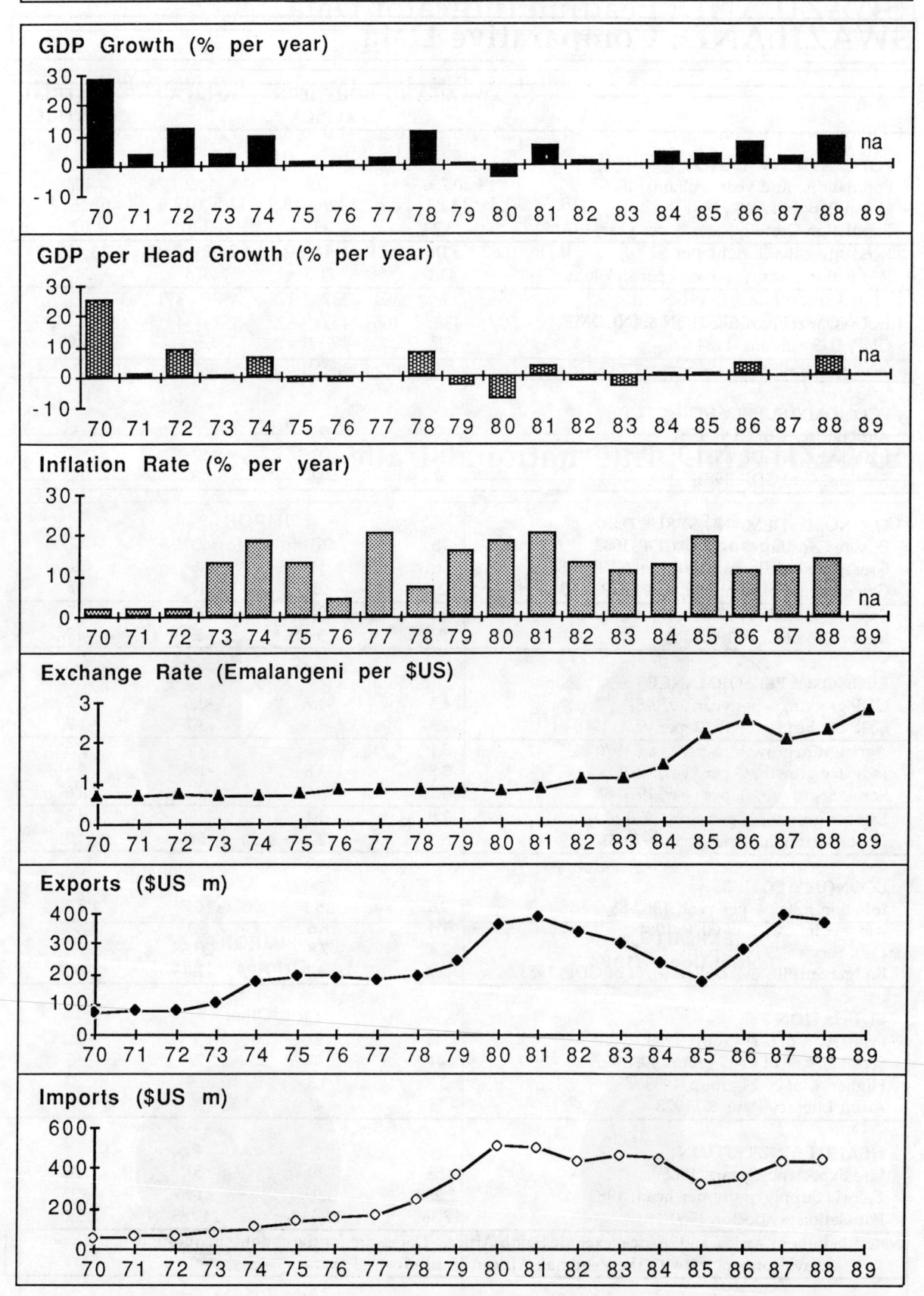

GDP Growth (% per year)

GDP per Head Growth (% per year)

Inflation Rate (% per year)

Exchange Rate (Emalangeni per $US)

Exports ($US m)

Imports ($US m)

Note: Leading indicator data for 1989 are based on the first half of 1989. 1989 exchange rate is for mid-1989.

SWAZILAND: Leading Indicator Data

	1980	1981	1982	1983	1984	1985	1986	1987	1988	1989
GDP Growth (% per year)	-4.8	6.7	1.2	-0.8	4.0	3.2	7.4	2.5	8.9	
GDP per head growth (% per year)	-8.0	3.5	-2.0	-4.0	0.8	0.0	4.2	-0.7	5.7	
Inflation (% per year)	18.7	20.7	13.2	11.2	12.9	19.9	11.5	12.6	14.6	
Exchange rate (E'geni per $US)	0.78	0.87	1.08	1.11	1.44	2.18	2.53	2.03	2.26	2.77
Exports, merchandise ($US m)	368	388	339	304	237	179	280	391	373	
Imports, merchandise ($US m)	517	507	433	465	447	323	352	431	440	

Note: Leading indicator data for 1989 are based on the first half of 1989. 1989 exchange rate is for mid-1989.

SWAZILAND: International Trade

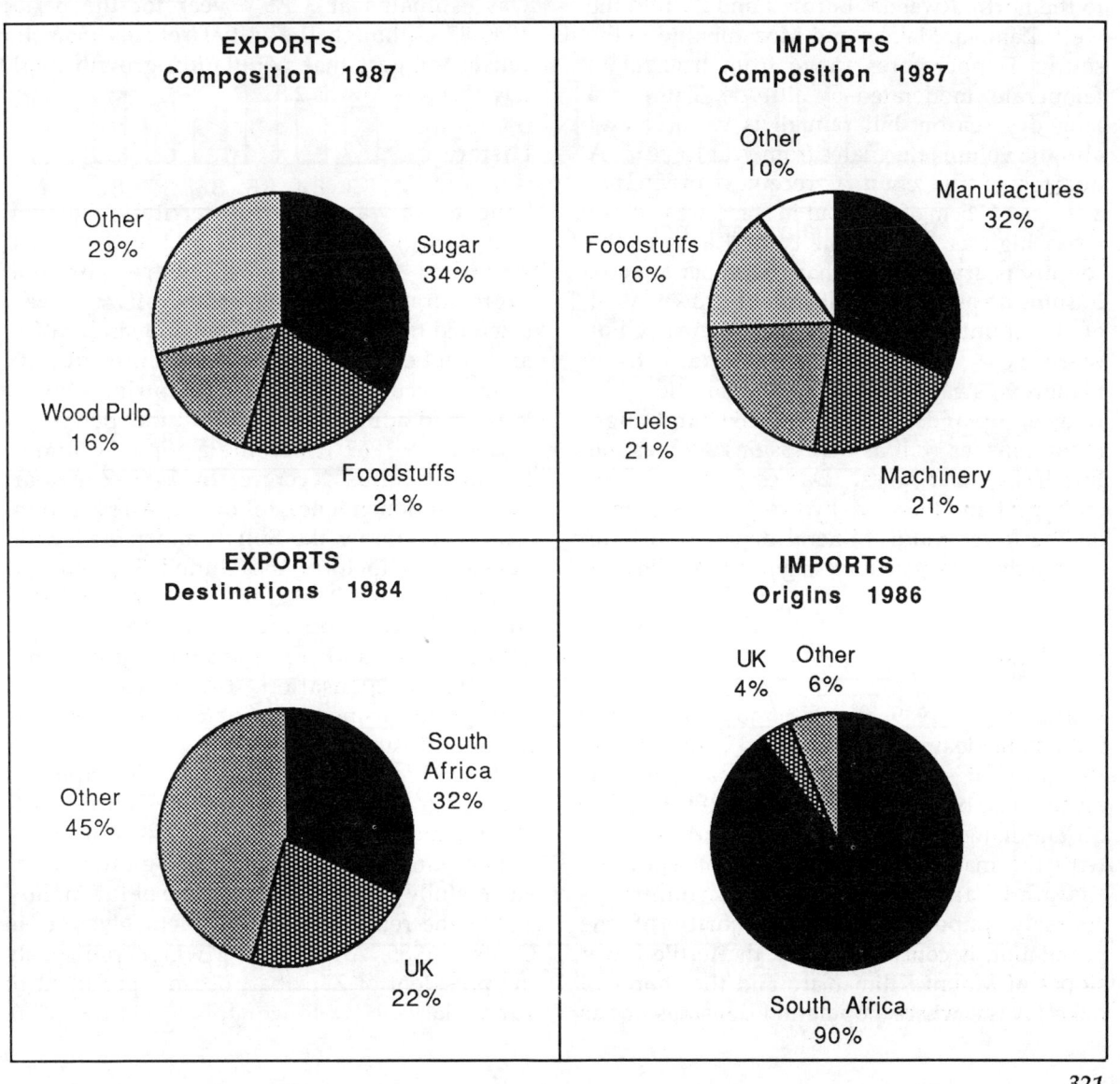

EXPORTS Composition 1987
- Sugar 34%
- Foodstuffs 21%
- Wood Pulp 16%
- Other 29%

IMPORTS Composition 1987
- Manufactures 32%
- Machinery 21%
- Fuels 21%
- Foodstuffs 16%
- Other 10%

EXPORTS Destinations 1984
- South Africa 32%
- UK 22%
- Other 45%

IMPORTS Origins 1986
- South Africa 90%
- UK 4%
- Other 6%

43 TANZANIA

Physical Geography and Climate

Tanzania, in the East African region, is a large coastal country lying just below the Equator and includes the islands of Pemba and Zanzibar. It is bounded by Kenya and Uganda to the north, Rwanda, Burundi and Zaïre to the west, Zambia, Malawi and Mozambique to the south. Temperatures range from tropical to temperate, moderated by altitude. There is a long dry season, but rainfall is variable and also the volume fluctuates from year to year. A quarter of the country receives an annual average of 75cm of rain, but in some areas it can be as high as 125cm. The central area of the country is dry with less than 50cm per annum. In some areas two harvests can be grown. Most of the country consists of high plateaux but there is a wide variety of terrain from mangrove swamps, coral reefs, plain, low hill ranges, uplands, volcanic peaks and high mountains, as well as depressions such as the Rift Valley and lakes. Dar es Salaam is the main port and there are hydro-electric schemes on the River Rufiji. Mineral deposits include diamonds, gold, gemstones, gypsum, kaolin and tin.

Population

Estimates for 1989 give a population of 25.5 million, made up of largely mixed Bantu groups, the largest being the Sukuma and the Nyamwezi, but there are at least 120 tribes. The official languages are Swahili and English, with the majority of the population speaking Kiswahili and Kisukuma. The country is sparsely populated. The majority of the population is concentrated on the fertile lower slopes of Mount Kilimanjaro and the shores of Lake Nyasa, where population densities are as high as 250 persons per square kilometre; whereas the average density for 1988 was 26 persons per square kilometre. The urban population in 1985 was around 14%, half the regional average, with a concentration around the port of Dar es Salaam. Population growth was estimated at 3.5% a year for the period 1980-87, although the initial returns from the census suggest that population growth might have been as low as 2.8%.

History

Tanganyika was a German territory, German East Africa, between 1890 and 1918, whilst Zanzibar and Pemba were British Protectorates. In 1896 the railway was extended from Tanga to Moshi and sisal, coffee and rubber plantations were introduced. Britain occupied Tanganyika during World War I and administered it until independence in 1961. In 1962 it became a republic under President Julius Nyerere. In 1963 Zanzibar gained Independence following a revolution which overthrew the Sultan and joined with Tanganyika to form the United Republic of Tanzania. In 1967 Tanzania nationalised financial and business enterprise and collectivised and reorganised agriculture. Despite compensation, there was severe damage to business confidence and a halt to foreign investment.

In 1977 the East African Community collapsed, and the border with Kenya was closed, not re-opening until 1983. In 1979, Tanzanian forces invaded Uganda and successfully overthrew the regime of Idi Amin.

On the retirement of President Nyerere, in October 1985, Ali Hassan Mwinyi, previously the president of Zanzibar, became president of Tanzania.

Tanzania

Stability

Since Independence in 1961, Tanzania has had single party rule under Julius Nyerere until 1985, when power was transferred to Ali Hassan Mwinyi. There has been negligible internal unrest since the revolution in Zanzibar in 1964, closely followed by a supressed army mutiny on the mainland. The invasion of Uganda in 1979, and a subsequent period of peace-keeping occupation had minimal disrupting effect on the Tanzanian economy.

Economic strategy underwent a profound change in 1967 when financial and business enterprises were taken into public ownership and a major re-organisation of the agricultural sector was introduced involving collective production and re-locating the population into villages. By 1977, the collectivisation of agriculture had virtually been abandoned. In 1986 Tanzania signed an agreement with the IMF which heralded the beginnings of a reversal of economic strategy to more encouragement for the private sector and reliance on market forces, rather than on planning and control.

Overall, Tanzania's stability has been moderate. The government has remained secure, but economic policy has undergone two changes of direction. It will take a period of consolidation of the Mwinyi administration and perseverance with liberalising reforms before international business and finance is confident about involvement in Tanzania again.

Economic Structure

GDP in 1987 was estimated at $US 3,080m, and this makes the economy roughly of average size in Africa. GNP per head in 1987 was $US 180, and this placed Tanzania among the poorer group of low-income countries.

Agriculture contributed 61% of GDP in 1987, and this was almost twice the average dependence on this sector in Africa. The industrial sector contributed 8% of GDP, of which 5% of GDP was manufacturing. The average in Africa in 1987 was for the industrial sector to contribute 28% of GDP. The services sector was responsible for 31% of GDP, below the African average of 40%.

On the demand side, private consumption was the equivalent of 98% of GDP in 1987, well above the African average of 72%. Government consumption was 8%, about half the allocation of resources to this sector that is typical of Africa. Gross domestic investment was 17% of GDP, while saving was -6%, with the level of investment being virtually achieved by virtue of a net inflow of resources, almost all aid, equivalent to 23% of GDP.

Exports were equivalent to 13% of GDP, and imports 36%. The percentage of GDP generated by exports is about half the average for Africa, while the dependence on imports is greater by about a sixth.

Coffee was the main merchandise export in 1987, comprising 31% of the total, with cotton next at 13%. Machinery was 19% of imports, with fuel 14%, manufactures 12%, foodstuffs 9% and metals 8%.

The adult literacy rate in 1980 was estimated at 79%, well above the African average of 39%. Primary enrolments were 69% in 1986, and this is slightly below the African average of 76%. The secondary enrolment rate was 3% and this is well below the average African rate of 16%, and one of the lowest secondary enrolment rates in the world. The higher education enrolment was less than half of 1%, and again one of the lowest rates anywhere.

Life expectancy in 1987 was 53 years, and this is slightly above the African average of 51. In 1980, there was one doctor for every 17,360 persons, and this was rather better than the African average of one doctor for every 24,000 persons. Average daily calorie supply per person was 2,192 in 1986, and this was slightly above the African average.

Overall, Tanzania is a low-income country, with heavy dependence on agriculture, a very small industrial sector, and a below average size for its service sector. Private consumption is a large proportion of GDP, and a significant proportion of this, as well as of investment, is sustained by net aid inflows. Exports are low and reliance on imports high. Educational provision is uneven, with high literacy rates co-existing with poor secondary and tertiary enrolment rates. Health provision is good, particularly in view of Tanzania's low-income status.

Tanzania

Economic Performance

GDP grew at 1.7% a year in the period 1980-87, and this was better than the 0.4% average for Africa. However, population growth estimated at 3.5% implied GDP per head falling at -1.8%, a slower rate of decline than in Africa generally.

The agriculture sector expanded at 3.8% a year in the period 1980-87, and this more than kept pace with the growth in population. However the industrial sector contracted at -2.4% a year, twice the rate of contraction experienced on average in Africa, while manufacturing declined at -3.5% a year. The services sector grew at 0.8% a year.

Export volumes declined at -7.4% a year in the 1980-87 period, rather worse than the African average of -1.0% a year. Import volumes contracted by -0.4% a year, sustained by substantial aid inflows, whereas for Africa as a whole, import volumes declined by -5.8% a year.

The average price level has expanded by 24.9% a year in the 1980-87 period, and this is rather worse than the 15.2% annual inflation average in Africa in the same period.

Overall, Tanzania has experienced poor economic performance in the 1980s with GDP per head falling significantly, and only in the agricultural sector has output kept pace with population growth. Export performance has been particularly poor, with volumes falling rapidly, and inflation has been twice the African average.

Recent Economic Developments

In June 1986, the Tanzanian government reached agreement with the IMF, after resisting IMF terms since 1979. The main measures involved a substantial devaluation of the currency, rises in producer prices for key export crops, and a budget which aimed to keep government spending constant in real terms. The agreement released $US 400m in frozen aid commitments and $US 130m in new aid, and in August 1986 Tanzania requested a stand-by facility from the IMF of $US 64m. These changes, together with the revelation that the government was to register individual landholdings, marked a decisive movement away from planning and government intervention in the economy, which had been the main feature of Tanzania's economic strategy since 1967.

As part of the reform programme, the government has ended the monopoly of the National Milling Corporation (NMC) in the purchase of domestic foodstuffs. The emergence of private traders forced the NMC in 1987 to raise its purchase prices by 50% for rice, 30% for maize and 25% for wheat to remain competitive. There is now strong pressure from farmers to wind up the other crop marketing boards, particularly those dealing with coffee, cotton, tea and sugar. Several key hotels are to be run by international corporations, and other hotels and tourist facilities have been made into private companies. Lonrho has indicated its willingness to invest in Tanzania, and is reported to be extending its tea holdings, and to be involved in brewing and tractor production.

Recent economic performance has shown an encouraging response to the reform programme, with economic growth estimated at 3.8% in 1986, the first time that output has kept pace with population growth since 1980 and this performance being sustained through to 1989. Inflation is reported to be running at 30% annually, which is about the level experienced throughout the 1980s, and given the substantial adjustments to the exchange rate, and rises in producer prices, it is a creditable achievement to have prevented an acceleration. Exports have shown little improvement as yet, and the response to the increased producer prices for coffee and tea will only become apparent over the next five years, as there is a lag between new plantings and the first crops. However, cotton which is an annual crop, reached record output levels with 390 thousand bales as compared with 168 thousand in 1985/6, a 132% increase, and realised $US 55m in sales. There is still a substantial balance of trade deficit, with exports in 1989 estimated at $US 380m and merchandise imports at $US 1,200m. The trade gap is covered by aid flows.

The exchange rate has continued to depreciate. It stood at TSh 17.50 = $US 1 in mid 1985 and by the end of 1989 had fallen to TSh 187.7 = $US 1. Nevertheless, the black market rate is reported to stand at TSh 260 = $US 1, and further depreciations are expected before the exchange rate reflects the market valuation of the shilling.

The current aid programme concentrates heavily on rehabilitation of infrastructure. There is a massive $US 700m scheme to improve roads coordinated by the World Bank. Other projects include a $US 8m programme to upgrade electricity supplies funded by the World Bank, and schemes to up-grade port berths in Dar es Salaam financed by Sweden and Norway. Railway rehabilitation, being undertaken by various donors, and the Petro-Canada International Assistance Corporation is engaged in a $US 27m oil exploration programme funded by the World Bank.

Economic Outlook and Forecasts

The forecasts assume that Tanzania maintains its record of political stability, and perseveres with a steady programme of liberalising economic reforms.

Coffee prices are expected to fall in real terms by -8.0% up to 1995, while cotton prices are forecast to remain virtually unchanged.

GDP is forecast to expand at around 4.5% a year up to 1995, allowing an increase in GDP per head of about 1.1% a year.

Export volumes are expected to expand at just over 2.0% a year, with the growth rate showing some acceleration over the period. Import volumes are expected to expand at slightly above 5.0% a year as donors continue to provide the balance of payments support necessary for the rehabilitation of infrastructure.

Inflation performance is expected to improve, but still to be close to 20% a year up to 1990.

Tanzania: Economic Forecasts
(average annual percentage change)

	Actual 1980-87	Forecast Base 1990-95	Forecast High 1990-95
GDP	1.7	4.5	4.8
GDP per Head	-1.8	1.1	1.4
Inflation Rate	24.9	16.0	19.0
Export Volumes	-7.4	2.2	2.5
Import Volumes	-0.4	5.1	5.5

TANZANIA: Comparative Data

	TANZANIA	EAST AFRICA	AFRICA	INDUSTRIAL COUNTRIES
POPULATION & LAND				
Population, mid year, millions, 1989	25.5	12.2	10.2	40.0
Urban Population, %, 1985	14	30.5	30	75
Population Growth Rate, % per year, 1980-86	3.5	3.1	3.1	0.8
Land Area, thou. sq. kilom.	945	486	486	1,628
Population Density, persons per sq kilom., 1988	26.0	24.2	20.4	24.3
ECONOMY: PRODUCTION & INCOME				
GDP, $US millions, 1986	4,020	2,650	3,561	550,099
GNP per head, $US, 1986	250	250	389	12,960
ECONOMY: SUPPLY STRUCTURE				
Agriculture, % of GDP, 1986	59	43	35	3
Industry, % of GDP, 1986	10	15	27	35
Services, % of GDP, 1986	31	42	38	61
ECONOMY: DEMAND STRUCTURE				
Private Consumption, % of GDP, 1986	89	77	73	62
Gross Domestic Investment, % of GDP, 1986	17	16	16	21
Government Consumption, % of GDP, 1986	8	15	14	17
Exports, % of GDP, 1986	10	16	23	17
Imports, % of GDP, 1986	25	24	26	17
ECONOMY: PERFORMANCE				
GDP growth, % per year, 1980-86	0.9	1.6	-0.6	2.5
GDP per head growth, % per year, 1980-86	-2.5	-1.7	-3.7	1.7
Agriculture growth, % per year, 1980-86	0.8	1.1	0.0	2.5
Industry growth, % per year, 1980-86	-4.5	1.1	-1.0	2.5
Services growth, % per year, 1980-86	2.9	2.5	-0.5	2.6
Exports growth, % per year, 1980-86	-9.8	0.7	-1.9	3.3
Imports growth, % per year, 1980-86	-1.3	0.2	-6.9	4.3
ECONOMY: OTHER				
Inflation Rate, % per year, 1980-86	21.5	23.6	16.7	5.3
Aid, net inflow, % of GDP, 1986	15.2	11.5	6.3	-
Debt Service, % of Exports, 1984	44.5	18	20.6	-
Budget Surplus (+), Deficit (-), % of GDP, 1979	-11.3	-3.0	-2.8	-5.1
EDUCATION				
Primary, % of 6-11 group, 1985	72	62	76	102
Secondary, % of 12-17 group, 1985	3	15	22	93
Higher, % of 20-24 group, 1985	3	1.2	1.9	39
Adult Literacy Rate, %, 1980	79	41	39	99
HEALTH & NUTRITION				
Life Expectancy, years, 1986	53	50	50	76
Calorie Supply, daily per head, 1985	2,316	2,111	2,096	3,357
Population per doctor, 1980	17,360	35,986	24,185	550

Notes: 'Southern Africa' and 'Africa' exclude South Africa. Dates are for the country in question, and do not always correspond with the Regional, African and Industrial averages.

TANZANIA: Leading Indicators

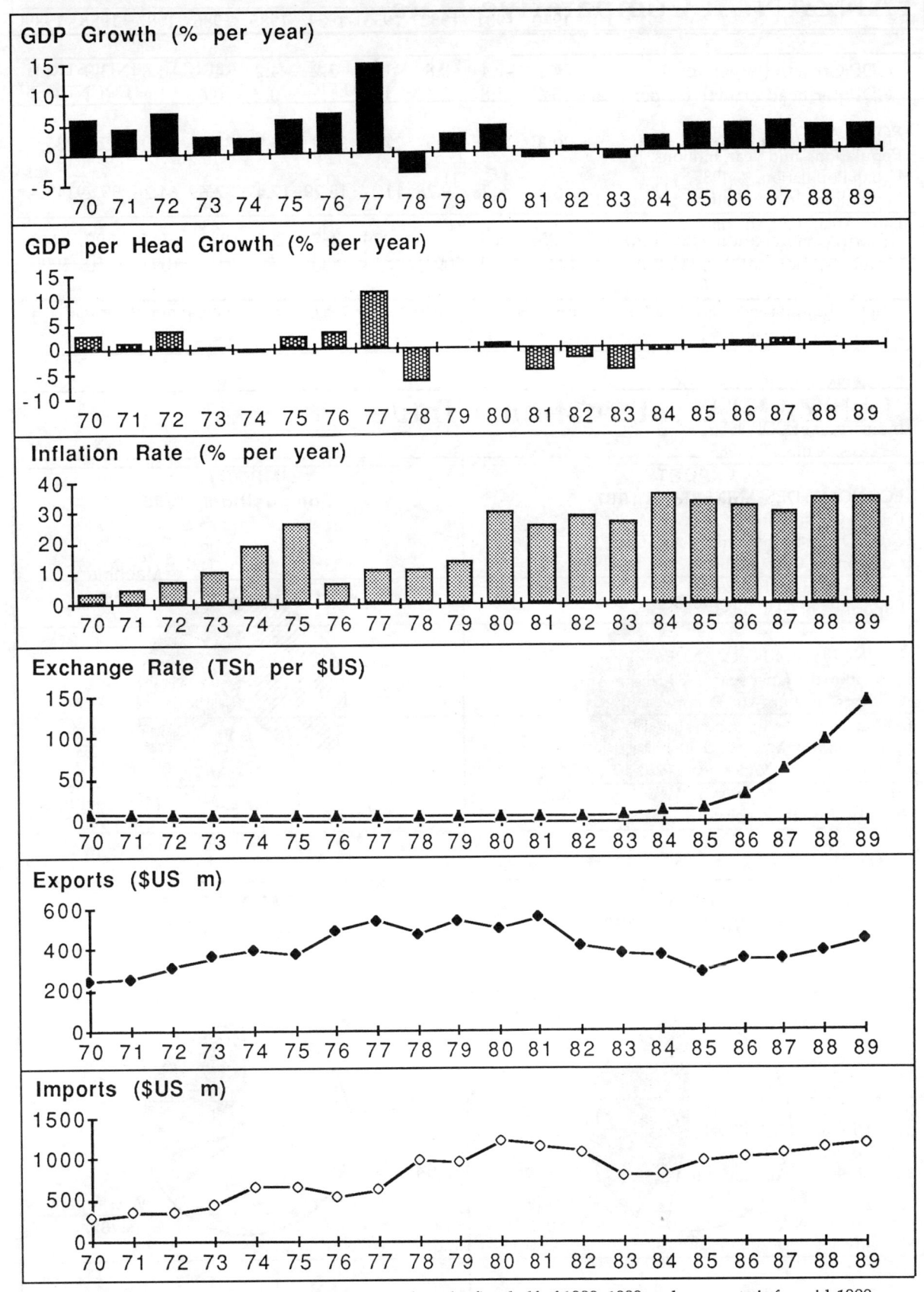

Note: Leading indicator data for 1989 are based on the first half of 1989. 1989 exchange rate is for mid-1989.

TANZANIA: Leading Indicator Data

	1980	1981	1982	1983	1984	1985	1986	1987	1988	1989
GDP Growth (% per year)	4.2	-1.4	0.8	-1.5	2.2	4.2	4.1	4.6	3.9	4.0
GDP per head growth (% per year)	0.6	-4.8	-2.6	-4.9	-1.2	-0.7	0.6	1.1	0.4	0.5
Inflation (% per year)	30.3	25.6	28.9	27.1	35.8	33.3	32.4	30.0	35.0	35.0
Exchange rate (Tsh per $US)	8.20	8.28	9.28	11.14	15.29	17.47	32.69	64.26	99.29	145.3
Exports, merchandise ($US m)	505	563	413	379	367	286	348	347	390	440
Imports, merchandise ($US m)	1220	1159	1093	819	839	999	1048	1092	1150	1200

Note: Leading indicator data for 1989 are based on the first half of 1989. 1989 exchange rate is for mid-1989.

TANZANIA: International Trade

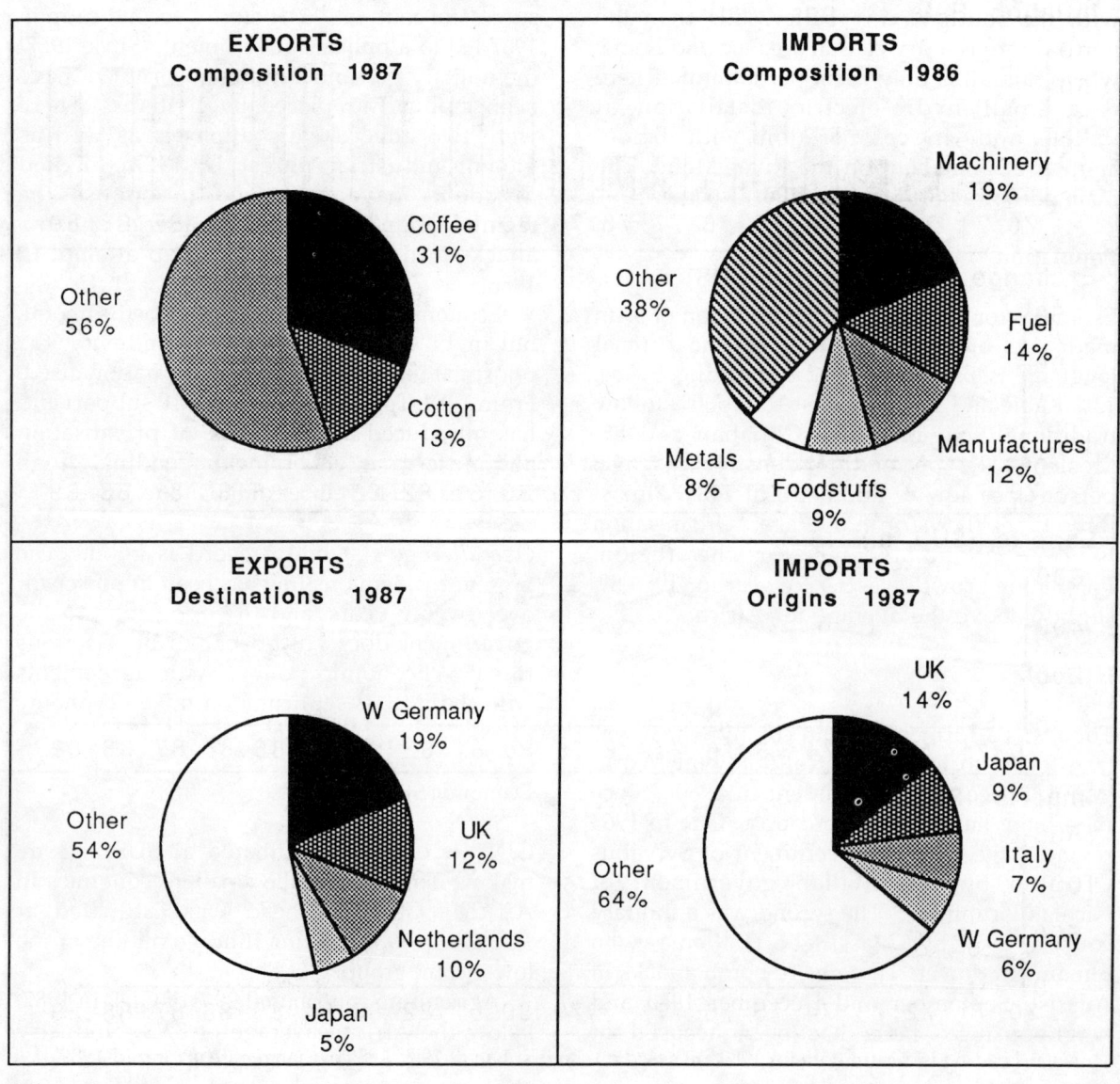

EXPORTS Composition 1987
- Coffee 31%
- Cotton 13%
- Other 56%

IMPORTS Composition 1986
- Machinery 19%
- Fuel 14%
- Manufactures 12%
- Foodstuffs 9%
- Metals 8%
- Other 38%

EXPORTS Destinations 1987
- W Germany 19%
- UK 12%
- Netherlands 10%
- Japan 5%
- Other 54%

IMPORTS Origins 1987
- UK 14%
- Japan 9%
- Italy 7%
- W Germany 6%
- Other 64%

44 TOGO

Physical Geography and Climate

Togo, in the West African region, is a small coastal state bounded by Ghana, Burkina Faso and Benin. The coast features lagoons and blocked estuaries with mangrove swamps. The vegetation is physically varied from rain forest in the south to savannah in the north, with the south having better soils than the north. There is a dry coastal area around Lomé, where rainfall is only 78cm per annum. There is a small hydro-electric installation at Palimé and, in co-operation with Benin, another at Nangbeto on the River Mono. The main ports are Port Lomé and Port Kpémé.

Population

Estimates for 1989 suggest a population of 3.4m made up of Ewe and Kabne. The official language is French with 22% speaking Evebe, 13% Kabie and 10% Wacibe. About 60% follow traditional religions, 25% Christian and 8% Moslem. The population density was 59.6 persons per square kilometre in 1989, almost three times the African average. Urbanisation is 23%, about average for the region. Population growth was 3.4% a year for 1980-89, slightly above the average for Africa.

History

Togo was a former German colony which was transferred to French control following World War I. It became independent in 1960. Since then there have been two coups. One in 1963 replaced the civilian government of Sylvanus Olympio by the civilian government of Nicolad Grunitzky. The second was a military coup in 1967 by Gnassingbe Eyadéma, who remains in power. There were bomb attacks in August, September and December 1985 and neighbouring Ghana has been accused of harbouring opponents of the regime. To appease Ghana, Ghanaian exiles opposed to President Rawlings have been deported.

Stability

Since independence in 1960, Togo had seven years of civilian rule, albeit interrupted by a successful coup in 1963, before a second coup in 1967 led to a military government. Since 1967, the military personnel in the government have been gradually replaced by civilians. There were two alleged coup attempts as well as violent industrial unrest in 1977. Exiled dissidents have continued to oppose the regime, and in 1985 there were a series of bomb attacks in the capital, and a coup attempt in 1986.

Economic strategy has been mostly liberal, but in 1974 the country's major industry, the phosphate mines, were partially nationalised. From 1979, Togo has relied on IMF support and has introduced a programme of privatisation and austerity in government spending. Togo has remained a member of the Franc Zone.

Overall,Togo's stability record is good. The present Eyadéma regime has been in power for over twenty years, and the opposition to the government does not appear to be a serious threat. Economic policy, with agreements with the IMF, is set firmly on mixed economy lines.

Economic Structure

GDP in 1987 was estimated at $US 1,230m, making Togo one of the smaller economies in Africa. GNP per head was estimated at $US 290 in 1987, putting it in the middle of the low-income group.

Agriculture contributed 29% of GDP in 1987, below the African average of 33%. Industry

Togo

generated 18% of GDP, and, within industry, manufacturing contributed 7%, both those being below the African averages. This leads to a heavy dependence on the services sector, which generates 54% of GDP, whereas in Africa the average is 40%.

On the demand side, private consumption was equivalent to 74% of GDP in 1987, about the African average, with government consumption at 21% of GDP, higher than the African average of 15%. Gross domestic investment was 17% of GDP, and with domestic savings at 6%, there was a net inflow of resources, mostly aid, equivalent to 12% of GDP. Exports, at 31% of GDP, generated more of GDP than is typical in Africa where the average is 25%. Import dependence, at 43% of GDP, was significantly higher than the African average of 29%.

In 1987, the main merchandise exports were phosphates 29%, cocoa 10%, coffee 9% and cotton 8%. Merchandise imports in 1985 were made up of manufactures 46%, foodstuffs 21%, machinery 18% and fuel 7%.

The adult literacy rate was estimated at 18% in 1975, but this is expected to have improved to about 40% at present, close to the African average. In 1986, primary enrolment rates were 102%, with a 21% enrolment rate in secondary education, and both these are significantly better than the African averages. The higher education enrolment rate at 2% is about the African norm.

Life expectancy in 1987 was 53 years, fairly typical of Africa. Provision of doctors at one for every 8,720 persons and nurses at one for every 1,240 in 1984 was three times better than the average for Africa for doctors, and almost twice as good for nurses. Daily calorie supply per person at 2,207 in 1986 was about 10% better than the African average.

Overall, Togo has one of the smaller economies in the medium-size group, with incomes per head in the low-income category. The economy has a large services sector. On the demand side, there is a large government sector and aid inflows sustain investment in the face of low savings rates. There is considerable reliance on export earnings and heavy reliance on imports. Educational and health provisions are above average.

Economic Performance

Togo's GDP declined by -0.5% a year between 1980 and 1987, whereas in all Africa it has expanded by 0.4%. GDP per head has declined by 3.9% a year, faster than in Africa generally. The agricultural sector grew by 0.8%, well below the population growth rate of 3.4% a year. The industrial sector declined by -1.6% compared with Africa's decline of -1.2% a year. The service sector also declined by -0.7% a year, whereas in Africa it expanded by 1.2% a year.

Export volumes fell at -3.0% in the period 1980-87, faster than the decline in Africa. Import volumes fell at -4.6% a year, and this contraction was slower than in Africa where import volumes fell at -5.8% a year.

Inflation averaged 6.6% a year in the period 1980-87, and this was under half the rate of Africa generally.

Overall, Togo has experienced poor economic performance in the 1980s, with falls in GDP, GDP per head and in all sectors except agriculture. Export volumes have declined steadily, and only in price stability has Togo's record been above average.

Recent Economic Developments

Togo has pursued policies of austerity in government spending in the 1980s, and has introduced a programme of privatisation of state-owned enterprises. Short-run balance of payments support from the IMF has been utilised since 1980. Togo received an enhanced structural adjustment facility from the IMF in May 1989 of $US 48m over three years. Twelve enterprises have been privatised covering plastics, dairying, oil exploration, steel, agricultural processing, quarrying, detergents, fats, oils, salt and textiles. Plans are in progress to privatise galvanising, transport, recording and confectionery enterprises. The current structural adjustment programme is expected to emphasise liberalisation of foreign investment regulations and reforms of agricultural pricing.

After a period of contracting GDP, 1981-83, Togo began to recover with positive rates of

GDP growth, with estimated expansion of GDP by 4.7% in 1985 and 3.1% in 1986. There was a slowing in 1987, but a recovery to over 4.0% a year in 1988 and 1989. Inflation performance improved up to 1984 when average prices fell by -3.5%, from close to a 20% increase in the price level in 1981. There has been a return to positive inflation subsequently, with the rate in 1989 estimated at 2.5% a year.

Togo's export revenues have fallen since 1980 when they stood at $476, being estimated at $US 340m in 1989. The reduction in export revenues caused by falling prices for phosphates and cocoa, and poor production performance in the coffee sector, has limited the ability to import, with the value of merchandise imports falling from $524m in 1980 to an estimated $US 370m in 1989.

Togo is estimated to have close to $1,000m in outstanding external debt at the end of 1989. Togo has been unwilling to request debt re-schedulings since 1985, but the pressure debt-service has put on the ability to import essential spares, fuels and inputs has led to a re-scheduling, over 20 years with a 16 year grace period. Until export earnings show a substantial recovery, Togo is expected to need regular debt re-schedulings.

Togo has been a member of the Franc Zone since independence, and since 1980 the CFA franc depreciated from CFA 211.3= $US 1 in 1980 to CFA 449.3=$US 1 in 1985. Subsequently, the weakness of the dollar has seen an appreciation to CFA 289.3=$US 1 at the end of 1989. This appreciation has led to speculation that the CFA franc might be devalued against the French franc.

France continues to be the major source of concessionary finance, and French and World Bank funds are involved in a $US 33.2m scheme to improve coffee and cocoa production. France is also providing $US 6.9m to combat coastal erosion and $US 2.7m for telecommunications. The United Nations is contributing $US 4.3m to infrastructure. The World Bank is allocating $US 9.7m to agricultural extension, $US 11.4m to encourage small-scale enterprises and is considering a $US 100m roads project.

Private foreign investment is involved in the privatisation of the textile sector where US and South Korean firms are involved in a joint venture in which the World Bank's International Finance Corporation has some equity. The privatised steel sector is embarking on a $US 2m project to produce galvanised pylons for export in West Africa.

Economic Outlook and Forecasts

The forecasts assume that the stability of the country remains good, and that Togo continues to benefit from the discipline bestowed by Franc Zone membership. It is assumed that the liberalising economic reform programme remains in place.

Prices of phosphates are expected to increase in real terms by 10% up to 1995. Cocoa prices, however, are projected to fall by -19%. Cotton prices are expected to remain virtually unchanged.

GDP is expected to maintain the positive growth rates achieved in the late 1980s, and to expand at an average rate of just over 4.0% a year up to 1995. This will allow GDP per head to expand at just over 1.0% a year.

Under the influence of improved producer prices, export volumes are projected to show positive growth of just over 3.0% a year. This, together with aid flows, should allow import volumes to expand at around 3.5% a year.

Inflation is expected to be low, at under 3% a year.

Togo: Economic Forecasts
(average annual percentage change)

	Actual	Forecast Base	High
	1980-87	1990-95	1990-95
GDP	-0.5	4.2	4.5
GDP per Head	-3.9	1.1	1.4
Inflation Rate	6.6	3.0	2.0
Export Volumes	-3.0	3.2	3.5
Import Volumes	-4.6	3.5	3.7

TOGO: Comparative Data

	TOGO	WEST AFRICA	AFRICA	INDUSTRIAL COUNTRIES
POPULATION & LAND				
Population, mid year, millions, 1989	3.4	12.3	10.2	40.0
Urban Population, %, 1985	23	28.7	30	75
Population Growth Rate, % per year, 1980-86	3.4	3.2	3.1	0.8
Land Area, thou. sq. kilom.	57	384	486	1,628
Population Density, persons per sq kilom., 1988	57.9	31.1	20.4	24.3
ECONOMY: PRODUCTION & INCOME				
GDP, $US millions, 1986	980	4,876	3,561	550,099
GNP per head, $US, 1986	250	510	389	12,960
ECONOMY: SUPPLY STRUCTURE				
Agriculture, % of GDP, 1986	32	40	35	3
Industry, % of GDP, 1986	20	26	27	35
Services, % of GDP, 1986	48	34	38	61
ECONOMY: DEMAND STRUCTURE				
Private Consumption, % of GDP, 1986	71	77	73	62
Gross Domestic Investment, % of GDP, 1986	28	13	16	21
Government Consumption, % of GDP, 1986	15	13	14	17
Exports, % of GDP, 1986	33	18	23	17
Imports, % of GDP, 1986	47	20	26	17
ECONOMY: PERFORMANCE				
GDP growth, % per year, 1980-86	-1.1	-2.03	-0.6	2.5
GDP per head growth, % per year, 1980-86	-4.4	-4.7	-3.7	1.7
Agriculture growth, % per year, 1980-86	1.7	1.3	0.0	2.5
Industry growth, % per year, 1980-86	-2.2	-3.6	-1.0	2.5
Services growth, % per year, 1980-86	-2.3	-2.5	-0.5	2.6
Exports growth, % per year, 1980-86	-6.6	-4.0	-1.9	3.3
Imports growth, % per year, 1980-86	-10.0	-12.9	-6.9	4.3
ECONOMY: OTHER				
Inflation Rate, % per year, 1980-86	6.7	13.0	16.7	5.3
Aid, net inflow, % of GDP, 1986	18.5	3.7	6.3	-
Debt Service, % of Exports, 1986	32.5	21.4	20.6	-
Budget Surplus (+), Deficit (-), % of GDP, 1986	-5.1	-3.4	-2.8	-5.1
EDUCATION				
Primary, % of 6-11 group, 1985	95	75	76	102
Secondary, % of 12-17 group, 1985	21	24	22	93
Higher, % of 20-24 group, 1985	2	2.6	1.9	39
Adult Literacy Rate, %, 1975	18	30	39	99
HEALTH & NUTRITION				
Life Expectancy, years, 1986	53	50	50	76
Calorie Supply, daily per head, 1985	2,221	2,105	2,096	3,357
Population per doctor, 1981	21,140	16,199	24,185	550

Notes: 'Southern Africa' and 'Africa' exclude South Africa. Dates are for the country in question, and do not always correspond with the Regional, African and Industrial averages.

TOGO: Leading Indicators

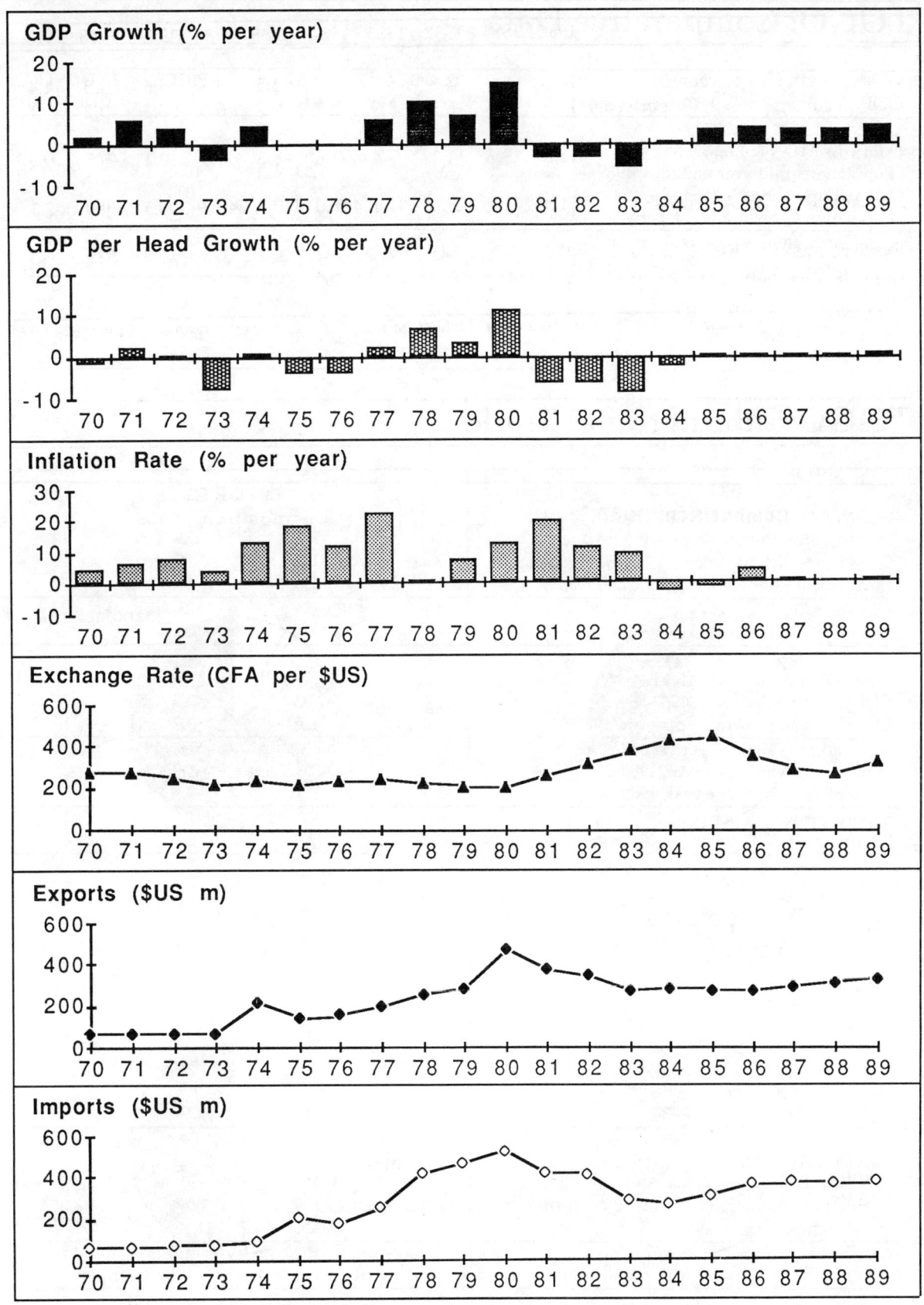

GDP Growth (% per year)

GDP per Head Growth (% per year)

Inflation Rate (% per year)

Exchange Rate (CFA per $US)

Exports ($US m)

Imports ($US m)

Note: Leading indicator data for 1989 are based on the first half of 1989. 1989 exchange rate is for mid-1989.

TOGO: Leading Indicator Data

	1980	1981	1982	1983	1984	1985	1986	1987	1988	1989
GDP growth (% per year)	14.5	-3.4	-3.3	-5.7	0.5	3.5	3.8	3.6	3.5	4.4
GDP per head growth (% per year)	11.3	-6.6	-6.5	-8.9	-2.7	0.2	0.5	0.3	0.2	1.1
Inflation (% per year)	12.3	19.7	11.1	9.4	-3.6	-1.8	4.1	0.1	-0.5	0.9
Exchange rate (CFA per $US)	211.3	271.7	328.6	381.1	437.0	449.3	346.3	300.5	279.9	327.3
Exports, merchandise ($US m)	476	378	345	274	291	282	273	298	319	340
Imports, merchandise ($US m)	524	414	408	289	263	304	355	362	360	366

Note: Leading indicator data for 1989 are based on the first half of 1989. 1989 exchange rate is for mid-1989.

TOGO: International Trade

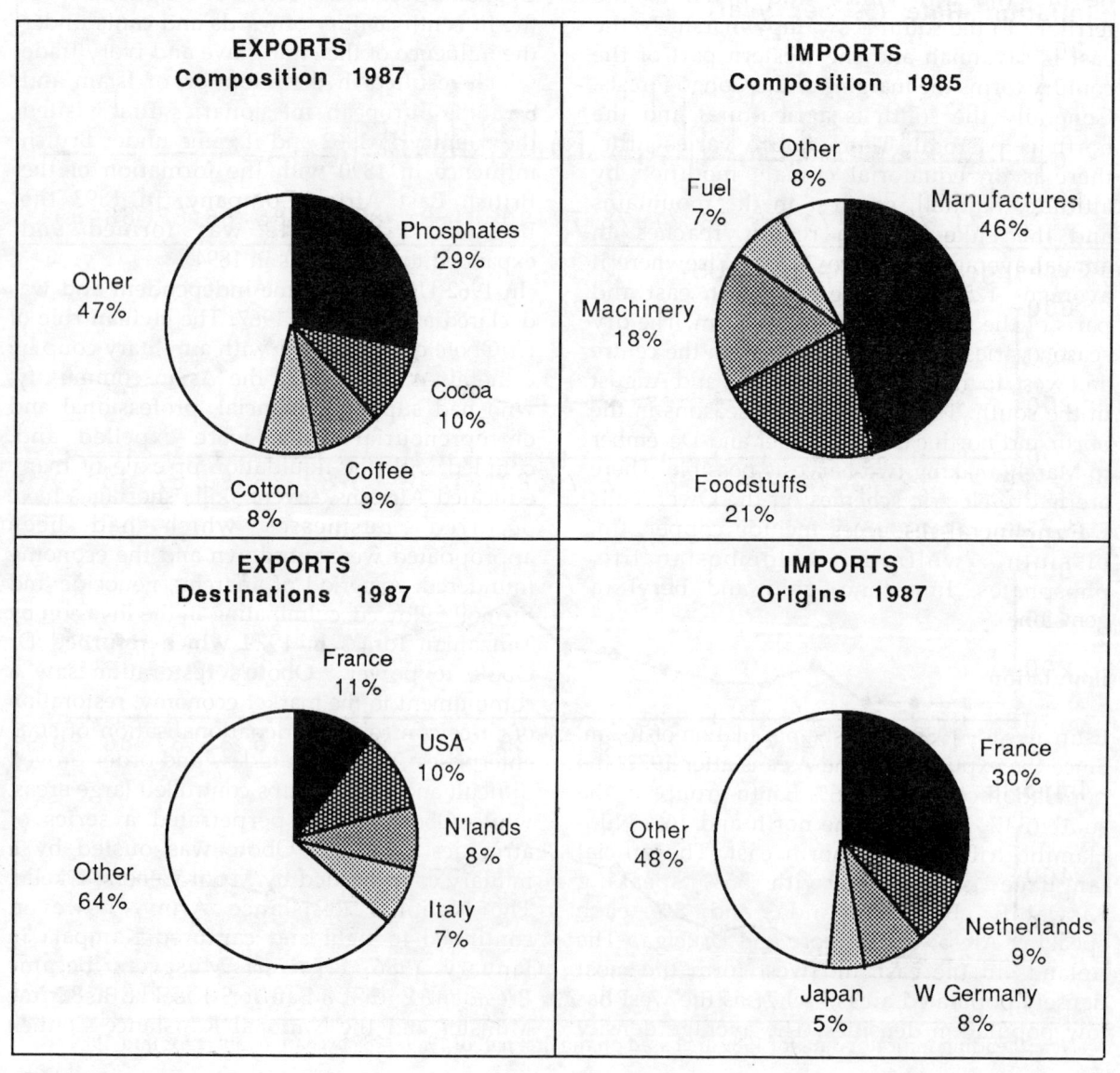

EXPORTS Composition 1987
- Phosphates 29%
- Cocoa 10%
- Coffee 9%
- Cotton 8%
- Other 47%

IMPORTS Composition 1985
- Manufactures 46%
- Foodstuffs 21%
- Machinery 18%
- Fuel 7%
- Other 8%

EXPORTS Destinations 1987
- France 11%
- USA 10%
- N'lands 8%
- Italy 7%
- Other 64%

IMPORTS Origins 1987
- France 30%
- Netherlands 9%
- W Germany 8%
- Japan 5%
- Other 48%

45 UGANDA

Physical Geography and Climate

Uganda, in the East African region, is a medium-sized landlocked state bordered by Sudan, Kenya, Tanzania, Rwanda and Zaïre It forms part of the central African plateau, dropping to the White Nile Basin in the north, the chief river. Lake Kionga and Lake Albert lie in the Rift Valley and much of the territory to the south is swampy marsh. To the east is savannah and the western part of the country forms the margins of the Congo forests. Generally the south is agricultural and the north is pastoral. Temperature varies little; there is an equatorial climate modified by altitude. Rainfall, greatest in the mountains and the Lake Victoria region, reaches an annual average of up to 200cm. Elsewhere it averages 125cm but the dry north east and parts of the south receive up to 75cm. The dry season varies between one month in the centre and west, to the months of June, July and August in the south. There are two dry seasons in the north and north east in October and December to March, making two harvests possible. There are hydro-electric schemes on the Owen Falls Dam. Mineral resources include copper, tin, bismuth, wolfram, colombo-tanalite, phosphates, limestone, gold and beryl (a gemstone).

Population

Estimates for 1989 suggest a population of 16.7m. Since the expulsion of the Asians after 1972, the population consists of 66% Bantu groups in the south, 33% Nilotic in the north and 16% Nilo-Hamitic tribes in the north east. The official language is English, with 35% speaking Kiswahili, 16% Oluganda and 8% each speaking Atesa, Orunankore and Uruciga. The uplands in the east and west form the most densely populated areas, whereas the west has low population densities. The average density of population was 71 per square kilometre in 1989. Urbanisation, at 10%, as a result of the turmoil of recent years, is well below the African average. Population growth at 3.1% a year for 1980-87 is about the African average.

History

Uganda developed a series of kingdoms from the fifteenth century onwards and came under the influence of the Arab slave and ivory trade which resulted in the adoption of Islam and Swahili. European missionaries first visited the country in 1862 and it came under British influence in 1890 with the formation of the British East Africa Company. In 1893 the Buganda Protectorate was formed and expanded to other areas in 1894.

In 1962 Uganda became independent and was declared a republic in 1967. The civilian rule of Dr Obote ended in 1971 with a military coup by General Amin. In 1972 the Asian community, who had supplied financial, professional and entrepreneurial skills, were expelled and, coupled with the liquidation or exile of many educated Africans, severe skills shortages have occurred. Businesses which had been appropriated were run down and the economy foundered. A period of anarchy, genocide and turmoil followed, culminating in the invasion by Tanzanian forces in 1979 which returned Dr Obote to power. Obote's restoration saw a commitment to the market economy, restoration of price controls and denationalisation of state enterprises. The return to law and order proved difficult and rebel groups controlled large areas, while Obote's army perpetrated a series of atrocities. In 1985 Obote was ousted by a military coup headed by Major General Okello. The National Resistance Army, however, continued to fight and captured Kampala in January 1986. Yoweri Museveni became President with Dr Samson Kisekka as Prime Minister and the National Resistance Council

Uganda

formed a joint military and civilian government.

Stability

Since Independence in 1963, Uganda had nine years of civilian government before a military coup in 1971 led to eight years of capricious and chaotic rule by General Amin. Amin was overthrown by Tanzanian intervention, but despite the return of Obote, political institutions were weak, and the economy suffered from disruption by lawless groups. In 1985 there was a military coup followed by a takeover in 1986 by the National Resistance Army of Yoweri Museveni, who had waged a guerrilla campaign since the return of Obote in 1980.

Economic strategy has fluctuated with Obote initially pledged to pursue a socialist development path, followed by the chaos of the Amin years which included the expulsion of the skilled Asian community. The restored Obote regime relied on market forces, but lack of security prevented any substantial progress. Museveni spent a year considering development options, but now appears to have cautiously committed the government to an IMF supported market-oriented strategy.

Overall, Uganda's stability has been very poor indeed. Under Museveni, internal security has improved and the economic environment will benefit if Museveni continues to pursue the present set of market-economy policies.

Economic structure

GDP in 1987 was estimated at $US 3,560m, making it roughly average for Africa in terms of economic size. GNP per head was estimated at $US 260, and this places Uganda roughly in the middle of the low-income group of countries.

Agriculture provides the bulk of GDP, contributing 76% in 1987, and this is over double the dependence on this sector than is typical elsewhere in Africa. The industry sector provides only 5% of GDP, and there was negligible production in industry outside manufacturing. The services sector contributed 19% of GDP, and the contributions of industry and services are very low compared with other African countries.

On the demand side, private consumption was equivalent to 88% of GDP in 1987, and this is high compared with other low-income countries in Africa. Government consumption was 7% of GDP, investment was 12%, and domestic saving was 5% of GDP, and these three categories are well below the African norm.

Exports are equivalent to 10% of GDP, and imports are 17%, the gap, which allows investment to be greater than domestic saving, is covered by the net inflow of aid. Exports are predominantly coffee, providing 98% of the total in 1987, with small amounts of cotton and tea also sold overseas. Imports in 1987 were mostly machinery (46%) and manufactures (38%), with fuel comprising 9% of the total.

The adult literacy rate was estimated at 55% in 1986, and this is above the African average. The turmoil and dislocation of the last two decades have prevented the collection of education statistics in recent years. In 1977, enrolment in primary education was 90%, and in secondary education it was 23%, and at that time this was a rather better provision than on average elsewhere in Africa. In 1984, the enrolment rate in higher education was 1%, and this is below the African average which is close to 2%.

Life expectancy in 1987 was 48, close to the African average of 51. In 1983, there were 21,900 persons per doctor and 2,060 persons per nurse, comparable with the average for the rest of Africa, and this is a considerable achievement given that many skilled professionals have migrated or been expelled since 1970. Average calorie supply, at 2,344 per person per day, is above the African average.

Overall, Uganda is a low-income country with an economy of about average size for the continent. The turmoil of the past two decades has led to a reliance on agriculture for subsistence, which in turn has caused consumption to be emphasised at the expense of saving and investment. Uganda takes limited advantage of the opportunities offered by trade. Educational and health provision are good considering the disruptions and migration of professionals which the country has experienced.

Uganda

Ecomomic Performance

GDP grew by 0.4% a year from 1980 to 1987, but when population growth is taken into account, GDP per head declined by -2.7% a year.

Agriculture, greatly affected by the security problems in rural areas, declined by -0.5% a year. This sector may take time to recover as the rural population gradually begins to live a normal lifestyle. The industrial sector, which like the agricultural sector has suffered from the destruction of much of Uganda's infrastructure, grew by 1.4% a year, while the service sector grew by 3.0%.

Export volumes increased by 2.7% a year, despite the domestic problems and large-scale smuggling of goods, and this compares favourably with the African figure of -1.0%. Import volumes were able to grow faster than exports, at 3.0% a year, as a result of aid inflows.

The inflation rate for 1980-87 was 95.2%, the highest in Africa, reaching 224% during 1987. Salaries have not risen at the same rate, thus encouraging corruption and poor work performance, particularly in the public sector.

Overall, Uganda has had some expansion of GDP in the 1980s, but GDP per head has fallen. Agricultural output declined, industrial output increased modestly, as did services output, exports and imports, although none of these categories expanded as fast as population growth. Price stability has been the worst in Africa.

Recent Economic Developments

In May of 1987 President Museveni appeared to end the period of indecision over Uganda's economic strategy when agreement was reached on a programme which would see the IMF supplying $US 76m and the World Bank $US 100m in balance of payments support. The currency was exchanged with a substantial devaluation and conversion tax, producer prices for export crops and import prices were raised. There have been large rises in domestic prices, but in June the government indicated that there would be no return to control of prices. Some of the properties returned to their former Asian owners were repossessed by the government after concern that the ownership claims were not valid, and this move is expected to discourage participation by Asian entrepreneurs. Administration and production continued to be hampered by shortages of skilled personnel. Economic performance appeared to show some recovery to positive growth rates in the three years 1981 to 1983, but there was a relapse to negative growth of around -5.0% in 1984 and 1985, with the contraction in GDP slowing to -1.0% in 1986. GDP per head rose in the 1981 to 1983 recovery, but this has been more than offset by the falls in the 1984 to 1986 period, before recovering in 1987 and 1988. Inflation has accelerated considerably, from around 20% annual increase in prices in 1980 to 100% in 1985, 175% in 1986, and a reported 300% in the first half of 1987. In 1989, the IMF agreed an enhanced structural adjustment facility of $US 199m, although $US 100m of this, due at the end of 1989, has been withheld as the bank is concerned over the size of the budget deficits which are fuelling inflation.

Export earnings, predominantly coffee, were lower in 1989 than in 1980. 1986 was an excellent year for coffee prices, but continuing smuggling diminished official receipts. It was hoped that more Draconian measures against smuggling and a 200% rise in produce prices would increase deliveries, but transport problems and payment delays appear to have offset these effects.

Uganda was estimated to have $US 1,800m of outstanding debt at the end of 1989. There are reported to be arrears in re-payments to some bi-lateral donors, and in June the government arranged for $100m of debt falling due in 1987 to be re-scheduled. Barter deals with Yugoslavia, Libya, Cuba and private organisations from UK, US and West Germany have been concluded since mid 1986, involving about $US 170m of trade.

The currency exchange introduced in May of 1987, together with a conversion tax of 80%, involved a substantial devaluation of the currency. Prior to the exchange, the Uganda shilling had been quoted at USh 1,350=$US 1, to which it had steadily depreciated from USh 7.47=$US 1 in 1980. The currency exchange was effected with 100 old shillings being exchanged for 1 new shilling, but with the conversion tax, this meant 130 old shillings were required for 1 new shilling. The exchange

Uganda

rate for the new currency was initially set at USh 60 = $US 1. Further devaluation of the currency has seen the rate depreciate to USh 370 = $US 1 by the end of 1989.

With improving security, and the apparent resolution of Uganda's economic strategy with the May 1987 agreement with the IMF, there has been a restoration of suspended aid commitments and substantial promises of new funds. The Islamic Development Bank has reactivated a $US 4.5m rice project, and $US 6.5m for roads. Kuwait has restored $US 10m for livestock and $US 10m balance of payments support, and Saudi Arabia is considering confirming $US 30m of projects. The African Development Fund and the Islamic Development Bank have committed $US 25.4m to a rice scheme. Italy and France are reported to be considering $US 130m of aid for rail rehabilitation, while the World Bank is involved in $US 67m of road projects. The Africa Development Bank is considering a $US 96m phosphate plant, the Arab Bank for Economic Development in Africa is committing $US 60m for roads and the African Development Fund is lending $US 4.9m for hospitals. China is involved in $US 21.3m of projects, Yugoslavia is lending $US 16.5m and Italy $US 15.1m for hotel rehabilitation. The EEC is allocating $US 3.9m for refugees, and the World Bank, Denmark, and the United Nations are involved in a $US 73m timber project. The World Bank, the Commonwealth Development Corporation and the UK are devoting $US 55m for electricity rehabilitation and work on the Owen Falls hydro-electric installations.

Economic Outlook and Forecasts

The forecasts assume that Uganda's stability continues to show steady improvement, and that the World Bank and IMF continue their support.

Coffee prices are expected to fall in real terms by -8.0% up to 1995.

GDP is forecast to expand at around the rate of population expansion, allowing slight annual increases in GDP per head. Export volumes are expected to expand as the agriculture sector recovers, but deteriorating terms of trade will limit the expansion of import volumes.

Pressure from Washington institutions to contain the budget deficit is expected to bring inflation under control with an annual rate of 20% being achieved by 1995.

Uganda: Economic Forecasts
(average annual percentage change)

	Actual 1980-87	Forecast Base 1990-95	Forecast High 1990-95
GDP	0.4	3.5	3.7
GDP per Head	-2.7	0.2	0.4
Inflation Rate	95.2	35.0	40.0
Export Volumes	2.7	4.2	4.5
Import Volumes	3.0	3.3	3.5

UGANDA: Comparative Data

	UGANDA	EAST AFRICA	AFRICA	INDUSTRIAL COUNTRIES
POPULATION & LAND				
Population, mid year, millions, 1989	16.7	12.2	10.2	40.0
Urban Population, %, 1985	7	30.5	30	75
Population Growth Rate, % per year, 1980-86	3.1	3.1	3.1	0.8
Land Area, thou. sq. kilom.	236	486	486	1,628
Population Density, persons per sq kilom., 1988	68.6	24.2	20.4	24.3
ECONOMY: PRODUCTION & INCOME				
GDP, $US millions, 1986	3,310	2,650	3,561	550,099
GNP per head, $US, 1986	230	250	389	12,960
ECONOMY: SUPPLY STRUCTURE				
Agriculture, % of GDP, 1986	76	43	35	3
Industry, % of GDP, 1986	6	15	27	35
Services, % of GDP, 1986	18	42	38	61
ECONOMY: DEMAND STRUCTURE				
Private Consumption, % of GDP, 1986	76	77	73	62
Gross Domestic Investment, % of GDP, 1986	14	16	16	21
Government Consumption, % of GDP, 1986	13	15	14	17
Exports, % of GDP, 1986	12	16	23	17
Imports, % of GDP, 1986	15	24	26	17
ECONOMY: PERFORMANCE				
GDP growth, % per year, 1980-86	0.7	1.6	-0.6	2.5
GDP per head growth, % per year, 1980-86	-2.3	-1.7	-3.7	1.7
Agriculture growth, % per year, 1980-86	-0.1	1.1	0.0	2.5
Industry growth, % per year, 1980-86	0.9	1.1	-1.0	2.5
Services growth, % per year, 1980-86	3.3	2.5	-0.5	2.6
Exports growth, % per year, 1980-86	4.4	0.7	-1.9	3.3
Imports growth, % per year, 1980-86	2.2	0.2	-6.9	4.3
ECONOMY: OTHER				
Inflation Rate, % per year, 1980-86	74.9	23.6	16.7	5.3
Aid, net inflow, % of GDP, 1986	5.7	11.5	6.3	-
Debt Service, % of Exports, 1986	6.5	18	20.6	-
Budget Surplus (+), Deficit (-), % of GDP, 1986	-2.8	-3.0	-2.8	-5.1
EDUCATION				
Primary, % of 6-11 group, 1977	90	62	76	102
Secondary, % of 12-17 group, 1977	23	15	22	93
Higher, % of 20-24 group, 1984	1	1.2	1.9	39
Adult Literacy Rate, %, 1986	55	41	39	99
HEALTH & NUTRITION				
Life Expectancy, years, 1986	48	50	50	76
Calorie Supply, daily per head, 1985	2,483	2,111	2,096	3,357
Population per doctor, 1981	21,270	35,986	24,185	550

Notes: 'Southern Africa' and 'Africa' exclude South Africa. Dates are for the country in question, and do not always correspond with the Regional, African and Industrial averages.

UGANDA: Leading Indicators

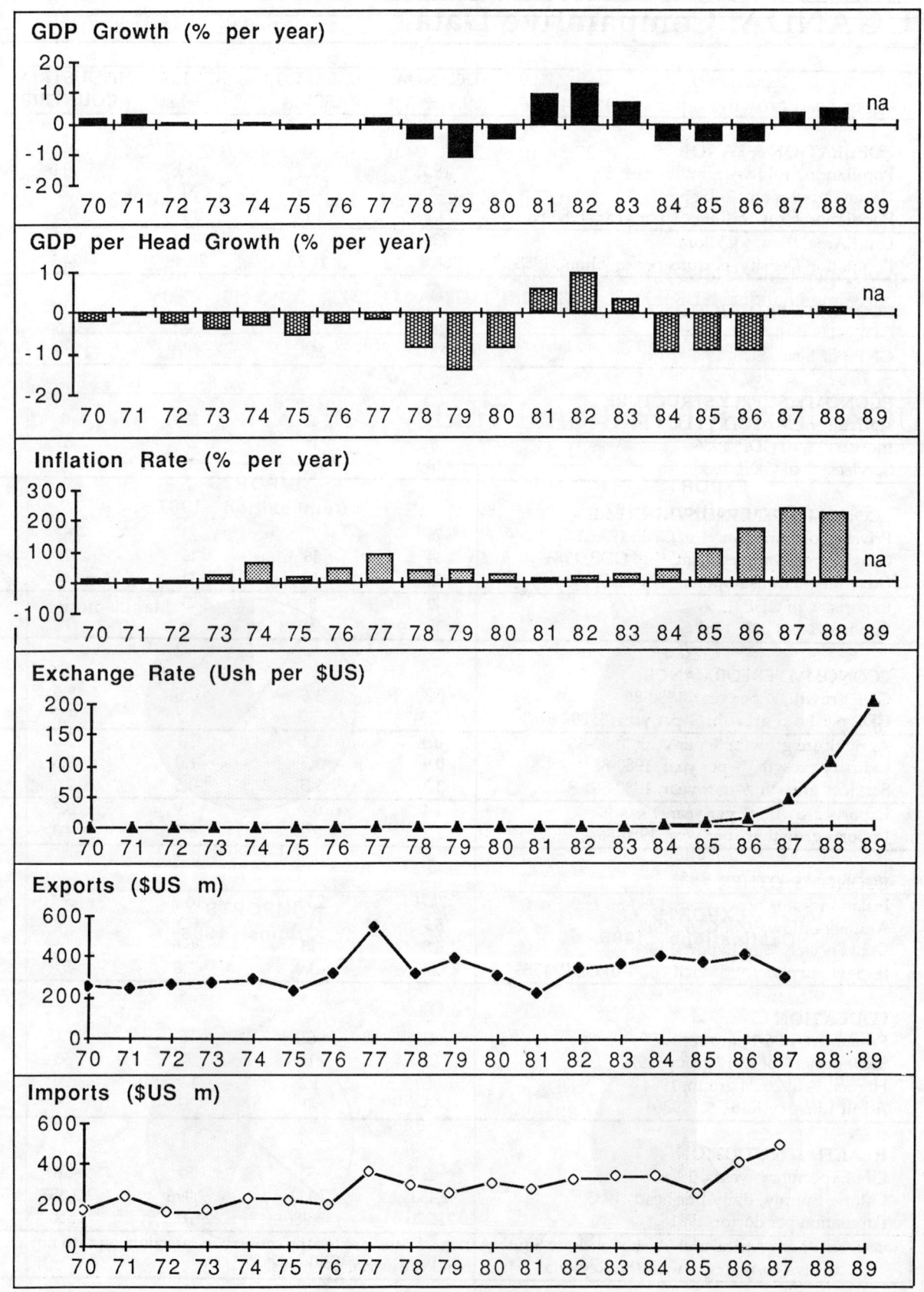

GDP Growth (% per year)

GDP per Head Growth (% per year)

Inflation Rate (% per year)

Exchange Rate (Ush per $US)

Exports ($US m)

Imports ($US m)

Note: Leading indicator data for 1989 are based on the first half of 1989. 1989 exchange rate is for mid-1989.

UGANDA: Leading Indicator Data

	1980	1981	1982	1983	1984	1985	1986	1987	1988	1989
GDP growth (% per year)	-5.2	9.5	13.1	7.1	-6.2	-6.0	-6.0	4.0	5.0	
GDP per head growth (% per year)	-8.4	6.3	9.9	3.9	-9.4	-9.2	-9.2	0.8	1.8	
Inflation (% per year)		27.8	10.6	21.4	24.0	38.3	106.7	168.0	238.0	224.0
Exchange rate (Ush per $US)		0.07	0.50	0.94	1.54	3.60	6.72	14.00	42.80	106.30	200.00
Exports, merchandise ($US m)		319	229	347	368	408	379	415	310	
Imports, merchandise ($US m)		318	284	338	343	342	264	413	500	

Note: Leading indicator data for 1989 are based on the first half of 1989. 1989 exchange rate is for mid-1989.

UGANDA: International Trade

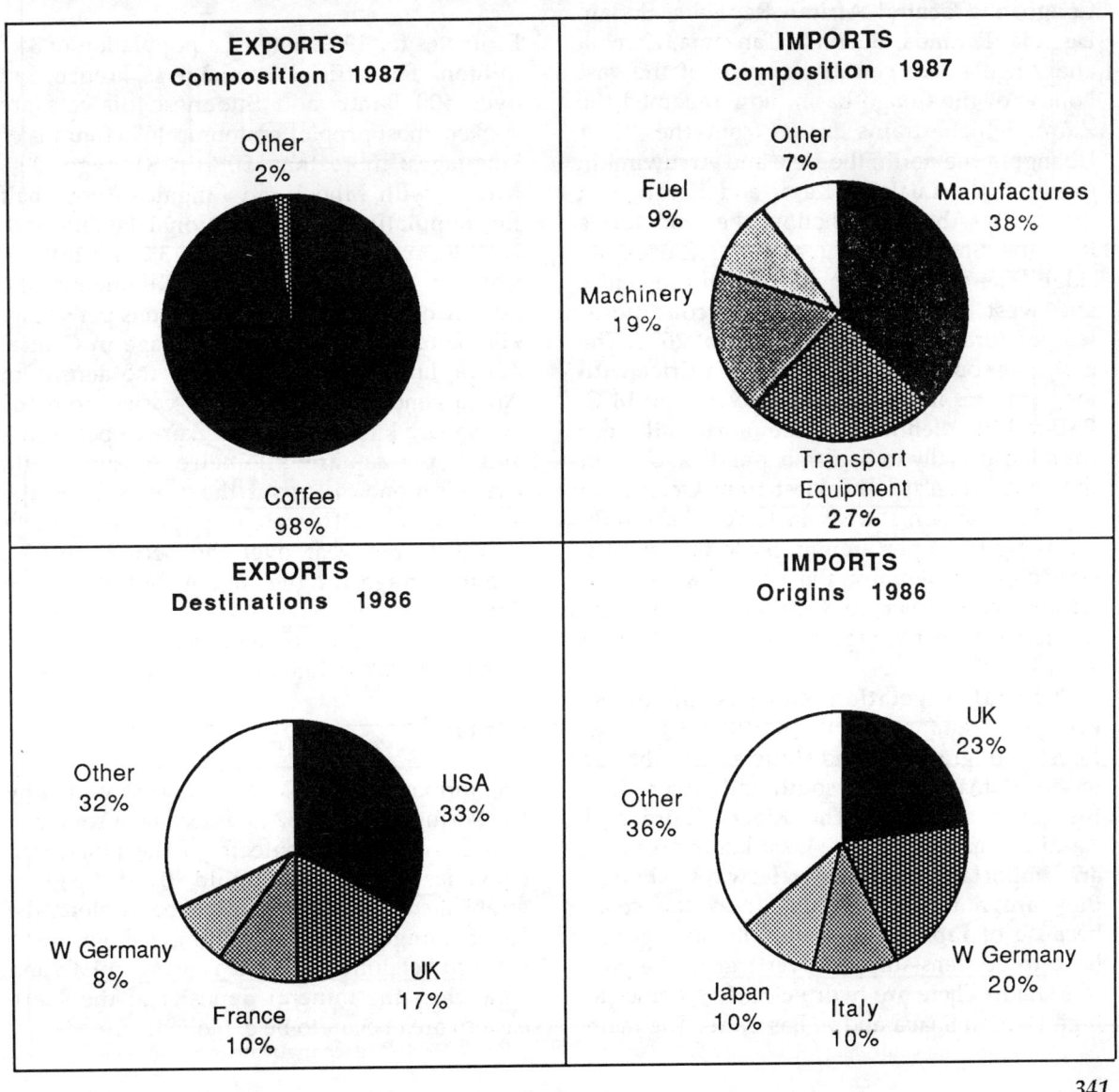

EXPORTS Composition 1987
- Coffee 98%
- Other 2%

IMPORTS Composition 1987
- Manufactures 38%
- Transport Equipment 27%
- Machinery 19%
- Fuel 9%
- Other 7%

EXPORTS Destinations 1986
- USA 33%
- Other 32%
- UK 17%
- France 10%
- W Germany 8%

IMPORTS Origins 1986
- Other 36%
- UK 23%
- W Germany 20%
- Japan 10%
- Italy 10%

46 ZAIRE

Physical Geography and Climate

Zaïre, in the Central African region, is the second largest sub-Saharan country in geographical area. It lies across the Equator, with coastal access, and is bounded by the Congo, the Central African Republic, Sudan, Uganda, Rwanda, Burundi, Tanzania, Zambia and Angola. The country consists of the vast hollow of the Congo basin, now renamed the Zaïre, which drains water from the River Ubangi in the north; the Uele and Arruwimi in the east; the Lualaba, Kasai and Kwango in the south. Above the hollow the land levels into the Shaba plateau and the Zaïre-Nile ridge. The climate is equatorial in the centre and west with hot and humid conditions, temperatures reaching an average of 26°C. The east and south east are cooler and drier with temperatures in the mountains averaging 18°C. Rainfall is plentiful in the north all year round, especially in September and October; in the south west it is heaviest from October to March, with some areas in lower Zaïre only receiving 80cm per annum. Since the country straddles the Equator the north has a drier winter, November to February, whilst the south has a drier winter, in the months May to August.

Natural vegetation consists of dense evergreen rainforest in the north and tropical forest with grassland and shrub savannah, due to fire damage, in the south and covers half the total landspace. The Rivers Zaïre and Kasai are navigable for 14,000 kilometres and are important national waterways, though they are not navigable as far as the coast because of rapids and waterfalls, and goods have to be trans-shipped overland to the port of Matadi. There are hydro-electric schemes at Inga Dam in Shaba and in Bas Zaïre. The main ports are Kinshasa, Boma and Matadi. Mineral resources include copper, cobalt, diamonds, gold, oil, manganese, silver, tungsten, uranium and zinc.

Population

Estimates for 1989 suggest a population of 34.7 million. The official language is French, but over 400 Bantu and Sudanese dialects are spoken; most people communicate in four major languages: more than 50% in Kikongo, 20% Kiteke, with Mbosi and Lingala. About half the population have traditional beliefs with 30% Roman Catholics and 13% Protestant. With an area of 2.3m square kilometres, the density of population is 14.4 persons per square kilometre, higher than the average in Central Africa, but about two-thirds of the density in Africa generally. The density varies from 100 per square kilometre in Bas Zaïre to between 1 and 3 per square kilometre in the south. Urbanisation was about the regional average at 38% in 1987. The rate of population growth was 3.1% per year over the period 1980-87, about average for the region, but above the African rate.

The population of the capital, Kinshasa, is about 2,654,000 while Lubumbashi has 543,000.

History

The former Belgian Congo was first visited by the Portuguese in 1482 and became a source of slaves from the seventeeth to the nineteenth centuries. In 1885 Leopold II of Belgium established a series of trading posts along the River Congo. The flourishing rubber trade induced Belgium to make it a colony in 1908 and from then the mineral deposits of the south eastern area began to be exploited.

Zaire

Independence was granted in 1960, but preparation was inadequate and the nationalist Prime Minister Patrice Lumumba and President Joseph Kasavubu were immediately faced with an army mutiny and the threat of secession by the mineral rich province of Katanga (now called Shaba) and South Kasai. United Nations forces intervened and following the assassination of Lumumba, a new government was formed under Cyrille Adoula.

By 1963 the UN forces and the army, led by Col. Joseph-Desiré Mobutu, had defeated both rebel groups. A new federal constitution was adopted and in 1964 Moïse Tshombe, leader of the Katanga rebellion, was invited to become Prime Minister. Further rebellions in Kwilu, Katanga and Kivu were crushed.

The elections were won by Tshombe's coalition, but the grouping collapsed following deadlock and executive power was passed to President Mobutu in 1965. In 1966 President Mobutu formed his own political organisation, Mouvement Populaire de la Révolution (MPR), and power began to concentrate in the Presidency. Union Minière, the main mining company, was nationalised in 1967. At the same time a referendum established Presidential government. Further elections endorsed Mobutu and in 1972 he took the title of Mobutu Sese Seko and inaugurated a programme of authenticity, which involved name changes and alterations to modes of dress. In Shaba (formerly Katanga) there were uprisings in 1977 and 1978 which were supressed with French and Moroccan help. About 75% of food, petroleum and chemicals are imported via South Africa. Despite scandals involving corruption and terrorist activity instigated by dissident exiles, the supression of opposition and further attacks by Shaba rebels in 1985, President Mobutu remains in power.

Stability

Since Independence in 1960 there was a period of considerable volatility until the coming to power of Mobutu Sese Seko in 1965. Subsequently there have been outbreaks of fighting in Shaba Province, and these have disrupted mining production.

Economic strategy up to 1983 was marked by haphazard intervention in the economy. In the early years of Independence there was an extension of government ownership, in particular the nationalisation of the mining sector in 1967. Since 1983 there have been moves to liberalise the economy, with the exchange rate floated, and plans to privatise state-owned enterprises and end the monopoly of marketing bodies.

Mobutu has been able to create a strong security network which has, together with frequent government reshuffles, prevented the growth of opposition within the country and pre-empted any coup attempts. Although there are a number of opposition groups outside Zaïre (mainly in Belgium), they appear too divided to become a serious threat.

Overall, Zaïre's stability record is moderate. The Mobutu regime appears secure but there have been two major changes in the direction of economic strategy and there is still haphazard intervention in the economy. It will be a while before business confidence in the private sector is fully restored under more liberal economic policies.

Economic Structure

In 1987 Zaïre had a GDP of $US 5,770m, the second largest in the region after Cameroon. However, together with Chad, Zaïre has the lowest standard of living in the region with a GNP per capita of only $US 150, one of the lowest in Africa, and Zaïre's rural population is particularly poor with most of the wealth being in the hands of a small urban élite.

The agricultural sector is proportionately smaller than in most other countries in Central Africa, generating only 32% of GDP. Industry, which includes the important mining sector, accounted for a further 33% (against a regional average of 25%), while the service sector generated the other 35% (below the regional average of 43%).

Private consumption was 73% of GDP, with government consumption 17%. The level of investment is low compared with the rest of Africa, at 13%, of which 10% is contributed by domestic saving.

Exports were equal to 33% of GDP with imports of 36%. The main exports are copper 42%, coffee 13% and diamonds 9%, while main

Zaïre

imports were machinery 40%, food produce 29% and fuels 24%.

The adult literacy rate was estimated at 55% in 1981, quite a bit better than in the region and in Africa generally. It was estimated in 1985 that 98% of primary-age children were in school with 57% of secondary-aged children being educated. Of the 20-24 year age group, 2% were in higher education. These figures make Zaïre's educational provision well above average.

Life expectancy of 52 years is about average. There were around 13,400 people per doctor and 1,700 per nurse in 1981. Zaïre would appear to have average health provision for the region, though many rural areas have limited health services. Average daily calorie supply per head in 1986 was just above the Africa norm.

Overall, Zaïre has a medium-sized economy but with very low average income. The agriculture and services sectors are small, and the industry sector, which includes mining, is large. Investment levels are low. There is about average trade dependency. Educational and health provisions appear good, although there is considerably better provision in the urban areas than in the rural.

Economic Performance

GDP grew by 1.6% a year over the period 1980-87, compared with a regional average of 4.2%, though this is still above the figure of 0.4% for the whole of Africa. GDP per head declined at -1.5% a year.

The 3.2% growth in the agricultural sector for this period is well above the regional rate of 0.5%. Although the industrial sector, which is largely dependent on mining, shows slightly better growth at 3.6%, it is however well below the regional average of 7.9% (though still above the African average of -1.2%). Services fell by -1.2% a year in Zaïre while the region as a whole showed growth of 3.2%.

While export volumes fell by -3.4% a year, import volumes only declined by -0.4%. Although regional export growth is 4.5%, this is largely a reflection of the growth of the oil industry in the countries along the coast of the Gulf of Guinea.

Inflation in Zaïre was 53.5% a year, 1980-87, well above the regional and African averages of 17% and 15.2% respectively.

Overall, Zaïre has managed to achieve expansion of GDP in the 1980s, but there has still been declining GDP per head, with agriculture and industry performing better than the services sector. Export volumes have fallen steadily, although imports have shown a smaller decline. The inflation rate has been one of the highest in Africa.

Recent Economic Developments

Zaïre began a programme of economic reform in 1983 which involved a float of the currency, the ending of price controls, reductions in government spending, the ending of state monopolies and the winding-up of some state-owned enterprises. There has been some faltering in this programme, and the IMF and World Bank have pressed for more rapid withdrawal from public sector involvement in manufacturing, mining and marketing. In October 1986 the President indicated that debt servicing would be limited to 20% of overall government spending or 10% of export earnings. This led to the withholding of World Bank lending, but this was restored by mid 1987 when further agreements with the IMF introduced plans to reduce projected public sector spending, increase tax revenues and petrol prices, and continue reforms of public sector enterprises. In August 1987, nine state-owned enterprises covering trading, housing, management training, savings, fishing, livestock, construction and steel, were dissolved. Zaïre has now revised its debt servicing proposals, and with re-schedulings these will be kept to 30% of government spending.

There have been some signs of improvement in economic performance since 1983. Economic growth has been positive, but below the rate of population growth. GDP was estimated to grow by 2.5% in 1989, but with population increasing at 3.1% a year, this has resulted in a decline in GDP per head. However, the rate of decline of GDP per head in the middle 1980s has been slower than in the latter part of the 1970s. Rehabilitation of Zaïre's mining sector is expected to be a long process, and will depend on improvements to the transport network, and are not expected to be completed before the end of 1990. There have been encouraging increases in diamonds marketed through official channels with the ending of the government monopoly in 1982 and the depreciation of the currency has discouraged smuggling although this had begun to increase again by 1987. Diamond sales rose by 80% in 1983 and 55% in 1984 and continued to rise in 1987.

Inflation performance has been variable, with annual rates of price increase in the 1980s between 20% and 75%. From 1983 to 1985, the inflation rate fell, from 77% to 24%, but then rose again to 47% in 1986. Inflation was estimated at around 60% for 1989.

A combination of falling prices for Zaïre's main mineral exports and poor production performance have resulted in falling export earnings, though copper prices have started to improve again recently. However, Gécamines has invested over $US 700m for repair and modernisation of capital equipment and in 1989 the World Bank granted them $US 20m for a technical assistance programme.

Zaïre's external debt was estimated at $US 7,800m in 1989, of which 28% is with commercial banks.

The currency has steadily depreciated in the 1980s, first by devaluations, and then after 1983 by a floating exchange rate. This has seen the exchange rate for Zaïre depreciate from Z 2.8 = $US 1 in 1980 to Z 119.6 = $US 1 in August 1987 and further devaluations brought it down to Z 435 = $US 1 by the end of 1989. The flexibility in the exchange rate has diminished the importance of the black market. The parallel rate now offers only a 20% premium over the official rate for foreign exchange, whereas before the float the premium was over 100%.

The IMF committed $US 230m in structural adjustment lending in mid 1987, and the World Bank $US 149m. Belgium has committed $US 24.1m of programme aid and Japan $US 15.9m. Project aid includes $US 167m from the African Development Bank for sugar rehabilitation, and $US 6.1m with a further $US 33.8m from the African Development Fund for hydro-electricity. The EEC is allocating $US 32.4m to the $US 700m rehabilitation of the state mining corporation, Gecamines. The EEC is also lending $US 11.9m for railway upgrading, while the World Bank, Belgium and the US are contributing $US 47.9m to an inland waterways scheme. Brazil is lending $US 65m, and the African Development Fund $US 9.7m for gold mine rehabilitation. The European Development Fund has allocated $US 11.4m to harbours and $US 20.9m to roads, the African Development Fund $US 2.5m to railways and education, and the Fund for Agricultural Development $US 7.6m to farming projects. The World Bank has committed $US 30.4m to rural and small industry schemes, $US 12m to technical assistance, and $US 11m to education. The EEC has allocated $US 45.2m to agricultural and rural schemes, and the European Development Fund $US 86m to roads, rural development and electrification.

Foreign investment has included a $US 20m commitment by Belgium's Petrofund for oil exploration, and a joint venture by Romania in a $US 17m manganese project.

Economic Outlook and Forecasts

The forecasts assume that Zaïre remains securely under the control of President Mobutu, and that although the current programme of liberalising reforms remains in place, the government will continue to find ways of delaying implementation.

Zaïre : Economic Forecasts
(average annual percentage change)

	Actual 1980-87	Forecast Base 1990-95	Forecast High 1990-95
GDP	1.6	2.2	2.4
GDP per Head	-1.5	-0.9	-0.7
Inflation Rate	53.5	48.0	42.0
Export Volumes	-3.4	3.0	3.5
Import Volumes	-0.4	0.2	0.5

Copper prices are projected to fall by -32% in real terms up to 1995. Coffee prices are projected to fall by -19%, and petroleum prices are expected to be virtually unchanged.

GDP is forecast to grow slightly faster up to 1995, but not faster than population expansion,

Zaire

and GDP per head will continue to fall.

The rehabilitation of mining and transport will allow export volumes to expand at around 3.0% a year. Import volumes are expected to expand very little as increased export volumes and aid offset adverse price movements.

Inflation is expected to remain high, close to 50% a year.

ZAIRE: Comparative Data

	ZAIRE	CENTRAL AFRICA	AFRICA	INDUSTRIAL COUNTRIES
POPULATION & LAND				
Population, mid year, millions, 1989	34.7	7.3	10.2	40.0
Urban Population, %, 1985	39	38.6	30	75
Population Growth Rate, % per year, 1980-86	3.1	3.0	3.1	0.8
Land Area, thou. sq. kilom.	2,345	638	486	1,628
Population Density, persons per sq kilom., 1988	14.4	11.1	20.4	24.3
ECONOMY: PRODUCTION & INCOME				
GDP, $US millions, 1986	6,020	4,146	3,561	550,099
GNP per head, $US, 1986	160	395	389	12,960
ECONOMY: SUPPLY STRUCTURE				
Agriculture, % of GDP, 1986	29	18	35	3
Industry, % of GDP, 1986	36	41	27	35
Services, % of GDP, 1986	35	41	38	61
ECONOMY: DEMAND STRUCTURE				
Private Consumption, % of GDP, 1986	81	62	73	62
Gross Domestic Investment, % of GDP, 1986	12	25	16	21
Government Consumption, % of GDP, 1986	7	14	14	17
Exports, % of GDP, 1986	33	36	23	17
Imports, % of GDP, 1986	33	36	26	17
ECONOMY: PERFORMANCE				
GDP growth, % per year, 1980-86	1.0	4.2	-0.6	2.5
GDP per head growth, % per year, 1980-86	-2.0	-0.8	-3.7	1.7
Agriculture growth, % per year, 1980-86	1.7	0.5	0.0	2.5
Industry growth, % per year, 1980-86	2.7	7.9	-1.0	2.5
Services growth, % per year, 1980-86	-0.7	3.2	-0.5	2.6
Exports growth, % per year, 1973-84	-4.3	4.5	-1.9	3.3
Imports growth, % per year, 1973-84	-1.2	0.4	-6.9	4.3
ECONOMY: OTHER				
Inflation Rate, % per year, 1980-86	54.1	17	16.7	5.3
Aid, net inflow, % of GDP, 1986	8.0	5.4	6.3	-
Debt Service, & of Exports, 1984	14.8	28	20.6	-
Budget Surplus (+), Deficit (-), % of GDP, 1973	-3.0	-1.4	-2.8	-5.1
EDUCATION				
Primary, % of 6-11 group, 1985	98	93	76	102
Secondary, % of 12-17 group, 1985	57	42	22	93
Higher, % of 20-24 group, 1985	2	1.9	1.9	39
Adult Literacy Rate, %, 1981	55	52	39	99
HEALTH & NUTRITION				
Life Expectancy, years, 1986	52	52	50	76
Calorie Supply, daily per head, 1985	2,151	2,115	2,096	3357
Population per doctor, 1981	13,430	13,835	24,185	550

Notes: 'Southern Africa' and 'Africa' exclude South Africa. Dates are for the country in question, and do not always correspond with the Regional, African and Industrial averages.

ZAIRE: Leading Indicators

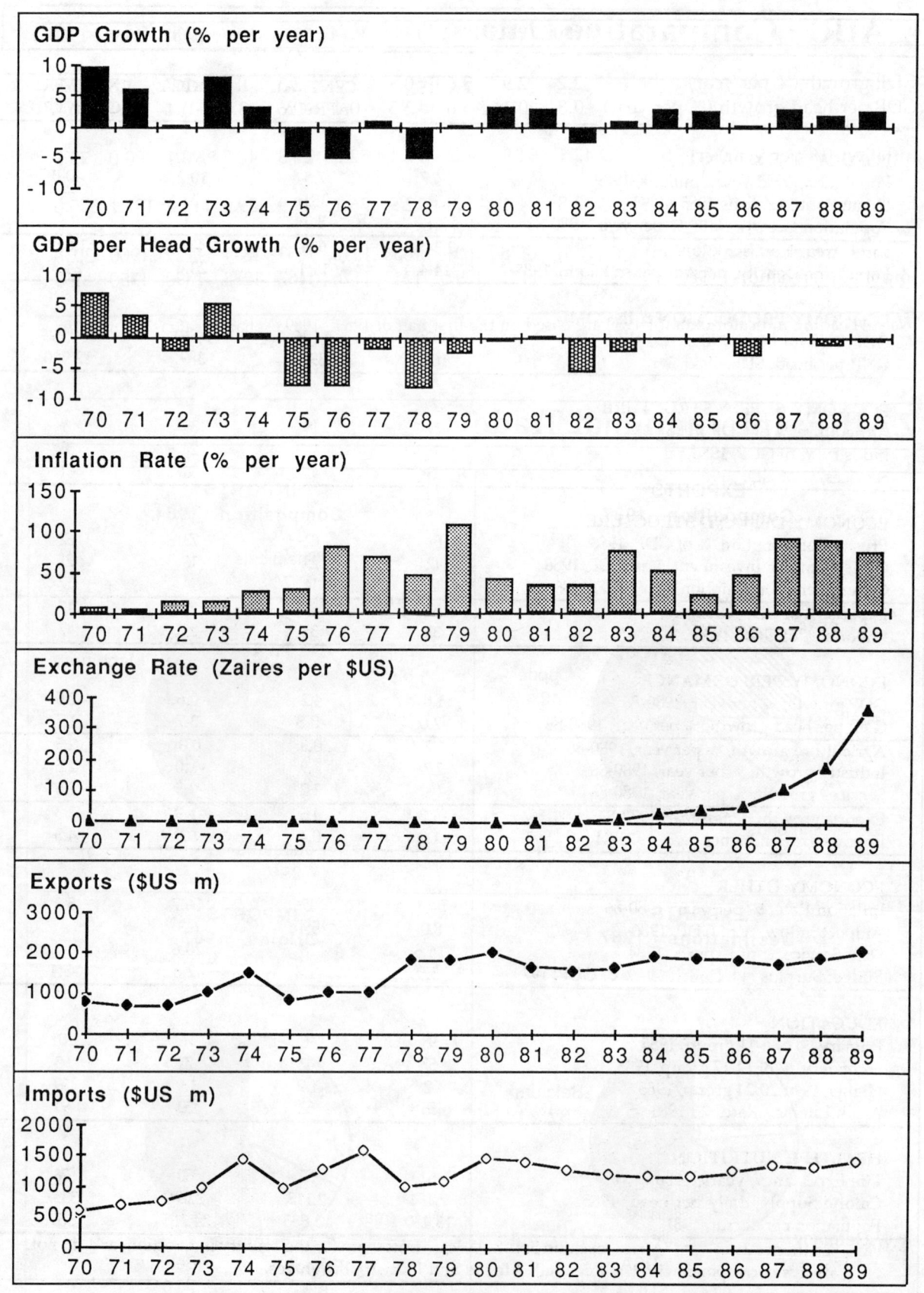

GDP Growth (% per year)

GDP per Head Growth (% per year)

Inflation Rate (% per year)

Exchange Rate (Zaires per $US)

Exports ($US m)

Imports ($US m)

Note: Leading indicator data for 1989 are based on the first half of 1989. 1989 exchange rate is for mid-1989.

ZAIRE: Leading Indicator Data

	1980	1981	1982	1983	1984	1985	1986	1987	1988	1989
GDP growth (% per year)	2.2	2.9	-2.6	0.8	2.7	2.5	0.1	2.7	1.8	2.5
GDP per head growth (% per year)	-0.8	-0.1	-5.6	-2.3	-0.4	-0.6	-3.0	-0.4	-1.4	-0.7
Inflation (% per year)	42.1	34.9	36.2	77.1	52.2	23.8	46.7	90.4	90.0	75.0
Exchange rate (Zaires per $US)	2.8	4.4	5.8	12.9	36.1	49.9	59.6	112.4	187.1	375.0
Exports, merchandise ($US m)		1678	1601	1686	1918	1853	1844	1729	1900	2050
Imports, merchandise ($US m)	1472	1421	1297	1213	1176	1187	1283	1390	1350	1450

Note: Leading indicator data for 1989 are based on the first half of 1989. 1989 exchange rate is for mid-1989.

ZAIRE: International Trade

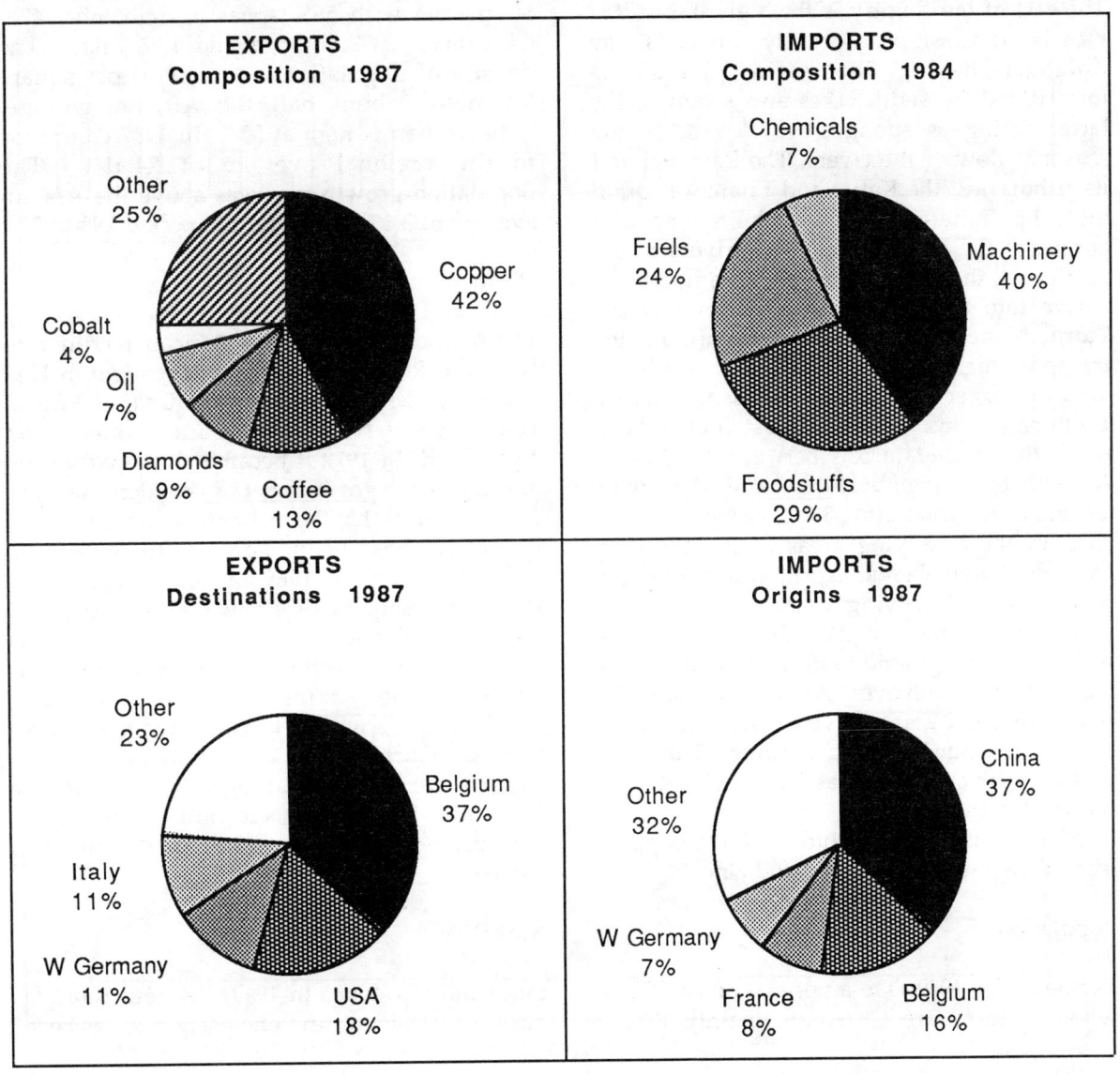

EXPORTS Composition 1987
- Copper 42%
- Other 25%
- Coffee 13%
- Diamonds 9%
- Oil 7%
- Cobalt 4%

IMPORTS Composition 1984
- Machinery 40%
- Foodstuffs 29%
- Fuels 24%
- Chemicals 7%

EXPORTS Destinations 1987
- Belgium 37%
- Other 23%
- USA 18%
- Italy 11%
- W Germany 11%

IMPORTS Origins 1987
- China 37%
- Other 32%
- Belgium 16%
- France 8%
- W Germany 7%

47 ZAMBIA

Physical Geography and Climate

Zambia, in the Southern Africa region, is a geographically large landlocked state bounded by Zaïre, Tanzania, Malawi, Mozambique, Zimbabwe, Botswana, Namibia and Angola. The east of the country is the highest, but it is mostly flat, sloping gently towards the Kalahari Basin. The main plateau is interrupted by sudd, lakes and swamps, the latter acting as sponges which modify the seasonal flow of the rivers. The Zambezi and its tributaries, the Kafue and Luangwa, drain into the Indian Ocean via the spectacular gorges of the Victoria Falls. The rivers of the north east, the Luapula and the Chambeshi drain into the Atlantic as tributaries of the Zaïre. None of the rivers are suitable for transport due to the large numbers of rapids and the larger lakes are not utilised. There are three seasons: April to August, cool and dry when the temperature is between 18°C-24°C; August to November, hot and dry with temperatures from 30°C-35°C, rising to above 35°C in the low lying areas of the south in October; November to April, warm and wet with frost in low-lying western areas and a temperature range of 24°C-30°C. Rainfall is highest on high ground to the north and west of the copperbelt with over 120cm per annum. The south west and Zambezi valley are driest with rainfall of around 75cm per annum. There are hydro-electric schemes at Kafne Gorge and Kariba North Bank.

Mineral deposits include cobalt, copper, lead, manganese, silver, gold and zinc.

Population

Estimates for 1989 give a population of 7.7m of whom a third are Christian, mainly Roman Catholic, with the majority following traditional beliefs. There are 73 different ethnic groups and over 80 languages. The main tribes being the Bemba in the northeast, the Nyanja in the east, the Tonga in the south and the Lozi in the west. English is the official languague with 56% speaking Icibemba, 42% Cinyanjaii, 23% Citonga and 17% Silozi. The density of population is low at 10 per square kilometre, about half the African average. Urbanisation is high at 53% in 1987 compared to the regional average of 24.1%. The population growth rate was above the African average at 3.6% a year for the period 1980-87.

History

The former British South African territory of Northern Rhodesia became independent in 1964 following the dissolution of the Central African Federation of North and South Rhodesia and Nyasaland. In 1972 it became a one-party state and under the government of President Kenneth Kaunda there has been political stability. The economy has been subject to extensive nationalization. In 1969 the mines came into state ownership, in 1974 mineral marketing, and the nationalisation of land was announced in 1977. Zambia has suffered from the unrest in the neighbouring states of Zimbabwe and Mozambique, which have disrupted export routes. Currently 50% of Zambia's mineral exports are shipped through South Africa. In December 1986 there were riots in the copper belt districts following rises in the price of maize.

Stability

Since Independence in 1964, Kenneth Kaunda has been President, and one-party rule has been in formal operation since 1972. Kaunda's popularity has remained high with the electorate, and sporadic student disturbances and civil unrest in the wake of recent economic reforms do not appear to be a serious threat to the regime. The only concern centres on the ability of the institutions of government to effect an orderly transfer of power to Kaunda's successor.

Economic strategy has undergone two major changes in direction, and at present is by no means coherent. The early years of Independence saw extension of public ownership and control in the economy, in particular the nationalisation of the copper mines in 1969. In 1982 the government introduced a reform and austerity programme which led to an auction system for foreign exchange and the phasing out of government subsidies. However, some of these policies were reversed early in 1987, although the Government still claims to be pursuing a structural adjustment programme.

Overall, Zambia's stability is good. The political security of the Kaunda regime is offset by the changes and uncertainty in economic strategy, and a present absence of coherence in economic policies.

Economic Structure

GDP in 1987 was estimated at $US 2,030m, making it one of the smaller medium-sized African economies. GNP per head was estimated at $US 250 putting it firmly in the middle of the range of low-income countries.

The agriculture sector is small, generating only 12% of GDP in 1987, whereas typically in Africa this sector would be almost three times larger. The industry sector generated 36% of GDP, of which manufacturing contributed 23%, and, apart from Zimbabwe, this is the largest manufacturing share in Africa. The services sector is large, at 52% of GDP, whereas the African average is 40%.

On the demand side, private consumption was euqivalent to 55% of GDP in 1987, and this is the lowest allocation to private consumption in Africa. Government consumption is 25% of GDP, and this is the highest in Africa. Investment is 15% of GDP, but domestic saving is 20%, so there was a net outflow of resources of 5% of GDP.

Export dependence is high at 47% of GDP, and only Mauritania in Africa has exports a greater percentage of GDP. The net resource outflow implies that imports are 42% of GDP, one of the highest import ratios in Africa, where the average is half Zambia's at 21%.

Merchandise exports are mostly copper at 91% of earnings, with cobalt providing 6% and zinc 2%. Merchandise imports are made up of manufactures 41%, machinery 39%, fuel 12% and foodstuffs 7%.

Adult literacy in 1985 was estimated at 76%, and this is well above the African average of 39%. The enrolment rate in 1986 was 104% in primary education and 19% in secondary, with 2% enrolment in higher education. The primary and secondary enrolment rates are better than the African averages, and the higher education enrolment is comparable.

Life expectancy at 53 years in 1987 was about the African average. There was one doctor for every 7,100 persons in 1983, and one nurse for every 740 persons. This is roughly three times the level of provision in Africa generally. Daily calorie supply per head was 2,343, roughly 10% above the African average.

Overall, Zambia is a medium-sized African economy, with a level of income per head that places it in the low-income category. There is a large manufacturing and industrial sector and a small agriculture sector. Allocation to private consumption is low, but to government consumption is high. There is heavy dependence on international trade. Educational and health provision is very good.

Economic Performance

GDP declined by -0.1% a year in the period 1980-87, whereas in Africa generally, GDP expanded by 0.4% over the same period. GDP per head declined at -3.7% a year, a rather faster rate of decline than the average for the continent of -2.8% a year.

The agriculture sector performed best, growing at 3.2% a year, whereas the industry sector has declined by -0.7% a year. Within the industry sector, manufacturing manged to expand at 0.8% a year. The services sector has contracted at -0.6% a year.

Zambia

Merchandise export volumes have contracted at -3.3% a year, and declining terms of trade have led to import volumes falling even faster at -6.2% a year. Both these rates are faster rates of decline than have been experienced on average in Africa in this period.

The inflation rate averaged 28.7% a year in the period 1980-87, and this is almost twice the rate experienced elsewhere in Africa.

Overall, Zambia's economic performance has been very poor in the 1980s, with falling GDP and falling GDP per head, and with only agriculture and manufacturing showing positive growth. Traded volumes have contracted and inflation has been quite high.

Recent Economic Developments

Since 1981 Zambia has been introducing elements of structural adjustment in its economic policies which aim to bring prices in line with scarcities, and a series of credits have been negotiated with the IMF to support this programme. However, Zambia's commitment to the series of reforms has not been particularly enthusiastic, and the IMF suspended payments in 1982 and 1985. The major disagreements have been over restrictions in government spending and adjustment in the exchange rate.

In October 1985 the government introduced an auction system for foreign exchange and in early 1986 ended control of interest rates, increased agricultural producer prices, abolished the state monopoly of agricultural marketing, ended import licensing, agreed to do away with food and fuel subsidies and introduced proposals for paying off accumulated trade debt arrears. The IMF responded with a two-year stand-by credit, and this was followed by a substantial re-scheduling of debt by the Paris Club of Zambia's creditor countries.

At the end of 1986, there was urban unrest in response to the increase in food prices caused by the ending of subsidies, and the subsidies were immediately restored. In January 1987 the proposals for meeting the trade payment arrears were frozen, the foreign exchange auction suspended, and in February interest rates were cut. Arrears of payments to the IMF built up, and further credits were suspended pending an agreement on restoring the programme of reforms.

In May of 1987, however, the Zambian government announced that it would not be seeking a further agreement with the IMF. A series of economic proposals were announced involving the limitation of debt servicing to 10% of export earnings, reduction of interest rates, and price controls on consumer goods introduced. The exchange rate was fixed and import licensing re-introduced. There was dismay at these reversals among the donor countries, but although the UK ended balance of payments support, other countries, notably the Scandinavian group, have continued project aid. Subsequently the government introduced price cuts of 10% in some commodities and announced rises in agricultural producer prices, although for the most part these did not compensate for inflation. Industrial and mining concerns began working at reduced capacity as a result of shortages of imported raw materials and fuel caused by the unavailability of foreign exchange.

In September 1989, however, Zambia came to an agreement with the IMF which committed the government to devaluing the currency, ending regulation of prices, raising interest rates, reducing food subsidies and cutting public spending. Inflation has steadily accelerated, from 12% in 1980 to 90% in 1989.

Falling copper prices and copper production have led to falling export revenue up to 1986. This peaked at $US 1,457m in 1980 but steadily declined to $US 692m in 1986. Since 1986, export revenues have recovered, rising to an estimated $US 1,350m in 1989. Import levels were sustained at around the $US 1,000m level between 1980 and 1982 by virtue of aid and credits. Falling export revenues and the ending of trade credit as a result of payment arrears, led to imports halveing to $US 518m in 1986 before recovering slightly to $US 790m in 1989.

Zambia's external debts are estimated in excess of $US 5b at the end of 1989, $5, giving rise to over $US 800m of debt servicing each year. In addition there are external trade, salary and pension arrears of $US 400m, and Over $US 1b of arrears with the IMF and the World Bank.

Zambia only began serious adjustment of the

Zambia

international value of the currency in 1983 with a 20% devaluation. Previously the black market gave a 50% premium to foreign exchange over the official rate. There was further depreciation until the introduction of the auctions in 1985. From ZK 0.93=$US 1 in 1987, the auction depreciated the kwacha to ZK 14.5=$US 1 before arrears in auction disbursement of foreign exchange led to a suspension of the system at the end of January 1987. Two tiers were temporarily introduced, with a priority rate of ZK 9=$US 1 and ZK 12=$US 1 for other transactions. In March the auction was restored with a ZK 15=$US 1 rate being set, with an official rate continuing in existence, and expected to be between the limits of ZK 9=$US 1 and ZK 12.5=$US 1. In May the kwacha was revalued to ZK 8=$US 1, and the shortage of foreign exchange at this price required the reintroduction of import licensing. The adjustments in September 1989 have led to rates of ZK 20.71=$US 1 at the end of 1989.

Project aid has continued, albeit at a reduced level, and can be expected to increase with the 1989 IMF agreement. Current projects include $US 5.7m from Japan for water supply and $US 8.2m for agriculture. Italy has donated $US 40m for rehabilitation of the oil pipeline from Tanzania and $US 38.2m for dam construction. Sweden is supplying $US 72.5m for education, health, agriculture and rural projects, Canada $US 23.9m for rail rehabilitation and China $US 13.4m for roads. The European Investment Bank is allocating $US 11.8m to small enterprises and the EEC $US 45m for fertilisers.

Private foreign investment includes $US 7.1m by Intercontinental Hotels with a loan from the World Bank's International Finance Corporation. Heinz has invested $US 1.5m in cooking oil production, and Ireland's Mosstock International is to invest $US 10m in wheat and cotton production.

Economic Outlook and Forecasts

The forecasts assume that Zambia maintains its political stability under President Kaunda, and that the government persists with the current adjustment programme.

Copper prices are projected to fall in real terms by -32% up to 1995. However, expanding copper production should offset this price fall and maintain foreign exchange earnings. Greater aid flows and more sympathetic treatment of external debts, together with the possibility of improved relations in the region with South Africa, should lead to expansion of the economy up to 1995.

GDP is projected to expand at around 1.5% a year, not fast enough, however, to prevent GDP per head falling at -2.0% a year.

Export volumes are expected to expand steadily at around 4.5% a year, but worsening terms of trade, despite increased aid flows, will prevent import volumes rising at much more than 1.2% a year.

Inflation is expected to remain high as the effects of devaluation and high interest rates increase costs, only falling toward the end of the forecast period. Over the period, the annual average rate of price increase is expected to be around 40%.

Zambia: Economic Forecasts
(average annual percentage change)

	Actual 1980-87	Forecast Base 1990-95	Forecast High 1990-95
GDP	-0.1	1.5	1.8
GDP per Head	-3.7	-2.0	-1.7
Inflation Rate	28.7	44.0	42.0
Export Volumes	-3.3	4.5	4.9
Import Volumes	-6.2	1.2	1.5

ZAMBIA: Comparative Data

	ZAMBIA	SOUTHERN AFRICA	AFRICA	INDUSTRIAL COUNTRIES
POPULATION & LAND				
Population, mid year, millions, 1989	7.7	6.0	10.2	40.0
Urban Population, %, 1985	39	23.8	30	75
Population Growth Rate, % per year, 1980-86	3.5	3.1	3.1	0.8
Land Area, thou. sq. kilom.	753	531	486	1,628
Population Density, persons per sq kilom., 1988	9.8	11.3	20.4	24.3
ECONOMY: PRODUCTION & INCOME				
GDP, $US millions, 1986	1,660	2,121	3,561	550,099
GNP per head, $US, 1986	300	383	389	12,960
ECONOMY: SUPPLY STRUCTURE				
Agriculture, % of GDP, 1986	11	25	35	3
Industry, % of GDP, 1986	48	33	27	35
Services, % of GDP, 1986	41	42	38	61
ECONOMY: DEMAND STRUCTURE				
Private Consumption, % of GDP, 1986	62	67	73	62
Gross Domestic Investment, % of GDP, 1986	15	14	16	21
Government Consumption, % of GDP, 1986	25	21	14	17
Exports, % of GDP, 1986	46	34	23	17
Imports, % of GDP, 1986	48	36	26	17
ECONOMY: PERFORMANCE				
GDP growth, % per year, 1980-86	-0.1	-3.7	-0.6	2.5
GDP per head growth, % per year, 1980-86	-3.5	-6.6	-3.7	1.7
Agriculture growth, % per year, 1980-86	2.8	-6.8	0.0	2.5
Industry growth, % per year, 1980-86	-0.7	-3.8	-1.0	2.5
Services growth, % per year, 1980-86	-0.5	-0.9	-0.5	2.6
Exports growth, % per year, 1980-86	-2.1	-5.1	-1.9	3.3
Imports growth, % per year, 1980-86	-7.3	-2.5	-6.9	4.3
ECONOMY: OTHER				
Inflation Rate, % per year, 1980-86	23.3	18.1	16.7	5.3
Aid, net inflow, % of GDP, 1986	31.2	8.6	6.3	-
Debt Service, % of Exports, 1984	11.4	7.8	20.6	-
Budget Surplus (+), Deficit (-), % of GDP, 1986	-16.3	-4.1	-2.8	-5.1
EDUCATION				
Primary, % of 6-11 group, 1984	103	95	76	102
Secondary, % of 12-17 group, 1984	19	18	22	93
Higher, % of 20-24 group, 1985	2	1.3	1.9	39
Adult Literacy Rate, %, 1980	44	47	39	99
HEALTH & NUTRITION				
Life Expectancy, years, 1986	53	50	50	76
Calorie Supply, daily per head, 1985	2,126	1,998	2,096	3,357
Population per doctor, 1981	7,800	26,118	24,185	550

Notes: 'Southern Africa' and 'Africa' exclude South Africa. Dates are for the country in question, and do not always correspond with the Regional, African and Industrial averages.

ZAMBIA: Leading Indicators

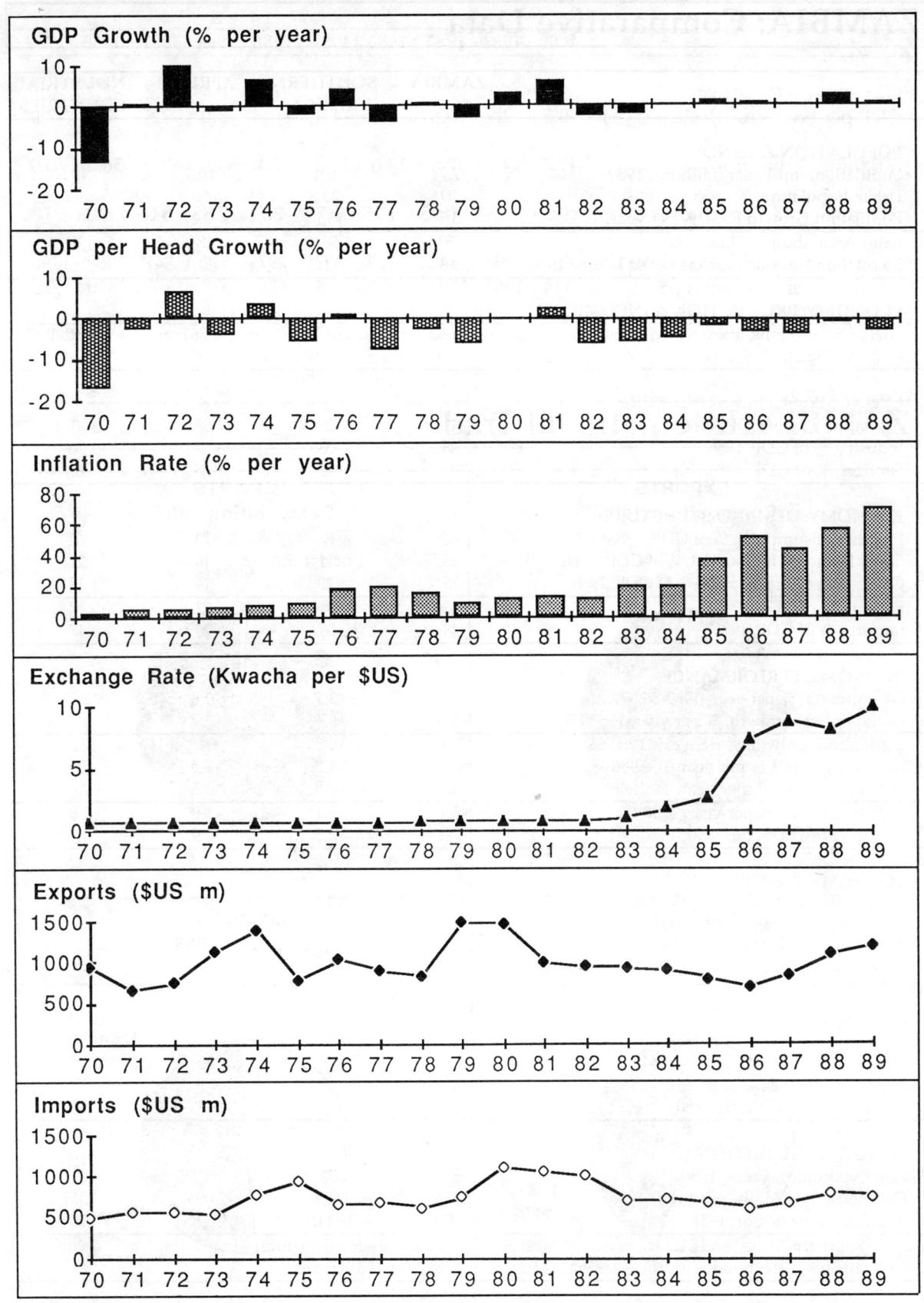

GDP Growth (% per year)

GDP per Head Growth (% per year)

Inflation Rate (% per year)

Exchange Rate (Kwacha per $US)

Exports ($US m)

Imports ($US m)

Note: Leading indicator data for 1989 are based on the first half of 1989. 1989 exchange rate is for mid-1989.

ZAMBIA: Leading Indicator Data

	1980	1981	1982	1983	1984	1985	1986	1987	1988	1989
GDP growth (% per year)	3.0	6.1	-2.7	-2.1	-0.5	1.4	0.4	-0.2	2.7	0.5
GDP per head growth (% per year)	-0.2	2.9	-6.0	-5.4	-4.8	-1.9	-3.0	-3.6	-0.7	-2.9
Inflation (% per year)	11.7	14.0	12.5	19.6	20.0	37.4	51.6	43.0	56.0	70.0
Exchange rate (Kwacha per $US)	0.79	0.87	0.93	1.25	1.79	2.71	7.33	8.89	8.22	10.00
Exports, merchandise ($US m)	1457	996	942	923	893	797	692	847	1080	1190
Imports, merchandise ($US m)	1114	1065	1004	711	734	685	622	702	810	770

Note: Leading indicator data for 1989 are based on the first half of 1989. 1989 exchange rate is for mid-1989.

ZAMBIA: International Trade

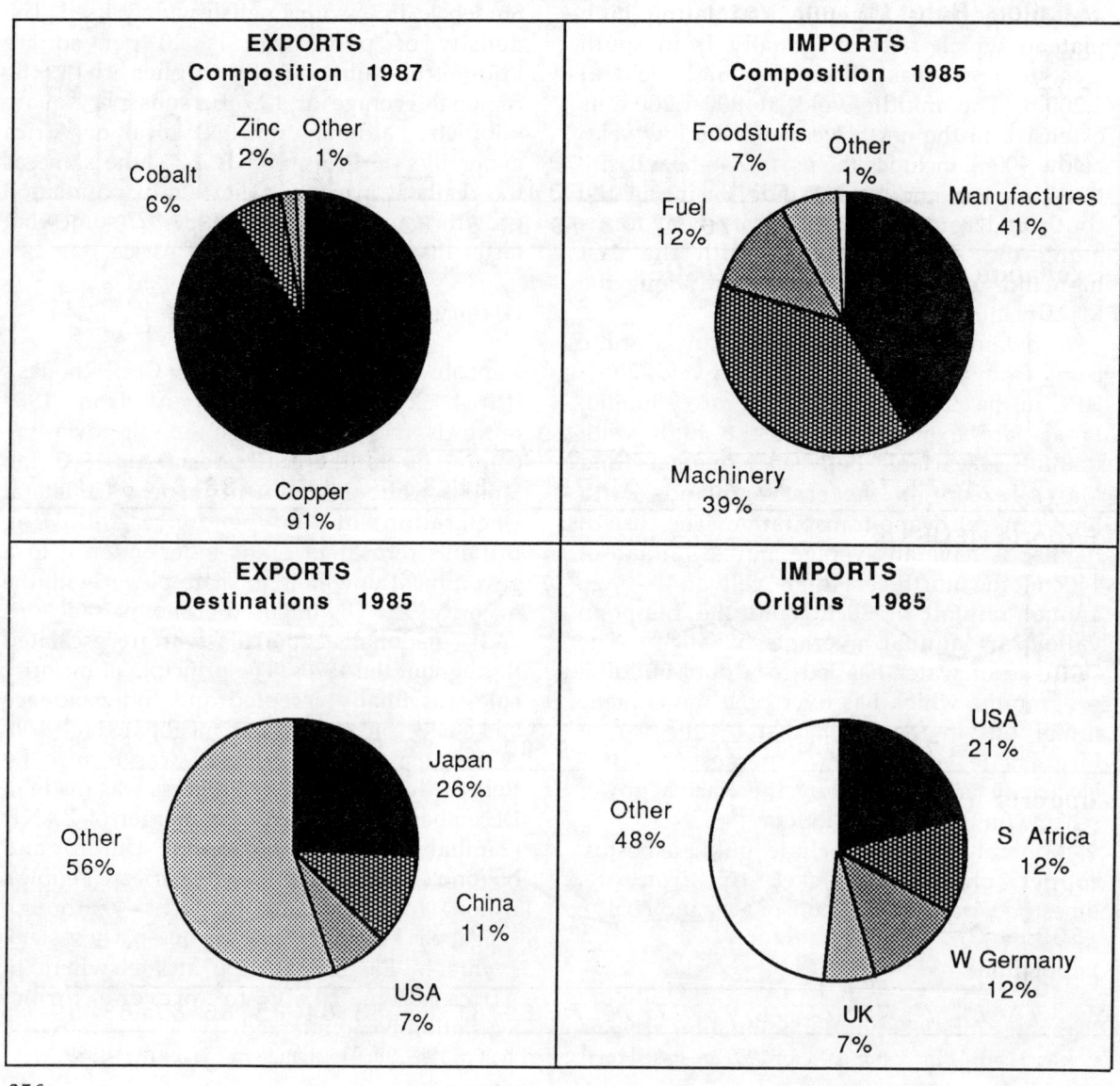

EXPORTS Composition 1987
- Copper 91%
- Cobalt 6%
- Zinc 2%
- Other 1%

IMPORTS Composition 1985
- Manufactures 41%
- Machinery 39%
- Fuel 12%
- Foodstuffs 7%
- Other 1%

EXPORTS Destinations 1985
- Japan 26%
- China 11%
- USA 7%
- Other 56%

IMPORTS Origins 1985
- USA 21%
- S Africa 12%
- W Germany 12%
- UK 7%
- Other 48%

48 ZIMBABWE

Physical Geography and Climate

Zimbabwe, in the Southern Africa region, is a medium-sized landlocked country bounded by Zambia, Mozambique, South Africa, Botswana and Namibia. Between the Rivers Zambezi and Limpopo is a gentle undulating high plateau which rises diagonally from south west to north east, becoming high veld at 1,200m. The middle veld at 900-1,200m is extensive in the north west and the low veld, below 900m, includes the north Zambezi Basin, the Limpopo and Save-Runde Basins in the south and south east. An escarpment forms along the Zambezi trough with the east highlands rising to 1,800m-2,600m along the Mozambique border.

The temperature is modified by altitude being from 13°C-22°C on the high veld, 20°C-30°C in the Zambezi valley, but rarely humid. Frosts are experienced on the high veld. Rainfall is variable between November and March, except in the east highlands, with high rates of evapo-transpiration. The eastern highlands have an average annual rainfall of 140cm, the north-east high velt an average annual rainfall of 80cm, and the Limpopo valley an annual average of 40cm. The shortage of water has led to a dam-building programme which has overcome the climatic limitations in the south east by the use of large-scale irrigation schemes. Hydro-electricity is generated by the Kariba power scheme on the Zambezi River.

Mineral deposits include gold, asbestos, copper, chromium, nickel, tin, iron ore, limestone, iron pyrites, phosphates and coal.

Population

Estimates for 1989 put the population at 9.7m. Most follow traditional beliefs, with around 20% Christians. The population is 68% Shona and 15% Ndebele with the minor groups, the Tonga, Sena, Hlengwe, Venda, Kalanga and Sotho making up 5%, Europeans 2% and Asians 0.5%. The major languages are Shona 67%, Sindebele 15%, with English widely used. The density of population is 20 per square kilometre, substantially higher than the regional average of 12.7 persons per square kilometre, although fairly typical of Africa generally. Urbanisation is 24%, the same as the regional average. The rate of population growth was 3.7% a year 1980-87, somewhat faster than the general rate in Africa.

History

Zimbabwe was administered by Cecil Rhodes's British South Africa Company from 1889 onwards. In 1923 it became the self-governing colony of Southern Rhodesia. In 1963 Ian Smith's white government made a Unilateral Declaration of Independence following Britain's refusal to grant independence to a government unwilling to share power with the majority. The UN imposed economic sanctions and a nationalist guerrilla warfare escalated throughout the 1970s. The principle of majority rule was finally accepted and Independence and a new constitution were negotiated in 1979. A new majority government was formed by Robert Mugabe in 1980. Progress was made in December 1985 towards the merger of ZANU (Zimbabwe African National Union) and Nkomo's ZAPU (Zimbabwe African People's Union) to form a single party. Although Mugabe is known to favour a one-party system, events in Eastern Europe and elswhere in Africa seem likely to prevent formal legislation being adopted.

Zimbabwe

Stability

Independence in 1980 was preceded by seven years of guerrilla activities by the liberation forces which resulted in widespread disruption of the economy, particularly in rural areas. Although security has improved under the government of Prime Minister Robert Mugabe, there has continued to be insurgent activity in Matabeleland, and tension between Joshua Nkomo's supporters and the government. In August 1987, the 20 parliamentary seats reserved for the 33,000 white community by the 1979 independence settlement were abolished. Economic strategy has involved a commitment by the Mugabe government to socialist policies, but there has been only modest implementation of these principles. Resettlement of land has been gradual, and the bulk of the mining sector has remained in private hands.

Overall, the stability of Zimbabwe has been poor, with the continuity in economic policies offset by the uncertainties and disruption of the run-up to Independence and subsequent inability to secure consensus among regional and minority groups.

Economic Structure

GDP was estimated to be $US 5,240m in 1987, by far the largest in the region (except for South Africa). At the same time GNP per head was estimated to be $US 580, third biggest in the region (excuding South Africa) after Botswana and Swaziland.

Agriculture generates only 11% of GDP, well below the regional average of 25%. In Africa as a whole, it averages 35%. On the other hand, industry generates 43% of GDP well above the regional and African averages of 33% and 27% respectively. Zimbabwe has a large manufacturing sector, generating 31% of GDP, the largest in Africa. The service sector generates 46% of GDP and this is just above the regional average of 42%, while the African average is 38%.

On the demand side, private consumption is equivalent to 59% of GDP, low in comparison with the African average of 72%. Government consumption is 20% of GDP, and this is about a third higher than the Africa norm. Gross domestic investment is 18% of GDP, fairly typical of Africa, but a domestic savings rate of 22% of GDP indicates a net outflow of resources equivalent to 4% of GDP.

Exports made up 27% of GDP, and while this is greater than average for Africa, the regional figure is 34%. Imports represent 24% of GDP, compared with 36% in the region and 26% for Africa.

Gold made up 19% of export earnings in 1987, followed by tobacco 18%, metals 14% and cotton 5%. Imports in 1987 comprised machinary 32%, manufactures 20%, chemicals 15% and energy 12%.

The adult literacy rate was estimated at 74% in 1985, almost double the average for Africa. Because of the disruption to education caused by the civil war and the policies of the pre-independence government, the number of people in primary education represents 129% of the 6-11 age group with many people attending adult education classes. This is well above the regional average of 95%. In Africa as a whole the average is only 76%. Of the 12-17 age group, 46% attend secondary schools, more than double the regional and African averages of 18% and 22% respectively. Higher education enrollment is high with 4% of the 20-24 age group attending such institutions, compared with 1.3% and 1.9% each in the region and Africa.

Life expectancy is 58 years, above the regional average of 50 years. Health provision is generally good. There is one doctor for every 6,700 people, nearly four times better than the regional average. There is also one nurse per thousand people, three times better than average for the region. Daily calorie supply in 1986 was 2,143, about the African norm.

Overall, Zimbabwe has a medium-sized economy, with income levels at the lower end of the middle-income range. The economy is well industrialised with a large manufacturing sector. Trade dependence is at roughly the norm for Africa. Educational and health provision is very good.

Economic Performance

GDP growth between 1980 and 1987 averaged 2.4% a year though this was restricted by serious drought between 1982 and 1987.

Despite the drought, growth in the

agricultural sector was still 2.3%. Industrial growth was only 1.4% with manufacturing restricted by shortages of foreign exchange for raw materials, components, spare parts and replacement plant, despite having some of the best infrastructure in Africa. Service sector growth was more encouraging at 3.3%. All three sectors showed growth rates well above the regional averages.

Export volume growth for the period was a disappointing 0.9% a year and depended heavily on mineral output, especially during the drought. Import volumes decreased by -6.8% as falling terms of trade and higher debt service payments have caused foreign currency allocations to be more tightly restricted. Reluctance to become dependent on South African ports while transport routes through Mozambique are hampered by security problems has been a major logistics problem.

Inflation averaged 12.4% and this is below the African average and good for a country without externally-imposed monetary discipline such as experienced in the Franc Zone group.

Overall, Zimbabwe's economic performance has been creditable in view of the drought and instability in the region. The economy has managed to expand although GDP per head has fallen. Export volumes have expanded despite falling import volumes. Inflation has been modest.

Recent Economic Developments

Zimbabwe has made gradual changes to the structure of the economy since Independence in 1980. The exchange rate underwent adjustment between 1983 and 1985, but remains overvalued and foreign exchange is rationed, causing problems for the industrial and mining sectors. There was a premium of foreign exchange in the black market of over 100% in mid-1989. World Bank concern over the size of the budget deficit has led to delay in arranging sector adjustment loans for the mining and agriculture sector, although lending has been made for industry sector adjustment. The government has steadily increased its participation in the economy by buying into various private sector enterprises.

After a surge in growth which marked Independence and the ending of sanctions, the economy grew slowly between 1982 and 1984. There was good growth in 1985 with excellent weather raising agricultural output. There was a return to more modest GDP expansion after 1986 and 1987 when GDP was falling, with good recovery in 1988 and 1989. GDP per head has fallen in every year from 1982 with the exception of 1985 and 1988. Inflation was over 20% in 1983 and 1984, but fell to 8.4% in 1985 before rising to 15% in 1989.

Merchandise export revenues have declined from a peak of $US 1,446m in 1980 to $US 1,087m in 1985 and then recovered to over $US 1,650m in 1989. Falling export revenue has limited the ability to import, with imports lagging a year behind exports. Merchandise imports peaked at $US 1,535m in 1981, fell to $US 1,068m in 1986, but recovered to $US 1,450m in 1989. Zimbabwe's external debt has risen from $US 519m in 1979 to over $US 2b in 1989. Zimbabwe's middle-income status means higher interest payments and shorter maturity periods, giving rise to high annual debt service payments. In addition, Zimbabwe has sought to ease constraints on imports by arranging short-term bridging loans. Debt service comprises over 30% of total export earnings on goods and services, and this is a constraint on domestic output as it limits the ability to purchase essential imports.

The exchange rate remained fairly constant until the end of 1982, and then a series of devaluations have seen the Zimbabwe dollar steadily depreciate to half its value by 1985 and by a further 29% up to the end of 1989. The exchange rate has subsequently remained virtually unchanged.

The main concessionary lending to Zimbabwe has included $US 4.9m from the EEC for export promotion, a commitment of $US 15.3m over 3 years from the UK for general balance of payments support, and $US 11.2m from Canada for agriculture, power and health. The Arab Bank for Economic Development in Africa is lending $US 9.2m and the Commonwealth Development Corporation $US 24m to meat production projects. The World Bank is allocating $US 3.5m to power projects together with $US 17m from the European Investment Bank. Finland is lending $US 6.6m for paper, mining, power and forestry schemes.

Zimbabwe

The private foreign sector has expanded, despite the difficulties created by foreign exchange shortage and unreliable supplies of imports. In 1989, Zimbabwe intoduced a new investment code and actively encouraged foreign investment.

Rio-Tinto, British American Tobacco, Delta, Distillers and Circle Cement all report increased profits. The government has pursued a steady policy of extending its involvement in the economy by purchasing shares in private sector enterprises, and it has taken holdings in Mardon, Kenning, Astra and Delta. Sweden's Mitre Nobel is involved in a $US 5m joint venture, UK's Cluff Gold is investing up to $US 12m in mining, Hoechst is setting up a $US 2.4m chemicals operation and UK's Aberfoyle Holdings is establishing a $US 90m palm oil project.

Economic Outlook and Forecasts

The forecasts assume that Zimbabwe maintains its stability and that political developments in South Africa ease Zimbabwe's transport problems. It is assumed that the government continues to be cautious over introducing liberalising economic reforms.

The price of gold is projected to fall by -22.0% in real terms up to 1995, and tobacco prices are expected to fall by -12.0%. Cotton prices are expected to remain more-or-less unchanged.

GDP is forecast to maintain the recovery observed in the late 1980s to grow at just over 3.0% up to 1995, allowing a slight improvement in GDP per head of around 0.2%.

Export volumes are forecast to expand at around 2.5% a year, but average terms of trade movements are expected to prevent import volumes expanding at anything more than 1.0% a year.

The inflation rate is expected to be slightly higher than in the 1980s, averaging around 15.0% a year.

Zimbabwe: Economic Forecasts
(average annual percentage change)

	Actual 1980-87	Forecast Base 1990-95	Forecast High 1990-95
GDP	2.4	3.2	3.3
GDP per Head	-1.3	0.2	0.3
Inflation Rate	12.4	15.5	15.0
Export Volumes	0.9	2.5	2.8
Import Volumes	-6.8	0.7	1.0

ZIMBABWE: Comparative Data

	ZIMBABWE	SOUTHERN AFRICA	AFRICA	INDUSTRIAL COUNTRIES
POPULATION & LAND				
Population, mid year, millions, 1989	9.7	6.0	10.2	40.0
Urban Population, %, 1985	27	23.8	30	75
Population Growth Rate, % per year, 1980-86	3.7	3.1	3.1	0.8
Land Area, thou. sq. kilom.	391	531	486	1,628
Population Density, persons per sq kilom., 1988	24.04	11.3	20.4	24.3
ECONOMY: PRODUCTION & INCOME				
GDP, $US millions, 1986	4,940	2,121	3,561	550,099
GNP per head, $US, 1986	620	383	389	12,960
ECONOMY: SUPPLY STRUCTURE				
Agriculture, % of GDP, 1986	11	25	35	3
Industry, % of GDP, 1986	46	33	27	35
Services, % of GDP, 1986	43	42	38	61
ECONOMY: DEMAND STRUCTURE				
Private Consumption, % of GDP, 1986	62	67	73	62
Gross Domestic Investment, % of GDP, 1986	18	14	16	21
Government Consumption, % of GDP, 1986	19	21	14	17
Exports, % of GDP, 1986	26	34	23	17
Imports, % of GDP, 1986	24	36	26	17
ECONOMY: PERFORMANCE				
GDP growth, % per year, 1980-86	2.6	-3.7	-0.6	2.5
GDP per head growth, % per year, 1980-86	-1.1	-6.6	-3.7	1.7
Agriculture growth, % per year, 1980-86	3.4	-6.8	0.0	2.5
Industry growth, % per year, 1980-86	0.8	-3.8	-1.0	2.5
Services growth, % per year, 1980-86	3.7	-0.9	-0.5	2.6
Exports growth, % per year, 1980-86	-2.7	-5.1	-1.9	3.3
Imports growth, % per year, 1980-86	-6.7	-2.5	-6.9	4.3
ECONOMY: OTHER				
Inflation Rate, % per year, 1980-86	13.0	18.1	16.7	5.3
Aid, net inflow, % of GDP, 1986	4.2	8.6	6.3	-
Debt Service, % of Exports, 1984	19.9	7.8	20.6	-
Budget Surplus (+), Deficit (-), % of GDP, 1986	-7.0	-4.1	-2.8	-5.1
EDUCATION				
Primary, % of 6-11 group, 1985	131	95	76	102
Secondary, % of 12-17 group, 1985	43	18	22	93
Higher, % of 20-24 group, 1985	3	1.3	1.9	39
Adult Literacy Rate, %, 1981	69	47	39	99
HEALTH & NUTRITION				
Life Expectancy, years, 1986	58	50	50	76
Calorie Supply, daily per head, 1985	2,114	1,998	2,096	3,357
Population per doctor, 1981	7,100	26,118	24,185	550

Notes: 'Southern Africa' and 'Africa' exclude South Africa. Dates are for the country in question, and do not always correspond with the Regional, African and Industrial averages.

ZIMBABWE: Leading Indicators

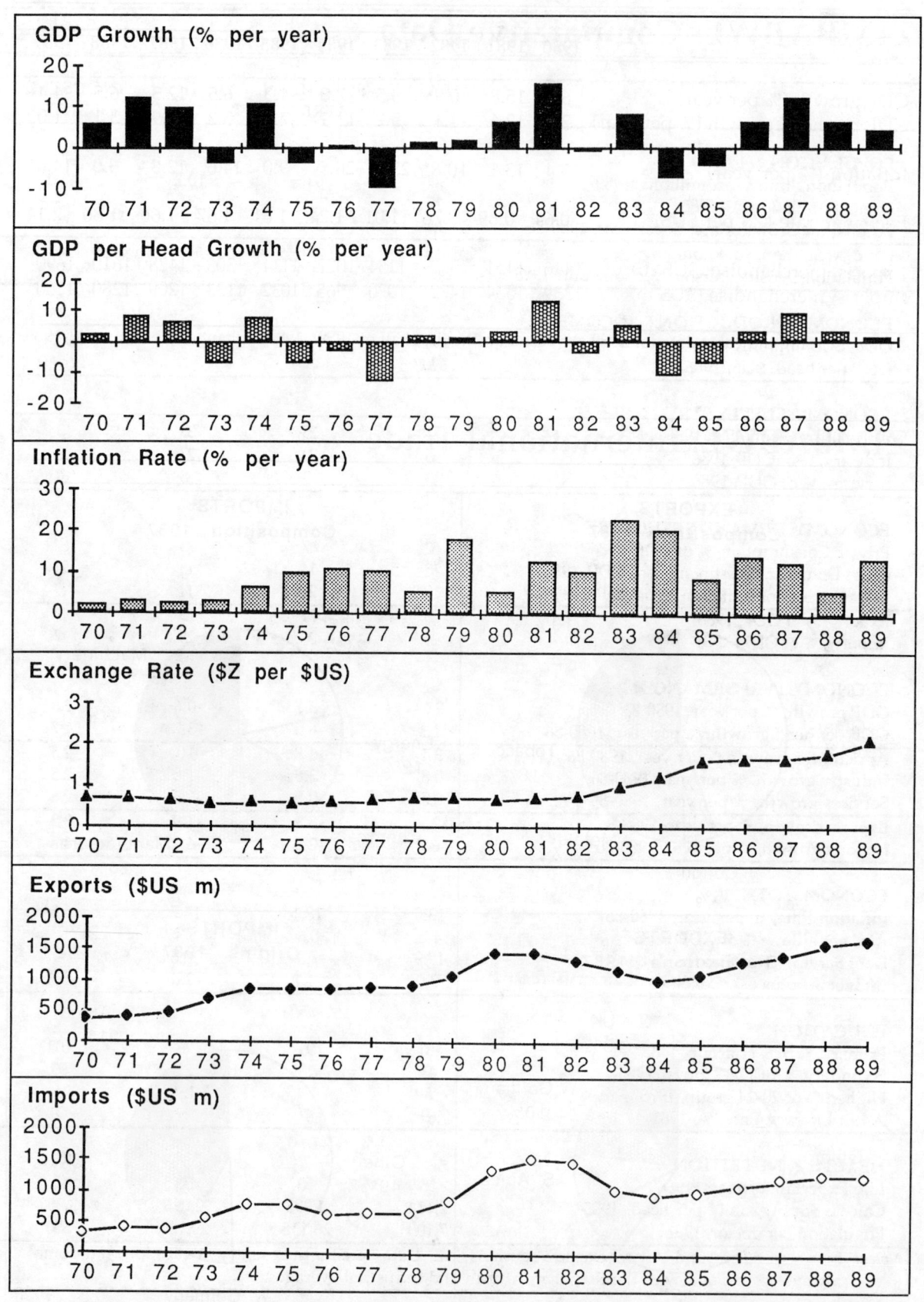

Note: Leading indicator data for 1989 are based on the first half of 1989. 1989 exchange rate is for mid-1989.

ZIMBABWE: Leading Indicator Data

	1980	1981	1982	1983	1984	1985	1986	1987	1988	1989
GDP growth (% per year)	6.1	15.8	-0.9	8.5	-7.8	-4.3	6.5	12.3	6.5	4.4
GDP per head growth (% per year)	2.9	12.6	-4.1	5.2	-11.1	-7.3	3.2	8.9	3.1	1.0
Inflation (% per year)	5.4	13.1	10.6*	23.1	20.5	9.0	14.6	12.8	6.0	14.0
Exchange rate (Z$ per $US)	0.64	0.69	0.76	1.01	1.24	1.61	1.67	1.66	1.80	2.14
Exports, merchandise ($US m)	1446	1451	1312	1154	1022	1114	1303	1428	1610	1690
Imports, merchandise ($US m)	1339	1534	1472	1070	965	1032	1133	1209	1280	1260

Note: Leading indicator data for 1989 are based on the first half of 1989. 1989 exchange rate is for mid-1989.

ZIMBABWE: International Trade

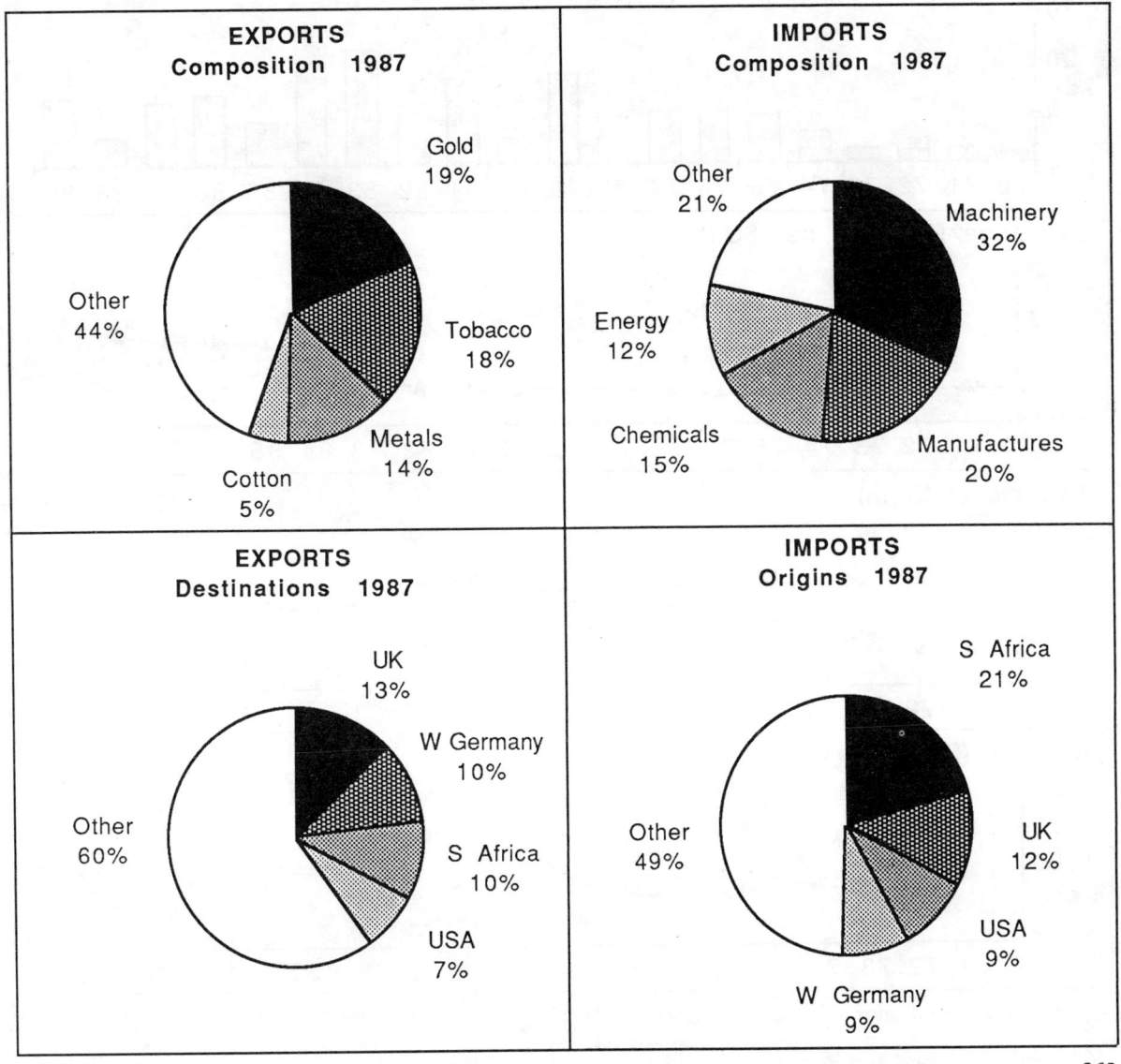

EXPORTS Composition 1987
- Gold 19%
- Tobacco 18%
- Metals 14%
- Cotton 5%
- Other 44%

IMPORTS Composition 1987
- Machinery 32%
- Manufactures 20%
- Chemicals 15%
- Energy 12%
- Other 21%

EXPORTS Destinations 1987
- UK 13%
- W Germany 10%
- S Africa 10%
- USA 7%
- Other 60%

IMPORTS Origins 1987
- S Africa 21%
- UK 12%
- USA 9%
- W Germany 9%
- Other 49%